BRAIDED RIVERS

Other publications of the International Association of Sedimentologists

SPECIAL PUBLICATIONS

35 Fluvial Sedimentology VII
Edited by M.D. Blum, S.B. Marriott and
S.F. Leclair
2005, 589 pages, 319 illustrations

34 Clay Mineral Cements in Sandstones
Edited by R.H. Worden and S. Morad
2003, 512 pages, 246 illustrations

33 Precambrian Sedimentary Environments
A Modern Approach to Ancient Depositional Systems
Edited by W. Altermann and P.L. Corcoran
2002, 464 pages, 194 illustrations

32 Flood and Megaflood Processes and Deposits
Recent and Ancient Examples
Edited by I.P. Martini, V.R. Baker and G. Garzón
2002, 320 pages, 281 illustrations

31 Particulate Gravity Currents
Edited by W.D. McCaffrey, B.C. Kneller and
J. Peakall
2001, 320 pages, 222 illustrations

30 Volcaniclastic Sedimentation in Lacustrine Settings
Edited by J.D.L. White and N.R. Riggs
2001, 312 pages, 155 illustrations

29 Quartz Cementation in Sandstones
Edited by R.H. Worden and S. Morad
2000, 352 pages, 231 illustrations

28 Fluvial Sedimentology VI
Edited by N.D. Smith and J. Rogers
1999, 328 pages, 280 illustrations

26 Carbonate Cementation in Sandstones
Edited by S. Morad
1998, 576 pages, 297 illustrations

25 Reefs and Carbonate Platforms in the Pacific and Indian Oceans
Edited by G.F. Camoin and P.J. Davies
1998, 336 pages, 170 illustrations

24 Tidal Signatures in Modern and Ancient Sediments
Edited by B.W. Flemming and A. Bartholomä
1995, 368 pages, 259 illustrations

23 Carbonate Mud-mounds
Their Origin and Evolution
Edited by C.L.V. Monty, D.W.J. Bosence, P.H. Bridges and B.R. Pratt
1995, 543 pages, 330 illustrations

3 The Seaward Margin of Belize Barrier and Atoll Reefs
Edited by N.P. James and R.N. Ginsburg
1980, 203 pages, 110 illustrations

1 Pelagic Sediments on Land and Under the Sea
Edited by K.J. Hsu and H.C. Jenkyns
1975, 448 pages, 200 illustrations

REPRINT SERIES

4 Sandstone Diagenesis: Recent and Ancient
Edited by S.D. Burley and R.H. Worden
2003, 648 pages, 223 illustrations

FIELD GUIDE SERIES

1 A Field Guide to the Neogene Sedimentary Basins of the Almeria Province, SE Spain
Edited by A.E. Mather, J.M. Martin,
A.M. Harvey and J.C. Braga
2001, 368 pages, 170 illustrations

SPECIAL PUBLICATION NUMBER 36 OF THE INTERNATIONAL ASSOCIATION OF SEDIMENTOLOGISTS

Braided Rivers: Process, Deposits, Ecology and Management

EDITED BY

Gregory H. Sambrook Smith, James L. Best, Charlie S. Bristow and Geoff E. Petts

SERIES EDITOR

Ian Jarvis
School of Earth Sciences and Geography
Centre for Earth and Environmental Science Research
Kingston University
Penrhyn Road
Kingston-upon-Thames KT1 2EE
UK

The second international conference on Braided Rivers was supported by the following sponsors, to whom we are extremely grateful:
International Association of Sedimentologists
British Sedimentological Research Group
British Geomorphological Research Group
BP

Blackwell Publishing

© 2006 International Association of Sedimentologists
and published for them by
Blackwell Publishing Ltd

BLACKWELL PUBLISHING
350 Main Street, Malden, MA 02148-5020, USA
9600 Garsington Road, Oxford OX4 2DQ, UK
550 Swanston Street, Carlton, Victoria 3053, Australia

The right of Gregory H. Sambrook Smith, James L. Best, Charlie S. Bristow and
Geoff E. Petts to be identified as the Authors of the Editorial Material in this Work
has been asserted in accordance with the UK Copyright, Designs, and Patents Act 1988.

All rights reserved. No part of this publication may be reproduced, stored in a
retrieval system, or transmitted, in any form or by any means, electronic, mechanical,
photocopying, recording or otherwise, except as permitted by the UK Copyright,
Designs, and Patents Act 1988, without the prior permission of the publisher.

First published 2006 by Blackwell Publishing Ltd

1   2006

Library of Congress Cataloging-in-Publication Data

Braided rivers : process, deposits, ecology and management / edited by
Gregory H. Sambrook Smith . . . [et al.].
   p. cm.
  Includes bibliographical references and index.
  ISBN-13: 978-1-4051-5121-4 (pbk. : alk. paper)
  ISBN-10: 1-4051-5121-8 (pbk. : alk. paper)
  1. Braided rivers.   I. Sambrook Smith, Gregory H.

GB1205.B67 2006
333.91′62—dc22
                                                                   2005034253

A catalogue record for this title is available from the British Library.

Set in 10.5/12.5 pt Palatino
by Graphicraft Limited, Hong Kong
Printed and bound in Singapore
by Markono Print Media Pte Ltd

The publisher's policy is to use permanent paper from mills that operate a sustainable
forestry policy, and which has been manufactured from pulp processed using acid-free
and elementary chlorine-free practices. Furthermore, the publisher ensures that the
text paper and cover board used have met acceptable environmental accreditation
standards.

For further information on
Blackwell Publishing, visit our website:
www.blackwellpublishing.com

Contents

Acknowledgements, vi

Braided rivers: where have we come in 10 years? Progress and future needs, 1
Greg Sambrook Smith, Jim Best, Charlie Bristow and Geoff Petts

Depositional models of braided rivers, 11
John S. Bridge and Ian A. Lunt

A sedimentological model to characterize braided river deposits for hydrogeological applications, 51
Peter Huggenberger and Christian Regli

Scaling and hierarchy in braided rivers and their deposits: examples and implications for reservoir modelling, 75
Sean Kelly

Approaching the system-scale understanding of braided river behaviour, 107
Stuart N. Lane

Cellular modelling of braided river form and process, 137
A.P. Nicholas, R. Thomas and T.A. Quine

Numerical modelling of alternate bars in shallow channels, 153
A. Bernini, V. Caleffi and A. Valiani

Methods for assessing exploratory computational models of braided rivers, 177
Andrea B. Doeschl, Peter E. Ashmore and Matt Davison

Bed load transport in braided gravel-bed rivers, 199
Christian Marti and Gian Reto Bezzola

Sediment transport in a microscale braided stream: from grain size to reach scale, 217
P. Meunier and F. Métivier

Morphological analysis and prediction of river bifurcations, 233
Guido Zolezzi, Walter Bertoldi and Marco Tubino

Braided river management: from assessment of river behaviour to improved sustainable development, 257
Hervé Piégay, Gordon Grant, Futoshi Nakamura and Noel Trustrum

Bank protection and river training along the braided Brahmaputra–Jamuna River, Bangladesh, 277
Erik Mosselman

Morphological response of the Brahmaputra–Padma–Lower Meghna river system to the Assam earthquake of 1950, 289
Maminul Haque Sarker and Colin R. Thorne

Use of remote-sensing with two-dimensional hydrodynamic models to assess impacts of hydro-operations on a large, braided, gravel-bed river: Waitaki River, New Zealand, 311
D. Murray Hicks, U. Shankar, M.J. Duncan, M. Rebuffé and J. Aberle

Effects of human impact on braided river morphology: examples from northern Italy, 327
Nicola Surian

Ecology of braided rivers, 339
Klement Tockner, Achim Paetzold, Ute Karaus, Cécile Claret and Jürg Zettel

Riparian tree establishment on gravel bars: interactions between plant growth strategy and the physical environment, 361
Robert A. Francis, Angela M. Gurnell, Geoffrey E. Petts and Peter J. Edwards

Index, 381

Acknowledgements

We would like to thank the following reviewers: P.E. Ashmore, P.J. Ashworth, N. Asselman, P.A. Carling, F. Cioffi, P. Conaghan, T.J. Coulthard, R.I. Ferguson, C.R. Fielding, R.A. Francis, A. Gardiner, A.J.G. Gerrard, D.J. Gilvear, H.-P. Hack, T.B. Hoey, J. Hooke, P. Huggenberger, F.M.R. Hughes, D.L. Jacoby, B. Jones, M.G. Kleinhans, S.N. Lane, I.A. Lunt, F.J. Magilligan, C. Marti, B.W. McArdell, D. Moreton, E. Mosselman, A.B. Murray, A.P. Nicholas, C. Paola, J.D. Phillips, J.M. Pugh, M. Sterling, R.E. Thomas, B.R. Turner, J. West and D. Whited.

Braided rivers: where have we come in 10 years? Progress and future needs

GREG SAMBROOK SMITH*, JIM BEST†, CHARLIE BRISTOW‡ and GEOFF PETTS*

*School of Geography, Earth and Environmental Sciences, University of Birmingham, Edgbaston, Birmingham B15 2TT, UK (Email: g.smith.4@bham.ac.uk)
†Earth and Biosphere Institute, School of Earth and Environment, University of Leeds, Leeds LS2 9JT, UK
‡Research School of Geological and Geophysical Sciences, Birkbeck College and University College London, Gower Street, London WC1E 6BT, UK

INTRODUCTION

In 1992, the first conference on Braided Rivers was held at the Geological Society of London (Best & Bristow, 1993). It sought to bring together a wide range of practitioners, primarily sedimentologists, geomorphologists and engineers, to review the state-of-the-art in research, and discuss common problems and future research directions. At the time of the first conference, considerable effort had been devoted to the investigation of single-channel meandering channels, as well as more generic considerations of coarse-grained rivers, as championed by the benchmark series of gravel-bed rivers conferences that had begun in the 1980s. However, with the backdrop of the onset of management plans for one of the world's greatest braided rivers, the Brahmaputra–Jamuna (see Mosselman, this volume, pp. 277–288 and Sarker & Thorne, this volume, pp. 289–310, for updates), there was a consensus that research into braided rivers was not as advanced as for single channels, and several of the papers at the 1992 meeting outlined future avenues for research (e.g. Bridge, 1993; Bristow & Best, 1993; Ferguson, 1993). A decade on, after the second conference on braided rivers held at the University of Birmingham in 2003, it is thus appropriate to ask how much progress we have made during this period and what are the future research needs? The purpose of this brief introductory paper to the contributions in this book is to outline some key developments, place the chapters that follow into some kind of context and, perhaps most importantly, highlight areas where little progress has been made or where recent research has shown that more work is required. The topics identified are not, of course, exhaustive, but reflect the flavour of discussions at the second conference, and hopefully will serve as a backdrop to the contributions that follow.

DYNAMICS

It is fair to say that study of the dynamics of braided rivers has been revolutionized in the past decade, largely through the development of new techniques for measuring flow and bed morphology. Three broad techniques can be highlighted as having produced significant results and hold huge future promise:

1 flow quantification using acoustic Doppler current profiling (ADCP);
2 direct measurement of bed morphology using multibeam echo sounding (MBES);
3 remote sensing of braided river morphology using synoptic digital photogrammetric and airborne laser survey methods, from which digital elevation models (DEMs) can be developed (e.g. Lane, 2000, this volume, pp. 107–135; Westaway et al., 2000, 2001; Hicks et al., this volume, pp. 311–326).

Significant progress has been made in each of these areas and these now offer, for the first time, the possibility of quantifying flow process, bed morphology and channel change in a wide range of types and scale of braided rivers.

Richardson et al. (1996), Richardson & Thorne (1998, 2001) and McLelland et al. (1999) have applied ADCP to the study of flow around braid bars in the Brahmaputra–Jamuna, and examined the nature of secondary flows associated with braid bars. Acoustic Doppler current profiling technology, which has largely been transferred from oceanographic studies, has thus opened up the opportunity to study the fluid dynamics of large rivers (see Best et al., in press) for examples in the Brahmaputra–Jamuna). Application of ADCP within large braided rivers can achieve quantification of the mean flow field in complex bed geometries such as that associated with large sand-dune fields (e.g. Parsons et al., 2005), and at-a-point monitoring can offer the opportunity to examine turbulence characteristics (Barua & Rahman, 1998; Roden, 1998), although careful consideration must be given to sampling volume and time scales in relation to the scale of turbulence. Additionally, the recent development of shallow-water ADCP (see http://www.sontek.com/product/asw/aswov.htm) now opens up the opportunity of achieving flow-field quantification in smaller braided rivers. Recent work (Rennie et al., 2002; Rennie & Villard, 2003; Kostaschuk et al., 2004) has even begun to suggest that ADCP may be used to estimate bedload grain velocities over sand beds, and this may offer future potential for quantifying sediment transport rates.

Multibeam echo sounding technology has improved rapidly in the past few years, and systems are now available that can achieve accuracies of 5–10 mm in bed height and cover large swath widths relatively rapidly (e.g. see Bartholomä et al., 2004; Wilbers, 2004; Parsons et al., 2005). Multibeam echo sounding promises the ability to obtain high accuracy repeat surveys of the deepest and most turbid braided rivers, and achieve the long sought after aim of being able to accurately quantify channel change on a range of scales from the bedform to barform and whole channel. Again, these systems can be used with ADCP to examine sediment transport rates through quantification of bedform and bed morphological change (e.g. Abraham & Pratt, 2002; Bartholomä et al., 2004).

Photogrammetric quantification of the surface morphology of braided rivers has also evolved rapidly in the past decade (see Lane, 2000, this volume, pp. 107–135; Westaway et al., 2000, 2001; Lane et al., 2003, this volume, pp. 107–135), and DEMs can be produced to quantify morphological change and estimate sediment transport rates, something that has always been problematic in braided rivers. Additionally, it has been demonstrated (Lane et al., 2003) that the levels of error using these methods are similar to those of traditional ground surveying and are far superior where cross-section spacing exceeds 100 m. However, perhaps the real breakthrough is that these methods can be used over a much greater spatial scale than traditional ground surveys. For example, Lane et al. (2003) used DEMs that covered an area 1 km wide and 3.5 km long on the braided Waimakariri River (see Lane, this volume, pp. 107–135). Using traditional ground-based surveying, study reaches have been typically only of the order of hundreds of metres in length (e.g. Goff & Ashmore, 1994).

The improved quality of the morphological data now available will enable small-scale processes to be studied over much wider areas. This provides the opportunity for much greater collaboration with those undertaking numerical modelling, as data will become available at a much higher resolution, which is more appropriate for both gridding input and validation of some of the numerical modelling techniques outlined below. The challenges that lie ahead for the further development of DEM techniques are to try and apply them both at increased temporal scales and during high discharge events. This may restrict this type of study to rivers whose water may be relatively clear during flood. However, if DEMs could be constructed through flood events, this would greatly enhance our understanding of the dynamics of morphological change, as well as providing techniques for channel management. Additionally, all three of the methods outlined above hold significant promise for combining with improved techniques to study the subsurface (see 'Deposits' below) and provide a fuller understanding of the links between process, form and deposit.

When reviewing the research conducted over the past decade on both channel and bar dynamics, it is apparent that although braided rivers are composed of roughly equal numbers of confluences and diffluences, research has tended to be dominated by studies of the confluence zone. This is a clear gap in understanding, as diffluences also exert an important influence on the routing of sediment and water downstream, and hence the overall dynamics of the braidplain evolution. In recent years, some progress has begun to be made in this area (e.g. Richardson & Thorne, 2001; Bolla

Pittaluga et al., 2003; Federici & Paola, 2003; Zolezzi et al., this volume, pp. 233–256), but it is apparent that much further work is required to detail the dynamics of diffluences at a range of scales, and to elucidate how flow and sediment routing may be connected between confluence and diffluence nodes. In particular, there is an urgent need for comprehensive field studies of these diffluence zones utilizing the techniques outlined above to complement physical and numerical modelling work.

DEPOSITS

A decade ago, Bridge (1993, p. 13) concluded that 'existing braided-river facies models are virtually useless as interpretive and predictive tools'; it is thus pertinent to ask what has been achieved since the 1992 conference, and if we are any nearer to better predictive models of the subsurface character of braided rivers (see Bridge & Lunt, this volume, pp. 11–50)? One significant area of advance in this time has been the development and widespread adoption of ground penetrating radar (GPR) for use alongside the more traditional methods of trenching, outcrop description and coring. The application of GPR has enabled significant advances in the three-dimensional description of the architecture of braided river deposits and their facies. However, whilst being extensively utilized by those studying modern deposits (e.g. Bridge et al., 1998; Best et al., 2003; Skelly et al., 2003; Lunt et al., 2004) and Quaternary sediments (e.g. Huggenberger, 1993; Beres et al., 1999), GPR has not been widely adopted by those studying ancient deposits, who often still rely on traditional techniques of outcrop observation. The more limited application of GPR to ancient braided river deposits is perhaps principally due to the problems generated by diagenesis and rock fracture that can overprint and obscure primary depositional fabrics (see Bristow & Jol, 2003, and references therein), although some studies have successfully employed GPR within ancient fluvial sediments (e.g. Corbeanu et al., 2001; Truss, 2004).

Despite the significant progress made through the use of GPR, many issues remain unresolved. For instance, only a handful of modern braided rivers have received detailed study using GPR, and only one of these rivers has been gravel bed (Lunt & Bridge, 2004; Lunt et al., 2004; Bridge & Lunt, this volume, pp. 11–50). Studies of ancient gravelly braided alluvium (e.g. Huggenberger, 1993; Beres et al., 1999; Huggenberger & Regli, this volume, pp. 51–74) have thus been useful, although there is still a relatively poor characterization and understanding of both the similarities and differences in the depositional facies as a function of bed grain size. Additionally, most GPR data are still presented in a largely descriptive and qualitative way, and thus, although this has permitted a better understanding of facies relationships within and between different braided rivers (e.g. Sambrook Smith et al., 2005), these data have rarely been collected, presented or used in a way that make them useful for modelling or undertaking statistical analyses to assess the differences/similarities between rivers.

Additionally, although progress has been made towards establishing explicit links between formative processes and depositional product (e.g. Ashworth et al., 2000; Best et al., 2003; Lunt et al., 2004), it is clear that far more such studies, over a range of sizes and types of braided rivers, are required. Specifically, the new techniques outlined above (ADCP, MBES and photogrammetry) hold great promise for quantifying holistic braided river change, at a range of scales from the bedform to channel belt, and could be undertaken in conjunction with GPR surveys over longer time periods to link process and product more precisely than has hitherto been possible. Such an approach appears a priority. A final key problem is that the GPR studies mentioned above have generally concentrated on a few active bars within each river studied, and there is thus very little information on facies relationships across the entire braidplain (i.e. including channels). Such information is, however, critical when assessing preservation potential and applying such data to the ancient record, and there is thus a clear need for data to be collected from not just active bars but all areas of the braidplain. The study of abandoned sections of the braidplain may prove useful in this respect, and may require the use of other geophysical techniques such as electrical resistivity (Baines et al., 2002) or time domain reflectometry (Truss, 2004), given that GPR does not perform well in clay-rich sediments such as may be found in older deposits of the braidplain. These techniques may also allow us to piece together more precisely the links between sub-surface flow and alluvial architecture (e.g. Bridge & Tye, 2000; Truss, 2004; Huggenberger & Regli, this volume, pp. 51–74).

NUMERICAL MODELLING

A decade ago, little progress had been made in the numerical modelling of either flow dynamics or sediment transport within braided rivers. However, given the difficulty of obtaining reliable field-based datasets on flow and sediment transport in braided rivers, the derivation of robust numerical models was a clear priority that needed to be addressed, as identified by Ferguson (1993, p. 85) who stated that 'numerical modelling is potentially a much better way of investigating dynamic interactions in evolving braid units'. Since this statement, progress has been made in three broad areas:

1 the application of commercially available computational fluid dynamics (CFD) software;
2 the development of cellular models to investigate planform evolution and characteristics;
3 initial attempts to produce full physically-based numerical models of aspects of braided river evolution.

Computational fluid dynamics models involve the construction of a grid or mesh to represent the flow domain over bed topography, and the Navier–Stokes equations are then calculated for each cell in the grid in three dimensions, such that rules relating to the conservation of mass and momentum are not broken. Some success has been achieved with this approach where it has been applied to either channel confluences (e.g. Bradbrook *et al.*, 1998) or flow around an individual bar (e.g. Lane & Richards, 1998; Nicholas & Sambrook Smith, 1999). However, there are two outstanding issues that need to be resolved before such models can be applied further. First, since braided rivers are characterized by varying topography and multiple channels, the geometry at a range of scales from grain to channel is complex. Such topography requires schemes to account for complex flow both near the bed and within the bed (an important consideration in streambed ecology), such as those being advanced by Lane *et al.* (2002, 2004) and Hardy *et al.* (2005), which can better capture near-bed flow, including the numerous areas of separated flow. Second, modelling of braided river change requires linkage to sediment transport algorithms (see Marti & Bezzola, this volume, pp. 199–215; Meunier & Metivier, this volume, pp. 217–231) and simulation of new bed topography, thus demanding that a new mesh is constructed for each change in topography following erosion or deposition. This problem has also begun to be addressed by Lane *et al.* (2002) who have developed a method for using CFD that does not require structured grids with boundary-fitted coordinates. Although only tested on a small scale, this new methodology has great potential to begin to develop full physically based models of braided river flow and sediment transport dynamics. Some progress in this area has also begun with attempts to model the evolution of alternate bars (see Bernini *et al.*, this volume, pp. 153–175). Since alternate bars may represent one route in the first stages of bar growth (e.g. see model of Bridge, 1993), if these forms can be modelled, it could lead to the successful development of models for far more complex bed morphology. However, this remains some way off since the modelling conducted to date has been restricted to fairly simple channel geometries. The further development of these models also requires, as always, more computational power, although multinode parallel processing now offers greater potential than ever before.

Another area where progress has been made in modelling braided rivers in the past decade is the development of cellular models, as first outlined by Murray & Paola (1994, 1997) and more recently modified by others (e.g. Nicholas, 2000; Thomas & Nicholas, 2002; Lane, this volume, pp. 107–135; Nicholas *et al.*, this volume, pp. 137–150). The cellular modelling approach operates in a relatively simple way, starting with a network of cells with a random variation in their elevation. Water is then routed to adjacent cells on the basis of the bed gradient between cells, with more discharge flowing between cells with a greater bed slope. The movement of sediment can then also be routed such that the transport rate is related to the discharge moving between two cells, with lateral flows of water and sediment being accommodated in the model. Although this approach does not capture or simulate the process dynamics or real physics of braided river flow, Murray & Paola (1994, 1997) demonstrated that this relatively simple approach can replicate some of the gross planform features of braided rivers. The work highlights that the presence or absence of appreciable local redeposition of sediment appears to be a key control of braiding, as revealed in the early physical modelling studies of Leopold & Wolman (1957). This cellular automata approach was further developed by Nicholas (2000) to take account of more variable conditions of channel width and discharge, and generated improved estimates of bedload yield when compared with previous approaches. A significant result of the study of Nicholas (2000) was that, even at relatively low flows, the model predicted higher rates of sediment transport than other models, a reflection of how areas of deep water, and higher transport rates, still occur within the braidplain even when average depths are low. An increased braiding intensity may thus lead to higher rates of sediment transport. The further development of these approaches represents an area where progress should be possible over the next decade, as discussed by Nicholas (2005), Lane (this volume, pp. 107–135) and Nicholas et al. (this volume, pp. 137–150).

The development of numerical models to simulate the deposits of braided rivers has perhaps been more limited. Instead of the more physically based approaches adopted by geomorphologists and engineers in modelling using a CFD approach, focus has largely concentrated on geometrical models that 'attempt to produce spatial patterns similar to those observed in the field using empirically derived geometrical relationships' (Webb & Anderson, 1996, p. 534). This approach has the advantage that it requires less computational power and so can potentially be used to address issues on a broader temporal and spatial scale. However, some have questioned how closely such approaches represent the actual deposits of braided rivers (see Bridge & Lunt, this volume, pp. 11–50) and there is clearly more progress to be made in this respect. A more explicit linking of the physically based approaches used to study surface processes with likely depositional sequences forms one of the key frontiers in predicting the architecture of braided river alluvium and forms a key area for research between geomorphologists, applied mathematicians, engineers and Earth scientists.

SCALE

Issues relating to the scale of braided rivers are fundamental and were identified as a key research area by Bristow & Best (1993, p. 3) who wrote 'The issue of scaling depositional form and formative process across this range [laboratory flume to 20 km wide braidplains] of braided channel sizes is rarely addressed yet is central when applying results and models from one channel size to a system of a completely different magnitude'. There has been significant research into various aspects of scale over the past decade, perhaps most notably the papers by Sapozhnikov & Foufoula-Georgiou (1997; Foufoula-Georgiou & Sapozhnikov, 2001) on the scale invariant aspects of braided river surface morphology and evolution. This work concludes that braided rivers display statistical scale invariance in their morphology but dynamic scaling in their evolution. These conclusions were suggested to be valid for a wide range and type of braided rivers, and only break down where there are strong geological controls on the river channel pattern. The importance of this work is that it suggests that the results of morphological studies from a small section of a braided river can be applied to a larger one, or that the results obtained from physical modelling experiments can be applied to a field prototype. The application of these issues with respect to the modelling of braided rivers is discussed by Doeschl et al. (this volume, pp. 177–197). The challenge now remains to investigate whether hydrological variables also display the same type of scale invariance.

Progress has also been made in investigating the scale invariance of braided river deposits, a more complex problem since the three-dimensionality of the form must be considered, together with preservation potential, rather than solely the two-dimensional planform as discussed above. Relationships between the scale of bedforms and their preserved stratasets have been proposed (Bridge, 1997; Bridge & Best, 1997; Leclair et al., 1997) and there is now a need to extend this approach to cover larger features such as unit and compound bars. From a broader perspective, Sambrook Smith et al. (2005) compared, in a qualitative sense, facies from three different sandy braided rivers spanning three orders of magnitude in scale. They concluded that these rivers did exhibit a degree of scale invariance although a range of other factors, such as discharge regime, local bar and channel topography, the channel width:depth ratio and abundance of vegetation, also needed to be taken into account. The discussions by Bridge & Lunt (this volume, pp. 11–50) and Kelly (this volume, pp. 75–106) develop the issue of scale further within the context of braided river deposits.

ECOLOGY

In 1999, Ward et al. (1999, p. 71) stated that 'The role of islands has been almost totally ignored by stream ecologists'. However, in recent years there has been a major re-evaluation of some of the paradigms of river ecology in the context of braided rivers (e.g. Ward et al., 1999, 2001, 2002; Tockner et al., 2000; Stanford & Ward, 2001; Tockner, this volume, pp. 339–359), together with a growing new understanding of how ecological and geomorphological processes interact in such areas as braid bar formation and evolution (e.g. Edwards et al., 1999; Kollmann et al., 1999; Gurnell et al., 2000, 2001; Johnson, 2000; Petts et al., 2000; Gurnell & Petts, 2002; Van der Nat et al., 2003; Francis et al., this volume, pp. 361–380). This body of work represents substantive progress in new areas over the past decade, as evidenced by the fact that there were no contributions on ecological processes at the first Braided Rivers conference in 1992. Progress in this area has also been characterized by an interdisciplinary approach, primarily between ecologists and geomorphologists, and the results of this research have led to development of channel management based on sound ecological, as well as geomorphological, principles (see Piégay et al., this volume, pp. 257–275; Surian, this volume, pp. 327–338); the need for, and utility of, such approaches are shown, for example, by the fact that they are a key component of new legislation such as the European Union Water Framework Directive.

The basic assumption underpinning many key concepts within river ecology, such as the River Continuum Concept (Vannote et al., 1980), the Serial Discontinuity Concept (Ward & Stanford, 1983) and the Flood Pulse Concept (Junk et al., 1989), is that rivers are essentially single-channel systems. Such an assumption makes these concepts difficult, if not impossible, to apply to braided rivers that characteristically have great spatial and temporal variability at a wide range of scales in properties such as flow depth, grain size, bar stability and the connectivity between channels. It is this variability in habitat that contributes towards the high biodiversity typical of braided rivers. For example, Tockner et al. (2000) modified the Flood Pulse Concept and coined the term 'flow pulse' to represent the importance of below-bankfull events in connecting different areas within a river system. This is especially important for braided rivers, where the multiple bars and channels can all be at slightly different elevations, and where modest changes in discharge can thus lead to significant differences in connectivity between the different parts of the river. The nature of this connectivity will determine the exchange of solids (e.g. organic material and siliciclastic sediment) and organisms across the braidplain; the complexity inherent within these relationships has only just begun to be identified and studied.

The problems associated with flow stage inherent within the idea of the 'flow pulse' are not unique within river ecology. Geomorphologists have long debated how flow stage can influence the processes of bar evolution, and whether it is valid to relate low-flow observations of exposed bars to high-flow conditions when bars may be submerged and thus often difficult to observe. Likewise, the calculation of the braiding index of a river or measurement of bar dimensions will all be influenced by flow stage. Bristow & Best (1993, p. 2) stated that 'little data exists for the comparison of bar and channel morphology at different flow stages', and it is still apparent that such data could prove a useful collaborative venture for both ecologists and geomorphologists alike. The new remote sensing techniques currently being used to quantify braided river topography (see Lane, this volume, pp. 107–135; Hicks et al., this volume, pp. 311–326) promise greatly improved levels of accuracy in this respect and will find strong uses in ecological applications. For example, sequential images could be used to accurately quantify patterns of flow depth, bar area and bar edge length that can be related to ecological surveys conducted over the same time period; these can then be used to predict temporally varying connectivity and its ecological consequences across the braidplain. Use of this knowledge in a predictive sense requires closer collaboration between ecologists and Earth surface scientists seeking to model braided river systems. Again, although the numerical modelling of braided rivers is still in its infancy (see above), there is great potential to use such numerical models to predict patterns of change within a system of relevance to ecologists, as outlined by Richards *et al.* (2002).

SUMMARY

Perhaps the most influential positive trend in the past decade has been the increased level of communication and collaboration between the range of disciplines that have researched braided rivers. It is thus apparent that much progress has been made in meeting the challenge of Bristow & Best (1993) for a greater discourse between braided river scientists. This continued willingness to work in an interdisciplinary manner will be central when meeting the challenges of the next decade, and applying this knowledge in areas as diverse as channel management and sustainable river ecology to the characterization of ancient alluvium forming hydrocarbon reservoirs. Key areas of research that appear ripe for study, and some of which are begun in the contributions within this volume, include:

1 Quantification of the full range of feedbacks and links operating between the geomorphological and ecological components of the braidplain.
2 Increasing the temporal resolution at which geomorphological data are collected, with particular emphasis on flood events and long-term monitoring of channel evolution.
3 Providing a more explicit link between the processes of bedform, bar and channel evolution and their resultant deposits.
4 Using and applying sedimentological data in a more quantitative way to enable the development of statistical and numerical models of deposits.
5 Conducting the above studies in a range of braided river environments, scales and grain sizes to properly assess the degree to which current models (whether ecological, geomorphological or sedimentological) can be applied.

ACKNOWLEDGEMENTS

We are extremely grateful to the International Association of Sedimentologists (IAS), British Sedimentological Research Group (BSRG), British Geomorphological Research Group (BGRG) and BP for their sponsorship of the second international conference on Braided Rivers, and to the IAS and Blackwell for their support of this volume. Ian Jarvis is thanked for his advice and careful handling of the manuscripts. Much of our interest and research concerning braided rivers has been part of larger collaborative research programmes over the past decade involving a range of researchers, and we are grateful in particular to Phil Ashworth, John Bridge, Frank Ethridge, Angela Gurnell, Ian Lunt, Stuart Lane, Oscar Orfeo, Dan Parsons, Chris Simpson, Ray Skelly and Klement Tockner for their role in these field campaigns and discussions on many ideas within this paper. Our work on braided rivers over the past decade has been enabled and supported by grants from the UK Natural Environment Research Council for which we are extremely grateful: grants to GSS and JB (GR9/04273 with Ashworth and NER/A/S/2003/00538 with Ashworth and Lane); JB (NER/A/S/2001/00445 and NER/B/S/2003/00243 with Lane and Parsons) and the Royal Society (with Lane and Parsons); JB and CB (GR9/02034 with Ashworth); GP (GR9/03249 with Gurnell). JB is also grateful for award of a Leverhulme Trust Research Fellowship that aided the completion of this paper and editing of some of the papers in this volume.

REFERENCES

Abraham, D. and Pratt, T. (2002) *Quantification of Bed-load Transport using Multibeam Survey Data and Traditional Methods.* ERDC/CHL CHETN-VII-4, US Army Engineer Research and Development Center, Vicksburg, MS. http://chl.wes.army.mil/library/publications/chetn

Ashworth, P.J., Best, J.L., Roden, J.E., Bristow, C.S. and Klaassen, G.J. (2000) Morphological evolution and dynamics of a large, sand braid-bar, Jamuna River, Bangladesh. *Sedimentology*, **47**, 533–555.

Baines, D., Smith, D.G., Froese, D.G., Bauman, P. and Nimeck, G. (2002) Electrical resistivity ground imaging (ERGI): a new tool for mapping the lithology and geometry of channel-belts and valley-fills. *Sedimentology*, **49**, 441–449.

Bartholomä, A., Ernstern, V.B., Flemming, B.W. and Bartholdy, J. (2004) Bedform dynamics and net sediment transport paths over a flood-ebb cycle in the Grådyb channel (Denmark), determined by high-resolution multibeam echosounding. *Geogr. Tidsskr. Dan. J. Geogr.*, **104**, 45–55.

Barua, D.K. and Rahman, K.H. (1998) Some aspects of turbulent flow structure in large alluvial rivers. *J. Hydraul. Res.*, **36**, 235–252.

Beres, M., Huggenberger, P., Green, A.G. and Horstmeyer, H. (1999) Using two- and three-dimensional georadar methods to characterize glaciofluvial architecture. *Sediment. Geol.*, **129**, 1–24.

Best, J.L. and Bristow, C.S. (Eds) (1993) *Braided Rivers*. Special Publication 75, Geological Society Publishing House, Bath, 419 pp.

Best, J.L., Ashworth, P.J., Bristow, C.S. and Roden, J.E. (2003) Three-dimensional sedimentary architecture of a large, mid-channel sand braid bar, Jamuna River, Bangladesh. *J. Sediment. Res.*, **73**, 516–530.

Best, J., Ashworth, P., Sarker, M.H. and Roden, J. (in press) The Brahmaputra–Jamuna River, Bangladesh. In: *Large Rivers* (Ed. A. Gupta). Wiley, Chichester.

Bolla Pittaluga, M., Repetto, R. and Tubino, M. (2003) Channel bifurcation in braided rivers: equilibrium configurations and stability. *Water Resour. Res.*, **39**, 1046–1059.

Bradbrook, K.F., Biron, P.M., Lane, S.N., Richards, K.S. and Roy, A.G. (1998) Investigation of controls on secondary circulation in a simple confluence geometry using a three-dimensional numerical model. *Hydrol. Process.*, **12**, 1371–1396.

Bridge, J.S. (1993) The interaction between channel geometry, water flow, sediment transport and deposition in braided rivers. In: *Braided Rivers* (Eds J.L. Best and C.S. Bristow), pp. 13–71. Special Publication 75, Geological Society Publishing House, Bath.

Bridge, J.S. (1997) Thickness of sets of cross strata and planar strata as a function of formative bedwave geometry and migration, and aggradation rate. *Geology*, **25**, 971–974.

Bridge, J.S. and Best, J.L. (1997) Preservation of planar laminae arising from low-relief bed waves migrating over aggrading plane beds: comparison of experimental data with theory. *Sedimentology*, **44**, 253–262.

Bridge, J.S. and Tye, R.S. (2000) Interpreting the dimensions of ancient fluvial channel bars, channels, and channel belts from wireline logs and cores. *Am. Assoc. Petrol. Geol. Bull.*, **84**, 1205–1228.

Bridge, J.S., Collier, R. and Alexander, J. (1998) Large-scale structure of Calamus River deposits (Nebraska, USA) revealed using ground-penetrating radar. *Sedimentology*, **45**, 977–986.

Bristow, C.S. and Best, J.L. (1993) Braided rivers: perspectives and problems. In: *Braided Rivers* (Eds J.L. Best and C.S. Bristow), pp. 1–9. Special Publication 75, Geological Society Publishing House, Bath.

Bristow, C.S. and Jol, H.M. (Eds) (2003) *Ground Penetrating Radar in Sediments*. Special Publication 211, Geological Society Publishing House, Bath, 338 pp.

Corbeanu, R.M., Soegaard, K., Szerbiak, R.B., Thurmond, J.B., McMechan, G.A., Wang, D., Snelgrove, S., Forseter, C.B. and A. Menitove (2001) Detailed internal architecture of a fluvial channel sandstone determined from outcrop cores, and 3-D ground-penetrating radar: examples from the middle Cretaceous Ferron Sandstone, east-central Utah. *Am. Assoc. Petrol. Geol. Bull.*, **85**, 1583–1608.

Edwards, P.J., Kollmann, J., Gurnell, A.M., Petts, G.E., Tockner, K. and Ward, J.V. (1999) A conceptual model of vegetation dynamics on gravel bars of a large Alpine river. *Wetlands Ecol. Manage.*, **7**, 141–153.

Federici, B. and Paola, C. (2003) Dynamics of channel bifurcations in noncohesive sediments. *Water Resour. Res.*, **39**, 1162.

Ferguson, R.I. (1993) Understanding braiding processes in gravel-bed rivers: progress and unsolved problems. In: *Braided Rivers* (Eds J.L. Best and C.S. Bristow), pp. 73–87. Special Publication 75, Geological Society Publishing House, Bath.

Foufoula-Georgiou, E. and Sapozhnikov, V. (2001) Scale invariances in the morphology and evolution of braided rivers. *Math. Geol.*, **33**, 273–291.

Goff, J.R. and Ashmore, P.E. (1994) Gravel transport and morphological change in braided Sunwapta River, Alberta, Canada. *Earth Surf. Process. Landf.*, **19**, 195–212.

Gurnell, A.M. and Petts, G.E. (2002) Island-dominated landscapes of large floodplain rivers, a European perspective. *Freshwat. Biol.*, **47**, 581–600.

Gurnell, A.M., Petts, G.E., Hannah, D.M., Smith, B.P.G., Edwards, P.J., Kollmann, J., Ward, J.V. and Tockner, K. (2000) Wood storage within the active zone of a large European gravel-bed river. *Geomorphology*, **34**, 55–72.

Gurnell, A.M., Petts, G.E., Hannah, D.M., Smith, B.P.G., Edwards, P.J., Kollman, J., Ward, J.V. and Tockner, K. (2001) Riparian vegetation and island formation along the gravel-bed Fiume Tagliamento, Italy. *Earth Surf. Process. Landf.*, **26**, 31–62.

Hardy, R.J., Lane, S.N., Lawless, M.R., Best, J.L., Elliot, L. and Ingham, D.B. (2005) Development and testing of a numerical code for treatment of complex river channel topography in three-dimensional CFD models with structured grids. *J. Hydraul. Res*, **43**, 1–13.

Huggenberger, P. (1993) Radar facies: recognition of facies patterns and heterogeneities within Pleistocene Rhine gravels, NE Switzerland. In: *Braided Rivers* (Eds J.L. Best and C.S. Bristow), pp. 163–176. Special Publication 75, Geological Society Publishing House, Bath.

Johnson, W.C. (2000) Tree recruitment and survival in rivers: influence of hydrological processes. *Hydrol. Process.*, **14**, 3051–3074.

Junk, W.J., Bayley, P.B. and Spinks, R.E. (1989) The flood-pulse concept in river-floodplain systems. *Can. Spec. Publ. Fish. Aquat. Sci.*, **106**, 110–127.

Kollmann, J., Vieli, M., Edwards, P.J., Tockner, K. and Ward, J.V. (1999) Interactions between vegetation development and island formation in the Alpine river Tagliamento. *Appl. Vegetat. Sci.*, **2**, 25–36.

Kostaschuk, R.A., Best, J.L., Villard, P.V., Peakall, J. and Franklin, M. (2004) Measuring flow velocity and sediment transport with an acoustic Doppler current profiler. *Geomorphology*, doi: 10.1016/j.geomorph.2004.07.012.

Lane, S.N. (2000) The measurement of river channel morphology using digital photogrammetry. *Photogram. Rec.*, **16**, 937–957

Lane, S.N. and Richards, K.S. (1998) Two-dimensional modelling of flow processes in a multi-thread channel. *Hydrol. Process.*, **12**, 1279–98.

Lane, S.N., Hardy, R.J., Elliott, L. and Ingham, D.B. (2002) High-resolution numerical modelling of three-dimensional flows over complex river bed topography. *Hydrol. Process.*, **16**, 2261–2272.

Lane, S.N., Westaway, R.M. and Hicks, D.M. (2003) Estimation of erosion and deposition volumes in a large gravel-bed, braided river using synoptic remote sensing. *Earth Surf. Process. Landf.*, **28**, 249–71.

Lane, S.N., Hardy, R.J., Elliot, L. and Ingham, D.B. (2004) Numerical modelling of flow processes over gravelly-surfaces using structured grids and a numerical porosity treatment. *Water Resour. Res.*, doi:10.1029/2002WR001934.

Leclair, S.F., Bridge, J.S. and Wang, F. (1997) Preservation of cross-strata due to migration of subaqueous dunes over aggrading and non-aggrading beds: comparison of experimental data with theory. *Geosci. Can.*, **24**, 55–66.

Leopold, L.B. and Wolman, M.G. (1957) River channel patterns—braided, meandering and straight. *US Geol. Surv. Prof. Pap.*, **282B**, 39–85.

Lunt, I.A. and Bridge, J.S. (2004) Evolution and deposits of a gravely braid bar Sagavanirktok River, Alaska. *Sedimentology*, **51**, 1–18.

Lunt, I.A., Bridge, J.S. and Tye, R.S. (2004) A quantitative, three-dimensional depositional model of gravely braided rivers. *Sedimentology*, **51**, 377–414.

McLelland, S.J., Ashworth, P.J., Best, J.L., Roden, J.E. and Klaassen, G.J. (1999) Flow structure and transport of sand-grade suspended sediment around an evolving braid-bar, Jamuna River, Bangladesh. In: *Fluvial Sedimentology VI* (Eds N.D. Smith and J. Rogers), pp. 43–57. Special Publication 28, International Association of Sedimentologists. Blackwell Science, Oxford.

Murray, A.B. and Paola, C. (1994) A cellular model of braided rivers. *Nature*, **371**, 54–57.

Murray, A.B. and Paola, C. (1997) Properties of a cellular braided stream model. *Earth Surf. Process. Landf.*, **22**, 1001–1025.

Nicholas, A.P. (2000) Modelling bedload yield in braided gravel bed rivers. *Geomorphology*, **36**, 89–106.

Nicholas, A.P. (2005) Cellular modelling in fluvial geomorphology. *Earth Surf. Process. Landf.*, **30**, 645–649.

Nicholas, A.P., and Sambrook Smith, G.H. (1999) Numerical simulation of three-dimensional flow hydraulics in a braided channel. *Hydrol. Process.*, **13**, 913–929.

Parsons, D.R., J.L. Best, R.J. Hardy, R. Kostaschuk, S.N. Lane and O. Orfeo (2005) The morphology and flow fields of three-dimensional dunes, Rio Paraná, Argentina: results from simultaneous multibeam echo sounding and acoustic Doppler current profiling. *J. Geophys. Res. Earth Surf*, **110**, FO4503, doi:10.1029/2004JF000231.

Petts, G.E., Gurnell., A.M., Gerrard, A.J., Hannah, D.M., Hansford, B., Morrissey, I., Edwards, P.J., Kollmann, J., Ward, J.V., Tockner, K., and Smith, B.P.G. (2000) Longitudinal variations in exposed riverine sediments: a context for the ecology of the Fiume Tagliamento, Italy. *Aquat. Conserv.*, **19**, 249–266.

Rennie, C.D., Millar, R.G. and Church, M.A. (2002) Measurement of bedload velocity using an acoustic Doppler current profiler. *J. Hydraul. Eng.*, **128**, 473–483.

Rennie, C.D. and Villard, P.V. (2003) Bedload measurement in both sand and gravel using an aDcp. *16th Canadian Hydrotechnical Conference October 22–24, 2003*. Canadian Society for Civil Engineers, Burlington, ON. 10 pp.

Richards, K., Brasington, J. and Hughes, F. (2002) Geomorphic dynamics of floodplains: ecological implications and a potential modelling strategy. *Freshwat. Biol.*, **47**, 559–579.

Richardson, W.R.R. and Thorne, C.R. (1998) Secondary currents around braid bar in Brahmaputra River, Bangladesh. *J. Hydraul. Eng.*, **124**, 325–328.

Richardson, W.R. and Thorne, C.R. (2001) Multiple thread flow and channel bifurcation in a braided river: Brahmaputra–Jamuna River, Bangladesh. *Geomorphology*, **38**, 185–196.

Richardson, W.R.R., Thorne, C.R. and Mahmood, S. (1996) Secondary flow and channel changes around a bar in the Brahmaputra River, Bangladesh. In:

Coherent Flow Structures in Open Channels (Eds P.J. Ashworth, S.J. Bennett, J.L. Best and S.J. McLelland), pp. 520–543. Wiley, Chichester.

Roden, J.E. (1998) *Sedimentology and dynamics of mega-sand dunes, Jamuna River, Bangladesh.* Unpublished PhD thesis, University of Leeds, 310 pp.

Sambrook Smith, G.H., Ashworth, P.J., Best, J.L., Woodward, J. & Simpson, C.J. (2005) The morphology and facies of sandy braided rivers: some considerations of spatial and temporal scale invariance. In: *Fluvial Sedimentology VII* (Eds M.D. Blum, S.B. Marriott and S.F. Leclair), pp. 145–158. Special Publication 35, International Association of Sedimentologists. Blackwell Publishing, Oxford.

Sapozhnikov, V.B. and Foufoula-Georgiou, E. (1997) Experimental evidence of dynamic scaling and indications of self-organized criticality in braided rivers. *Water Resour. Res.*, **32**, 1109–1112.

Skelly, R.L., Bristow, C.S. and Ethridge, F.G. (2003) Architecture of channel-belt deposits in an aggrading shallow sandbed braided river: the lower Niobrara River, northeast Nebraska. *Sediment. Geol.*, **158**, 249–270.

Stanford, J.A. and Ward, J.V. (2001) Revisiting the serial discontinuity concept. *Regul. River. Res. Manage.*, **17**, 303–310.

Thomas, R. and Nicholas, A.P. (2002) Simulation of braided river flow using a new cellular routing scheme. *Geomorphology*, **43**, 179–95.

Tockner, K., Malard, F. and Ward, J.V. (2000) An extension of the flood pulse concept. *Hydrol. Process.*, **14**, 2861–2883.

Truss, S. (2004) *Characterisation of sedimentary structure and hydraulic behaviour within the unsaturated zone of the Triassic Sherwood Sandstone aquifer in North East England.* Unpublished PhD thesis, University of Leeds, 252 pp.

Van der Nat, D., Tockner, K., Edwards, P.J., Ward, J.V. and Gurnell, A.M. (2003) Habitat change in braided flood plains (Tagliamento, NE-Italy). *Freshwat. Biol.*, **48**, 1799–1812.

Vanotte, R.L., Minshall, G.W., Cummins, K.W., Sedell, J.R. and Cushing, C.E. (1980) The river continuum concept. *Can. J. Fish. Aquat. Sci.*, **37**, 130–137.

Ward, J.V. and Stanford, J.A. (1983) The serial discontinuity concept of lotic ecosystems. In: *Dynamics of Lotic Ecosystems* (Eds T.D. Fontaine and S.M. Bartell), Ann Arbor Science, Ann Arbor, MI, 347–356.

Ward, J.V., Tockner, K., Edwards, P.J., Kollmann, J., Bretschko, G., Gurnell, A.M., Petts, G.E. and Rossaro, B. (1999) A reference system for the Alps: the 'Fiume Tagliamento'. *Regul. River. Res. Manage.*, **15**, 63–75.

Ward, J.V., Tockner, K., Uehlinger, U. and Malard, F. (2001) Understanding natural patterns and processes in river corridors as the basis for effective river restoration. *Regul. River. Res. Manage.*, **17**, 311–323.

Ward, J.V., Tockner, K., Arscott, D.B. and Claret, C. (2002) Riverine landscape diversity. *Freshwat. Biol.*, **47**, 517–539.

Webb, E.K. and Anderson, M.P. (1996) Simulation of preferential flow in three-dimensional, heterogeneous conductivity fields with realistic internal architecture. *Water Resour. Res.*, **32**, 533–545.

Westaway, R.M., Lane, S.N. and Hicks, D.M. (2000) The development of an automated correction procedure for digital photogrammetry for the study of wide, shallow, gravel-bed rivers. *Earth Surf. Process. Landf.*, **25**, 209–226.

Westaway, R.M., Lane, S.N. and Hicks, D.M. (2001) Airborne remote sensing of clear water, shallow, gravel-bed rivers using digital photogrammetry and image analysis. *Photogram. Eng. Remote Sens.*, **67**, 1271–1281.

Wilbers, A. (2004) *The development and hydraulic roughness of subaqueous dunes.* PhD thesis, Faculty of Geosciences, Utrecht University, Netherlands Geographical Studies 323, 224 pp.

Depositional models of braided rivers

JOHN S. BRIDGE and IAN A. LUNT[1]

Department of Geological Sciences, Binghamton University, P.O. Box 6000, Binghamton NY 13902-6000, USA

ABSTRACT

Depositional models of braided rivers are necessary for rational interpretation of ancient deposits, and to aid in the characterization of subsurface deposits (e.g. aquifers, hydrocarbon reservoirs). A comprehensive depositional model should represent bed geometry, flow and sedimentary processes, and deposits accurately, quantitatively, and in detail. Existing depositional models of braided rivers do not meet these requirements, and there are still many misconceptions about braided rivers and their deposits that need to be expunged. Over the past decade, there have been major advances in our understanding of braided rivers, making it possible to develop new and improved depositional models. First, the use of groundpenetrating radar in combination with cores and trenches has allowed detailed description of the different scales of deposit in braided rivers that vary widely in channel size and sediment size. Second, the study of braided channel geometry and kinematics has been facilitated by the use of aerial photographs taken at short time intervals. In some cases, these photographs have been analysed using digital photogrammetry to produce digital elevation models. Third, water flow and sediment transport at the all-important flood stages have been studied using new equipment and methods (e.g. acoustic Doppler current profilers, positioning using differential global positioning systems). However, such high-flow studies are rare, which is one reason why there has not been much progress in development of realistic theoretical models for the interaction between bed topography, water flow, sediment transport, erosion and deposition. Laboratory experimental studies of braided rivers have continued to be useful for examining the controls on channel geometry and dynamics, but have not been able to generate all of the different types and scales of strata observed in natural braided river deposits.

New depositional models for sand-bed and gravel-bed braided rivers are presented here based mainly on studies of natural rivers. They comprise:

1 maps showing idealized active and abandoned channels, compound bars and lobate unit bars;
2 cross-sections showing large-scale inclined strata and their internal structures, associated with migration of compound bars, unit bars and their superimposed bedforms;
3 vertical logs of typical sedimentary sequences through different parts of compound bar deposits and channel fills.

Compound bars migrate laterally and downstream, associated mainly with accretion of lobate unit bars. Abandoned channels are mainly filled with unit-bar deposits. The geometry of the different scales of strataset is related to the geometry and migration of the bedform associated with deposition of the strataset. In particular, the length-to-thickness ratio of stratasets is similar to the wavelength-to-height ratio of associated bedforms. Furthermore, the wavelength and height of bedforms such as dunes and bars are related to channel depth and width. Therefore, the thickness of a particular scale of strataset (e.g. medium-scale cross sets, large-scale sets of inclined strata)

---

[1]Present address: Earth and Biosphere Institute, School of Earth and Environment, University of Leeds, Leeds, West Yorkshire, LS2 9JT, UK (Email: ilunt@earth.leeds.ac.uk).

will vary with river dimensions. These relationships between the dimensions of stratasets, bedforms and channels mean that the depositional models can be applied to channels of all scales. However, realistic models of the spatial distribution and degree of preservation of channel bars and fills within channel belts need to be developed.

Keywords Braided rivers, depositional model, fluvial deposits, sedimentary structures.

INTRODUCTION

Braided rivers and their deposits are important components of the Earth's surface, now and in the past. The deposits of ancient braided rivers are indicators of past Earth surface environments, and may contain significant reserves of water and hydrocarbons (Martin, 1993; Bal, 1996; Anderson et al., 1999). Depositional models based on knowledge of modern braided rivers are necessary to allow rational interpretation of ancient deposits. Depositional models can also aid in the prediction of the nature of subsurface deposits where data (cores, well logs, seismic) are sparse. Ideally, a depositional model must represent landforms and sedimentary processes accurately, must contain detailed, three-dimensional sedimentary information (including the various superimposed scales of deposits), must be quantitative, and should have some predictive value. A depositional model should also provide parameters (e.g. permeability, porosity) relevant to modelling fluid flow through aquifers and hydrocarbon reservoirs. Existing depositional models for braided rivers (e.g. Miall, 1977, 1992, 1996; Bluck, 1979; Ramos et al., 1986; Bridge, 1993, 2003; Collinson, 1996) do not meet these ideals, because the nature and origin of modern river deposits are generally not known in detail. This is due partly to difficulties in:

1 describing deposits in three-dimensions below the water table;
2 studying depositional processes during the all-important high-flow stages, and over large time and space scales (see reviews by Bridge, 1985, 1993, 2003).

Over the past decade or so, some of these difficulties have been overcome by:

1 use of ground-penetrating radar (GPR) in combination with coring and trenching to describe all of the different scales of deposits in detail and in three dimensions (Jol & Smith, 1991; Gawthorpe et al., 1993; Huggenberger, 1993; Siegenthaler & Huggenberger, 1993; Alexander et al., 1994; Huggenberger et al., 1994; Jol, 1995; Bridge et al., 1995, 1998; Beres et al., 1995, 1999; Leclerc & Hickin, 1997; Asprion & Aigner, 1997, 1999; Van Overmeeren, 1998; Vandenberghe & Van Overmeeren, 1999; Bristow et al., 1999; Ekes & Hickin, 2001; Wooldridge, 2002; Regli et al., 2002; Best et al., 2003; Bristow & Jol, 2003; Skelly et al., 2003; Woodward et al., 2003; Lunt & Bridge, 2004; Lunt et al., 2004a,b; Sambrook Smith et al., 2005);
2 study of channel deposits in frozen rivers, allowing easy access to the whole channel belt, and the procurement of undisturbed cores of unconsolidated gravel (Lunt et al., 2004a,b; Lunt & Bridge, 2004);
3 study of the evolution of channel geometry using sequences of aerial photographs taken at short time intervals (Lane et al., 1998, 2001; Stoijic et al., 1998; Wooldridge, 2002; Lunt & Bridge, 2004).

These studies have allowed the construction of a new generation of depositional models for braided rivers (Lunt et al., 2004a,b; this paper).

However, studies of processes of fluid flow, sediment transport, erosion and deposition in braided rivers during flood stages are rare (Bridge & Gabel, 1992; Richardson et al., 1996; Richardson & Thorne, 1998; McLelland et al., 1999). Therefore, there has been little progress in the generation of realistic empirical models that link bed topography, fluid flow, sediment transport, erosion and deposition in braided rivers. The same goes for theoretical models. Braided-river processes and deposits have also been studied in laboratory flumes (Ashmore, 1982, 1991, 1993; Ashworth et al., 1999, 2004; Moreton et al., 2002; Sheets et al., 2002; Ashworth et al., 2004). Advantages of experimental studies are that the environment is manageable and channel-forming flow conditions can be studied. However, scaling problems must be overcome. Although experimental studies have been useful for elucidating the controls on braided channel geometry and dynamics, they have not been able to generate all of the superimposed scales of bedforms and associated strata observed in natural rivers. For example, in the experiments described by Ashworth *et al.* (1999), only relatively large-scale strata associated with bar accretion and channel filling were discernible.

The purpose of this paper is to discuss this new information on the geometry, flow and sedimentary processes, and deposits of braided rivers, and to show how this information has led to the development of new depositional models for braided rivers. The organization of this paper is as follows:

1 discussion of recent studies of braided river processes and deposits;
2 discussion of the usefulness of experimental and theoretical studies to development of depositional models;
3 presentation of new depositional models and their implications for interpreting and predicting the nature of ancient deposits.

STUDIES OF MODERN BRAIDED RIVER PROCESSES AND DEPOSITS

Methods

The most significant advance in studies of the deposits of modern braided rivers is the use of ground-penetrating radar (GPR) in combination with coring and trenching. Ground-penetrating radar reflection data are acquired in real time by moving transmitting and receiving antennae together along lines in a grid. The GPR reflections are due to changes in dielectric permittivity (and less commonly due to magnetic permeability), related to the amount and type of pore-filling material, sediment texture and composition (Van Dam & Schlager, 2000; Kowalsky *et al.*, 2001, 2004; Van Dam *et al.*, 2002; Lunt *et al.*, 2004a,b). As reflections are primarily caused by changes in pore-water saturation or volume, which are closely related to sediment texture and composition, reflections give a record of sedimentary strata. The amplitude of radar reflections depends on contrasts in radar velocity between adjacent strata and on stratal thickness compared with the wavelength of the transmitted pulse (Kowalsky *et al.*, 2001). A particular reflection amplitude can be caused by different combinations of sediment texture (sands adjacent to sandy gravel, open-framework gravel adjacent to sandy gravel). Above the water table, reflections are primarily related to variations in water saturation. Below the water table, reflections are related mainly to variations in porosity and sediment composition. In permafrost, reflections are related to porosity and relative proportions of water and ice in the pores.

Central frequencies of antennae commonly used in sedimentological studies are 50 to 1000 MHz. For a given subsurface sediment type, the depth of penetration decreases, and the resolution of sedimentary strata increases, with increasing antenna frequency (Jol, 1995; Woodward *et al.*, 2003). For example, 100 MHz antennae may have a penetration depth of the order of 10 m, and be able to resolve strata 0.3 m thick, whereas 450 MHz antennae may have a depth of penetration of several metres and be able to resolve strata 0.1 m thick. In order to distinguish all scales of stratification, it is desirable to use a range of antenna frequencies (50–1000 MHz). As an example, cross strata that are centimetres to decimetres thick, within cross sets of the order of many decimetres thick, can be discerned using high-frequency antennae. With low frequency antennae, only the set boundaries would be represented, giving the appearance of planar strata. If the set thickness of cross strata is of the order of only a few decimetres, even the high frequency antennae will only pick up the set boundaries. As a result, GPR has been most useful for imaging the larger scales of strata in channel belts. Furthermore, radar facies cannot be clearly related to sedimentary facies unless their three-dimensional geometry is investigated using a range of antenna frequencies, the antenna frequency is stated, and reasons for variation in reflection amplitude are understood. For example, concave-upward reflections in radar profiles may be associated with confluence scours, main channel fills, small cross-bar channels adjacent to unit bars, scours upstream of obstacles such as logs or ice blocks, or bases of trough cross strata. Interpretation of the origin of such reflections will depend on their three-dimensional geometry and an understanding of adjacent strata. The central antenna frequency and vertical exaggeration of GPR profiles need to be considered when making sedimentological interpretations. In addition, vertical variation in radar wave velocity must be considered when converting two-way travel time to depth in order to obtain the correct geometry of reflections.

Interpretation of the origin of the observed deposits has been accomplished from knowledge of, or inferences about, the nature of bed topography, flow and sedimentary processes during the flood stages when most erosion and deposition occurs. However, such high-flow data are very sparse. In the case of the Calamus River, data collection was facilitated by the operation of measuring equipment from bridges built across the channel (Bridge & Gabel, 1992; Gabel, 1993). In the case of the Brahmaputra River, high-stage flow data (and suspended sediment concentration) were collected using an acoustic Doppler current profiler operated from a boat, positioned using differential global positioning systems (GPS) (Richardson et al., 1996; Richardson & Thorne, 1998; McLelland et al., 1999).

Knowledge of the nature of channel and bedform migration is also essential for interpreting the origin of river deposits. This has been accomplished by repeat surveys using depth sounders, differential GPS, and using aerial photographs taken at frequent intervals. Recently, it has been possible to survey the bed topography of rivers and assess sediment transport rates using digital photogrammetry in combination with geographical information system (GIS) based digital elevation models (Lane et al., 1994, 1995, 1998, 2001; Martin & Church, 1996; Ashmore & Church, 1998; Stojic et al., 1998; Chandler, 1999; Westaway et al., 2000; Lane et al., 2003). These techniques have a vertical accuracy of the order of 0.1 m in clear, shallow water, but cannot be used in turbid opaque rivers.

Results

Only river studies where there are extensive new data on the channel geometry, flow and sedimentary processes, and deposits are discussed here. These studies were conducted on the Calamus River in Nebraska, USA (Bridge et al., 1998), the Brahmaputra/Jamuna River in Bangladesh (Best et al., 2003), the Sagavanirktok River on the North Slope of Alaska (Lunt et al., 2004a,b; Lunt & Bridge, 2004), and the Niobrara River in Nebraska, USA (Skelly et al., 2003). These studies are from sand-bed and gravel-bed rivers, and cover channel-belt widths from 200 m to 10 km. Note, for future reference, that sand-bed rivers commonly contain some gravel, and gravel-bed rivers always contain appreciable quantities of sand. These field studies, and some theoretical and experimental studies, form the main basis for the depositional models presented.

Sagavanirktok

The deposits of the gravelly Sagavanirktok River on the North Slope of Alaska were described using cores, wireline logs, trenches and GPR profiles (Lunt et al., 2004a,b; Lunt & Bridge, 2004). Study of the Sagavanirktok River when it was frozen allowed access to the whole channel belt and procurement of undisturbed cores of unconsolidated gravel. Porosity and permeability were determined from core samples, and the age of the deposits was determined using optically stimulated luminescence (OSL) dating. The origin of the deposits was inferred from:

1 interpretation of channel and bar formation and migration, and channel filling, using annual aerial photographs;
2 observations of water flow and sediment transport during floods;
3 observations of bed topography and sediment texture at low-flow stage.

The Sagavanirktok River contains compound braid bars and point (side) bars. Compound braid bars originate by growth of lobate unit bars and by chute cutoff of point bars (Fig. 1). Compound bars migrate downstream and laterally, associated with accretion of successive unit bars. The upstream ends of compound bars may be sites of erosion or accretion. During floods, most of the active riverbed is covered with sinuous-crested dunes, with minor proportions of bedload sheets (Fig. 2). Transverse ribs and ripples occur rarely in very shallow water. A channel segment may become abandoned if its upstream end becomes blocked by a channel bar. Channels that are becoming abandoned and filled contain lobate unit bars.

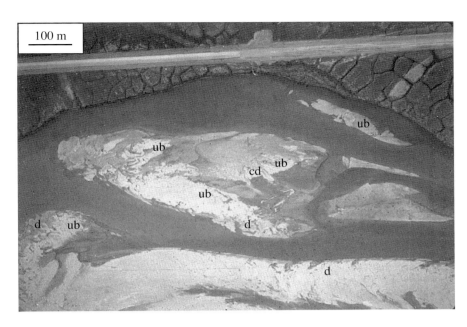

Fig. 1 Aerial photograph of the gravel-bed Sagavanirktok River, Alaska, with different scales of bed form indicated: ub, lobate unit bar; d, dunes superimposed on bar surfaces; cd, small channels and depositional lobes associated with the edges of unit bars. The photograph shows a single compound braid bar with flow from left to right. The bar tail is a relatively low area (dark colour) and contains a relatively large cross-bar channel. The compound braid bar grew by accretion of unit bars on its upstream end, sides and downstream end.

Recognition of different scales of bedform (e.g. bars, dunes), their migration patterns, and associated stratification was essential for building a model of gravelly braided rivers (Lunt et al., 2004a,b: Figs 1 to 3). Channel-belt deposits are composed of deposits of compound braid bars and point bars and large channel fills (Fig. 3A–C). Channel belt deposits are up to 7 m thick and 2.4 km wide, and are composed mainly of gravels, with minor sands and sandy silts (Fig. 3A). A channel belt may be capped by a decimetre thick, sandy–silty soil horizon, unless eroded by later channel belts. Migration of compound bars within channel belts forms compound sets of large-scale inclined strata composed of simple sets of large-scale inclined strata (formed by unit bars) and small channel fills (formed by cross-bar channels). Compound sets are hundreds of metres long and wide, metres thick, have basal erosion surfaces, and terminate laterally in large channel fills (Fig. 3B & C). Set thickness and vertical trends in grain size depend on the bed geometry and surface grain size of compound bars and the nature of bar migration.

Relatively thick, fining-upward sequences form as bar-tail regions migrate downstream into a curved channel or confluence scour. Grain size may increase towards the top of a thick fining-upward sequence where bar-head lobes migrate over the bar-tail. Relatively thin compound sets of large-scale strata with no vertical grain-size trend are found in riffle regions. Compound sets are composed mainly of sandy gravel, but open-framework gravel is common near set-bases (Fig. 3A). Reconstructing the origin and evolution of compound-bar deposits from only recent aerial photographs or cores is impossible. It is also impossible and unnecessary to determine from core whether a compound bar was a point bar or a braid bar (see Lunt & Bridge, 2004). A complete understanding of the evolution of a compound-bar deposit is only possible with a combination of frequent aerial photography, orthogonal GPR profiles and cores. Large channel fills are also composed mainly of simple sets of large-scale inclined strata (Fig. 3C), but are capped with sandy strata containing small- and medium-scale cross sets, and planar strata. These deposits are also generally sandier in the downstream parts of a channel fill. Between two and five compound bar deposits or large channel fills occur within the thickness of a channel-belt deposit (Fig. 3A).

Migration of unit bars forms simple sets of large-scale inclined strata (Fig. 3D). These sets are decimetres to metres thick, tens of metres long and wide, and generally fine upwards, although they may show no grain-size trend. Between three and seven simple sets of large-scale strata make

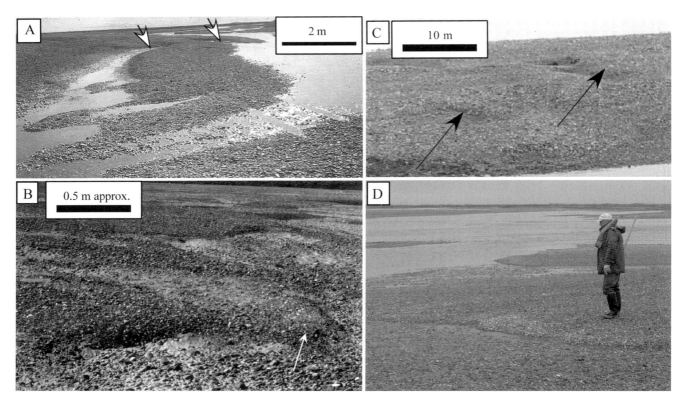

Fig. 2 Photographs of different scales of typical bedforms in the Sagavanirktok River. (A) Lobate unit bar with its side cut by small cross-bar channels and depositional lobes in the adjacent swale. Tyre tracks occur in cross-bar channel nearest to observer. This unit bar is in the tail region of a compound bar. The fronts of two other unit bars (arrows) occur in the (upstream) head region of the compound bar. (B) Dunes on surface of compound bar. Dunes are 0.5 m high and have crestline lengths of around 10 m. Modification by falling-stage flow indicated by arrow. (C) Dunes on the exposed side of a unit bar. Flow is from right to left. (D) Bedload sheets.

up the thickness of a compound large-scale set. Open-framework gravels are common at the base of a simple large-scale set. The large-scale strata are generally inclined at <10°, but may reach the angle of repose at the margin of the unit bar.

Small channel fills (cross-bar channels) are made up of small- and medium-scale cross sets, and planar strata. They are tens of metres long, metres wide, and decimetres thick, and occur at the tops of simple sets of large-scale strata, especially where simple large-scale sets occur at the top of compound large-scale sets.

Dune migration forms medium-scale sets of cross strata. Medium-scale cross sets are decimetres thick and wide, up to 3 m long, and contain isolated open-framework gravel cross-strata and sandy trough drapes. The thickness of medium-scale cross sets decreases upwards in compound bar deposits. Planar strata are formed by bedload sheets and may be made up of sandy gravel, open-framework gravel or sand. Individual strata are centimetres thick and decimetres to metres long and wide. Imbricated pebbles and pebble clusters are commonly found at the base of planar strata. Small-scale sets of cross strata, formed by ripples, generally occur in channel fills, as trough drapes, and as overbank deposits. Small-scale cross sets are centimetres thick and long, are always composed of sand, and may contain organic remains, root traces or burrows where they occur in channel fills or overbank deposits.

The geometry of the different scales of strataset in the Sagavanirktok River is related to the geometry and migration of their formative bedforms. In particular, the wavelength:height ratio of bedforms is similar to the length:thickness ratio of their associated deposits (Fig. 4). These relationships seem to apply to other sandy and gravelly

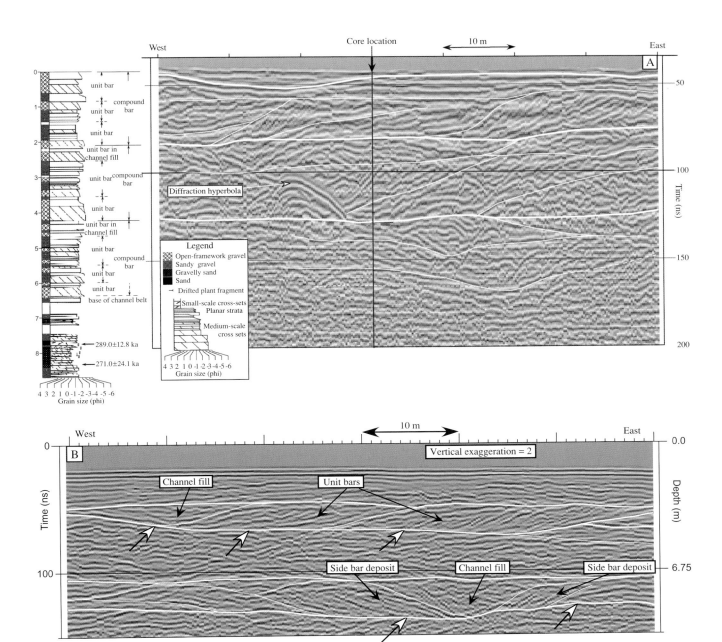

Fig. 3 Sagavanirktok channel deposits. (A) Example of 110 MHz GPR profile (across stream view) and core log through the channel belt. Core log and GPR profile are shown at the same vertical scale. Core log shows three compound-bar deposits each comprising three or four unit-bar deposits. Internal structure of unit-bar deposits is medium-scale cross strata (from dunes) and planar strata (from bedload sheets). The base of the active channel-belt (not visible on the GPR profile) is 6.4 m below the surface and overlies burrowed and rooted sands containing small-scale and medium-scale cross sets. The GPR profile has a vertical exaggeration of 5:1 and shows the two-dimensional geometry of simple sets of large-scale inclined strata due to unit bars (set bases marked by thin white lines) and compound sets of large-scale inclined strata due to compound bars (set bases marked with thick white lines). From Lunt *et al.* (2004b). (B) Example of 110 MHz GPR profile (across stream view) showing compound large-scale inclined strata associated with lateral accretion of a braid bar (upper part). The basal erosion surface is marked by arrows, and unit-bar deposits and a channel fill are bordered by white lines. The lower part of the profile shows accretion of side bars (bordered by white lines) towards a central fill of a confluence scour zone. From Bridge (2003).

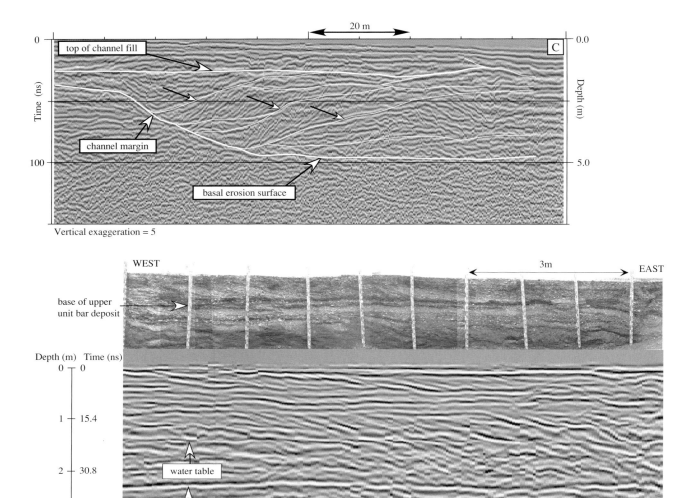

Fig. 3 (*cont'd*) (C) Example of 110 MHz GPR profile (across stream view) showing channel filling with unit bar deposits (bases arrowed) that accreted onto the western margin of a compound bar. Large-scale set boundaries are represented by white lines. From Bridge (2003). (D) Trench photomosaic and 450 MHz GPR profile through two superimposed unit bar deposits, cut oblique to flow direction. Large-scale inclined strata dipping to the east are formed by migration of unit bars. These strata vary in inclination laterally and reach the angle of repose in places. The large-scale inclined strata are composed internally of sets of medium-scale cross strata (formed by dunes) or planar strata (formed by bedload sheets). From Bridge (2003). Depth scale for GPR profile converted from TWTT using a radar velocity of 0.13 m ns^{-1}.

braided river deposits (Fig. 4). Furthermore, the wavelength and height of bedforms such as dunes and bars are related to channel depth and width (Bridge, 2003). Therefore, the thickness of a particular scale of strata set (i.e. medium-scale cross sets and large-scale sets of inclined strata) will vary with the scale of the palaeoriver. These relationships between the dimensions of stratasets, bedforms and channels mean that the depositional model developed from the Sagavanirktok River can be applied to other fluvial deposits.

Calamus

The sandy Calamus River was studied in a reach that is transitional between meandering and

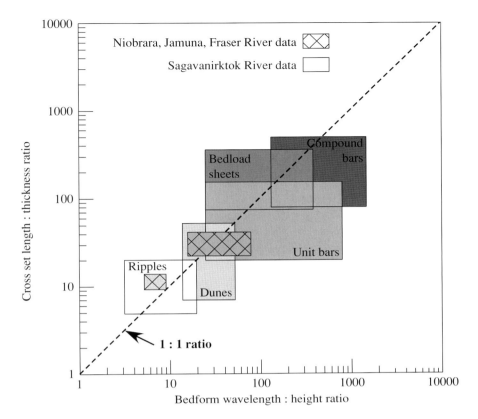

Fig. 4 Bedform length-to-height ratio plotted against corresponding strataset length-to-thickness ratio. The bedform dimensions were taken from reinterpreted aerial photograph measurements or descriptions, and preserved cross-set dimensions from reinterpreted GPR profiles or descriptions. Bedform measurements for the Sagavanirktok River were taken at low flow stage, and may not represent formative bedform dimensions. Measurements from other rivers were made from aerial photographs, trenches, river bed profiles and GPR profiles. Data sources are: Best *et al.* (2003), Lunt *et al.* (2004b), Skelly *et al.* (2003) and Wooldridge (2002).

braided, such that the average number of channels per cross-valley transect (a braiding index) is between 1 and 2 (Bridge *et al.*, 1986, 1998; Bridge & Gabel, 1992; Gabel, 1993). The interaction between channel geometry, water flow, sediment transport and deposition around a braid bar on the Calamus River was studied using measurements made over a large discharge range from catwalk bridges (Bridge & Gabel, 1992; Gabel, 1993). In the curved channels on either side of the braid bar, the patterns of flow velocity, depth, water surface topography, bed shear stress, and the rate and mean grain size of bedload transport rate are similar to those in single-channel bends. Because of the low sinuosity of these channels, topographically induced across-stream water flow was stronger than curvature-induced secondary circulation (as also found by McLelland *et al.* (1999) in the much larger Brahmaputra/Jamuna River). A theoretical model of bed topography, flow and bedload transport in bends (Bridge, 1992) agreed well with Calamus River data, and was subsequently used as the basis for simple numerical models for deposition in braided channels (Bridge, 1993). In the diffluence zone upstream of the Calamus braid bar, the depth-averaged flow velocity converges over the deeper areas (talweg) and diverges over topographic highs (bars).

The evolution of the channel geometry in the Calamus River reach was studied by Bridge *et al.* (1986), and further analysed by Bridge *et al.* (1998). Braid bars form either from chute cut-off of point bars or from growth of lobate unit bars in mid-channel, as is the case in most braided rivers (e.g. Brahmaputra, Sagavanirktok, discussed here). Growth of braid bars is mainly by an increase in width and downstream length (i.e. lateral and downstream accretion). Lateral accretion is not necessarily symmetrically distributed on each side of the braid bar. The upstream ends of braid bars may experience erosion or deposition. A curved channel may become abandoned as a side bar or point bar grows into its entrance. During the early stages of filling, the channel contains relatively small unit bars, especially at the upstream end.

Vibracores and GPR profiles show that channel bar deposits reflect:

1 the distribution of grain size and bedforms on the streambed during high flow stages;
2 the mode of bar growth and migration.

The geometry and orientation of large-scale inclined strata reflect mainly lateral and downstream migration of the bar, and accretion in the form of lobate unit bars and scroll bars (Bridge *et al.*, 1998). Within the large-scale strata are mainly medium-scale cross strata associated with the migration of dunes, which cover most of the bed at high flow stages. Small-scale cross strata from ripple migration occur in shallow water near banks. Bar sequences have an erosional base and generally fine upwards except for those near the bar head, which show little vertical variation in mean grain size. Channel-fill deposits are similar to channel-bar deposits, except that the large-scale inclined strata are concave upward in cross-channel view, and the deposits are generally finer grained in the downstream part of the channel fill (Bridge *et al.*, 1998). Figure 5 shows a model of the channel geometry, mode of channel migration, and deposits of the Calamus River reach studied.

Brahmaputra/Jamuna

There have been many studies of the geometry, flow and sedimentary processes, and patterns of erosion and deposition of the sandy Brahmaputra/Jamuna River (Coleman, 1969; Bristow, 1987, 1993; Bristow *et al.*, 1993; Thorne *et al.*, 1993; Mosselman *et al.*, 1995; Richardson *et al.*, 1996; Best & Ashworth, 1997; Richardson & Thorne, 1998; McLelland *et al.*, 1999; Ashworth *et al.*, 2000; Best *et al.*, 2003).

Detailed studies of water flow in the Brahmaputra/Jamuna include those by Richardson *et al.* (1996), Richardson & Thorne (1998) and McLelland *et al.* (1999). McLelland *et al.* (1999) measured near-bed and near-surface velocity vectors, and suspended sediment concentrations, in the channels around a compound braid bar at high and low flow stages. However, as the time interval between measurements was several months, and bedload transport rate was not measured, the sediment continuity equation could not be used to quantitatively relate the velocity and sediment transport fields to the nature of channel and bar erosion and deposition. Richardson *et al.* (1996) and Richardson & Thorne (1998) measured flow patterns in the diffluence zone upstream of a braid bar. Complicated patterns of convergence and divergence of the depth-averaged flow velocity were associated with bars. The primary flow tends to diverge away from bar troughs and converge towards bar crests, giving rise to associated secondary circulation cells. The pattern of flow in curved channel segments is similar to that in single-river bends. These flow patterns measured by Richardson *et al.* (1996) and Richardson & Thorne (1998) differed from those measured by McLelland *et al.* (1999), particularly in the nature of the secondary circulation.

Braid bars in the Brahmaputra/Jamuna originate from mid-channel unit bars and by chute cut-off of point bars, as seen in other braided rivers. Once formed, braid bars grow episodically by lateral and downstream accretion. The upstream parts of braid bars may be sites of accretion or erosion, depending on flow stage and geometry. Braid bars in the Brahmaputra/Jamuna are typically up to 15 m high, 1.5 to 3 km long, and 0.5 to 1 km wide (Coleman, 1969; Bristow, 1987; Ashworth *et al.*, 2000). The amount of lateral and downstream accretion on bars during a monsoonal flood can be of the order of kilometres. The accretion on braid bars is normally associated with migration of unit bars. Unit bars (called sand waves by Coleman, 1969) are commonly several metres high, but can exceed 10 m. Although lobate unit bars within channels and on compound braid bars are figured prominently by Coleman (1969), Ashworth *et al.* (2000) and Best *et al.* (2003), Ashworth *et al.* (2000) and Best *et al.* (2003) argued that unit bars occur less commonly in the main channels of the Brahmaputra/Jamuna than in other sand-bed rivers. This may be a reflection of the fact that unit bars may not be prominent at low flow stage (as shown by Coleman, 1969), and would be difficult to observe in aerial photographs taken at high flow stage when the water is extremely turbid. During high flow stages, the channels and bars are covered with dunes, although there may be restricted areas of upper-stage plane beds and ripples near bar tops. In the deeper parts of the channels, dunes are typically 3 to 4 m high, but are 0.5 to 1 m high near bar tops. The 'accretionary dune front' described by Ashworth *et al.* (2000) and Best *et al.* (2003) is probably the front of a lobate bar-head unit bar (Fig. 6).

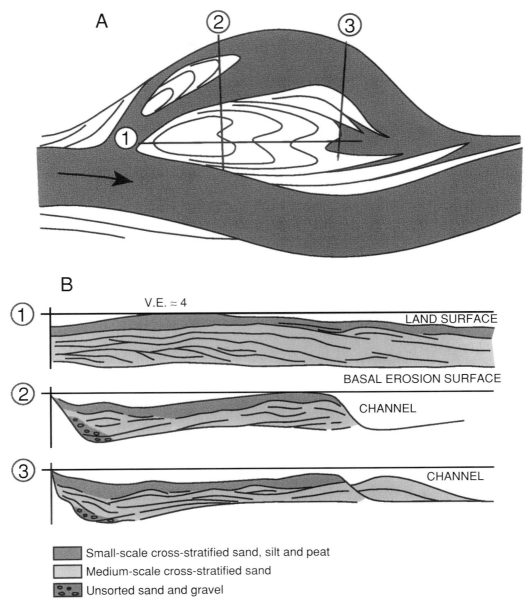

Fig. 5 Model of channel geometry, mode of channel migration, and deposits of the Calamus River. The map shows accretion topography on a compound braid bar, suggesting bar growth mainly by incremental lateral and downstream accretion. Relatively small unit bars occur within the filling channel. The cross sections show large-scale inclined strata associated with the accretion of unit bars (convex upward patterns), the lateral and downstream accretion of the compound braid bar, and channel filling. Lower bar deposits are composed mainly of medium-scale cross strata due to dune migration. Upper bar deposits and channel fills are composed mainly of small-scale cross strata (due to ripples), plus bioturbated silt and peat. The overall sedimentary sequence generally fines upward from very coarse or coarse sand to fine or very fine sand. Modified from Bridge *et al.* (1998).

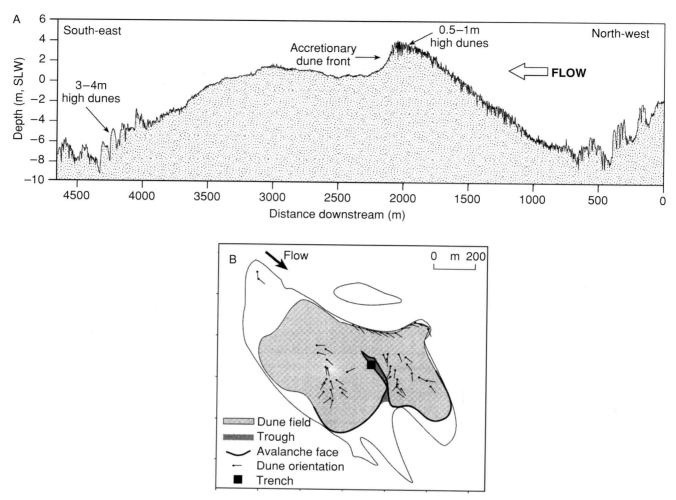

Fig. 6 (A) Along-stream topographic profile of a Brahmaputra/Jamuna compound braid bar at high flow stage on 12 August 1994 (from Ashworth et al., 2000). Note decrease in dune height with water depth. The 'accretionary dune front' marked is probably the front of a unit bar. As the vertical exaggeration of profile is about 75, this bar front is not at the angle of repose. (B) Map of the compound bar in A at low flow stage on 13 March 1995 (from Ashworth et al., 2000). The compound bar is covered with dunes, and the 'avalanche faces' of two lobate unit bars are separated by a trough. The bar-tail region contains three scroll (unit) bars.

Bristow (1993) described the deposits in the upper 4 m of channel bars exposed in long cut banks during low flow stage. He described the deposits of sinuous-crested dunes (trough cross strata), ripples (small-scale cross strata) and upper stage plane beds (planar strata). These stratasets were arranged into larger scale sets of strata indicative of seasonal deposition on bars (Fig. 7). The orientation of these flood-generated stratasets (large-scale inclined strata) indicated bar migration by upstream, downstream and lateral accretion. Bristow (1993) cited evidence for migration of small dunes over larger dunes at various stages of a monsoonal flood; however, it is likely that some of the larger, metres-high bedforms are actually unit bars (Fig. 7). The fills of cross-bar channels are common in these upper bar deposits (Fig. 7). These data are incorporated in the depositional model below.

Best et al. (2003) recognized the following types of cross (inclined) strata in their GPR profiles, trenches and cores through a compound braid bar (Fig. 8):

1 large-scale (sets up to 8 m thick, cross strata dipping at the angle of repose or less), associated with deposition on bar margins;

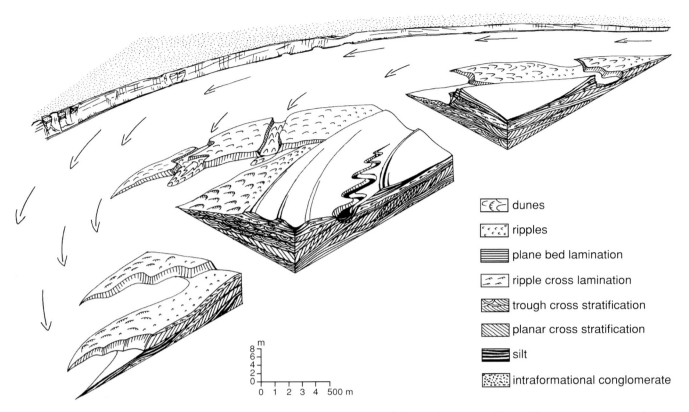

Fig. 7 The model of Bristow (1993) for compound bar-top deposits of the Brahmaputra River. The upstream part of the bar may be erosional or accretionary. The central part shows vertical and lateral accretion. Unit bars (including scroll bars) can be seen accreting onto the compound braid bar, giving rise to ridge-and-swale topography. These bars are covered with sinuous-crested dunes and in places are cut by small channels. Internal structures include trough cross strata from sinuous-crested dunes, small-scale cross strata from ripples, and planar strata from deposition on upper-stage plane beds. 'Planar cross strata' are associated with accretion of unit bars according to Bristow.

2 medium-scale (1 to 4 m thick, cross strata dipping at the angle of repose or less), associated with migration of large dunes;
3 small-scale (sets 0.5 to 2 m thick) due to dunes migrating over bar flanks.

Set thickness of small-scale cross strata decreases upwards. Cross strata associated with ripples (called ripple cross stratification) occur locally at the top of the bar. The terminology of large, medium and small scales of cross strata is different from that used in this paper. Furthermore, the ranges of set thickness in these categories overlap, so that it is difficult to know how to classify any specific cross set. Best et al. (2003) did not explicitly recognize the distinction between unit bars and compound bars, and the different scales of deposits associated with them. Cross sets associated with dunes that have a mean height of 3 m are likely to have a mean set thickness of the order of 1 m (Leclair & Bridge, 2001). Therefore, most of the medium-scale cross sets (with thickness of 1–4 m) are unlikely to be formed by dunes, but rather by unit bars (as recognized by Best et al. (2003, p. 521), and implied in fig. 11B of their paper).

Figure 9 indicates that vertical accretion deposits on the bar top pass laterally into both upstream and lateral accretion deposits, and that both bar margin slipface deposits and vertical accretion deposits in the channel pass laterally into downstream/oblique accretion deposits. There are clearly terminology problems here. In fact, if vertical deposition occurs on any mound-like form (i.e. a braid bar), it would appear that there are components of accretion in the upstream, lateral and downstream directions. If such vertical accretion on

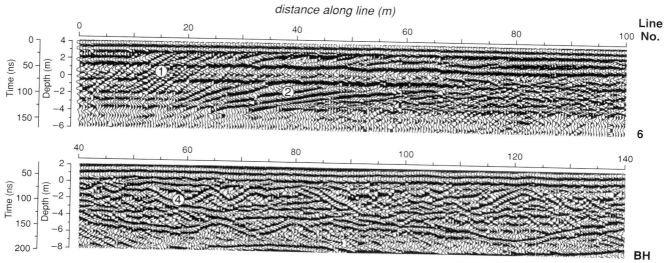

Fig. 8 Examples of 100 MHz GPR profiles through Brahmaputra compound braid bar deposits (from Best *et al.*, 2003). Vertical exaggeration is 1.58. Profile 6, across stream through the downstream part of the braid bar (location on Fig. 9), shows large-scale inclined strata associated with unit bar migration. The spatially variable dip angles (up to angle of repose) are typical of unit bar deposits. Profile BH, alongstream through the upstream part of the bar, shows concave upward reflections that were interpreted by Best *et al.* (2003) as trough cross sets up to 2 m high associated with migration of large sinuous-crested dunes. However, Fig. 6A shows that dunes in this region are only 0.5 to 1 m high. Lower in this profile, 3 m thick inclined strata are ascribed to downstream accretion of a bar head.

a braid bar is followed by local erosion on the bar top, the original vertical deposition may not even be discernible. As the braid bar studied by Best *et al.* (2003) experienced both lateral and downstream growth simultaneously, a lateral accretion deposit will also be a downstream accretion deposit. Therefore, terminology based on two-dimensional vertical profiles can be very misleading. The data of Best *et al.* (2003) are reinterpreted below and incorporated in the new depositional model.

Niobrara

Skelly *et al.* (2003) interpreted the following sedimentary facies from radar facies using 100 and 200 MHz GPR antennae: trough cross beds associated with small three-dimensional dunes; planar cross beds associated with small two-dimensional dunes; horizontal to low-angle planar strata or cross-set boundaries associated with large three-dimensional dunes (or linguoid bars); channel-shaped erosion surfaces associated with secondary channels (and filled with deposits of two-dimensional and three-dimensional dunes); sigmoidal strata associated with accretion on braid bar margins. Skelly *et al.* (2003) also interpreted superimposed sequences, 1 to 1.5 m thick, as due to stacking of braid bars, and most bar sequences fined upward. These depositional patterns were associated with high-stage bar construction and upstream, lateral and downstream bar accretion, and low-stage dissection of the bars by small channels. Unfortunately, the uppermost parts of the radar profiles were not linked to the geometry and migration of specific extant channels and bars. The distinction between unit bars and compound bars was not made. However, Fig. 10A (Skelly *et al.*, 2003, fig. 10) shows an example of a lateral transition from low-angle strata to angle of repose strata to low-angle strata, as observed in unit-bar deposits elsewhere. These deposits were interpreted by Skelly *et al.* (2003) as due to small two-dimensional dunes. These unit-bar deposits accreted onto the western margin of a compound bar deposit (Fig. 10B: Skelly *et al.*, 2003; fig. 11) and the dimensions of these unit bar deposits are consistent with those in other rivers (Fig. 4).

Discussion of recent studies of braided river deposits

The new data reviewed above have resulted in greatly improved understanding of braided rivers

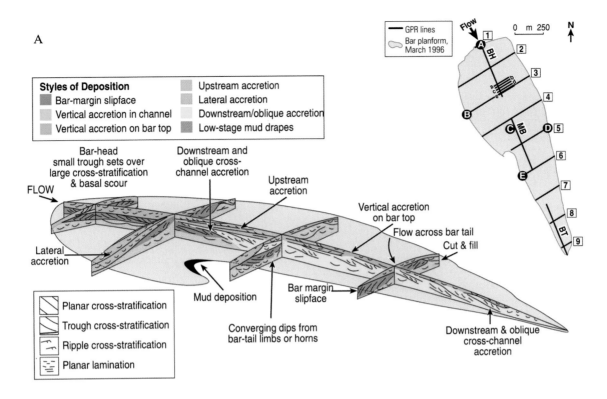
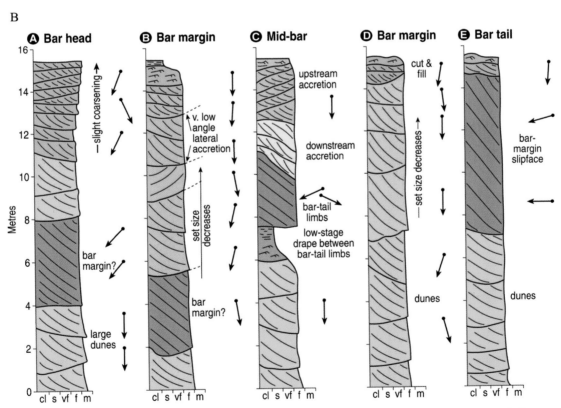

Fig. 9 The depositional model of Best *et al.* (2003) for a Brahmaputra braid bar. (A) Three-dimensional diagram of the principal styles of deposition. The bar is 3 km long, 1 km wide and 12–15 m high. (B) Schematic sedimentary logs at five localities within the braid bar (see inset in A for location), illustrating the characteristic sedimentary structures, large-scale bedding surfaces, and styles of deposition. Arrows depict approximate flow directions for sedimentary structures, with flow down the page indicating flow parallel to the mean flow direction. See text for discussion of this model.

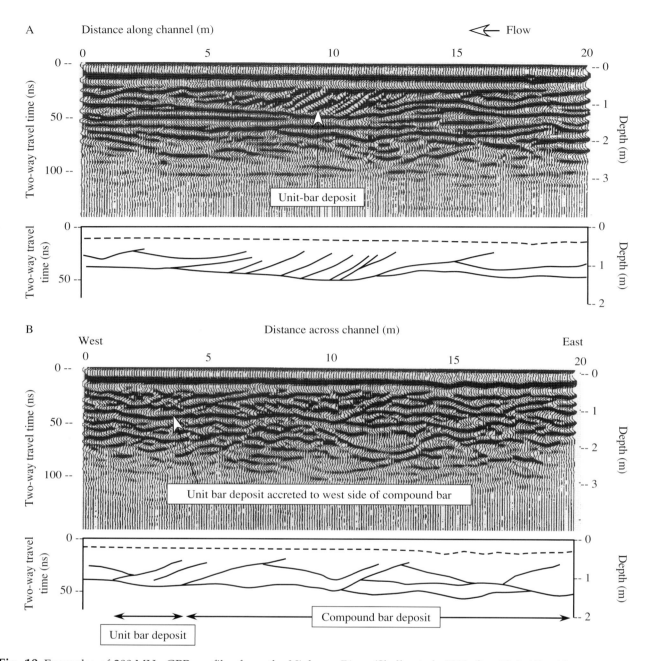

Fig. 10 Examples of 200 MHz GPR profiles from the Niobrara River (Skelly *et al.*, 2003, figs 10 & 11) with reinterpretation.

and their deposits. It is clear from these data that there is no truth to the common claims that:

1 braided rivers transport relatively coarse sediment as bedload and have unstable banks;
2 discharge variability is very large for braided rivers.

There is remarkable similarity in geometry, flow and sedimentary processes, and deposits despite differences in river scale and grain size: e.g. unit bars and dunes at high flow stage; braid bar geometry, evolution and migration (as also suggested by Sambrook Smith *et al.*, 2005). Differences between gravel-bed and sand-bed rivers are: more bedload sheets and associated planar strata in gravel-bed rivers, increasing as mean grain size increases; more abundant ripples (small-scale cross strata) and upper stage plane beds (planar strata) in sand bed rivers, increasing as mean grain size decreases.

There appears to be a perception in the literature that dunes and associated cross strata are uncommon in gravelly rivers. However, it is now well known that dunes occur in gravel-bed rivers (Dinehart, 1989, 1992; Carling, 1999; Kleinhans *et al.*, 2002; Lunt *et al.*, 2004a,b). One reason why gravel dunes have not been more widely reported is that gravel-bed rivers are rarely studied during peak flood conditions. Also, gravel dunes commonly have low heights on the upper parts of bars where they can be observed easily at low flow stage. Cross stratification formed by dunes is common in gravelly fluvial deposits (Bridge & Jarvis, 1976, 1982; Gustavson, 1978; Dawson, 1989; Siegenthaler & Huggenberger, 1993; Lunt *et al.*, 2004a,b), but is perhaps not as easily seen as in sands. Massive (structureless) fluvial gravel is widely reported, but cannot be explained easily without appeal to a gravity flow origin. This is rarely observed in modern rivers except in the case of bank collapse.

Unit bars and associated sets of planar cross strata have been specifically associated with braided rivers (Collinson, 1970; Smith, 1971, 1972, 1974; Bluck, 1976,1979; Cant, 1978; Cant & Walker, 1976, 1978; Blodgett & Stanley, 1980; Crowley, 1983). However, unit bars occur in meandering rivers also (McGowen & Garner, 1970; Bluck, 1971; Jackson, 1976; Levey, 1978; Bridge *et al.*, 1995). Furthermore, it has become increasingly apparent that most of the internal structure of unit bars is not planar cross strata, but is due to the bedforms (dunes, ripples, bedload sheets) migrating over them (Collinson, 1970; Jackson, 1976; Nanson, 1980; Bridge *et al.*, 1986, 1995, 1998; Ashworth *et al.*, 2000; Best *et al.*, 2003; Lunt *et al.*, 2004a,b). It appears that angle-of-repose (planar) cross strata are restricted to the margins of unit bars, particularly where the bars migrate into relatively deep, slow-moving water (Best *et al.*, 2003; Lunt *et al.*, 2004a,b). Some workers have distinguished the deposits of 'bar cores' and 'bar margins' (Eynon & Walker, 1974; Ashworth *et al.*, 1999; Moreton *et al.*, 2002; Best *et al.*, 2003), although there may be gradations between the core and margins of both unit and compound bars.

Detailed understanding of modern river deposits requires description of channel bed topography, channel and bar movements (i.e. changes in topography), and nature of deposits. Channel bed topography has traditionally been measured using standard topographic surveying techniques and depth soundings. Relatively accurate surveying of bed topography, with vertical and horizontal resolution of centimetres, is now possible with GPS (Brasington *et al.*, 2000). Digital photogrammetry in combination with GIS-based digital elevation models has also been developed for measuring subaerial topography, although these methods are not yet as accurate as other methods (Lane *et al.*, 1994, 1995, 1998, 2001; Chandler, 1999). Digital photogrammetry has also been used to determine the topography of subaqueous parts of shallow, clear rivers at low flow-stage by making corrections for the effects of light refraction in water (Westaway *et al.*, 2000; Lane *et al.*, 2001). Topographic measurements are accurate to ± 0.15 m on exposed surfaces and ± 0.25 m for subaqueous surfaces (Lane *et al.*, 2003). Although these measurements are not as accurate as ground-based topographic measurements, they allow estimation of rates of channel change over relatively large areas. However, these methods may not be used for investigating subaqueous bed topography in turbid rivers at high flow stage.

Studies of flow and sediment transport on large braided rivers are not common because of measurement difficulties during floods. The flow and sediment transport in curved channels adjacent to braid bars may be similar to those in curved single channels (Bridge & Gabel, 1992; Bridge, 1993). The flow and sediment transport in confluence zones has also been likened to that in back-to-back meanders (reviews in Bridge, 1993, 2003). This could be valid for symmetrical confluences, but the analogy can be taken only so far (Bradbrook *et al.*, 2000). The curvature of the converging channels in river confluences may give rise to super-elevation of the water surface in mid-channel and spiral flow with near-bed flow components towards the outer banks (Mosley, 1976; Ashmore, 1982; Ashmore & Parker, 1983; Ashmore *et al.*, 1992; Bridge, 1993; Rhoads & Kenworthy, 1995, 1998; Rhoads, 1996; Rhoads & Sukhodolov, 2001; Sukhodolov & Rhoads,

2001). The maximum water-surface super-elevation and the magnitude of the transverse components of the spiral flow tend to increase as the confluence angle increases (or radius of curvature of joining channels decreases). However, even with symmetrical confluences, the amount of water-surface super-elevation is not as great as would be expected for a single meander of the same geometry. The pattern of flow in confluences is also dependent upon flow acceleration associated with reduced cross-sectional area of the conjoined channels, and with anisotropic turbulence associated with flow separation downstream from mouth-bar crests, inequality in the depths of confluent channels, and the enhanced turbulence of the mixing layer between the joining streams (Best, 1987, 1988; Best & Roy, 1991; Biron et al., 1993a,b, 1996a,b, 2002; McLelland et al., 1996; Rhoads, 1996; De Serres et al., 1999). These influences on confluence flow have been modelled numerically with reasonable success by Bradbrook et al. (1998, 2000, 2001). Despite these complications to the flow pattern in confluences, the maximum bedload sediment size and transport rate occur in confluence scour zones where flow velocity and bed shear stress are greatest (Davoren & Mosley, 1986; Best, 1987, 1988; Ashworth et al., 1992a,b; Ferguson et al., 1992; Rhoads, 1996).

Flow and sediment transport in flow divergence zones upstream of braid bars are not known well (but see Mosley, 1976; Ashworth & Ferguson, 1986; Davoren & Mosley, 1986; Ashmore et al., 1992; Ferguson et al., 1992; Bridge and Gabel, 1992; Richardson et al., 1996; Richardson and Thorne, 1998). The maximum velocity locus is in mid-channel in the upstream parts of this zone, and splits downstream such that each thread of maximum velocity is close to the upstream tip of the braid bar downstream. There may be relatively complicated patterns of convergence and divergence of depth-averaged velocity in this zone, associated with bar-scale bed topography (Bridge & Gabel, 1992; Richardson et al., 1996). With large diffluence angles, zones of flow separation with vertical axes may occur near the outer banks of the channel entrances, enhancing deposition there. Recent experimental and theoretical studies of diffluence dynamics have concentrated on the stability of the divergent channels at different sediment transport stages (Bolla-Pittaluga et al., 2003; Frederici & Paola, 2003).

LABORATORY STUDIES

Braided-river geometry and processes have been studied extensively in laboratory flumes (Schumm & Khan, 1972; Ashmore, 1982, 1991, 1993; Ouchi, 1985; Schumm et al., 1987; Hoey & Sutherland, 1991; Germanowski & Schumm, 1993; Leddy et al., 1993; Wood et al., 1993; Ashworth et al., 1994, 1999; Koss et al., 1994; Warburton & Davies, 1994; Bryant et al., 1995; Peakall et al., 1996; Ashworth, 1996; Gran & Paola, 2001; Paola et al., 2001; Heller et al., 2001; Moreton et al., 2002; Sheets et al., 2002; Cazanacli et al., 2002; Frederici & Paola, 2003). These small-scale experimental studies have been useful for elucidating the broad patterns of water flow and sediment transport, the kinematics of channel and bar migration, and the effects of riparian vegetation, tectonism, base-level change, aggradation and degradation on braided stream geometry and dynamics. Studies focusing on the nature of deposits are relatively recent (Ashworth et al., 1999; Heller et al., 2001; Paola et al., 2001; Moreton et al., 2002; Sheets et al., 2002). Ashworth et al. (1999) distinguished relatively coarse-grained deposits (channel fills, bar cores and bar margins) from fine-grained deposits (channel fills, bar tops and bar margins) (see Fig. 11). Moreton et al. (2002) also recognized relatively coarse-grained channel fills (including bar deposits showing evidence of lateral accretion), fine-grained channel fills, overbank splays (planar stratified), floodplain fines (laminated) and 'erosional remnants'. Sheets et al. (2002) also recognized coarse-grained channel fills (including lateral accretion deposits), fine-grained channel fills and overbank sheet deposits. However, it has not been possible to generate all of the superimposed scales of bedforms and associated strata observed in natural rivers (such as stratasets associated with migration of ripples and dunes). As a result of this, comparisons of experimental deposits with natural river deposits have been restricted to only the largest scales of strataset, associated with channel bars and channel fills (Ashworth et al., 1999). Channel fills appear to be much more prominent in some experimental deposits (Sheets et al., 2002) compared with natural deposits, probably due to unrealistically high rates of channel switching and abandonment.

Detailed experimental studies of the flow and sediment movement in confluences have been undertaken mainly in fixed bed channels (Mosley, 1976;

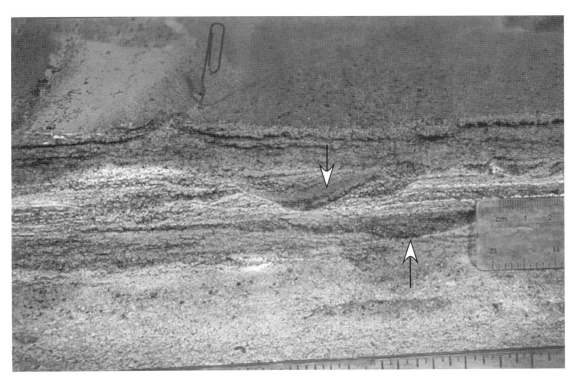

Fig. 11 Example of braided river deposits from a flume experiment undertaken in Leeds, UK, by Ashworth *et al.* (1999). The arrowed channel fills and adjacent bar deposits are composed of coarse sand. These channel deposits cut into light-coloured, horizontally laminated, fine sand to silt, deposited on tops of channel bars.

Best, 1987, 1988; Best & Roy, 1991; Biron *et al.*, 1996a,b; McLelland *et al.*, 1996; Huang *et al.*, 2002). The lack of a mobile bed, with no erosion and deposition, means that these studies have limited application to depositional models. Recent experimental studies of diffluence zones in braided rivers (Frederici & Paola, 2003) have examined the geometry, flow and sediment transport in the downstream channels, and whether or not one of the braided channels tends to be filled.

THEORETICAL STUDIES

Channel bars

Theoretical models of the interaction between bed topography, water flow, sediment transport, erosion and deposition in braided rivers are not well developed. Quantitative, dynamic, three-dimensional depositional models of channel bars have been developed only recently, and such models are at a rudimentary stage (Willis, 1989; Bridge, 1993).

These models require prediction of the interaction between bed topography, water flow, sediment transport rate, mean grain size of bedload, and bed forms within channels of prescribed geometry (Bridge, 1977, 1992). The flow conditions are assumed to be steady and bankfull, with the bed topography, water flow and sediment transport in equilibrium. The models apply to either single channel bends with an associated point bar, or two channel bends separated by a braid bar (Bridge & Gabel, 1992). The planforms of the channels are sine-generated curves, and features such as unit bars and cross-bar channels are not considered. The channels and associated bars must be put in a dynamic context by allowing them to migrate by bank erosion and bar deposition, and to change geometry in time. Net vertical deposition is not allowed over the time spans considered in the models. Despite being simplified, these models give important insights into the nature of channel bar deposits that could not come from the previous static one-dimensional and two-dimensional models (review in Bridge, 2003):

1 As channels migrate by lateral and downstream migration, the deposits from different parts of channel bars become vertically superimposed (e.g. bar head deposits overlying bar tail deposits, and bar tail deposits overlying confluence scour deposits).
2 Systematic spatial variations in the thickness of channel bar deposits, and the inclination and orientation of large-scale strata, are due to bed topography and the mode of channel migration. For example, it is common for channel bar deposits to thicken (by up to a factor of two) and for large-scale strata to steepen, towards a cut bank (channel-belt margin) or confluence scour.
3 Lateral and vertical variation in grain size and sedimentary structures are controlled by the bed topography, flow, sediment transport and bed forms, and by the mode of channel migration. Channel-bar deposits normally fine upwards, but they also commonly show little vertical variation in grain size. Some channel-bar deposits coarsen at the top if bar head deposits are preserved.

This type of model was used to interpret the geometry and hydraulics of curved channels associated with ancient braided rivers in the Miocene Siwaliks of northern Pakistan (Willis, 1993; Khan *et al.*, 1997; Zaleha, 1997). Willis (1993) was able to quantitatively reconstruct the width, depth, mean velocity, slope, wavelength and sinuosity of individual channel segments in these Siwalik deposits, and, because of the excellent exposures, to estimate channel-belt widths and braiding index. Channel bars were interpreted to have migrated mainly by downstream translation and bend expansion, but also by channel switching within the channel belts. Rates of channel migration were estimated at up to the order of channel width per seasonal flood period.

The models of Bridge and Willis predict the geometry, grain size and sedimentary structure of the deposits of single point bars or braid bars. However, they do not consider the somewhat complicated flow structures at channel diffluences and confluences. It is necessary to develop theoretical models for flow and sediment transport in these regions. Although there are numerical models of turbulent flow in confluences that agree fairly well with observed flows (Bradbrook *et al.*, 1998, 2000, 2001; Lane & Richards, 1998; Nicholas & Sambrook Smith, 1999; Lane *et al.*, 2000; Huang *et al.*, 2002), they do not describe the interaction between flow, sediment transport and bed topography. In addition, theoretical models for flow and sediment transport at diffluences (Bolla-Pittaluga *et al.*, 2003) are very simplistic and cannot predict detailed patterns of sediment transport and deposition. Finally, there are no quantitative models for the flow, sediment transport and deposition in abandoned channel fills.

Channel belts

Numerical simulation of the nature of channel deposits within channel belts is in its infancy. The spatial distribution of the deposits of individual channel bars and fills could not be included easily in the bar depositional models of Willis (1989) and Bridge (1993) because it is very difficult to predict how individual channel segments and bars will migrate and become preserved within channel belts. It is necessary to develop models for the deposits of several adjacent bars and channel fills within braided channel belts. Tetzlaff (1991) proposed a very simple empirical model for the geometry, lateral channel migration and aggradation of meandering and braided channels within channel belts. Murray & Paola (1994, 1997) modelled braided channel belts using a grid of cells, the top surfaces of which formed a uniform downstream slope but with random perturbations in elevation superimposed. Water discharge moving down the grid is distributed between adjacent cells such that more discharge moves down the steeper slopes. Sediment discharge between cells is a power function of either water discharge or discharge–slope product. The result of the model is that hollows are scoured and high areas receive deposits (due to the non-linear dependence of sediment transport rate on discharge), so that channels and bars form. Murray & Paola (1994, 1996, 1997) claim that their model can produce braided channel patterns that are similar to those in real rivers. A recent development of this cellular routing approach is by Thomas & Nicholas (2002). These types of models were not developed to predict the nature of deposits.

The sediment routing model of fluvial deposition, SEDSIM (Tetzlaff & Harbaugh, 1989), is based on solution of the simplified equations of motion of water and sediment. It is claimed to be able to simulate flow, sediment transport, erosion and deposition in river channel bends, braided rivers, alluvial fans, and deltas. However, there are fundamental problems with the basic assumptions and construction of this model, particularly in the treatment of sediment transport, erosion and deposition (see also North, 1996). The SEDSIM model has not been tested by detailed comparison with real-world data (see also Paola, 2000). Apparently, small changes in input parameters can result in large and unpredictable changes in output. Tetzlaff & Priddy (2001, and Griffiths *et al.* (2001) described development of SEDSIM into STRATSIM, in order to simulate a range of fluvial and marine sedimentary processes acting over time intervals of hundreds of thousands of years. Unfortunately, few details are given of the workings of these models, nor of how or whether they were tested against data from natural sedimentary environments (Paola, 2000). Despite the shortcomings of these models, they have been used to simulate the general character of fluvial hydrocarbon reservoirs.

Object-based stochastic models have been used to distribute channel deposits within channel belts (Tyler *et al.*, 1994; Webb, 1994, 1995; Deutsch & Wang, 1996; Webb & Anderson, 1996; Holden *et al.*, 1998; Deutsch and Tran, 2002). Webb (1994, 1995) and Webb & Anderson (1996) modelled a braided network of river channels as a random walk. Channel shapes were assigned using hydraulic geometry equations, and sedimentary facies within the channel fills were assigned using a calculated Froude number. A three-dimensional pattern of channel fills was produced by repeated simulation of channel networks with a fixed aggradation rate superimposed. Hydraulic conductivity values were assigned to the various lithofacies types in order to explore the effects of three-dimensional lithofacies heterogeneity on ground water flow (Webb & Anderson, 1996; Anderson *et al.*, 1999). In the approaches of Tyler *et al.* (1994), Deutsch & Wang (1996) and Holden *et al.* (1998), individual channel deposits within channel belts are represented by a series of sinuous channels superimposed in space. None of these approaches correctly represents the nature of channel deposits in channel belts, which are in fact composed predominantly of parts of channel bars with relatively minor volumes of channel fills. It is necessary to define shapes of these objects properly, and the information shown in Fig. 4 will assist in the scaling of objects.

There are also theoretical models for the distribution of channel belts within floodplain alluvium (review by Bridge, 2003). However, this scale of fluvial deposition is beyond the scope of this paper.

NEW DEPOSITIONAL MODELS AND THEIR USE

The new qualitative depositional models shown in Figs 12 and 13 comprise:

1 maps showing idealized active and abandoned channels, compound bars and lobate unit bars;
2 cross sections showing large-scale inclined strata and their internal structures, associated with migration of compound bars, unit bars and their superimposed bedforms;
3 vertical logs of typical sedimentary sequences through different parts of compound bar deposits and channel fills.

The bar head regions of the compound bars have formed by accretion of the fronts of lobate unit bars, and their bar tail regions have formed by accretion of the sides of lobate unit bars (i.e. scroll bars). Thus, compound bar growth has been mainly by lateral and downstream accretion. The abandoned channel is being filled with unit bar deposits, and its upstream end was blocked by a compound point bar. The cross sections and vertical logs differ somewhat between gravel-bed rivers (Fig. 12) and sand-bed rivers (Fig. 13). Remember that gravel-bed rivers normally have abundant sand, and most sand-bed rivers also contain some gravel.

The new depositional models show:

1 the relationship between different scales of bedforms and channels, and corresponding scales of strataset;
2 the spatial relationship between the scales of strataset;
3 the distribution of sediment texture within stratasets;
4 length:thickness ratios for all scales of strataset.

As the length and thickness of stratasets can be related to the length and height of formative bedforms, and some of these are in turn related to

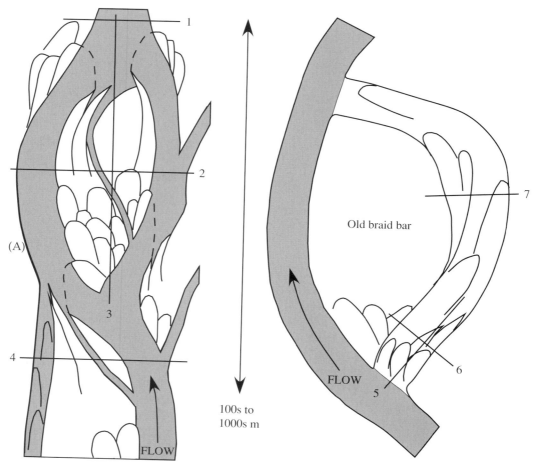

Fig. 12 Depositional model for gravel-bed rivers. (A) Map on left shows idealized active channels (stippled), compound bars and lobate unit bars. The bar-head regions of compound bars have formed by accretion of the fronts of lobate unit bars, and their bar-tail regions have formed by accretion of the sides of lobate unit bars (i.e. scroll bars). Thus, compound bar growth has been mainly by lateral and downstream accretion. Locations of cross sections 1 to 4 are shown. Map on right shows idealized abandoned channel (containing lobate unit bars) adjacent to a braid bar. The upstream end of the abandoned channel was blocked by a compound point (side) bar. Locations of cross sections 5 to 7 are shown.

depth and width of channels, these models can be generalized to all scales of river. The spatial distribution and degree of preservation of channel bars and channel fills within channel belts are discussed, but detailed models are not yet available. The models can be used to interpret ancient braided river deposits, and to aid in the building of three-dimensional sedimentary models of specific hydrocarbon reservoirs and aquifers (for flow and reactive contaminant transport simulation), given limited subsurface data (e.g. borehole logs, cores, geophysical profiles). These aspects are expanded upon below.

Description of the depositional model for gravel-bed braided rivers

A depositional model for gravelly braided rivers is based on the Sagavanirktok River data and pre-existing published information on gravelly braided rivers (Lunt et al., 2004a,b; Fig. 12). Stratasets are hierarchically organized within fluvial deposits, such that cosets of medium-scale cross strata (from dunes), planar strata (from bedload sheets) and small-scale cross strata (from ripples) make up simple sets of large-scale strata, simple sets of large-scale strata (from unit bars) and small

(B) 1 Across-stream view of compound side bars adjacent to a confluence scour

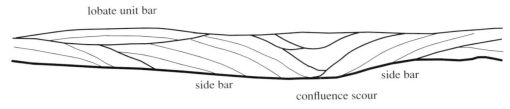

2 Across-stream view of compound braid bar that migrated over a confluence

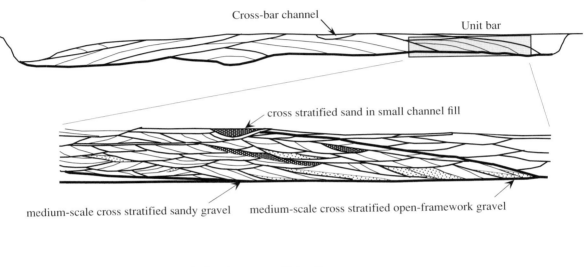

3 Across-stream view of compound point bar that accreted laterally

4 Along-stream view through compound bar that migrated laterally and downstream

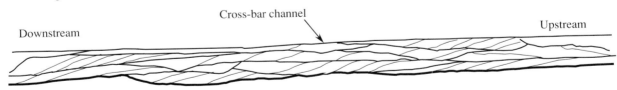

Fig. 12 (*cont'd*) (B) Cross sections identified in Fig. 12A showing large-scale inclined strata associated with migration of compound bars and unit bars. Thin lines bound simple large-scale strata, medium thickness lines represent bases of sets of simple large-scale strata (due to unit-bar migration), and thick lines represent bases of compound sets of large-scale strata (due to migration of compound bars). Vertical exaggeration of cross sections is approximately 5: therefore, apparent stratal dips are exaggerated. Large-scale strata generally dip at less than 12°, but simple large-scale strata approach the angle of repose in places.

5 Along-stream view through upstream end of large channel fill: lateral and downstream growth of compound bar

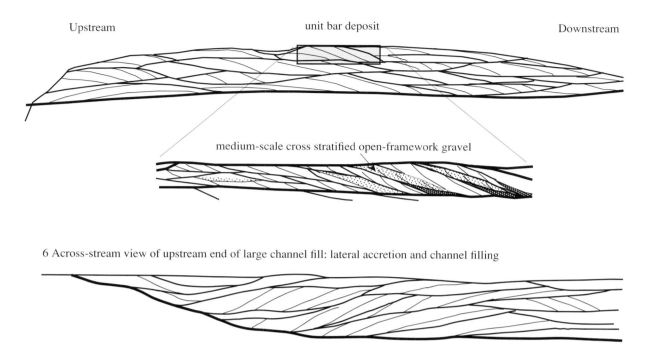

6 Across-stream view of upstream end of large channel fill: lateral accretion and channel filling

7 Across-stream view of downstream end of large channel fill: scroll bar accretion and channel filling

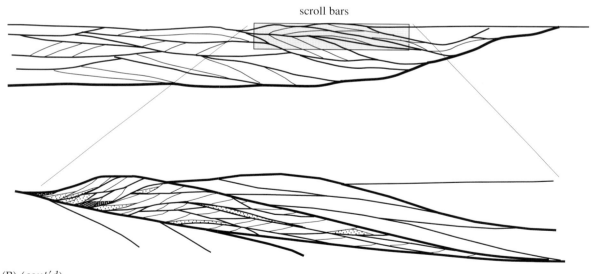

Fig. 12 (B) (*cont'd*)

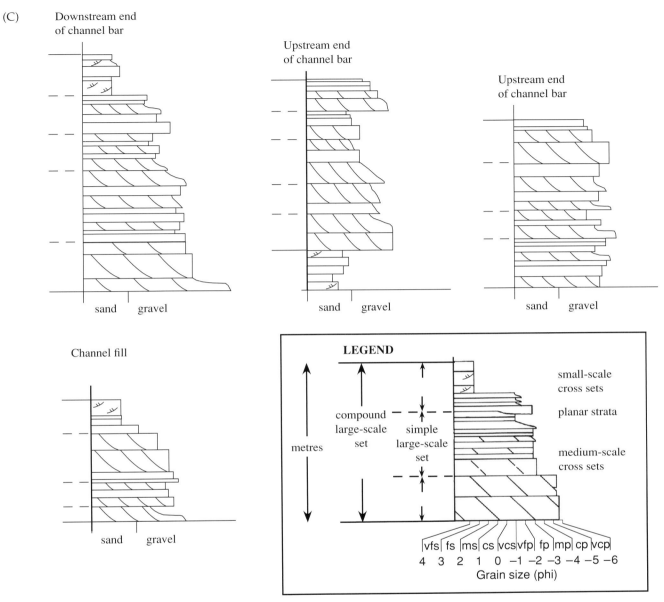

Fig. 12 (*cont'd*) (C) Vertical logs of typical sedimentary sequences through different parts of compound bar deposits and channel fills.

channel fills (from cross-bar channels) make up compound sets of large-scale inclined strata and large channel fills, and compound sets of large-scale strata (from compound bars) and large channel fills (from main channels) make up channel-belt deposits. The spatial distribution of different scales of strataset is predicted in the model (Fig. 12), as explained below.

Sandy small-scale cross sets occur mostly as overbank deposits capping channel-belts. They also occur in the upper parts of channel fills, and form drapes in the troughs of unit bars and dunes. Such trough drapes occur randomly within stratasets. Sandy small-scale cross strata make up a relatively small proportion of gravelly channel-belt deposits.

Sandy planar strata occur in channel fills, bar-tail swales and as overbank deposits. Planar strata in sandy gravel and open-framework gravel occur in small amounts within simple sets of large-scale

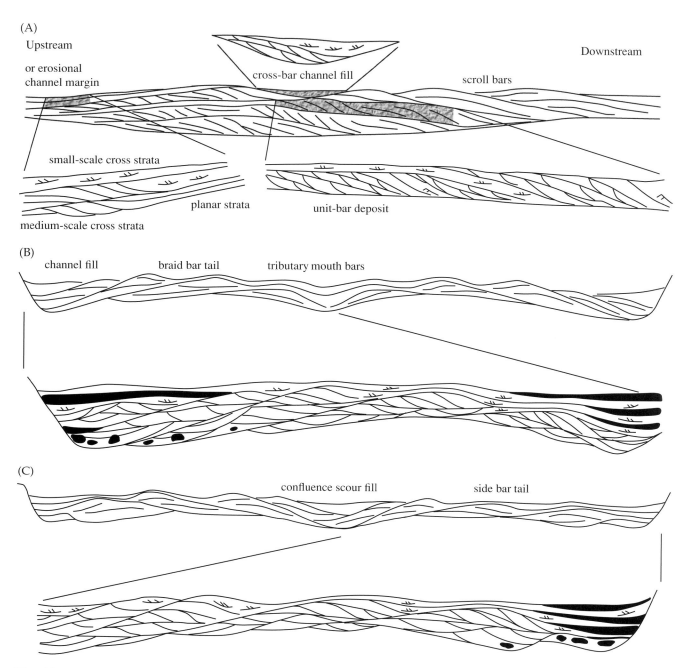

Fig. 13 Depositional model for sand-bed rivers, based on the same channel geometry and mode of channel and bar migration as for gravel-bed rivers. (A) Along-stream section (flow to right) along the axis of a compound braid bar showing large-scale inclined strata associated with migration of the compound bar and superimposed unit bars. Vertical exaggeration is between 5 and 10, so stratal inclinations are exaggerated. Insets (shaded) are expanded to show more sedimentological detail. The upstream inset (expanded lower left) shows medium-scale cross strata (from dunes) passing laterally and vertically into planar strata (from upper stage plane beds) or small-scale cross strata (from ripples). The set boundaries dip upstream, reflecting the geometry of the braid bar. The lower right inset is a section through the deposits of a unit bar. Simple large-scale inclined strata (due to unit bar migration) enclose sets of medium-scale cross strata (due to dune migration). The inclination of the simple large-scale strata reaches the angle of repose in places. Sets of small-scale cross strata formed by ripples represent deposition during falling flow stage. The upper inset represents the fill of a cross-bar channel, containing medium-scale and small-scale cross strata.

preservation of truncated compound bars within single channel belts can be assessed using:

$$s_m = lr/c + 0.82\beta \quad (1)$$

where s_m is the mean thickness of truncated bars (sets of compound large-scale inclined strata), l is the downstream distance between successive bar troughs, c is downstream migration rate of bars, and r is the long-term vertical deposition rate (Bridge, 1997; Leclair et al., 1997; Leclair & Bridge, 2001). Using data from the Sagavanirktok River, the vertical deposition rate in the channel belt (averaged over the past 6000 yr) is 1.7 mm yr^{-1}. This deposition rate is only a crude estimate of the average vertical deposition rate of the entire channel belt. The mean downstream migration rate (averaged over 52 yr) of compound bars is 3.55 m yr^{-1}, the mean compound bar length is 810 m, and the mean thickness of truncated compound bars is 1.94 m (as measured from 110 MHz GPR profiles). These data result in values of β of 1.9 m, using Eq. 1. Independent estimates of β can be made using the probability density function (PDF) of compound bar heights (Paola & Borgman, 1991). The simplest method is to fit an exponential curve to the tail of the probability density function:

$$f(h) = ae^{-ah} \quad (2)$$

where h is bar height and $a = 1/\beta$. Based on somewhat limited data, this results in values of β between 2.4 and 2.6 m, depending on which part of the tail is used for curve fitting. Using a β value of 2.5 m, together with the mean vertical deposition rate, migration rate and bar lengths quoted above, Eq. 1 predicts the mean truncated bar thickness to be 2.43 m. This predicted value is quite close to the mean truncated bar thickness measured from 110 MHz GPR profiles of 1.94 m. The first term on the right-hand side of Eq. 1 is much smaller than the second term, which implies that deposition rate does not have a significant impact on degree of preservation of bar deposits. The equivalent values for unit bar deposits cannot be derived for the Sagavanirktok River, because there are not sufficient data to create a PDF of unit bar heights. More data need to be collected and analysed before it can be assumed that Eq. 1 can be applied to all scales of bedform, from ripples to compound bars.

However, channels and compound bars do not migrate solely by episodic downstream migration: they also migrate within a channel belt by channel diversion and channel filling. Based upon the lessons learned from alluvial architecture models (Bridge & Leeder, 1979; Bridge & Mackey, 1993; Mackey & Bridge, 1995; Heller & Paola, 1996), the degree of truncation of channel bar and fill deposits within channel belts should increase with channel bar thickness (scour depth) and frequency of channel diversion, and decrease with channel-belt width/channel width and deposition rate. However, the effects of increased diversion frequency may be counteracted by associated increased deposition rate. Indeed, Ashworth et al. (1999) maintain that the architecture of channel bars and channel fills within the braided channel belts they studied is independent of deposition rate, and is controlled by variations in channel scour depths.

Use of new depositional models for characterization of aquifers and hydrocarbon reservoirs

There have been numerous studies of ancient braided river deposits, and some of these deposits have been used as analogues for hydrocarbon reservoirs or aquifers (Ramos et al., 1986; Cuevos Gozalo & Martinius, 1993; Dreyer, 1993; Dreyer et al., 1993; Miall, 1994; Doyle & Sweet, 1995; Olsen et al., 1995; Liu et al., 1996; Robinson & McCabe, 1997; Anderson et al., 1999; Asprion & Aigner, 1999; Hornung & Aigner, 1999; Bridge et al., 2000). However, many interpretations of ancient river deposits may not be reliable, because the depositional models used as the basis for interpretation have not been adequate until relatively recently.

These new depositional models should remove much of the uncertainty in interpreting the origin of fluvial strata, in establishing the lateral continuity of different types and scales of strata, and in predicting the three-dimensional distribution of sediment properties in fluvial channel deposits. This will greatly improve predictions of groundwater flow and contaminant transport in fluvial aquifers, and fluid flow in hydrocarbon reservoirs. Procedures for incorporating these models into reservoir/aquifer characterization are given by Lunt et al. (2004a).

Most hydrogeological models are unable to account realistically for the influence of different types, scales and spatial distributions of sedimentary strata on fluid flow and contaminant transport in fluvial aquifers (Webb 1994; Bierkens & Weerts, 1994; Jussel et al., 1994; Scheibe & Freyberg, 1995; Bierkens, 1996; Anderson et al., 1999). Geostatistical modelling techniques (Carle et al., 1998; Fogg et al., 1998; Rauber et al., 1998; Weissmann & Fogg, 1999; Ritzi et al., 2000) used in combination with these new depositional models should improve predictions of fluid flow and contaminant transport in the subsurface. It should be remembered, however, that it is virtually impossible to distinguish deposits of braided and meandering rivers using data from only cores and wireline logs (Bridge & Tye, 2000; Bridge, 2001).

FUTURE CHALLENGES

The new depositional models presented here, based on studies of natural rivers, laboratory experiments and theory, are detailed, quantitative and generally applicable. The models are presented as cross sections, parallel and normal to flow direction, through untruncated channel bars and channel fills. It is possible to construct such cross sections in any orientation with the new information available. Data are also available on the spatial distribution of channel bars and channel fills within channel belts, including the degree of erosional truncation of these deposits. However, there is not yet a good enough understanding of channel and bar migration to produce a general model for the spatial arrangement of channel bars and fills within channel belts. The scaling relationships between migrating bedforms, specifically unit bars and compound bars, and associated stratasets need to be developed. Future models should also have comprehensive palaeocurrent orientations. There remains a pressing need for studies of depositional processes during floods, over large areas, and over long time spans, in order to allow water flow and sediment transport to be linked to erosion and deposition, channel and bar migration, and the nature of deposits. Efficient methods to monitor changes in topography of large areas of channel bed in turbid braided rivers in flood must be developed. Such studies of braided rivers during flood are part of the basis for developing theoretical models of braided river processes. Existing theoretical models do not predict braided river deposition adequately, and much more effort is required in this area. Finally, future progress requires consistent use of terminology, and dispelling of the many misconceptions about braided rivers and their deposits.

ACKNOWLEDGEMENTS

We appreciate the detailed editorial comments of Phil Ashworth, Jim Best and Peter Huggenberger. Jim Best, Phil Ashworth, Greg Sambrook Smith and Colin Wooldridge also kindly provided us with manuscripts prior to their publication.

REFERENCES

Alexander, J., Bridge, J.S., Leeder, M.R., Collier, R.E.Ll. and Gawthorpe, R.L. (1994) Holocene meander belt evolution in an active extensional basin, southwestern Montana. *J. Sediment. Res.*, **B64**, 542–559.

Anderson, M.P., Aiken, J.S., Webb, E.K. and Mickelson, D.M. (1999) Sedimentology and hydrogeology of two braided stream deposits. *Sediment. Geol.*, **129**, 189–199.

Ashmore, P.E. (1982) Laboratory modelling of gravel braided stream morphology. *Earth Surf. Proc.*, **7**, 2201–2225.

Ashmore, P.E. (1991) How do gravel-bed rivers braid? *Can. J. Earth Sci.*, **28**, 326–341.

Ashmore, P. (1993) Anabranch confluences kinetics and sedimentation processes in gravel-braided streams. In: *Braided Rivers* (Eds J.L. Best and C.S. Bristow), pp. 129–146. Special Publication 75, Geological Society Publishing House, Bath.

Ashmore, P.E. and Church, M. (1998) Sediment transport and river morphology: a paradigm for study. In: *Gravel-bed Rivers in the Environment* (Eds P.C. Klingeman, R.L. Beschta, P.D. Komar and J.B. Bradley), pp. 115–148. Water Resources Publications, Highlands Ranch, Colorado.

Ashmore, P.E. and Parker, G. (1983) Confluence scour in coarse braided streams. *Water Resour. Res.*, **19**, 392–402.

Ashmore, P.E., Ferguson, R.I., Prestegaard, K.L., Ashworth, P.J. and Paola, C. (1992) Secondary flow in anabranch confluences of a braided, gravel-bed stream. *Earth Surf. Proc. Landf.*, **17**, 299–311.

Ashworth, P.J. (1996) Mid-channel bar growth and its relationship to local flow strength and direction. *Earth Surf. Proc. Landf.*, **21**, 103–123.

Ashworth, P.J. and Ferguson, R.I. (1986) Interrelationships of channel processes, changes and sediments in a proglacial braided river. *Geogr. Ann.*, **68A**, 361–371.

Ashworth, P.J., Ferguson, R.I., Ashmore, P.E., Paola, C., Powell, D.M. and Prestegaard, K.L. (1992a) Measurements in a braided river chute and lobe: II. Sorting of bedload during entrainment, transport and deposition. *Water Resour. Res.*, **28**, 1887–1896.

Ashworth, P.J., Ferguson, R.I. and Powell, M.D. (1992b) Bedload transport and sorting in braided channels. In: *Dynamics of Gravel-bed Rivers* (Eds P. Billi, C.R. Hey, C.R. Thorne and P. Tacconi), pp. 497–515, Wiley, Chichester.

Ashworth, P.J., Best, J.L., Leddy, J.O. and Geehan, G.W. (1994) The physical modeling of braided rivers and deposition of fine-grained sediment. In: *Process Models and Theoretical Geomorphology* (Ed. M.J. Kirkby), pp. 115–139. Wiley, Chichester.

Ashworth, P.J., Best, J.L., Peakall, J. and Lorsong, J.A. (1999) The influence of aggradation rate on braided alluvial architecture: field study and physical scale modelling of the Ashburton River gravels, Canterbury Plains, New Zealand. In: *Fluvial Sedimentology VI* (Eds N.D. Smith and J. Rogers), pp. 333–346. Special Publication 28, International Association of Sedimentologists. Blackwell Science, Oxford.

Ashworth, P.J., Best, J.L., Roden, J.E., Bristow, C.S. and Klaassen, G.J. (2000) Morphological evolution and dynamics of a large, sand braid-bar, Jamuna River, Bangladesh. *Sedimentology*, **47**, 533–555.

Ashworth, P.J., Best, J.L. and Jones, M. (2004) Relationship between sediment supply and avulsion frequency in braided rivers. *Geology*, **32**, 21–24.

Asprion, U. and Aigner, T. (1997) Aquifer architecture analysis using ground-penetrating radar: Triassic and Quaternary examples. *Environ. Geol.*, **31**, 66–75.

Asprion, U. and Aigner, T. (1999) Towards realistic aquifer models; three-dimensional georadar surveys of Quaternary gravel deltas (Singen Basin, SW Germany). *Sediment. Geol.*, **129**, 281–297.

Bal, A.A. (1996) Valley fills and coastal cliffs buried beneath an alluvial plain: evidence from variation of permeabilities in gravel aquifers, Canturbury Plains, New Zealand. *J. Hydrol. (NZ)*, **35**, 1–27

Beres, M., Green, A.G., Huggenberger, P. and Horstmeyer, H. (1995) Mapping the architecture of glaciofluvial sediments with three-dimensional georadar. *Geology*, **23**, 1087–1090.

Beres, M., Huggenberger, P., Green, A.G. and Horstmeyer, H. (1999) Using two- and three-dimensional georadar methods to characterize glaciofluvial architecture. *Sediment. Geol.*, **129**, 1–24.

Best, J.L. (1987) Flow dynamics at river channel confluences: implications for sediment transport and bed morphology. In: *Recent Developments in Fluvial Sedimentology* (Eds F.G. Ethridge, R.M. Flores and M.D. Harvey), pp. 27–35. Special Publication 39, Society of Economic Paleontologists and Mineralogists, Tulsa, OK.

Best, J.L. (1988) Sediment transport and bed morphology at river channel confluences. *Sedimentology*, **35**, 481–498.

Best, J.L. and Ashworth, P.J. (1997) Scour in large braided rivers and the recognition of sequence stratigraphic boundaries. *Nature*, **387**, 275–277.

Best, J.L. and Roy, A.G. (1991) Mixing-layer distortion at the confluence of channels of different-depth. *Nature*, **350**, 411–413.

Best, J.L., Ashworth, P.J., Bristow, C. and Roden, J. (2003) Three-dimensional sedimentary architecture of a large, mid-channel sand braid bar, Jamuna River, Bangladesh. *J. Sediment. Res.*, **73**, 516–530.

Bierkens, M.F.P. (1996) Modeling hydraulic conductivity of a complex confining layer at various spatial scales. *Water Resour. Res.*, **32**, 2369–2382.

Bierkens, M.F.P. and Weerts, H.J.T. (1994) Block hydraulic conductivity of cross-bedded fluvial sediments. *Wat. Resour. Res*, **30**, 2665–2678.

Biron, P., Best, J.L. and Roy, A.G. (1996a) Effects of bed discordance on flow dynamics at open channel confluences. *J. Hydraul. Eng., ASCE*, **122**, 676–682.

Biron, P., DeSerres, B., Roy, A.G. and Best, J.L. (1993a) Shear layer turbulence at an unequal depth channel confluence. In: *Turbulence: Perspectives on Flow and Sediment Transport* (Eds N.J. Clifford, J.R. French and J. Hardisty), pp. 197–214. Wiley, Chichester.

Biron, P., Roy, A.G., Best, J.L. and Boyer, C.J. (1993b) Bed morphology and sedimentology at the confluence of unequal depth channels. *Geomorphology*, **8**, 115–129.

Biron, P., Roy, A.G. and Best, J.L. (1996b) Turbulent flow structure at concordant and discordant open-channel confluences. *Exp. Fluids*, **21**, 437–446.

Biron, P.M., Richer, A., Kirkbride, A., Roy, A.G. and Han, S. (2002) Spatial patterns of water surface topography at a river confluence. *Earth Surf. Proc. Landf.*, **27**, 913–928.

Blodgett, R.H. and Stanley, K.O. (1980) Stratification, bedforms and discharge relations of the Platte River system, Nebraska. *J. Sediment. Petrol*, **50**, 139–148.

Bluck, B.J. (1971) Sedimentation in the meandering River Endrick. *Scott. J. Geol.*, **7**, 93–138.

Bluck, B.J. (1976) Sedimentation in some Scottish rivers of low sinuosity. *Trans. R. Soc. Edinb.*, **69**, 425–456.

Bluck, B.J. (1979) Structure of coarse grained braided stream alluvium. *Trans. R. Soc. Edinb.*, **70**, 181–221.

Bolla Pittaluga, M., Repetto, R. and Tubino, M. (2003) Channel bifurcations in braided rivers: equilibrium configurations and stability. *Water Resour. Res.*, **39**, 1046.

Bradbrook, K.F., Biron, P.M., Lane, S.N., Richards, K.S. and Roy, A.G. (1998) Investigation of controls on secondary circulation in a simple confluence using a three-dimensional numerical model. *Hydrol. Process.*, **12**, 1371–1396.

Bradbrook, K.F., Lane, S.N. and Richards, K.S. (2000) Numerical simulation of three-dimensional, time averaged flow structure at river channel confluences. *Water Resour. Res.*, **36**, 2731–2746.

Bradbrook, K.F., Lane, S.N., Richards, K.S., Biron, P.M. and Roy, A.G. (2001) Role of bed discordance at asymmetrical river confluences. *J. Hydraul. Eng., ASCE*, **127**, 351–368.

Brasington, J., Rumsby, B.T. and McVey, R.A. (2000) Monitoring and modelling morphological change in a braided gravel-bed river using high resolution GPS-based survey. *Earth Surf. Proc. Landf.*, **25**, 973–990.

Bridge, J.S. (1977) Flow, bed topography, grain size and sedimentary structure in open channel bends: a three-dimensional model. *Earth Surf. Proc.*, **2**, 401–416.

Bridge, J.S. (1985) Paleochannel patterns inferred from alluvial deposits: a critical evaluation. *J. Sediment. Petrol.*, **55**, 579–589.

Bridge, J.S. (1992) A revised model for water flow, sediment transport, bed topography and grain size sorting in natural river bends. *Water Resour. Res.*, **28**, 999–1023.

Bridge, J.S. (1993) The interaction between channel geometry, water flow, sediment transport and deposition in braided rivers. In: *Braided Rivers* (Eds J.L. Best and C.S. Bristow), pp. 13–71. Special Publication 75, Geological Society Publishing House, Bath.

Bridge, J.S. (1997) Thickness of sets of cross strata and planar strata as a function of formative bedwave geometry and migration, and aggradation rate. *Geology*, **25**, 971–974.

Bridge, J.S. (2001) Characterization of fluvial hydrocarbon reservoirs and aquifers: Problems and solutions. *Rev. Sed. Asoc. Argentina*, **8**, 87–114.

Bridge, J.S. (2003) *Rivers and Floodplains*. Blackwell, Oxford.

Bridge, J.S and Gabel, S.L. (1992) Flow and sediment dynamics in a low sinuosity, braided river: Calamus River, Nebraska Sandhills. *Sedimentology*, **39**, 125–142.

Bridge, J.S. and Jarvis, J. (1976) Flow and sedimentary processes in the meandering river South Esk, Glen Clova, Scotland. *Earth Surf. Proc.*, **1**, 303–336.

Bridge, J.S. and Jarvis, J. (1982) The dynamics of a river bend: a study in flow and sedimentary processes. *Sedimentology*, **29**, 499–541.

Bridge, J.S. and Leeder, M.R. (1979) A simulation model of alluvial stratigraphy. *Sedimentology*, **26**, 617–644.

Bridge, J.S. and Mackey, S.D. (1993) A revised alluvial stratigraphy model. In: *Alluvial Sedimentation* (Eds M. Marzo and C. Puidefabregas), pp. 319–337. Special Publication 17, International Association of Sedimentologists. Blackwell Scientific Publications, Oxford.

Bridge, J.S. and Tye, R.S. (2000) Interpreting the dimensions of ancient fluvial channel bars, channels, and channel belts from wireline-logs and cores. *Am. Assoc. Petrol. Geol. Bull.*, **84**, 1205–1228.

Bridge, J.S., Smith, N.D., Trent, F., Gabel, S.L. and Bernstein, P. (1986) Sedimentology and morphology of a low-sinuosity river: Calamus River, Nebraska Sand Hills. *Sedimentology*, **33**, 851–870.

Bridge, J.S., Alexander, J., Collier, R.E.L., Gawthorpe, R.L. and Jarvis, J. (1995) Ground-penetrating radar and coring used to document the large-scale structure of point-bar deposits in 3-D. *Sedimentology*, **42**, 839–852.

Bridge, J.S., Collier, R.E.Ll. and Alexander, J. (1998) Large-scale structure of Calamus river deposits revealed using ground-penetrating radar. *Sedimentology*, **45**, 977–985.

Bridge, J.S., Jalfin, G.A. and Georgieff, S.M. (2000) Geometry, lithofacies and spatial distribution of Cretaceous fluvial sandstone bodies, San Jorge Basin, Argentina: Outcrop analog for the hydrocarbon-bearing Chubut Group. *J. Sediment. Res.*, **70**, 341–359.

Bristow, C.S. (1987) Brahmaputra River: channel migration and deposition. In: *Recent Developments in Fluvial Sedimentology* (Eds F.G. Ethridge, R.M. Flores and M.D. Harvey), pp. 63–74. Special Publication 39, Society of Economic Paleontologists and Mineralogists, Tulsa, OK.

Bristow, C.S. (1993) Sedimentary structures exposed in bar tops in the Brahmaputra River, Bangladesh. In: *Braided Rivers* (Eds J.L. Best and C.S. Bristow), pp. 277–289. Special Publication 75, Geological Society Publishing House, Bath.

Bristow, C.S., Best, J.L. and Roy, A.G. (1993) Morphology and facies models of channel confluences. In: *Alluvial Sedimentation* (Eds M. Marzo and C. Puidefabregas), pp. 91–100. Special Publication 17, International Association of Sedimentologists. Blackwell Scientific Publications, Oxford.

Bristow, C.S. and Jol, H.M. (Eds) (2003) *Ground Penetrating Radar in Sediments*. Special Publication 211, Geological Society Publishing House, Bath, 330 pp.

Bristow, C.S., Skelly, R.L. and Ethridge, F.G. (1999) Crevasse splays from the rapidly aggrading, sand-bed, braided Niobrara River, Nebraska: effect of base-level rise. *Sedimentology*, **46**, 1029–1047.

Bryant, M., Falk, P. and Paola, C. (1995) Experimental study of avulsion frequency and rate of deposition. *Geology*, **23**, 365–369.

Cant, D.J. (1978) Bedforms and bar types in the South Saskatchewan River. *J. Sediment. Petrol*, **48**, 1321–1330.

Cant, D.J. and Walker, R.G. (1976) Development of a braided fluvial facies model for the Devonian Battery Point Sandstone, Quebec, Canada. *Can. J. Earth Sci.*, **13**, 102–119.

Cant, D.J. and Walker, R.G. (1978) Fluvial processes and facies sequences in the sandy braided South Saskatchewan River, Canada. *Sedimentology*, **25**, 625–648.

Carle, S.F., Labolle, E.M., Weissmann, G.S., Van Brocklin, D. and Fogg, G.E. (1998) Conditional simulation of hydrofacies architecture: a transition probability/Markov approach. In: *Hydrogeologic Models of Sedimentary Aquifers* (Eds G.S. Fraser and J.M. Davis), pp. 147–170. Concepts in Hydrogeology and Environmental Geology 1, Society of Economic Paleontologists and Mineralogists,Tulsa, OK.

Carling, P.A. (1999) Subaqueous gravel dunes. *J. Sediment. Res.*, **69**, 534–545.

Cazanacli, D., Paola, C. and Parker, G. (2002) Experimental steep, braided flow: application to flooding risk on fans. *J. Hydraul. Eng., ASCE*, **128**, 322–330.

Chandler, J.H. (1999) Effective application of automated digital photogrammetry for geomorphological research. *Earth Surf. Proc. Landf.*, **24**, 51–63.

Coleman, J.M. (1969) Brahmaputra River: Channel processes and sedimentation. *Sediment. Geol.*, **3**, 129–239.

Collinson, J.D. (1970) Bedforms of the Tana River, Norway. *Geogr. Ann.*, **52A**, 31–56.

Collinson, J.D. (1996) Alluvial sediments. In: *Sedimentary Environments and Facies*, 3rd edn (Ed. H.G. Reading), pp. 37–82. Blackwell Science, Oxford.

Crowley, K.D. (1983) Large-scale bed configurations (macroforms), Platte River Basin, Colorado and Nebraska: Primary structures and formative processes. *Geol. Soc. Am. Bull.*, **94**, 117–133.

Cuevas Gozalo, M.C. and Martinius, A.W. (1993) Outcrop data-base for the geological characterization of fluvial reservoirs: an example from distal fluvial fan deposits in the Loranca Basin, Spain. In: *Characterization of Fluvial and Aeolian Reservoirs* (Eds C.P. North and D.J. Prosser), pp. 79–94. Special Publication 73, Geological Society Publishing House, Bath.

Davoren, A. and Mosley, M.P. (1986) Observations of bedload movement, bar development and sediment supply in the braided Ohau River. *Earth Surf. Proc. Landf.*, **11**, 643–652.

Dawson, M. (1989) Flood deposits present within the Severn Main Terrace. In: *Floods: Hydrological, Sedimentological and Geomorphological Implications* (Eds K. Beven and P. Carling), pp. 253–264. Wiley, Chichester.

DeSerres, B., Roy, A.G., Biron, P.M. and Best, J.L. (1999) Three-dimensional structure at a river channel confluence with discordant beds. *Geomorphology*, **26**, 313–335.

Deutsch, C.V. and Tran, T.T. (2002) FLUVSIM: a program for object-based modeling of fluvial depositional systems. *Comp. Geosci.*, **28**, 525–535.

Deutsch, C.V. and Wang, L. (1996) Hierarchical object-based stochastic modeling of fluvial reservoirs. *Math. Geol.*, **28**, 857–880.

Dinehart, D.L. (1989) Dune migration in a steep, coarse-bedded stream. *Water Resour. Res.*, **25**, 911–923.

Dinehart, D.L. (1992) Evolution of coarse gravel bed forms: field measurements at flood stage. *Water Resour. Res.*, **28**, 2667–2689.

Doyle, J.D. and Sweet, M.L. (1995) Three-dimensional distribution of lithofacies, bounding surfaces, porosity and permeability in a fluvial sandstone—Gypsy Sandstone of northern Oklahoma. *Am. Assoc. Petrol. Geol. Bull.*, **79**, 70–96.

Dreyer, T. (1993) Quantified fluvial architecture in ephemeral stream deposits of the Esplugafreda Formation (Palaeocene), Tremp-Graus Basin, northern Spain. In: *Alluvial Sedimentation* (Eds M. Marzo and C. Puidefabregas), pp. 337–362. Special Publication 17, International Association of Sedimentologists. Blackwell Scientific Publications, Oxford.

Dreyer, T., Falt, L.-M., Hoy, T., Knarud, R., Steel, R. and Cuevas, J.-L. (1993) Sedimentary architecture of field analogues for reservoir information (SAFARI): a case study of the fluvial Escanilla Formation, Spanish Pyrenees. In: *The Geological Modelling of Hydrocarbon Reservoirs and Outcrop Analogues* (Eds S. Flint and I.D. Bryant), pp. 57–80. Special Publication 15, International Association of Sedimentologists. Blackwell Scientific Publications, Oxford.

Ekes, C. and Hickin, E.J. (2001) Ground penetrating radar facies of the paraglacial Cheekye Fan, southwestern British Columbia. *Sediment. Geol.*, **143**, 199–217.

Eynon, G. and Walker, R.G. (1974) Facies relationships in Pleistocene outwash gravels, southern Ontario; a model for bar growth in braided rivers. *Sedimentology*, **21**, 43–70.

Ferguson, R.I., Ashmore, P.E., Ashworth, P.J., Paola, C. and Prestegaard, K.L. (1992) Measurements in a braided river chute and lobe: I. Flow pattern, sediment transport and channel change. *Water Resour. Res.*, **28**, 1877–1886.

Fogg, G.E., Noyes, C.D. and Carle, S.F. (1998) Geologically based model of heterogeneous hydraulic conductivity in an alluvial setting. *Hydrogeol. J.*, **6**, 131–143.

Frederici, B. and Paola, C. (2003) Dynamics of channel bifurcations in noncohesive sediments. *Water Resour. Res.*, **39**, 1162.

Gabel, S.L. (1993) Geometry and kinematics of dunes during steady and unsteady flows in the Calamus River, Nebraska, USA. *Sedimentology*, **40**, 237–269.

Gawthorpe, R.L., Collier, R.E. Ll., Alexander, J., Bridge, J.S. and Leeder, M.R. (1993) Ground penetrating radar: application to sandbody geometry and heterogeneity studies. In: *Characterization of Fluvial and Aeolian Reservoirs* (Eds C.P. North and D.J. Prosser), pp. 421–432. Special Publication 73, Geological Society Publishing House, Bath.

Germanowski, D. and Schumm, S.A. (1993) Changes in braided river morphology resulting from aggradation and degradation. *J. Geol.*, **101**, 451–466.

Gran, K. and Paola, C. (2001) Riparian vegetation controls on braided stream dynamics. *Water Resour. Res.*, **37**, 3275–3283.

Griffiths, C.M., Dyt, C., Paraschivoiu, E. and Liu, K. (2001) SEDSIM in hydrocarbon exploration. In: *Geologic Modeling and Simulation: Sedimentary Systems* (Eds D.F. Merriam and J.C. Davis), pp. 71–97. Kluwer Academic/Plenum Publishers, Dordrecht.

Gustavson, T.C. (1978) Bedforms and stratification types of modern gravel meander lobes, Nueces River, Texas. *Sedimentology*, **25**, 401–426.

Heller, P.L. and Paola, C. (1996) Downstream changes in alluvial architecture: an exploration of controls on channel-stacking patterns. *J. Sediment. Res.*, **B66**, 297–306.

Heller, P.L., Paola, C., Hwang, I-G., John, B. and Steel, R. (2001) Geomorphology and sequence stratigraphy due to slow and rapid base-level changes in an experimental subsiding basin (XES 96–1). *Am. Assoc. Petrol. Geol. Bull.*, **85**, 817–838.

Hoey, T. and Sutherland, A.J. (1991) Channel morphology and bedload pulses in braided rivers: a laboratory study. *Earth Surf. Proc. Landf.*, **16**, 447–462.

Holden, L., Hauge, R., Skare, Ø. and Skorstad, A. (1998) Modeling of fluvial reservoirs with object models. *Math. Geol.*, **30**, 473–496.

Hornung, J. and Aigner, T. (1999) Reservoir and aquifer characterization of fluvial architectural elements: Stubensandstein, Upper Triassic, southwest Germany. *Sediment. Geol.*, **129**, 215–280.

Huang, J., Weber, L. and Lai, Y.G. (2002) Three-dimensional study of flows in open-channel junctions. *J. Hydraul. Eng., ASCE*, **128**, 268–280.

Huggenberger, P. (1993) Radar facies: recognition of characteristic braided river structures of the Pleistocene Rhine gravel (NE part of Switzerland). In: *Braided Rivers* (Eds J.L. Best and C.S. Bristow), pp. 163–176. Special Publication 75, Geological Society Publishing House, Bath.

Huggenberger, P., Meier, E. and Pugin, A. (1994) Ground-probing radar as a tool for heterogeneity estimation in gravel deposits: advances in data-processing and facies analysis. *J. Applied Geophys.*, **31**, 131–184

Jackson, R.G. (1976) Large-scale ripples of the Lower Wabash River. *Sedimentology*, **23**, 593–623.

Jol, H.M. (1995) Ground penetrating radar antennae frequencies and transmitter powers compared for penetration depth, resolution and reflection continuity. *Geophysical Prospecting*, **43**, 693–709.

Jol, H.M. and Smith, D.G. (1991) Ground penetrating radar of northern lacustrine deltas. *Can. J. Earth Sci.*, **28**, 1939–1947.

Jussel, P., Stauffer, F. and Dracos, T. (1994) Transport modeling in heterogeneous aquifers: 1. Statistical description and numerical generation of gravel deposits. *Water Resour. Res.*, **30**, 1803–1817.

Khan, I.A., Bridge, J.S., Kappelman, J. and Wilson, R. (1997) Evolution of Miocene fluvial environments, eastern Potwar plateau, northern Pakistan. *Sedimentology*, **44**, 221–251.

Kleinhans, M.G., Wilbers, A.W.E., De Swaaf, A. and Van Den Berg, J.H. (2002) Sediment supply-limited bedforms in sand-gravel rivers. *J. Sediment. Res.*, **72**, 629–640.

Koss, J.E., Ethridge, F.G. and Schumm, S.A. (1994) An experimental study of the effects of base-level change on fluvial, coastal plain and shelf systems. *J. Sediment. Res.*, **B64**, 90–98.

Kowalsky, M.B., Dietrich, P., Teutsch, G. and Rubin, Y. (2001) Forward modeling of ground-penetrating radar data using digitized outcrop images and multiple scenarios of water saturation. *Water Resour. Res.*, **37**, 1615–1625.

Kowalsky, M.B., Rubin, Y. and Dietrich, P. (2004) The use of ground-penetrating radar for characterizing sediments under transient conditions. In: *Aquifer Characterization* (Eds J.S. Bridge, and D. Hyndman), pp. 107–127. Special Publication 80, Society of Economic Paleontologists and Mineralogists, Tulsa, OK.

Lane, S.N. and Richards, K.S. (1998) High resolution, two-dimensional spatial modeling of flow processes in a multi-thread channel. *Hydrol. Process.*, **12**, 1279–1298.

Lane, S.N., Chandler, J.H. and Richards, K.S. (1994) Developments in monitoring and terrain modeling small-scale river-bed topography. *Earth Surf. Proc. Landf.*, **19**, 349–368.

Lane, S.N., Richards, K.S. and Chandler, J.H. (1995) Morphological estimation of the time-integrated bedload transport rate. *Water Resour. Res.*, **31**, 761–772.

Lane, S.N., Richards, K.S. and Chandler, J.H. (Eds) (1998) *Landform Monitoring, Modelling and Analysis*. Wiley, Chichester.

Lane, S.N., Bradbrook, K.F., Richards, K.S., Biron, P.M. and Roy, A.G. (2000) Secondary circulation cells in river channel confluences: measurement artifacts or coherent flow structures? *Hydrol. Process.*, **14**, 2047–2071.

Lane, S.N., Chandler, J.H. and Porfiri, K. (2001) Monitoring river channel and flume surfaces with digital photogrammetry. *J. Hydraul. Eng., ASCE*, **127**, 871–877.

Lane, S.N., Westaway, R.M, and Hicks, D.M. (2003) Estimation of erosion and deposition volumes in a large gravel-bed, braided river using synoptic remote sensing. *Earth Surf. Proc. Landf.*, **28**, 249–271.

Leclair, S.F. and Bridge, J.S. (2001) Quantitative interpretation of sedimentary structures formed by river dunes. *J. Sediment. Res.*, **71**, 713–716.

Leclair, S.F., Bridge, J.S. and Wang, F. (1997) Preservation of cross-strata due to migration of subaqueous dunes over aggrading and non-aggrading beds: comparison of experimental data with theory. *Geosci. Can.*, **24**, 55–66.

Leclerc, R.F. and Hickin, E.J. (1997) The internal structure of scrolled floodplain deposits based on ground-penetrating radar, North Thompson River, British Columbia. *Geomorphology*, **21**, 17–38.

Leddy, J.O., Ashworth, P.J. and Best, J.L. (1993) Mechanisms of anabranch avulsion within gravel-bed rivers: observations from a physical scale model. In: *Braided Rivers* (Eds J.L. Best and C.S. Bristow), pp. 119–127. Special Publication 75, Geological Society Publishing House, Bath.

Levey, R.A. (1978) Bedform distribution and internal stratification of coarse-grained point bars, Upper Congaree River, South Carolina. In: *Fluvial Sedimentology* (Ed. A.D. Miall), pp. 105–127. Memoir 5, Canadian Society of Petroleum Geologists, Calgary.

Liu, K., Boult, P., Painter, S. and Paterson, L. (1996) Outcrop analog for sandy braided stream reservoirs: permeability patterns in the Triassic Hawkesbury Sandstone, Sydney Basin, Australia. *Am. Assoc. Petrol. Geol. Bull.*, **80**, 1850–1866.

Lunt, I.A. and Bridge, J.S. (2004) Evolution and deposits of a gravelly braid bar and a channel fill, Sagavanirktok river, Alaska. *Sedimentology*, **51**, 415–432.

Lunt, I.A., Bridge, J.S. and Tye, R.S. (2004a) Development of a 3-D depositional model of braided river gravels and sands to improve aquifer characterization. In: *Aquifer Characterization* (Eds J.S. Bridge, and D. Hyndman), pp. 139–169. Special Publication 80, Society of Economic Paleontologists and Mineralogists, Tulsa, OK.

Lunt, I.A., Bridge, J.S. and Tye, R.S. (2004b) A quantitative, three-dimensional depositional model of gravelly braided rivers. *Sedimentology*, **51**, 377–414.

Mackey, S.D. and Bridge, J.S. (1995) Three dimensional model of alluvial stratigraphy: theory and application. *J. Sediment. Res.*, **B65**, 7–31.

Martin, J.H. (1993) A review of braided fluvial hydrocarbon reservoirs: the petroleum engineer's perspective. In: *Braided Rivers* (Eds J.L. Best and C.S. Bristow), pp. 333–367. Special Publication 75, Geological Society Publishing House, Bath.

Martin, Y. and Church, M. (1996) Bed-material transport estimated from channel surveys-Vedder River, British Columbia. *Earth Surf. Proc. Landf.*, **20**, 247–261.

McGowen, J.H. and Garner, L.E. (1970) Physiographic features and stratification types of coarse grained point bars: Modern and ancient examples. *Sedimentology*, **14**, 77–111.

McLelland, S.J., Ashworth, P.J. and Best, J.L. (1996) The origin and downstream development of coherent flow structures at channel junctions. In: *Coherent Flow Structures in Open Channels* (Eds P.J. Ashworth, S.J. Bennett, J.L. Best and S.J. McLelland), pp. 459–490. Wiley, Chichester.

McLelland, S.J., Ashworth R.J., Best, J.L., Roden J. and Klaassen, G.J. (1999) Flow structure and transport of sand-grade suspended sediment around an evolving braid bar, Jamuna River, Bangladesh. In: *Fluvial Sedimentology VI* (Eds N.D. Smith and J. Rogers), pp. 43–57. Special Publication 28, International Association of Sedimentologists. Blackwell Science, Oxford.

Miall, A.D. (1977) A review of the braided river depositional environment. *Earth Sci. Rev.*, **13**, 1–62.

Miall, A.D. (1992) *Alluvial deposits*. In: *Facies Models: Response to Sea Level Change* (Eds R.G. Walker and N.P. James), pp. 119–142. Geological Association of Canada, St John's, Newfoundland.

Miall, A.D. (1994) Reconstructing fluvial macroform architecture from two-dimensional outcrops: examples from the Castlegate Sandstone, Book Cliffs, Utah. *J. Sediment. Res.*, **B64**, 146–158.

Miall, A.D. (1996) *The Geology of Fluvial Deposits*. Springer-Verlag, New York, 582 pp.

Moreton, D.J., Ashworth, P.J. and Best, J.L. (2002) The physical scale modelling of braided alluvial architecture and estimation of subsurface permeability. *Basin Res.*, **14**, 265–285.

Mosley, M.P. (1976) An experimental study of channel confluences. *J. Geol.*, **84**, 535–562.

Mosselman, E., Huisink, M., Koomen, E. and Seijmonsbergen, A.C. (1995) Morphological changes in a large braided sand-bed river. In: *River Geomorphology* (Ed. E.J. Hickin), pp. 235–249. Wiley, Chichester.

Murray, A.B. and Paola, C. (1994) A cellular model of braided rivers. *Nature*, **371**, 54–57.

Murray, A.B. and Paola, C. (1996) A new quantitative test of geomorphic models, applied to a model of braided streams. *Water Resour. Res.*, **32**, 2579–2587.

Murray, A.B. and Paola, C. (1997) Properties of a cellular braided stream model. *Earth Surf. Proc. Landf.*, **22**, 1001–1025.

Nanson, G.C. (1980) Point bar and floodplain formation of the meandering Beatton River, northeastern British Columbia, Canada. *Sedimentology*, **27**, 3–29.

Nicholas, A.P. and Sambrook Smith, G.H. (1999) Numerical simulation of three-dimensional flow hydraulics in a braided channel. *Hydrol. Process.*, **13**, 913–929.

North, C.P. (1996) The prediction and modelling of subsurface fluvial stratigraphy. In: *Advances in Fluvial Dynamics and Stratigraphy* (Eds P.A. Carling and M.R. Dawson), pp. 395–508. Wiley, Chichester.

Olsen, T., Steel, R., Hogseth, K., Skar, T. and Roe, S.-L. (1995) Sequential architecture in a fluvial succession: sequence stratigraphy in the Upper Cretaceous Mesaverde Group, Price Canyon, Utah. *J. Sediment. Res.*, **B65**, 265–280.

Ouchi, S. (1985) Response of alluvial rivers to slow active tectonic movement. *Geol. Soc. Am. Bull.*, **96**, 504–515.

Paola, C. (2000) Quantitative models of sedimentary basin filling. *Sedimentology*, **47**(Supplement 1), 121–178.

Paola, C. and Borgman, L. (1991) Reconstructing random topography from preserved stratification. *Sedimentology*, **38**, 553–565.

Paola, C., Mullin, J., Ellis, C., Mohrig, D.C., Swenson, J.B., Parker, G., Hickson, T., Heller, P.L., Pratson, L., Syvitski, J., Sheets, B. and Strong, N. (2001) Experimental stratigraphy. *GSA Today*, **11**, 4–9.

Peakall, J., Ashworth, P. and Best, J. (1996) Physical modeling in fluvial geomorphology: principles, applications and unresolved issues. In: *The Scientific Nature of Geomorphology*. (Eds B.L. Rhoads, and C.E. Thorn), pp. 221–253. Wiley, Chichester.

Ramos, A., Sopeña, A. and Perez-Arlucea, M. (1986) Evolution of Bundsandstein fluvial sedimentation in the Northwest Iberian Ranges (Central Spain). *J. Sediment. Petrol*, **56**, 862–875.

Rauber, M., Stuaffer, F., Huggenberger, P. and Dracos, T. (1998) A numerical three-dimensional conditioned/unconditioned stochastic facies type model applied to a remediation well system: *Water Resour. Res.*, **34**, 2225–2233.

Regli, C, Huggenberger, P. and Rauber, M. (2002) Interpretation of drill core and georadar data of coarse gravel deposits. *J. Hydrol.*, **255**, 234–252.

Rhoads, B.L. (1996) Mean structure of transport effective flows at an asymmetrical confluence when the main stream is dominant. In: *Coherent Flow Structures in Open Channels* (Eds P.J. Ashworth, S.J. Bennett, J.L. Best and S.J. McLelland), pp. 491–517. Wiley, Chichester.

Rhoads, B.L. and Kenworthy, S.T. (1995) Flow structure at an asymmetrical stream confluence. *Geomorphology*, **11**, 273–293.

Rhoads, B.L. and Kenworthy, S.T. (1998) Time-averaged flow structure in the central region of a stream confluence. *Earth Surf. Proc. Landf.*, **23**, 171–191.

Rhoads, B.L. and Sukhodolov, A.N. (2001) Field investigation of three-dimensional flow structure at stream confluences:1 Thermal mixing and time-averaged velocities. *Water Resour. Res.*, **37**, 2393–2410.

Richardson, W.R.R. and Thorne, C.R. (1998) Secondary currents and channel changes around a braid bar in the Brahmaputra River, Bangladesh. *J. Hydraul. Engrg., ASCE*, **124**, 325–328.

Richardson, W.R.R., Thorne, C.R. and Mahmood, S. (1996) Secondary flow and channel changes around a bar in the Brahmaputra River, Bangladesh. In: *Coherent Flow Structures in Open Channels* (Eds P.J. Ashworth, S.J. Bennett, J.L. Best and S.J. McLelland), pp. 520–543. Wiley, Chichester.

Ritzi, R.W., Dominic, D.F., Slesers, A.J., Greer, C.B., Reboulet, E.C., Telford, J.A., Masters, R.W., Klohe, C.A., Bogle, J.L. and Means, B.P. (2000) Comparing statistical models of physical heterogeneity in buried-valley aquifers: *Water Resour. Res.*, **36**, 3179–3192.

Robinson, J.W. and McCabe, P.J. (1997) Sandstone-body and shale-body dimensions in a braided fluvial system: Salt Wash Sandstone Member (Morrison Formation), Garfield County, Utah. *Am. Assoc. Petrol. Geol. Bull.*, **81**, 1267–1291.

Sambrook Smith, G.H., Ashworth, P.J., Best, J.L., Woodward, J. and Simpson, C.J. (2005) The morphology and facies of sandy braided rivers: some considerations of scale invariance. In: *Fluvial Sedimentology VII* (Eds M.D. Blum, S.B. Marriott and S.F. Leclair), pp. 145–158. Special Publication 35, International Association of Sedimentologists. Blackwell Publishing, Oxford.

Scheibe, T.D. and Freyberg, D.L. (1995) Use of sedimentological information for geometric simulation of natural porous media structure: *Water Resour. Res.*, **31**, 3259–3270.

Schumm, S.A. and Khan, H.R. (1972) Experimental study of channel patterns. *Bull. Geol. Soc. Am.*, **83**, 1755–1770.

Schumm, S.A., Mosley, M.P. and Weaver, W.E. (1987) *Experimental fluvial Geomorphology*. Wiley, New York.

Sheets, B.A., Hickson, T.A. and Paola, C. (2002) Assembling the stratigraphic record: depositional patterns and time-scales in an experimental alluvial basin. *Basin Res.*, **14**, 287–301.

Siegenthaler, C. and Huggenberger, P. (1993) Pleistocene Rhine gravel : deposits of a braided river system with dominant pool preservation. In: *Braided Rivers* (Eds J.L. Best and C.S. Bristow), pp. 147–162. Special Publication 75, Geological Society Publishing House, Bath.

Skelly, R.L., Bristow, C.S. and Ethridge, F.G. (2003) Architecture of channel-belt deposits in an aggrading shallow sandbed braided river: the lower Niobrara River, northeast Nebraska. *Sediment. Geol.*, **158**, 249–270.

Smith, N.D. (1971) Transverse bars and braiding in the Lower Platte River, Nebraska. *Geol. Soc. Am. Bull*, **82**, 3407–3420.

Smith, N.D. (1972) Some sedimentological aspects of planar cross-stratification in a sandy braided river. *J. Sediment. Petrol*, **42**, 624–634.

Smith, N.D. (1974) Sedimentology and bar formation in the upper Kicking Horse River, a braided outwash stream. *J. Geol.*, **81**, 205–223.

Stojic, M., Chandler, J.H., Ashmore, P. and Luce, J. (1998) The assessment of sediment transport rates by automated digital photogrammetry. *Photogram. Eng. Remote Sens.*, **64**, 387–395.

Sukhodolov, A.N. and Rhoads, B.L. (2001) Field investigation of three-dimensional flow structure at stream confluences: 2. Turbulence. *Wat. Resour Res.*, **37**, 2411–2424.

Tetzlaff, D. (1991) The combined use of sedimentary process modeling and statistical simulation in reservoir characterization. SPE Paper 22759, presented at the *SPE Annual Technical Conference and Exhibition*, 6–9 October, Dallas, TX, pp. 937–942.

Tetzlaff, D.M. and Harbaugh, J.W. (1989) *Simulating Clastic Sedimentation*. New York, Van Nostrand Reinhold.

Tetzlaff, D. and Priddy, G. (2001) Sedimentary process modeling: from academia to industry. In: *Geologic Modeling and Simulation: Sedimentary Systems* (Eds D.F. Merriam and J.C. Davis), pp. 45–69. Kluwer Academic/Plenum Publishers, Dordrecht.

Thomas, R. and Nicholas, A.P. (2002) Simulation of braided river flow using a new cellular routing scheme. *Geomorphology*, **43**, 179–195.

Thorne, C.R., Russell, A.P.G. and Alam, M.K. (1993) Platform pattern and channel evolution of Brahmaputra River, Bangladesh. In: *Braided Rivers* (Eds J.L. Best and C.S. Bristow), pp. 257–276. Special Publication 75, Geological Society Publishing House, Bath.

Tyler, K., Henriquez, A. and Svanes, T. (1994) Modeling heterogeneities in fluvial domains: a review of the influence on production profiles. In: *Stochastic modeling and Geostatistics* (Eds J.M. Yarus and R.L. Chambers), pp. 77–89. Computer Applications in Geology 3, American Association of Petroleum Geologists, Tulsa, OK.

Van Dam, R.L. and Schlager, W. (2000) Identifying the causes of ground-penetrating radar reflections using time-domain reflectometry and sedimentological analyses. *Sedimentology*, **47**, 435–449.

Van Dam, R.L., Van den Berg, E.H., van Heteren, S., Kasse, C., Kenter, J.A.M. and Groen, K. (2002) Influence of organic matter in soils on radar-wave reflection: sedimentological implications. *J. Sediment. Res.*, **72**, 341–352.

Vandenberghe, J. and Van Overmeeren, R.A. (1999) Ground-penetrating images of selected fluvial deposits in the Netherlands. *Sediment. Geol.*, **128**, 245–270.

Van Overmeeren, R.A. (1998) Radar facies of unconsolidated sediments in The Netherlands: A radar stratigraphy interpretation method for hydrogeology. *Appl. Geophys.*, **40**, 1–18.

Warburton, J. and Davies, T. (1994) Variability of bedload transport and channel morphology in a braided river hydraulic model. *Earth Surf. Proc. Landf.*, **19**, 403–421.

Webb, E.K. (1994) Simulating the three-dimensional distribution of sediment units in braided stream deposits. *J. Sediment. Res.*, **B64**, 219–231.

Webb, E.K. (1995) Simulation of braided channel topology and topography. *Water Resour. Res.*, **31**, 2603–2611.

Webb, E.K. and Anderson, M.P. (1996) Simulation of preferential flow in three-dimensional, heterogeneous conductivity fields with realistic internal architecture. *Water Resour. Res.*, **32**, 533–545.

Weissmann, G.S. and Fogg, G.E. (1999) Multi-scale alluvial fan heterogeneity modeled with transition probability geostatistics in a sequence stratigraphic framework. *J. Hydrol.*, **226**, 48–65.

Westaway, R.M., Lane, S.N. and Hicks, D.M. (2000) The development of an automated correction procedure for digital photogrammetry for the study of wide, shallow, gravel-bed rivers. *Earth Surf. Proc. Landf.*, **25**, 209–226.

Willis, B.J. (1989) Paleochannel reconstructions from point bar deposits: a three-dimensional perspective. *Sedimentology*, **36**, 757–766.

Willis, B.J. (1993) Ancient river systems in the Himalayan foredeep, Chinji village area, northern Pakistan. *Sediment. Geol.*, **88**, 1–76.

Wood, L.J., Ethridge, F.G. and Schumm, S.A. (1993) An expermental study of the influence of subaqueous shelf angles on coastal plain and shelf deposits. In: *Siliciclastic Sequence Stratigraphy* (Eds P. Weimer and H.W. Posamentier), pp. 381–391. Memoir 58, American Association of Petroleum Geologists, Tulsa, OK.

Woodward, J., Ashworth, P.J., Best, J.L., Sambrook Smith, G.H. and Simpson, C.J. (2003) The use and application of GPR in sandy fluvial environments: methodological considerations. In: *Ground Penetrating Radar in Sediments* (Eds C.S. Bristow and H.M. Jol), pp. 127–142. Special Publication 211, Geological Society Publishing House, Bath.

Wooldridge, C.L. (2002) *Channel bar radar architecture and evolution in the wandering gravel-bed Fraser and Squamish Rivers, British Columbia, Canada.* Unpublished MSc thesis, Department of Geography, Simon Fraser University, Burnaby, BC, Canada, 122 pp.

Zaleha, M.J. (1997) Fluvial and lacustrine palaeoenvironments of the Miocene Siwalik Group, Khaur area, northern Pakistan. *Sedimentology*, **44**, 349–368.

A sedimentological model to characterize braided river deposits for hydrogeological applications

PETER HUGGENBERGER and CHRISTIAN REGLI

Department of Geosciences, Applied and Environmental Geology, University of Basel, Bernoullistrasse 32, 4056 Basel, Switzerland
(Email: peter.huggenberger@unibas.ch)

ABSTRACT

Braided river deposits form important aquifers in many parts of the world, and their heterogeneity strongly influences groundwater flow and mass transport processes. To accurately characterize these coarse gravelly aquifers, it is important to understand the erosional and depositional processes that form these sediments. Moreover, it is important to evaluate the relative importance of various parameters that determine the preservation potential of different depositional elements over geological time scales. These objectives may be achieved by developing techniques that allow for the integration of different quality data into quantitative models. Information concerning sedimentary textures and the spatial continuity of sedimentary structures in braided river deposits, inherent in depositional facies descriptions, allows the spatial variability of hydrogeological properties (e.g. hydraulic conductivity and porosity) to be predicted. Depositional elements in gravel deposits can contain a restricted range of textures, which form a limited number of sedimentary structures. These depositional elements are bounded by erosional and/or lithological surfaces. The frequency, size and shape of different elements in a sedimentary sequence depend on several factors, including aggradation rate, channel belt mobility on the kilometre scale, gravel-sheet/scour activity at the scale of hundreds of metres and topographic position of the different elements within an evolving system. Preserved shape and size of the elements affect the correlation lengths and the standard deviations of the aquifer properties, such as hydraulic conductivity and porosity.

Different quality data sets that may be used in characterizing braided river deposits can be recognized in outcrop, boreholes and on ground-penetrating radar (GPR) sections. This paper proposes a means of integrating outcrop, borehole and GPR data into a stochastic framework of sedimentary structures and the distribution of hydraulic aquifer properties. Data integration results in variable degrees of uncertainty when assigning values to hydraulic properties and characterizing the geometry of sedimentary structures. An application of this approach is illustrated using a data set (400 m × 550 m) from the northeastern part of Switzerland at the confluence of the Rhine and Wiese rivers. The data set includes drill-core data from five boreholes and 14 GPR sections with a total length of 3040 m. The results of the variogram analysis provide the orientation of sedimentary structure types representing the main flow direction of the River Rhine in the lower part of the aquifer, and of the River Wiese in the upper part. The analysis also results in large ranges of spatial correlation, ranging from a few metres up to tens of metres for the different sedimentary structure types.

Keywords Gravel heterogeneity, sedimentological model, stochastic modelling, groundwater.

INTRODUCTION

A large proportion of European and North American aquifers were formed by former braided river systems. Accurate predictions of groundwater flow and transport behaviour within these sediments rely on detailed knowledge of the distribution of physical, chemical and biological aquifer properties. The complexity of depositional and erosional processes in braided river systems, however, leads to highly heterogeneous distributions of hydrogeological parameters such as hydraulic conductivity and porosity. Hydraulic conductivity variations over several orders of magnitude are of primary importance in groundwater flow and solute migration (Gelhar, 1986; Adams & Gelhar, 1992; Rehfeldt et al., 1993). Due to the complexity of the aquifer structures in these coarse gravelly deposits, and its consequences for hydraulic property distribution, stochastic modelling has rarely been applied to practical problems (Dagan, 2002).

Continuous three-dimensional information on the hydraulic properties of aquifers cannot be obtained easily in heterogeneous sediments. Consequently, different methods have been developed in order to map aquifer properties. Koltermann & Gorelick (1996) distinguished three main types of methods: (1) structure-imitating methods that use any combination of Gaussian and non-Gaussian statistical and geometric relationships in order to match observed sedimentary patterns; (2) process-imitating methods that consist of both aquifer calibration techniques, which solve the governing equations of fluid flow and transport, and geological process models that combine mass and momentum conservation principles with sediment transport equations; and (3) descriptive methods that use different field methods to translate geological facies models into hydrofacies models with characteristic aquifer properties.

All these methods have already been applied in coarse fluvial gravel deposits that are typical of braided river environments. Examples of each approach, including key information for the understanding of sedimentological facies description and aquifer characterization, as cited in the literature, are provided below.

1 Structure-imitating methods: Jussel et al. (1994) developed unconditioned, Gaussian stochastic models of glacio-fluvial deposits using statistical data derived from outcrop descriptions of gravel pits in northeastern Switzerland. Facies interpretations obtained from ground-penetrating radar (GPR) profiles by Rauber et al. (1998) from one of these gravel deposits (Hüntwangen, Switzerland) were used to condition stochastic models. Both groups of authors used their stochastic models to simulate a synthetic tracer transport experiment. Among non-Gaussian methods, indicator-based simulations offered the advantage of combining different types of observations, using 'hard' and 'soft' data (e.g. McKenna & Poeter, 1995). Regli et al. (2004) developed the software code GEOSSAV, which is capable of incorporating these different data types by including drill-core descriptions and GPR profiles, to produce a stochastic model that could be tested against tracer breakthrough responses in the 'Langen Erlen' well catchment zone (Regli et al., 2003).

Other non-Gaussian approaches include Boolean, or object-based methods, which are suited to randomly distributed sedimentary body shapes that are typical of fluvial deposits (Deutsch & Wang, 1996). Geological information has been used to develop geometrical and probabilistic rules that control the distribution, geometry, elongation axis and connectivity for each of the simulated objects or facies types. Another promising non-Gaussian approach has explored the influence of aquifer structure on groundwater flow regimes using transition probabilities (Markov chains; Carle et al., 1998; Fogg et al., 1998). Weissmann et al. (1999) presented a Markov chain geostatistical approach to create geologically plausible three-dimensional characterizations of a heterogeneous, fluvial aquifer system. The maps were conditioned using core, well-log and soil data from the Kings River alluvial fan aquifer system in California. Using this approach, pattern-imitating methods have simulated the architecture and lithology of fluvial deposits directly, based on geometric concepts of the observed geology. Webb (1994) and Webb & Anderson (1996) presented a simulation method that develops the three-dimensional distribution of sediment units in braided stream deposits. Their code, BCS-3D, combines a random-walk algorithm for braided topological networks with equations to describe hydraulic channel geometry, which were then used to estimate the channel shape and connectivity relations of a surface. Sediment aggradation in the third (vertical) dimension was obtained by stacking the generated geomorphological braided river surfaces using a constant offset, and the resulting sedimentary model was used to assign distributed hydraulic conductivities.

2 Process-imitating methods have not been used extensively for modelling fluvial sediments. Koltermann & Gorelick (1992) used large-scale process simulations to reconstruct the geological evolution of an alluvial fan in northern California over the past 600,000 yr. The reconstruction of the Alameda Creek fan highlighted the importance of climate change, tectonics and relative sea-level change on the depositional processes. Nonetheless, geological process models have rarely been applied to hydrogeological problems, largely because they use initial and boundary conditions that are largely unknown and consequently cannot be conditioned to measured values.

3 Descriptive methods produce maps of the architecture of fluvial deposits by the translation of geological or geophysical site-specific data into conceptual facies models. The quality of field observations is of prime importance. Lithofacies concepts of braided river deposits have been developed by Allen (1978), Miall (1985), Brierley (1991), Paola *et al.* (1992), Heller & Paola (1992), Bridge (1993, 2003) and Siegenthaler & Huggenberger (1993). Ashmore (1982, 1993) and Ashmore & Parker (1983) have demonstrated the importance of key elements such as braiding intensity and scouring on sedimentation using laboratory experiments and field observations of braided river systems (Bridge, 2003; Ashworth *et al.*, 1999, 2004). Anderson (1989) and Klingbeil *et al.* (1999) related the concept of lithofacies to hydrofacies in a fluvial environment, and Anderson *et al.* (1999) used extensive outcrop observations and sampling information for hydrofacies conceptualization. Generally, outcrops and boreholes offer a restricted image of the sedimentary architecture and grain-size distribution data. The additional application of shallow geophysical methods such as high-resolution GPR has been shown to be very effective in mapping the relevant hydrofacies (e.g. Huggenberger, 1993; Bridge *et al.*, 1995; Best *et al.*, 2003). Lunt *et al.* (2004) combined GPR surveys with data from trenches, cores, logs, and permeability and porosity measurements in a study of a three-dimensional depositional model of the Sagavanirktok River in Alaska. Techniques used in three-dimensional surveying offer the possibility of generating images of the spatial structure of fluvial aquifer architecture on a decimetric scale (Beres *et al.*, 1995, 1999; Asprion & Aigner, 1999; Heinz, 2001). Regli *et al.* (2002) coupled probability estimations of drill-core descriptions to radar facies types to define sedimentary structure types for the Langen Erlen aquifer.

Various braided river system depositional models are controversially debated in the literature. Controversies have arisen mainly as a result of the difficulty in finding simple relationships between morphological elements observed in modern braided river systems and the geometry of architectural elements of ancient braided river deposits. Such relations can be based only on two-dimensional outcrop observations combined with two-dimensional and three-dimensional GPR data sets that provide data to a limited depth, in conjunction with certain concepts concerning the preservation potential of former morphological elements in the sedimentary record. The prediction of the architecture of braided river deposits cannot be based on observations of recent braided river system morphology at low flow stage alone, since structures developing under very high flow conditions (catastrophic or landscape shaping events) give better information on the spatial extent of characteristic depositional elements. In terms of hydrogeological applications, it is important to remember that conceptual sedimentological models are based on their relevance to groundwater hydraulics. There are many indications from hydrogeological investigations (e.g. Flynn, 2003) that the role of highly permeable sediments, such as open-framework gravel, is often underestimated.

The objective of this paper is to derive a methodology to characterize braided river deposits for hydrogeological applications by using different quality geological information. This includes:

1 definition of sedimentary textures, structures and depositional elements together with a discussion of the links between surface morphology and the subsurface sedimentary structures;
2 presentation of a method by which outcrop, borehole and GPR data can be integrated in a structure-imitating aquifer model;
3 presentation of a case study of the application of subsurface modelling for the Rhine/Wiese sand and gravel aquifer near Basel, Switzerland (Fig. 1).

The stochastic simulations characterize possible spatial distributions of aquifer properties that can be used as an input for groundwater flow and transport simulations.

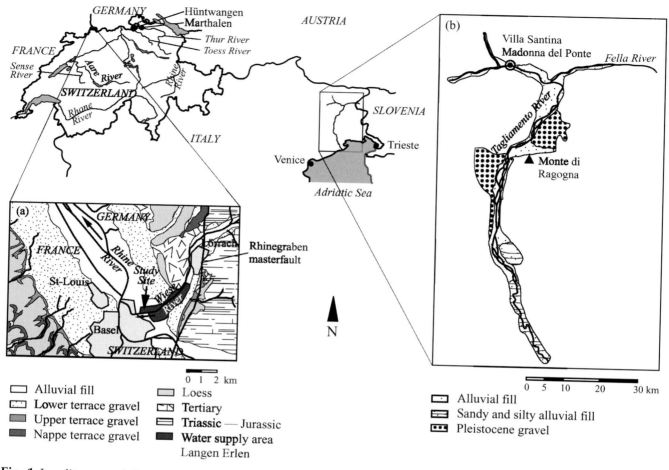

Fig. 1 Locality map of rivers and outcrops cited in the text. Insets show: (a) simplified geological overview of the Basel area (Switzerland) including the location of study site in the Basel water supply area (Langen Erlen), located within the ancient confluence of the River Rhine and River Wiese; (b) overview of the Tagliamento area, Friaul, northern Italy.

METHODS

In order to refine sedimentological models to permit characterization of braided river deposits to be extended to hydrogeological applications, different methods for modelling aquifer sedimentology and the subsurface have been applied. Following the hydrogeological facies concepts of Asprion & Aigner (1999), sedimentary textures, structures and depositional elements were defined. The preservation potential of depositional elements, and its significance for groundwater flow and transport in their influence on hydraulic conductivity and porosity, are used as key information in this type of classification. Important information was obtained by comparison of morphological elements, observed at the surface of modern braided river systems (e.g. River Tagliamento, Friaul, northeastern Italy), with preserved depositional elements that were investigated in gravel pits deposited by the Pleistocene Rhine River in Switzerland (Fig. 1). Moreover, the results from laboratory flume experiments, numerical models, field measurements and proposed fluvial processes (e.g. fluvial sediment sorting processes) can form the sedimentological facies relations described (Ashmore, 1982, 1993; Klingemann & Emmet, 1982; Van Dyke, 1982; Carling & Glaister, 1987; Carling, 1990; Tubino et al., 1999; Lanzoni, 2000).

Outcrop, borehole and GPR profiles represent data of different quality. Outcrop and laboratory investigations of representative samples concerning facies or hydraulic properties provide the most reliable data, and are considered as *hard data*. In contrast, drill-core and GPR data are less precise and considered to be *soft data*, and greater uncertainties are associated with these data values. The differences between hard and soft data required an interpretation method to be developed that allowed the integration of different quality datasets into proposed lithofacies schemes. The method provides probabilities that drill-core layer descriptions and radar facies types represent particular defined sedimentary structures.

Accurate modelling of subsurface parameter distributions becomes progressively more difficult with increasing heterogeneity and uncertainty in the spatial variability of available data. Uncertainty depends both on the quantity and the quality of available data. Stochastic simulation as described below implies sampling from conditional distributions, and the resulting spatial models consist of samples from a multivariate distribution characterizing the spatial phenomenon. The sequential indicator-based approach matches the sedimentary structures using conditioning data, such as drill-core and GPR information, with the spatial correlation of data values, depending on the sedimentological model as described in the following sections, and the geostatistical analysis of the available data verified in variogram analyses.

SEDIMENTARY TEXTURES, STRUCTURES AND DEPOSITIONAL ELEMENTS

Based on outcrop analyses of gravel pit exposures of sediments deposited during the Pleistocene by the Aare, Rhine, Rhône and Thur rivers in Switzerland (Fig. 1), and established facies schemes (Siegenthaler & Huggenberger, 1993; Jussel *et al.*, 1994; Rauber *et al.*, 1998; Regli *et al.*, 2002), a limited number of textural and structural sediment types, reflecting dominant depositional elements developing in braided river systems, were defined.

Sedimentary texture and structure types

Definition of sedimentary texture types is based on grain-size distribution and sediment sorting. In the Rhine gravel, sedimentary textures are easily recognizable in outcrop due to colour variations caused by the presence or absence of sand-fractions, silt and clay and the wetting/drying characteristics at the outcrops. Consequently, colour is used in field classifications of different gravel textures (Jussel *et al.*, 1994). Bearing in mind that colour differences are dependent on the source rock material, the sedimentary textures are considered as an important attribute in distinguishing different sedimentary structures in outcrop; such textures are also frequently used in drill-core descriptions. Investigations at different field locations and subsequent data interpretation allow the classification of seven different sedimentary texture types (Fig. 2a–g).

1 Open-framework gravel (OW, Figs 2a & 3a, c) is a well-sorted gravel, in which pore space is free of sand and silt, although clay and silt particles occasionally drape the pebbles and cobbles. Mean grain-size (d_m) and sorting coefficient ($s_c = d_{60}/d_{10}$) for OW are 0.015 m and 3.3, respectively (s_c corresponds to the slope of the grain-size distribution curve between d_{10} and d_{60} and is an indicator of sediment sorting (Swiss Standard Association, 1997)). Open-framework gravel corresponds to a well-sorted, poorly graded (high porosity) sedimentary texture type.
2 Bimodal gravel (BM, Figs 2b & 3a, c) consists of a matrix of well-sorted medium sand that fills interstices of a framework of well-sorted pebbles and occasional cobbles ($d_m = 0.021$ m, $s_c = 77$). Visual inspections of BM in outcrop give the impression that the grain-size and bed-thickness of BM are positively correlated (see also Steel & Thompson, 1983).
3 Grey gravel (GG, Figs 2c & 3b, d–f) is a poorly sorted gravel, containing coarse sand, granules, pebbles and, rarely, cobbles. Clay and silt particles never make up more than 5% of the deposit ($d_m = 0.007$ m, $s_c = 55$).
4 Brown gravel (BG, Figs 2d & 3d–e) is a poorly sorted gravel with sand and silt ($d_m = 0.021$ m, $s_c = 236$).
5 Silty gravel (SG, Fig. 2g) is a poorly sorted, brownish coloured, gravel, often containing up to 30% sand and nearly 20% silt and clay ($d_m = 0.002$ m, $s_c = 250$).
6 Sand lenses (SA, Fig. 3f) may be poorly sorted to well-sorted and do not contain significant silt or clay fractions.
7 Silt lenses (SI, Fig. 2f) consist of a poorly graded silt.

Sedimentary structure types are made up of one or a combination of two sedimentary texture

Fig. 2 Sedimentary textures types: (a) (OW) open-framework well-sorted gravel; (b) (BM) bimodal gravel; (c) (GG) poorly sorted, clean grey gravel; (d) (BG) poorly sorted, coarse brown gravel with silt fraction; (e) (SA) sand; (f) (SI) silt; (g) (SG) silty gravel. Photographs (a–c) and (e–g) from Pleistocene Rhine gravel at Hüntwangen, (d) from Sense River, Switzerland.

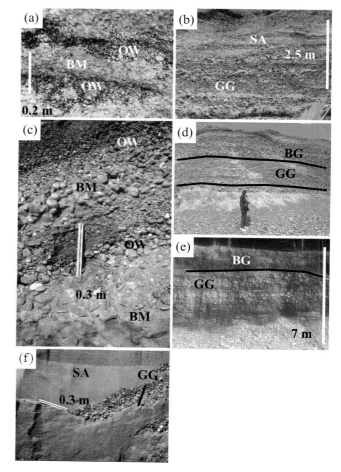

Fig. 3 Sedimentary structures types: (a), (c) (OW/BM) open-framework well-sorted/bimodal gravel couplet consisting of BM at the base and OW at the top; (b), (f) alternating beds of SA/GG; (d), (e) GG/BG and SG/GG. All samples from Pleistocene Rhine gravel, Hüntwangen, Switzerland.

types that may alternate. The seven sedimentary texture types identified may be grouped into nine sedimentary structure types (Fig. 3).

1 OW appears in fining upward beds that are one to several decimetres thick, and individual clasts are not oriented.

2 OW/BM couplets are fining-upwards sequences consisting of BM at the base and OW at the top. A transition from BM to OW is generally marked by a sharp boundary between the sand in the BM and the open pores of the OW. The pebbles, however, show a continuous fining-upwards sequence from BM to OW.

3 GG often forms thick sheets of poorly sorted sediments up to 2 m thick.

4 The bedding of BG is massive, and often forms beds up to 4–6 m thick.

5 & 6 GG and BG may also occur in alternating layers that are either horizontal (GG/BG-h) or inclined (GG/BG-i).

7 SG may occur in close relation to BG, as massive beds or associated with thin silt layers. However, the latter are not very frequently observed in outcrop.

8 SA may occur in different settings on a floodplain and with varying proportions of GG and SI.

9 SI mainly occurs as thin beds or near the top of gravel sequences, sometimes associated with gravels in smaller trough-shaped fills.

Depositional elements

Depositional elements (Fig. 4a–f) are defined based on the geometrical form of erosional and/or lithological boundaries and the character of the fill (e.g. structure types and geometry of planar or tangential foresets). The most abundant elements in the deposits investigated are: (1) scour or trough fills (Fig. 4d–f); (2) gravel sheets or gravel dunes that usually form thin gravel layers, although these may be up to more than 1 m thick (Fig. 4a, c); (3) bedload sheets (Fig. 4b, c) or traction carpets (Todd, 1989); and (4) overbank deposits (Siegenthaler & Huggenberger, 1993; Beres et al., 1999, fig. 3; Heinz, 2001).

1 Scour or trough fill deposits (Figs 3–5) can be composed of four different sedimentary structure types: (i) sets with OW/BM couplet cross-beds; (ii) sets with GG cross-beds; (iii) sets with SA cross-beds; and (iv), in the braided–meandering transition zone (which is marked by a decrease in slope and widening of the floodplain to the west of the Rhinegraben masterfault), sets with SG cross-beds (Regli et al., 2002). The cross-bedded sets are 0.1–0.5 m thick with trough-shaped, erosional concave-upward lower bounding surfaces. The erosional surface appears as a clear boundary that is discontinuously overlain by a lag of cobbles in some outcrops. In sections approximately perpendicular to the general flow direction, the erosional bounding surfaces may display the shape of circular arc segments, while the cross-beds are strongly curved and tangential to the lower set boundary (Huggenberger et al., 1988).

2 Horizontally bedded gravel sheets (Fig. 4a, c) mainly consist of GG with occasional alternations with single beds of BG. Individual beds range in thickness from <0.1 m to about 1 m thick, but are typically 0.1–0.3 m thick. Very thick beds are probably obscure low-angle cross-stratified sets. Single beds extend laterally from a few metres up to several tens of metres. The bedding is mostly vague and caused by: (i) variable sand fractions especially within the GG; (ii) vertical alternations of different textural types; (iii) thin interlayers of sand and pebbles; or (iv) discontinuous stringers or coarse pebbles and cobbles.

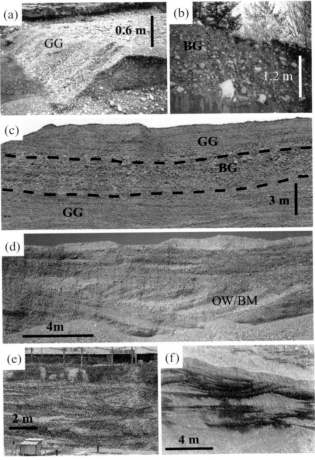

Fig. 4 Depositional elements: (a), (c) gravel sheets (a, flow from right to left, River Toess; c, section perpendicular to ancient mean flow direction, Pleistocene Rhine gravel, Hüntwangen); (b) bedload sheet (River Sense, Switzerland); (d–f) scour and trough-fill deposits (sections perpendicular to ancient mean flow direction, all outcrops from Pleistocene Rhine gravel; d, Hüntwangen, e, Basel, f, Marthalen). Note the concave-upward trough-shape and erosional lower bounding surfaces. Trough-fill deposits (d) consist of tangential bottom sets with gravel couplet cross-beds. (GG) Grey gravel; (BG) poorly sorted, coarse brown gravel with silt fraction; (OW/BM) open-framework well-sorted/bimodal gravel couplet.

3 Laterally extensive, massive and coarse-grained gravel sheets (Fig. 3d, e) are composed of BG. Single beds or sets of beds have thicknesses ranging from decimetres up to several metres. Laterally, thickness may vary abruptly and the beds may locally disappear or reduce to a discontinuous lag of coarse cobbles. A preferred orientation or imbrication of the clasts

Fig. 5 Pleistocene trough-fill dominated sequence in Rhine gravels at Marthalen, Switzerland, in a section perpendicular to the mean ancient flow direction. Note the dominance of filled trough structures and the absence of gravel-sheets. Type OW/BM makes up to more than 60% of the whole sequence.

is quite common, and the long axis of the clasts usually dips approximately 15° upstream.

4 Overbank deposits are composed of SA and SI and, compared with the active channel area, fine-grained sediments (fractions finer than fine sand) are more abundant. Floodplain deposits consist of massive to horizontally bedded medium to coarse silt and very fine sand, and in a few examples fine sand. Bioturbation may be intense in more recent sediments, and is mainly due to rooting by abundant vegetation. In Pleistocene sequences, however, only a small amount of organic matter may be observed due to the lack of a plant cover. Sandy beds with ripples or parallel laminae often occur near the active channel belt, mainly as a product of moderate to high magnitude floods (Huggenberger et al., 1998).

Relationship between depositional elements of ancient rivers and morphological elements of modern braided river systems

It is often difficult to estimate the maximum dimensions and orientation of depositional elements of braided river deposits preserved in the geological record. Although voluminous information is available from photographs of progressively excavated terrace walls (Jussel, 1992; Siegenthaler & Huggenberger, 1993), the maximum dimensions of depositional elements are commonly impossible to ascertain without two-dimensional and three-dimensional GPR data sets. Moreover, the results of a three-dimensional GPR study of a 70 m × 120 m × 15 m block have demonstrated that the dimensions of depositional elements such as trough fills can be significantly larger than the size of the block investigated (Beres et al., 1999).

Observations of morphological elements on the floodplain of the Tagliamento River (Fig. 1) immediately after a moderate magnitude flood (2000 $m^3 s^{-1}$) allowed study of both the geometrical form and extent of gravel sheets (Fig. 6) and the simultaneous progradation of a gravel sheet with the development of a scour fill in front of the gravel sheet (Fig. 7). This flood has allowed some important products of the high-flow stage to be documented that are generally overprinted by smaller events, such as development of smaller gravel sheets, or which are not accessible to observation (e.g. processes in scours). There are few examples documenting simultaneous gravel-sheet and scour development in rivers. However, Fay (2002) documented trough-shaped scour fills and related them to the November 1996 glacier-outburst flood, Skeidarársandur, southern Iceland. Their scour-fill geometry (Fay, 2002, fig. 9) strongly resembles structures found in the Rhine gravel deposits of northeastern Switzerland (Siegenthaler & Huggenberger, 1993, fig. 6).

Gravel sheets

The geometry of two gravel sheets that are more than 200 m long and 80 m wide is documented in Fig. 6. The gravel sheets are in concordance with the previous riverbed topography and developed at moderate magnitude flow, where the active channel belt was wholly inundated and the morphology of the floodplain was reshaped. The orientation of the longitudinal direction of the gravel sheet differs from the orientation of the channels developing at the end of a flood event. Irregularly sinuous, single-thread channels, including channel islands, dominate the morphological character of the fluvial system at low flow, and the angle between the longitudinal directions of the gravel sheets varies according to the intersection with these channels. In the present example, the angle is ~20–30°. This difference in transport direction marks the difference between the braiding character of the river system during flooding and

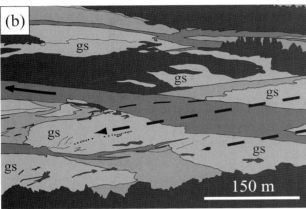

⟵ Flow direction at low flow conditions
⟵ - - Transport direction of gravel sheets at high flow conditions

Fig. 6 Observations of modern river analogues: (a) photograph and (b) interpretation of a view of the Tagliamento River from Monte di Ragogna (Friaul, Italy). Flow from right to left. Note the angle between the active channels under low-flow conditions and the long axis of gravel sheets (gs) developed during flood (high-flow) conditions. The low-flow channel in the central part of the photograph dissects the gravel sheets.

the channels at low-flow conditions (Leopold & Wolman, 1957; Church, 2002).

Simultaneous gravel-sheet and scour development

Whereas gravel sheets are at least partly preserved after floods (Lunt *et al.*, 2004, fig. 6), scours and related trough fills developing simultaneously during gravel-sheet propagation (Ashmore, 1982) are filled up with sediment deposited with declining flood energy. As trough fills are buried below the upper units of a fluvial channel, they

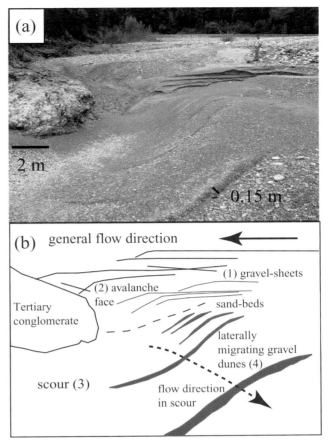

Fig. 7 Observations of modern river analogues: (a) photograph and (b) interpretation of relics developed by a scour-fill process at Madonna del Ponte, Villa Santina (Friaul, Italy). The photograph provides a three-dimensional view of a river section abandoned immediately after a flood event. The mean flow direction in the channel is from right to left. Within the scour, smaller gravel dunes have migrated across the lateral wall of the scour. Note the difference in elevation between the top of the gravel sheets and the deepest part of the scour pool (approximately 6 m). Gravel sheets (1), avalanche faces (2), transition zone to scour pool (3) and upward migrating gravel dunes (4) developed simultaneously.

generally can be portrayed only on GPR profiles (Huggenberger *et al.*, 1998). The development of sedimentary structures within a scour was observed at Madonna del Ponte (Villa Santina, Friaul, Fig. 1) immediately after a flood in 2000. Figure 7 gives a view perpendicular to the mean flow. The avalanche faces in front of the gravel sheets or dunes mark a transition from gravel sheets to the scour, the erosional base of which is at least 3–4 m below the top of the gravel sheets. Such transitions represent negative steps that can lead to the development of flow separation zones (Allen, 1984), which favour sediment sorting, such as cross-bedding and the development of gravel couplets (Carling & Glaister, 1987). In the scour itself, gravel dunes have been observed migrating across the lateral trough wall. Along the individual dune front (step), cross-bed sets with completely obscured foresets develop (e.g. Siegenthaler & Huggenberger, 1993, fig. 5). The maximum amplitude of the dune crests is in the order of 0.1–0.2 m. While the model of Carling & Glaister (1987) can explain planar, cross-bedded gravel couplets (OW/BM), this particular example from the Tagliamento River could illustrate the formation of tangential gravel couplets (e.g. Fig. 4d).

HYDROGEOLOGICAL CHARACTERIZATION OF BRAIDED RIVER DEPOSITS

Hydraulic properties of sedimentary structure types

The hydraulic properties of different sedimentary structure types have been determined from slightly or undisturbed sediment samples taken of individual sedimentary texture types from outcrops in the unsaturated zone (Huggenberger et al., 1988; Jussel et al., 1994). The results of hydraulic testing and geostatistical analysis display significant differences in the hydraulic properties of the different sedimentary structure types in comparison with the variability of the individual texture types. The mean and standard deviation of the measured hydraulic conductivities in unsaturated or slightly disturbed samples (Rauber et al., 1998) indicate that the effect of OW structures (geometric mean of hydraulic conductivity, K, 1×10^{-1} m s^{-1}; standard deviation of the natural logarithm of K, $\sigma_{\ln K}$, >0.8) on the groundwater velocity distribution is much more important than the influence of admixtures of other sedimentary structure types (K, 1×10^{-3} to 1×10^{-5} m s^{-1}; $\sigma_{\ln K}$, 0.4–0.8). The frequency of occurrence and the dimensions of the OW strata determine the correlation length and the standard deviation of the hydraulic conductivity in coarse gravel deposits. Stauffer & Rauber (1998) showed that OW strata in OW and OW/BM structures dominate these statistical parameters of the gravel deposits, although their total volumetric fraction was quite small (2.8%).

Recognition of sedimentary structure types from drill-core and GPR data

Outcrop, drill-core and geophysical information represent data of different quality and resolution at different scales. Due to the easy access and visibility of undisturbed sedimentary structures and textures, outcrop investigations provide very reliable hard data. Unfortunately, few outcrops are available for studies. Drilling may destroy sedimentary structures and can smear the interface with adjacent layers. Typically, drill-core layer descriptions are not very detailed and do not clearly indicate explicit texture or structure types, even if grain-size analyses are available (e.g. overlapping ranges of grain-size distributions of different sedimentary texture types; Jussel, 1992, fig. 2.5a–d). In addition, the quality of individual drill-core descriptions varies considerably depending on the geotechnical or sedimentological approach used. Consequently, drill-core data are normally considered as soft data.

Non-destructive GPR techniques permit depositional elements and sedimentary structures to be delineated based on the geometry of the bounding surfaces and on smaller internal reflections. However, the relationship between reflection patterns and sedimentary structure types is frequently ambiguous, and consequently the results of GPR studies are also considered as soft data.

The concept of recognizing sedimentary structure types using drill-core and GPR data is based on the lithofacies scheme developed for the Rhine gravel, described in Siegenthaler & Huggenberger (1993), Jussel et al. (1994), Rauber et al. (1998) and Regli et al. (2002). Drill-core and GPR data have also been interpreted and integrated into this lithofacies scheme (Fig. 8). The data interpretation includes a probability estimate that drill-core layer descriptions and reflection patterns represent defined sedimentary structure types.

Interpretation of drill-core data

Sedimentological drill-core descriptions of coarse gravel deposits provide information on the com-

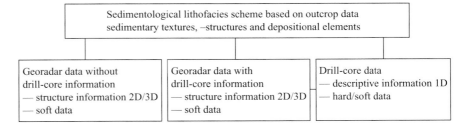

Fig. 8 Classification and integration of different quality data for characterizing braided river deposits.

position and texture of deposits. More specifically, details of grain-size categories, sorting, major constituent composition and the proportion of each grain-size fraction can be determined, as well as information about the colour, chemical precipitation, thickness of a deposit and its transition with the underlying layer (Table 1). Based on these data, the probability can be estimated that a drill-core layer description represents a defined sedimentary structure type (structure-type probability). The estimation of probability and the mathematical details of this method are described in Regli *et al.* (2002).

Differentiation of the probabilities of the sedimentary structure types follows an iterative process, whereby each type of information is considered. The iterative differentiation of sedimentary structure-type probabilities is illustrated using the example presented in Fig. 9 (Rhine/Wiese aquifer, Fig. 1). A drill-core layer described as grey, clean, poorly sorted gravel, with abundant medium and coarse sand, abundant pebbles and cobbles, normal thickness (0.25–2.5 m), and underlain by a layer of rather silty gravel, contains information (see Table 1) that allows the differentiation of the single structure-type probabilities. Starting with the initial probability values for sand, gravel, pebbles and cobbles (Regli *et al.*, 2002, table 2, row 13), the subsequent iterations result in variable differentiation of the structure-type probabilities for the same data set with different confidence levels. The vertical lines represent ranges of structure-type probabilities when different orders of iteration are chosen. In Fig. 9, two orders of iterations are presented; one starts with the information concerning the strongest relative weighting factor and then proceeds downward (Table 1, iteration number 1 to 13; data represented in Fig. 9 by symbols); the second order of iterations starts with the information on the weakest relative weighting factor and then

proceeds upward (Table 1, iteration number 13 to 1; data represented in Fig. 9 by the tops of the vertical lines). A confidence factor of 0.9 for the drill-core layer results in a GG classification with a probability of more than 70%. A factor of confidence of 0.9 reflects a high degree of confidence in the drill-core description (e.g. due to detailed drill-core analysis based on the Unified Soil Classification System (USCS: Wagner, 1957) and additional sedimentological attributes). A factor of confidence of 0.5 indicates a lower degree of confidence in the drill-core description (e.g. due to bad drill-core analysis or washing by flushing fluids during drilling).

Interpretation of GPR data

Ground-penetrating radar data (e.g. reflection profiles) provide two-dimensional images that permit the division of the subsurface into zones of more and less prominent reflection patterns. According to the interpretation concepts of Hardage (1987), Beres and Haeni (1991) and Huggenberger (1993), a radar facies type can be defined as a mappable sedimentary structure with a reflection pattern differing from those in adjacent structures. The geometry of the sedimentary structure types can be delineated by more continuous reflections and the types of reflection patterns within a certain visible structure. In addition, an angular unconformity between prominent reflections can be an indicator of an erosional surface, which separates different sedimentary units.

Transformation of the reflections from travel-time to depth requires information on the velocity distribution. The velocity field of the GPR waves is derived using the common midpoint method (CMP). Semblance velocity analysis (e.g. Beres *et al.*, 1999) shows that interval velocities in gravel deposits range from 0.07 m ns^{-1} to 0.11 m ns^{-1}

Table 1 Sedimentological information in drill-core layer descriptions and relative weighting factors with indications of the sedimentary structure type, for which the information is typical. OW, open framework well-sorted gravel; OW/BM, open framework well-sorted/bimodal gravel couplets; GG, grey gravel; BG, brown gravel; GG/BG-horizontal, alternating grey and brown gravel, horizontally layered; GG/BG-inclined, alternating grey and brown gravel, inclined; SG, silty gravel; SA, sand; SI, silt. The colour separation above and below −5.0 m arises from the different geology of the source areas: above −5.0 m the deposits consist exclusively of Wiese gravel, below −5.0 m the deposits consist of Rhine and subordinate quantities of Wiese gravel

Sedimentological information (Iteration number)		Typical sedimentary structure types	Relative weighting factor
Major constituent (1)	Silt	SI	1
	Sand	SA	
	Gravel	OW, OW/BM, GG, BG, GG/BG-h/-i, SG	
Quantity of (clay)–silt–(sand) (3)	Clean	OW, OW/BM, GG, GG/BG-h/-i, SA	0.7
	Little silt/silty	BG, GG/BG-h, GG/BG-i, SA	
	Little silt and sand	OW, OW/BM	
	Much silt/clay	SG	
Quantity of sand (4)	Little sand, 3–15%	OW/BM, BG, SG, SI	0.7
	Abundant sand, 16–30%	GG, BG, GG/BG-h/-i, SG	
	Much sand, 31–49%	GG, SG, SA	
Quantity of gravel (10)	Little gravel, 3–15%	SA, SI	0.25
	Abundant gravel, 16–30%	SA, SI	
	Much gravel, 31–49%	SA, SI	
Quantity of cobbles (11)	Few cobbles	OW/BM, GG, BG, GG/BG-h/-i, SG, SA	0.25
	Abundant cobbles	GG, BG, GG/BG-h/-i	
	Many cobbles	BG	
Fraction of sand (5)	Fine sand	BG, GG/BG-h/-i, SA, SI	0.55
	Medium sand	OW/BM, GG, BG, GG/BG-h/-i, SG, SA	
	Coarse sand	OW/BM, GG, SA	
Fraction of gravel (8)	Fine gravel	GG, SG, SA, SI	0.4
	Medium gravel	OW, OW/BM, GG, GG/BG-h/-i, SA	
	Coarse gravel	OW/BM, BG, GG/BG-h/-i	
Open framework gravel (2)	Open framework gravel	OW, OW/BM	0.85
	Fe-/Mg-precipitation	OW, OW/BM	
Sorting of sand (6)	Well sorted	OW/BM, GG, SA	0.55
	Poorly sorted	OW, BG, GG/BG-h/-i, SG, SA	
Sorting of gravel (9)	Well sorted	OW, OW/BM, SA	0.4
	Poorly sorted	GG, BG, GG/BG-h/-i, SG, SA	
Colour (7) (above −5 m)	Grey	OW, OW/BM, GG	0.55
	Brown	GG, BG, SG, SA, SI	
	Grey-brown	GG, BG, GG/BG-h/-i, SG, SA, SI	
Colour (7) (below −5 m)	Grey	OW, OW/BM, GG, SG, SA, SI	
	Brown	BG	
	Grey-brown	GG/BG-h/-i, SA	
Thickness of layer (12)	Thin, <0.25 m	OW	0.25
	Normal, 0.25–2.5 m	OW/BM, GG, BG, GG/BG-h/-i, SG, SA, SI	
	Thick, >2.5 m	BG	
Underlying layer (13)	SA	SI	0.1
	OW, OW/BM, GG	BG, GG/BG-h/-i, SG	
	BG, GG/BG-h/-i, SI, SG	OW, OW/BM, GG, SA	

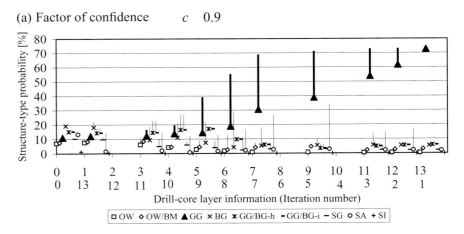

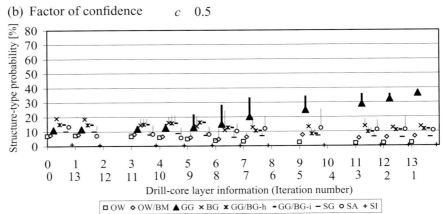

Fig. 9 Differentiation of sedimentary structure-type probabilities (%) based on the integration of sedimentological drill-core layer information and the factor of confidence in the drill-core description. Data shown are for layer 5 from borehole 1477 (see Regli *et al.*, 2002): (a) high confidence in drill-core description; (b) moderate confidence in drill-core description. The fine vertical lines represent ranges of structure-type probabilities when different orders of integration of layer information are chosen. See text for explanation.

(mean at 0.095 m ns^{-1}) for the different CMPs. Depending on the velocity field, linear or more complex velocity functions have to be considered for the transformation of the reflections from travel-time to depth. Due to the accuracy of the vertical resolution, which is equal to a quarter of the applied GPR wavelength, and the variance of the wave velocity in comparison to the resolution of the lithological units in drill cores, a constant velocity is assumed as a first approximation. For more complex velocity functions, calibration curves need to be considered, which transform individual boundaries from two-way travel-time to depths by using interpolation schemes for differing neighbouring velocity logs (Copty & Rubin, 1995).

The radar facies types can be calibrated based on interpreted drill cores located in the vicinity of GPR sections. The calibration process consists of assigning the calculated structure-type probabilities of the drill-core layer descriptions to the corresponding radar facies types. The mathematical details of this method are described in Regli *et al.* (2002). Figure 10a provides an example of oblique sigmoidal and oblique parallel radar facies types within a portion of a GPR section that incorporates borehole 1462 (Rhine/Wiese aquifer; see Regli *et al.*, 2002). The difference in thickness suggested by drill-core layers and the GPR structures depends on the resolution of the visual drill-core analysis (usually to a few centimetres) and the frequency used in GPR mapping (usually to a few decimetres for 50 MHz antennae).

Drill-core and GPR data processing

For the application of drill-core and GPR data to subsurface aquifer modelling, one-dimensional drill-core data and the two-dimensional images of GPR sections have to be transformed into point data. Data processing is necessary once facies analysis is performed. The lithofacies-based interpretation of drill-core and GPR data based on Regli *et al.*

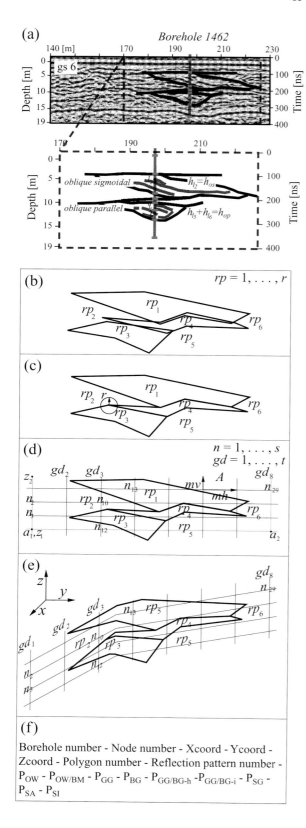

(2002) considers differences in data uncertainty and provides lithofacies probabilities for given points along boreholes and grid nodes with arbitrary mesh sizes along GPR sections. The processing steps for the GPR data are schematically shown in Fig. 10b–f. The procedure consists of (i) digitizing reflection pattern boundaries (Fig. 10b), (ii) snapping common points of neighbouring polygons (GPR structures; Fig. 10c), (iii) rasterizing polygons and generating nodes (grid-points; Fig. 10d), (iv) transforming relative coordinates into absolute coordinates (Fig. 10e) and (v) assigning data to nodes (Fig. 10f).

The digitization of the reflection patterns ($rp = 1, \ldots, r$) is carried out with digicps-3, an appended digitizing software for CPS-3 (Radian Corporation, 1992). Usually the points of neighbouring polygons are not coincident. For successive data processing steps, supplemental program routines were written in C. The snapping tool allows input of the radius (r) into which polygon points will be snapped. The rasterizing tool allows the input of the area of the GPR section ($A = a_1, a_2, z_1, z_2$) to be rasterized and the input of the horizontal (mh) and vertical (mv) mesh sizes between nodes. Nodes ($n = 1, \ldots, s$) with the same a coordinates (direction along GPR profile) but various z coordinates (depth) are grouped into a 'GPR borehole' ($gd = 1, \ldots, t$). The data density has to be chosen in such a way that GPR structures are clearly shown. The coordinate transformation tool changes relative two-dimensional coordinates (a,z) into absolute three-dimensional coordinates (x,y,z). Finally, the assigning tool allows the arrangement of all information such as data source (GPR or borehole),

Fig. 10 *(left)* A graphical summary of the steps used in radar facies analysis and data processing: (a) assignment of sedimentary structure-type probabilities from drill-core layer descriptions to corresponding radar facies types according to the proportion in thickness between drill-core layer and GPR structure; (b) digitizing GPR pattern boundaries; (c) snapping of common polygon points; (d) rasterizing polygons and generating nodes (grid-points); (e) transforming relative coordinates into absolute coordinates; (f) assigning information (e.g. physical parameters) to nodes. gs, GPR section; rp, reflection pattern; r, radius; n, node; gd, 'GPR borehole'; mh, horizontal mesh size; mv, vertical mesh size; A, area of GPR section (gs; $A: a_1,a_2,z_1,z_2$); P, probability of sedimentary structure type.

borehole number, node number, x, y and z coordinates, depth, polygon number, reflection pattern number, detail of whether a node is a surface node or not, probabilities of the sedimentary structure types OW, OW/BM, GG, BG, GG/BG-h, GG/BG-i, SG, SA, and SI to the corresponding nodes. The data are in a comma-separated-value (CSV) format, with each node in a separate line and each kind of information separated by a comma. Subsurface modelling software requires spatial coordinates (x,y,z) as well as data such as sedimentary structure types, probability details, hydraulic and geotechnical parameters obtained at the nodal location.

EXAMPLE OF STOCHASTIC AQUIFER SIMULATION USING HARD AND SOFT DATA

Simulation software and project integration

The mathematical tool GEOSSAV (Geostatistical Environment for Subsurface Simulation And Visualization) can be used for the integration of hard and soft data, and in the three-dimensional stochastic simulation and visualization of the distribution of geological structures and hydrogeological properties in the subsurface (Regli et al., 2004). The tool GEOSSAV can be used for data analysis, variogram computation of regularly or irregularly spaced data, and sequential indicator simulation of subsurface heterogeneities when employed as an interface to selected geostatistical programs from the Geostatistical Software Library, GSLIB (Deutsch & Journel, 1998). Simulations can be visualized by three-dimensional rendering and slicing perpendicular to the main coordinate axis. The data can then be exported into regular grid-based groundwater simulation systems (e.g. ASMWIN (Chiang et al., 1998); GMS (Environmental Modelling Systems Inc., 2002); PMWIN (Chiang & Kinzelbach, 2001)).

In the current study, GEOSSAV was used to generate the sedimentary structures and hydraulic aquifer properties in the vicinity of a drinking water well. The simulated aquifer properties were integrated into a 550 m × 400 m × 21 m, 11-layer finite-difference groundwater flow and transport model to simulate a river restoration pilot project near a well-field in a gravel aquifer supplying the city of Basel (Fig. 1). Groundwater simulation of this portion of the Rhine/Wiese aquifer is of interest since it includes simulations of changing well capture zones depending on the sedimentary structures present in the subsurface, hydrological variations, the operating regime of the drinking water wells, and the progress of river restoration activities (Regli et al., 2003). The aim of the following example is to demonstrate the different steps followed in the characterization of subsurface heterogeneity, depending on the proposed sedimentological model and the geostatistical analysis employed using the available data.

Field data of different quality

The study site is located at the ancient confluence where the River Rhine flowed to the northwest and its tributary, the River Wiese, flowed to the southwest (Fig. 1). The average discharge of the River Rhine at this location over the past 110 yr was 1052 m^3 s^{-1}, and the discharge of the Rhine is around 90 times larger than the average discharge of the Wiese (11.4 m^3 s^{-1}, based on measurements collected over the past 68 yr (Swiss Federal Office for Water and Geology, 2001)). Rhine and Wiese sediments are lithologically distinct from each other as the two rivers derived their sediments from two geologically different catchment areas. Within these two lithological units, a number of sedimentary structures are recognized that were generated by sedimentary processes operating in braided rivers. Lithofacies associated with the sedimentary structures at this location include (Regli et al., 2002): OW, OW/BM, GG, BG, GG/BG, as well as horizontally layered or inclined, and SG, SA and SI. Drill-core data from five boreholes and GPR data from 14 vertical GPR sections (total length of all sections is 3040 m; GPR line orientations within the field of 550 m × 400 m are given in Fig. 11) indicated that the unconfined aquifer consists of coarse unconsolidated Quaternary alluvial deposits. Tertiary marls underlie these gravels and are considered impermeable for the purposes of the model simulations. Aquifer thickness varies between 13 and 18 m. The lower 80% of the aquifer consists of Rhine gravel (primarily limestone) and the upper 20% of the aquifer consists of Wiese gravel (primarily silicates and limestone; Zechner et al., 1995). These lithological differences may be explained by the River Rhine reworking the Wiese gravel

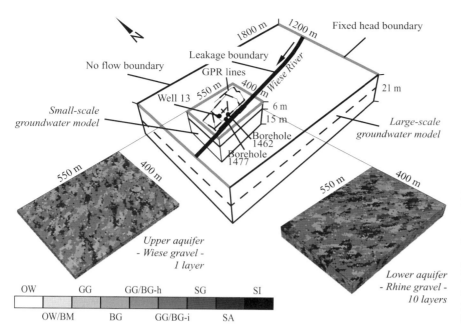

Fig. 11 Location of the small-scale groundwater model within a large-scale homogeneous groundwater model. The small-scale model relies on stochastically generated aquifer properties. OW, open framework well-sorted gravel; OW/BM, open framework well-sorted/bimodal gravel couplets; GG, grey gravel; BG, brown gravel; GG/BG-h, alternating grey and brown gravel, horizontally layered; GG/BG-i, alternating grey and brown gravel, inclined; SG, silty gravel; SA, sand; SI, silt.

at very high flow stages, whereas the uppermost part of the Wiese gravel sequence was preserved until the next shift of the active channel of the Rhine. The data sampled from the GPR sections are given for nodes with 5 m lateral spacing along the profile direction and 1 m vertical spacing.

Variogram computation

Modelling spatial variability of data is the key to any subsurface simulation. A variogram describes the spatial correlation of data as a function of the separation vector between two data points. The indicator variogram computation was based on the drill-core and GPR data described above, and was run separately for the lower part (Rhine gravel) and the upper part (Wiese gravel) of the aquifer. The indicator transform (Deutsch & Journel, 1998) at grid node locations was set to unity for structure types with the greatest probability values, but otherwise was zero. Experimental indicator variograms were calculated for various directions (azimuth, dip, plunge). Table 2 presents the resulting parameters for the nine sedimentary structure types identified at the study site. Azimuth (both dip and plunge are 0°) and the ranges corresponding to maximum and minimum horizontal and vertical spatial correlation distances characterize the geometric anisotropy of the sedimentary structure types.

The results of the variogram analysis provide the orientation of the sedimentary structure types representing the main flow direction of the River Rhine in the lower part, and of the River Wiese in the upper part of the aquifer, respectively. The relatively large ranges of spatial correlation, ranging from a few metres up to a few tens of metres for the different sedimentary structure types (Table 2), may be significantly influenced by the resolution of the GPR system and the density of the sampled data taken from the GPR sections. The sedimentary structures of the Rhine gravel were modelled as geostatistical structures that are around 20% and 45% larger in the horizontal and vertical directions respectively compared with the structures of the Wiese gravel.

Aquifer simulation

The aquifer structure was simulated by sequential indicator simulation (Deutsch & Journel, 1998; Regli *et al.*, 2004). The sequential indicator simulation principle is an extension of conditioning that includes all data available within the neighbourhood of a model cell, including the original data and all previously simulated values. Sequential

Table 2 Rhine gravel and Wiese gravel parameters used for the sequential indicator simulation to define the geometric anisotropy of the sedimentary structure types: OW, open framework well-sorted gravel; OW/BM, open framework well-sorted/bimodal gravel couplets; GG, grey gravel; BG, brown gravel; GG/BG-horizontal, alternating grey and brown gravel, horizontally layered; GG/BG-inclined, alternating grey and brown gravel, inclined; SG, silty gravel; SA, sand; SI, silt. Values in italics are estimates; the isotropic nugget constants of the sedimentary structure types are zero; the variogram models of the sedimentary structure types are exponential; the dip and plunge of the sedimentary structure types are zero degrees; the sill refers to the positive variance contribution in the variogram model (Deutsch & Journel, 1998)

	Parameter	Sedimentary structure type								
		OW	OW/BM	GG	BG	GG/BG horizontal	GG/BG inclined	SG	SA	SI
Wiese gravel	Volumetric fraction (%)	0.02	0.05	0.16	0.05	0.50	0.05	0.02	0.14	0.01
	Sill (–)	0.13	0.12	0.18	0.115	0.13	0.13	0.045	0.18	0.13
	Azimuth (°)	240	240	240	240	240	240	270	200	240
	Maximum horizontal range (m)	3	24	60	34	50	7	14	50	3
	Minimum horizontal range (m)	1.5	18	24	24	18	3	18	16	1.5
	Vertical range (m)	0.5	4	6	5	6	1	4	3	0.5
Rhine gravel	Volumetric fraction (%)	0.02	0.06	0.12	0.05	0.50	0.05	0.03	0.16	0.01
	Sill (–)	0.1	0.095	0.155	0.055	0.1	0.1	0.04	0.17	0.1
	Azimuth (°)	310	320	315	300	310	310	300	300	310
	Maximum horizontal range (m)	5	54	60	40	70	8	30	60	5
	Minimum horizontal range (m)	2	22	19	22	30	4	17	22	2
	Vertical range (m)	1	10	5	11	10	2	10	8	1

indicator simulations are processed in a number of steps. An initial step establishes a grid network and coordinate system, and this is followed by assigning data to the nearest grid node. Where more than one data point may be used at a node, the closest data point is assigned to the grid node. In a third step, a random path through all grid nodes is determined. For a node on a random path, adjacent data and previously simulated grid nodes are searched, to permit an estimation of the conditional distribution by indicator kriging. Based on this distribution, a simulated lithofacies is randomly drawn and set as hard data before the next node in the random path is selected and the process repeated. By using this approach, the simulation grid is built up sequentially. During the final step of the sequential indicator simulation, results are checked to ensure that orientations and sizes of the simulated sedimentary structures are in accordance with those observed.

The one sequential indicator simulation, using separate runs for the lower and the upper parts of the aquifer, is presented in Fig. 11. The regular model grid is defined by $110 \times 80 \times 10$ cells for the lower part and by $110 \times 80 \times 1$ cells for the upper part of the aquifer, respectively. Cell sizes are 5 m $\times$ 5 m $\times$ 1.5 m for the lower part, and 5 m $\times$ 5 m $\times$ 6 m for the upper part, respectively. Each simulated sedimentary structure-type distribution is termed an aquifer realization. In each model run, the resulting probability density functions of the sedimentary structure types deviate less than ±10% from the initial probability density functions, which represent the expected volumetric fractions of the sedimentary structure types over the entire model domain (see Table 2; Jussel et al., 1994). In order to determine statistical moments and their confidence limits by Monte Carlo type modelling, a minimum of 100 or 1000 runs are necessary. The changes in orientation and ranges of possible sedimentary structures, caused by the above-mentioned interactions of the two rivers over time, are recognized and included in the model by partitioning the aquifer vertically into two hydrostratigraphic units.

The sedimentary structures generated are characterized by randomly selecting hydraulic conductivity and porosity values provided from means and standard deviations calculated by Jussel et al. (1994). Files containing distributions of hydraulic conductivity and effective porosity values were generated and exported to a Modflow-based groundwater simulation system, in order to perform groundwater flow and transport simulations (PMWIN, Chiang & Kinzelbach, 2001). This process is described in detail in Regli et al. (2003). Groundwater flow and transport simulations are in accordance with field measurements including tracer breakthrough concentrations and groundwater head measurements. Each aquifer simulation represents various equiprobable representations of the subsurface with a variable degree of uncertainty in hydraulic parameter values and geometry of sedimentary structures, as based on the aquifer sedimentological and geostatistical analyses.

DISCUSSION

Sedimentary texture and structure types along with depositional elements of sedimentological models can be observed in almost all Pleistocene fluvial deposits in Switzerland as well as many other braided river systems (e.g. Steel & Thompson, 1983; Ramos et al., 1986; Heinz, 2001). The similarities in characteristic grain-size distributions and hydraulic conductivities of deposits from different Pleistocene rivers have been documented by Jussel (1992). Similarities in grain-size curves were also found when comparing the gravel-sheet deposits of modern rivers with samples observed in gravel pits or drill cores (Nägeli et al., 1996; Regli et al., 2002). Sedimentary texture and structure types of a former inactive floodplain can be observed only at very few locations in the uppermost metres of the Rhine valley deposits between the Alpine front and Basel. The sedimentary structure types observed in the former floodplain are largely irrelevant for hydrogeological applications because the water table is generally below these units.

It is uncertain whether the difference between bounding-surface models dominated by horizontal and trough-shaped boundary surfaces could be attributed to two different braided river systems, for the following reasons.

1 The dynamic character of braided rivers determines the preservation potential of the different depositional elements, and hence the spatial extent of the different sedimentary structure types in the geological record. In rivers with low aggradation rates and high bedload-sheet activities, scour fills have the highest chance of being preserved. An example of the dominance of OW/BM is documented in the Marthalen system (Fig. 5), where more than 60% of the deposits consist of scour fills. In rivers with high aggradation rates and low gravel-sheet activities, gravel sheets have a higher probability of being preserved.

2 Although it can be documented from many outcrops and GPR surveys that the horizontally dominated bounding surfaces are most abundant in the uppermost metres (Huggenberger et al., 1998; Carling et al., 2000), this may not be due to major differences in their depositional system. The dominance of horizontal bounding surfaces near the top of the deposits may be explained by (i) incision of the main channels of a stream (lowering of the base level) during normal runoff, including high flow stages, followed by (ii) broadening of the valley bottom by lateral erosion accompanied by minor deposition during occasional events of very high runoff (landscape shaping events) by the end of late Pleistocene. Such a shift would subsequently stop fluvial activity in the higher terrace levels. Consequently, when examining sedimentary structures in Pleistocene deposits, the dominance of horizontal bounding surfaces may be restricted to the uppermost part of the outcrop sections. This in turn implies that, when examining modern fluvial environments, an exploration depth of the geophysical method of at least the deepest scour is required to fully describe the bounding surfaces.

Bluck (1979), Steel & Thompson (1983), Lunt & Bridge (2004) and Lunt et al. (2004) provided descriptions of horizontally layered sediments overlaying tangentially dipping cross-beds, separated by erosional surfaces. Their observations suggest processes acting at two different vertical levels. Ashmore (1982), when studying the initiation of braiding in a straight channel in a laboratory flume, also found that each individual bar was associated with a local scour immediately upstream. The relative importance of scour pools in braided river deposits has also been shown in previous investigations by Ashmore (1993) and Bristow & Best (1993). Ashmore (1982) concluded that any attempt to explain the development of a braided pattern from alternating bars must account for scours, and that alternating bars and scours are interrelated. Scours in alluvial systems related to gravel sheets develop over distances that are up to several times bankfull channel width (Lewin, 1976; Tubino et al., 1999; Lanzoni, 2000; Church, 2002). The scours associated with gravel sheets develop and fill with the advance of gravel sheets, while the fill process is completed only during waning flow. Due to the turbulent, turbid water during high flow, the scour cut-and-fill process was not observed in the Madonna del Ponte outcrop in the Tagliamento River. However, scour structures can be portrayed by GPR profiling, and the sedimentary texture types in the central part of a gravel sheet have been shown to mainly consist of poorly to moderately sorted gravel ranging from coarse sand to cobbles with almost no silt and clay.

The example of the Tagliamento River provides some insight into the mechanisms of sediment sorting during flooding (high flow). Figure 7 illustrates a transition from an avalanche face of a gravel sheet into a scour. The transition represents a negative step, which may produce a flow separation zone (Best & Roy, 1991; Best, 1993). With an angle of about 90°, inclined gravel dunes migrating across the lateral scour wall produce smaller-scale secondary flow separation zones. The migration of several gravel dunes across the lateral trough wall might be explained by a sequence of sediment pulses, as recognized by Klingemann & Emmett (1982) and more recent flume experiments (e.g. Marti, 2002). However, a more continuous sediment supply cannot be excluded, although this has not been recorded under comparable experimental conditions. According to Carling & Glaister (1987) and Carling (1990), negative steps may lead to the formation of gravel couplets, which represent a fining-upward sequence of BM at the base and OW at the top. An advance of several migrating small gravel dunes across the lateral trough walls concurs with the results of numerical models of confluence dynamics (Bradbrook et al., 2000), which allow visualization of flow vectors when overpassing negative steps (e.g. the situation at the front of gravel dunes). Models of confluence dynamics have been compared with the results of laboratory experiments and field data (Rhoads, 1996; Rhoads & Kenworthy, 1998), and predictions of flow in cross-sectional planes portray a rotational flow and an upward component of the flow in the scour (Bradbrook et al., 2000). This flow pattern is consistent with the upward-migrating gravel dunes along the flank of the scour that produce tangential sets of OW/BM. A possible link between the processes described above and the resulting sedimentary structures in a vertical profile perpendicular to the mean ancient flow direction is presented in Fig. 4d. Contrary to that which might be expected from a first inspection of a trough fill consisting of sets with OW/BM cross-beds, these cross-beds could develop by migration of gravel dunes across the lateral trough walls.

Drill-core layer descriptions generally represent mixtures of different sedimentary structure types and often only sparse indications of OW strata can be identified in drill-core layer descriptions. Nevertheless, it is evident from many recent outcrop observations that OW and OW/BM occur frequently (Siegenthaler & Huggenberger, 1993). Consequently, hydrogeological models based on drill-core data may reproduce effective hydraulic conductivities, but underestimate their standard deviations. This underestimation may be critical when transport distances of particles or microorganisms in heterogeneous aquifers need to be evaluated. A fruitful subject of future research would be to investigate what percentage of OW strata are required to affect the hydrogeological, statistical characteristics of a deposit.

The ability to integrate different quality data, such as outcrop and drill-core analysis, geophysical methods and geostatistical data, into mathematical models offers a great potential for improving the accuracy of aquifer models. This is particularly relevant when considering the integration of drill-core and two-dimensional or three-dimensional structure information from GPR data into groundwater flow and transport models, especially when considering the variable degree of uncertainty associated with estimating the hydraulic properties and geometry of the sedimentary structures. Textural and structural interpretations of GPR data are fuzzy to some extent, and this results in a probability estimate that drill-core layer descriptions and radar facies types represent defined sedimentary structure types. These structure-type probabilities can be given for points along boreholes, and grid nodes with arbitrary mesh sizes along GPR sections. In order to derive maximum benefit from available textural and structural information, sequential indicator simulation techniques have been used to simulate the distributions of subsurface geological structures and their associated hydrogeological properties. Based on variogram modelling and the choice of the simulation routine, various equiprobable realizations have to be generated to provide an objective, statistically supported view of reality.

CONCLUSIONS

1 The approach used to characterize aquifer heterogeneity presented in this paper considers the differences between lithofacies-based interpretation of outcrop, drill-core and GPR data. The lithofacies scheme is based on observations of unsaturated sand and gravel outcrops and fluvio-dynamic interpretations of processes in a braided river. This approach is suitable for the interpretation of radar facies types. Moreover, the analysis of a large number of vertical sections in gravel pits and construction sites (e.g. Figs 1–5) in Switzerland, France and southern Germany has shown that trough-shaped erosional surfaces are the dominant sedimentary features. In contrast, horizontal gravel sheets were encountered less frequently than expected, based on modern analogues (e.g. Lunt *et al.*, 2004). The trough-fill deposits seen in gravel pits represent the infilling of ancient scour pools.

2 Observations and GPR experiments at different locations in the Pleistocene Rhine gravel deposits of Switzerland show that scour dimensions vary according to the palaeodischarge, as well as the width of the active channel belt or the valley (Beres *et al.*, 1999). Two-dimensional and three-dimensional GPR data may be used to estimate the scour-fill/channel-fill ratio and the proportion of other deposits present, and time slices of three-dimensional GPR experiments can provide maps of the ancient fluvial systems at different depths (Beres *et al.*, 1999). Both data types are of particular interest when studying aquifer heterogeneity.

3 Observations in gravel pits and on the River Tagliamento raise questions concerning the validity of the widely accepted assumption among sedimentologists that the description of modern analogues permits the essence of the heterogeneity of a deposit to be ascertained. By following this rationale, the role of scour fills containing open framework well-sorted (OW) and open framework well-sorted/bimodel (OW/BM) gravel might be at least underestimated, or even overlooked. When addressing many hydrogeological issues related to gravel aquifers, the most important structure types are OW and OW/BM. The extent of these highly permeable sediments is related to the dimensions of the depositional trough-fill deposits. A dominance of trough-fill deposits has also been recorded based on three-dimensional GPR investigations (Beres et al., 1999). Vertical GPR sections may be used to estimate the scour-fill/channel-fill ratio, and to characterize the geometry of the erosional surfaces of sedimentary structures and their radar facies types.

4 Application of the procedures outlined in this paper provides a means of integrating different quality data. The approach results in a simulated aquifer model that includes quantifiable uncertainty associated with the quantity and quality of data parameter values. This uncertainty must be considered in both assigning values to parameters and in the identification of particular structures. The application of stochastic methods based on reliable sedimentological models therefore provides an objective means of handling both hard and soft data associated with large-scale subsurface heterogeneities.

ACKNOWLEDGEMENTS

We thank K. Bernet and R. Flynn for reviewing the manuscript. Special thanks are addressed to J. Best, J. Bridge, I. Lunt, G. Weissman and J. West for valuable critiques and comments that have significantly improved the manuscript. This study was supported by the Swiss National Science Foundation, grant no. 2100-049272.96/1. The C routines referred to in this paper are available on request from the authors.

REFERENCES

Adams, E.E. and Gelhar, L.W. (1992) Field study of dispersion in a heterogeneous aquifer, 2, Spatial moment analysis. *Water Resour. Res.*, **28**, 3293–3308.

Anderson, M.P. (1989) Hydrogeologic facies models to delineate large-scale spatial trends in glacial and glaciofluvial sediments. *Geol. Soc. Am. Bull.*, **101**, 501–511.

Anderson, M.P., Aiken, J.S., Webb, E.K. and Mickelson, D.M. (1999) Sedimentology and hydrogeology of two braided stream deposits. *Sediment. Geol.*, **129**, 187–199.

Allen, J.R.L. (1978) Studies in fluviatile sedimentation: An exploratory quantitative model for the architecture of avulsion-controlled alluvial suits. *Sediment. Geol.*, **21**, 129–147.

Allen, J.R.L. (1984) Parallel lamination developed from upper-stage plane beds: a model based on the larger coherent structures of the turbulent boundary layer. *Sediment. Geol.*, **39**, 227–242.

Ashmore, P.E. (1982) Laboratory modelling of gravel braided stream morphology. *Earth Surf. Process.*, **7**, 2201–2225.

Ashmore, P. and Parker, G. (1983) Confluence scour in coarse braided streams. *Water Resour. Res.*, **19**, 392–402.

Ashmore, P. (1993) Anabranch confluence kinetics and sedimentation processes in gravel braided streams. In: *Braided Rivers* (Eds J.L. Best and C.S. Bristow), pp. 129–146. Special Publication 75, Geological Society Publishing House, Bath.

Ashworth, P.J., Best, J.L., Peakall, J. and Lorsong, J.A. (1999) The influence of aggradation rate on braided alluvial architecture: field study and physical scale modelling of the Ashburton River gravels, Canterbury Plains, New Zealand. In: *Fluvial Sedimentology VI* (Eds N.D. Smith and J. Rogers), pp. 333–346. Special Publication 28, International Association of Sedimentologists. Blackwell Science, Oxford.

Ashworth, P.J., Best, J.L. and Jones, M. (2004) The relationship between sediment supply and avulsion frequency in braided rivers. *Geology*, **32**, 21–24.

Asprion, U. and Aigner, T. (1999) Towards realistic aquifer models: three-dimensional georadar surveys of Quaternary gravel deltas (Singen Basin, SW Germany). *Sediment. Geol.*, **129**, 281–297.

Beres, M. and Haeni, F.P. (1991) Application of ground-penetrating radar methods in hydrogeologic studies. *Groundwater*, **29**, 375–386.

Beres, M., Green, A.G., Huggenberger, P. and Horstmeyer, H. (1995) Mapping the architecture of glaciofluvial sediments with three-dimensional georadar. *Geology*, **23**, 1087–1090.

Beres, M., Huggenberger, P., Green, A.G. and Horstmeyer, H. (1999) Using two- and three-dimensional georadar methods to characterise glaciofluvial architecture. *Sediment. Geol.*, **129**, 1–24.

Best, J.L. (1993) On the interactions between turbulent flow structure, sediment transport and bedform development: some considerations from recent experimental research. In: *Turbulence: Perspectives on Flow and Sediment Transport* (Eds N.J. Clifford, J.R. French and J. Hardisty), pp. 61–92. Wiley, Chichester.

Best, J.L. and Roy, A.G. (1991) Mixing layer distortion at the confluence of channels of different depth. *Nature*, **350**, 411–413.

Best, J.L., Ashworth, P.J., Bristow, C. and Roden, J. (2003) Three-dimensional sedimentary architecture of a large, mid-channel sand braid bar, Jamuna River, Bangladesh. *J. Sediment. Res.*, **73**, 516–530.

Bluck, B.J. (1979) Structure of coarse grained braided stream alluvium. *Trans. Roy. Soc. Edinb.*, **70**, 181–221.

Bradbrook, K.F., Lane, S.N. and Richards, K.S. (2000) Numerical simulation of three dimensional, time averaged flow structure at river channel confluences. *Water Resour. Res.*, **36**, 2731–2746.

Brierley, G.J. (1991) Bar sedimentology of the Squamish river, British Columbia: definition and application of morphostratigraphic units. *Journal of Sedimentary Petrology*, **61**, 211–225.

Bridge, J.S. (1993) The interaction between channel geometry, water flow, sediment transport and deposition in braided rivers. In: *Braided Rivers* (Eds J.L. Best and C.S. Bristow), pp. 13–71. Special Publication 75, Geological Society Publishing House, Bath.

Bridge, J.S. (2003) *Rivers and Floodplains*. Blackwell Scientific, Oxford, 504 pp.

Bridge, J.S., Alexander, J., Collier, R.E.L.L., Gawthorpe, R.L. and Jarvis, J. (1995) Ground-penetrating radar and coring to study the large-scale structure of point-bar deposits in three dimensions. *Sedimentology*, **42**, 839–852.

Bristow, C. and Best, J.L. (1993) Braided rivers: perspectives and problems. In: *Braided Rivers* (Eds J.L. Best and C.S. Bristow), pp. 1–9. Special Publication 75, Geological Society Publishing House, Bath.

Carle, S.F., LaBolle, E.M., Weissmann, G.S., Van Brocklin, D. and Fogg, G.E. (1998) Conditional simulation of hydrofacies architecture: A transition probability/Markov approach. In: *Hydrogeologic Models of Sedimentary Aquifers* (Eds G.S. Fraser and J.M. Davis), pp. 147–170. Concepts in Hydrogeology and Environmental Geology 1, Society of Economic Paleontologists and Mineralogists, Tulsa, OK.

Carling, P.A. (1990) Particle over-passing on depth-limited gravel bars. *Sedimentology*, **37**, 345–355.

Carling, P.A. and Glaister, M.S. (1987) Rapid depositions of sand and gravel mixtures downstream of a negative step: the role of matrix-infilling and particle overpassing in the process of bar-front accretion. *J. Geol. Soc. London*, **144**, 543–551.

Carling, P.A., Gölz, E., Orr, H.G. and Radecki-Pawlik, A. (2000) The morpholodynamics of fluvial sand dunes in the River Rhine, near Mainz, Germany. I. Sedimentology and morphology. *Sedimentology*, **47**, 227–252.

Chiang, W.-H. and Kinzelbach, W. (2001) *3D-Groundwater Modelling with PMWIN—a Simulation System for Modeling Groundwater Flow and Pollution.* Springer, Berlin, 346 pp.

Chiang, W.-H., Kinzelbach, W. and Rausch, R. (1998) *Aquifer Simulation Model for Windows—Groundwater Flow and Transport Modeling, an Integrated Program.* Gebrueder Borntraeger, Berlin, 137 pp.

Church, M. (2002) Geomorphic thresholds in riverine landscapes. *First International Symposium on Landscape Dynamics of Riverine Corridors*, Ascona, Switzerland, March, 2001. *Freshwat. Biol.*, **47**, 541–557.

Copty, N. and Rubin, Y. (1995) A stochastic approach to the characterisation of lithofacies from surface seismic and well data. *Water Resour. Res.*, **31**, 1673–1686.

Dagan, G. (2002) An overview of stochastic modelling of groundwater flow and transport: from theory to applications. *Eos (Trans. Am. Geophys. Soc.)*, **83/53**, 621–625.

Deutsch, C.V. and Wang, L. (1996) Hierarchical object-based stochastic modelling of fluvial reservoirs. *Math. Geol.*, **28**, 857–880.

Deutsch, C.V. and Journel, A.G. (1998) *GSLIB Geostatistical Software Library and User's Guide.* Oxford University Press, New York, 369 pp.

Environmental Modelling Systems Inc. (2002) *GMS: Groundwater Modelling System.* http://www.ems-i.com.

Fay, H. (2002) Formation of ice-block obstacle marks during the November 1996 glacier-outburst flood (Jökulhlaup), Skeidarársandur, southern Island. In: *Flood and Megaflood Processes and Deposits: Recent and Ancient Examples* (Eds I.P. Martini, V.R. Baker and G. Garzón), pp. 85–97. Special Publication 32, International Association of Sedimentologists. Blackwell Science, Oxford.

Flynn, R. (2003) *Virus transport and attenuation in perialpine gravel aquifers.* Unpublished PhD thesis, University of Neuchâtel, Switzerland, 178 pp.

Fogg, G.E., Noyes, C.D. and Carle, S.F. (1998) Geologically based model of heterogeneous hydraulic conductivity in an alluvial setting. *Hydrogeol. J.*, **6**, 131–143.

Gelhar, L.W. (1986) Stochastic subsurface hydrology from theory to applications. *Water Resour. Res.*, **22**, 135S–145S.

Hardage, B.A. (1987) *Seismic Stratigraphy.* Geophysical Press, London, 422 pp.

Heller, P. and Paola, C. (1992) The large-scale dynamics of grain-size variation in alluvial basins, 2: application to syntectonic conglomerate. *Basin Res.*, 4, 73–90.

Heinz, J. (2001) *Sediment. Geol. of Glacial and Periglacial Gravel Bodies (SW-Germany): Dynamic Stratigraphy and Aquifer Sedimentology.* Tübinger Geowissenschaftliche Arbeiten, C59, 102 pp.

Huggenberger, P. (1993) Radar facies: recognition of facies patterns and heterogeneities within Pleistocene Rhine gravels, NE Switzerland. In: *Braided Rivers* (Eds J.L. Best and C.S. Bristow), pp. 163–176. Special

Publication 75, Geological Society Publishing House, Bath.

Huggenberger, P., Siegentaler, C. and Stauffer, F. (1988) Grundwasserströmung in Schottern: Einfluss von Ablagerungsformen auf die Verteilung von Grundwasserfliessgeschwindigkeit. *Wasserwirtschaft*, **78**, 202–212.

Huggenberger, P., Höhn, E. and Beschta, R. (1998) Groundwater control on riparian/fluvial systems. *Freshwat. Biol.*, **40**, 407–425.

Jussel, P. (1992) *Modellierung des Transports gelöster Stoffe in inhomogenen Grundwasserleitern*. Unpublished PhD thesis, 9663. Eidgenössisch Technische Hochschule, Zürich, 323 pp.

Jussel, P., Stauffer, F. and Dracos, T. (1994) Transport modelling in heterogeneous aquifers: 1. statistical description and numerical generation of gravel deposits. *Water Resour. Res.*, **30**, 1803–1817.

Klingbeil, R., Kleineidam, S., Asprion, U., Aigner, T. and Teutsch, G. (1999) Relating lithofacies to hydrofacies: outcrop-based hydrogeological characterisation of Quaternary gravel deposits. *Sediment. Geol.*, **129**, 299–310.

Klingemann, P.C. and Emmett, W.W. (1982) Gravel bedload transport processes. In: *Gravel-bed Rivers* (Eds R.D. Hey, J.C. Bathurst and C.R. Thorne), pp. 145–169. Wiley, Chichester.

Koltermann, C.E. and Gorelick, S.M. (1992) Paleoclimate signature in terrestrial flood deposits. *Science*, **256**, 1775–1782.

Koltermann, C.E. and Gorelick, S.M. (1996) Heterogeneity in sedimentary deposits: A review of structure-imitating, process-imitating, and descriptive approaches. *Water Resour. Res.*, **32**, 2617–2658.

Leopold, L.B. and Wolman, M.G. (1957) River channel patterns, braided, meandering, straight. *US Geol. Surv. Prof. Pap.*, **282-B**.

Lanzoni, S. (2000) Experiments on bar formation in a straight flume 1. Uniform sediment. *Water Resour. Res.*, **36**(11), 3351–3363.

Lewin, J. (1976) Initiation of bed forms and meanders in coarse-grained sediment. *Geol. Soc. Am. Bull.*, **87**, 281–285.

Lunt, I.A. and Bridge, J.S. (2004) Evolution and deposits of a gravelly braid bar, Sagavanirktok River, Alaska. *Sedimentology*, **51**, 415–432.

Lunt, I.A., Bridge, J.S. and Tye, R.S. (2004) A quantitative, three-dimensional depositional model of gravelly braided rivers. *Sedimentology*, **51**, 377–414.

Marti, C. (2002) Morphodynamics of widenings in steep rivers. In: *River Flow 2002* (Eds D. Bousmar and Y. Zech), Vol. 2, pp. 865–873. A.A. Balkema, Lisse.

McKenna, S.A. and Poeter, E.P. (1995) Field example of data fusion in site characterization. *Water Resour. Res.*, **31**, 3229–3240.

Miall, A.D. (1985) Architectural-element analysis: a new method of facies analysis applied to fluvial deposits. *Earth Sci. Rev.*, **22**, 261–308.

Nägeli, M.W., Huggenberger, P. and Uehlinger, U. (1996) Ground penetrating radar for assessing sediment structures in the hyporheic zone of a perialpine river. *J. N. Am. Benthol. Soc.*, **15**, 353–366.

Paola, C., Heller, P. and Angevine, C. (1992) The large-scale dynamics of grain-size variation in alluvial basins, 1: Theory. *Basin Res.*, **4**, 73–90.

Radian Corporation (1992) *CPS-3, User's Manual*. Radian Corporation, Austin, London.

Ramos, A., Sopeña, A. and Perez-Arlucea, M. (1986) Evolution of Buntsandstein fluvial sedimentation in the north-west Iberian ranges (Central Spain). *J. Sediment. Petro.*, **56**, 862–875.

Rauber, M., Stauffer, F., Huggenberger, P. and Dracos, T. (1998) A numerical three-dimensional conditioned/unconditioned stochastic facies type model applied to a remediation well system. *Water Resour. Res.*, **34**, 2225–2233.

Regli, C., Huggenberger, P. and Rauber, M. (2002) Interpretation of drill-core and georadar data of coarse gravel deposits. *J. Hydrol.*, **255**, 234–252.

Regli, C., Rauber, M. and Huggenberger, P. (2003) Analysis of heterogeneity within a well capture zone, comparison of model data with field experiments: a case study from the river Wiese, Switzerland. *Aquat. Sci.*, **65**, 111–128.

Regli, C., Rosenthaler, L. and Huggenberger, P. (2004) GEOSSAV: a simulation tool for subsurface applications. *Comput. Geosci.*, **30**, 221–238.

Rehfeldt, K.R., Boggs, J.M. and Gelhar, L.W. (1993) Field study of dispersion in a heterogeneous aquifer, 3, Geostatistical analysis of hydraulic conductivity. *Water Resour. Res.*, **28**, 3309–3324.

Rhoads, B.L. (1996) Mean structure of transport-effective flows at an asymmetrical confluence when the main stream is dominant. In: *Coherent Flow Structures in Open Channels* (Eds P.J. Ashworth, S.J. Bennett, J.L. Best and S.J. McLelland), pp. 459–490. Wiley, Chichester.

Rhoads, B.L. and Kenworthy, S.T. (1998) Time-averaged flow structure in the central region of a stream confluence. *Earth Surf. Process. Landf.*, **23**, 171–191.

Siegenthaler, C. and Huggenberger, P. (1993) Pleistocene Rhine gravel: deposits of a braided river system with dominant trough preservation. In: *Braided Rivers* (Eds J.L. Best and C.S. Bristow), pp. 147–162. Special Publication 75, Geological Society Publishing House, Bath.

Stauffer, F. and Rauber, M. (1998) Stochastic macrodispersion models for gravel aquifers, *J. Hydraul. Res.*, **36**, 885–896.

Steel, R.J. and Thompson, D.B. (1983) Structures and textures in Triassic braided stream conglomerates ('Bunter Pebble Beds') in the Sherwood Sandstone Group, North Staffordshire, England. *Sedimentology*, **30**, 341–367.

Swiss Federal Office for Water and Geology (2001) *The Hydrological Yearbook of Switzerland 2000.* Geneva, 432 pp.

Swiss Standard Association (1997) *Identifikation der Lockergesteine—Labormethode mit Klassifikation nach USCS (SN 670008a).* Geneva, 7 pp.

Todd, S.P. (1989) Stream-driven, high-density gravelly traction carpets: possible deposits in the Trabeg Conglomerate formation, SW Ireland and some theoretical considerations on their origin. *Sedimentology*, **36**, 513–530.

Tubino, M., Repetto, R. and Zolezzi, G. (1999) Free bars in rivers. *J. Hydraul. Res.*, **37**, 759–775.

Van Dyke, M. (1982) *An Album of Fluid Motion.* Parabolic Press, Stanford, CA, 176 pp.

Wagner, A.A. (1957) The use of the Unified Soil Classification System by the Bureau of Reclamation. *Proceedings of the 4th International Conference on Soil Mechanics and Foundation Engineering*, Vol. 1, pp. 125–134.

Webb, E.K. (1994) Simulating the three-dimensional distribution of sediment units in braided-stream deposits. *J. Sediment. Res.*, **B64**, 219–231.

Webb, E.K. and Anderson, M.P. (1996) Simulation of preferential flow in three-dimensional, heterogeneous conductivity fields with realistic internal architecture. *Water Resour. Res.*, **32**, 533–545.

Weissmann, G.S., Carle, S.F. and Fogg, G.E. (1999) Three-dimensional hydrofacies modelling based on soil surveys and transition probability geostatistics. *Water Resour. Res.*, **35**, 1761–1770.

Zechner, E., Hauber, L., Noack, Th., Trösch, J. and Wülser, R. (1995) Validation of a groundwater model by simulating the transport of natural tracers and organic pollutants. In: *Tracer Technologies for Hydrological Systems* (Ed. Ch. Leibundgut), pp. 57–64. IAHS Publication 229, International Association of Hydrological Sciences, Wallingford.

Scaling and hierarchy in braided rivers and their deposits: examples and implications for reservoir modelling

SEAN KELLY

Danish Oil and Natural Gas, Agern Alle 24–26, 2970 Hørsholm, Denmark (Email: SEKE@dong.dk)

ABSTRACT

Globally, there are numerous significant hydrocarbon accumulations within braided fluvial reservoirs. Building successful subsurface models of braided fluvial reservoirs requires the inclusion of information from a wide variety of sources, including appropriate analogues (both modern and ancient) as well as specific field data. In addition to field data (primarily from boreholes), many published static modelling strategies for fluvial reservoirs use data collected from modern rivers or two-dimensional outcrops to infer three-dimensional reservoir architecture and geometry. However, there has been little attempt to evaluate the statistical significance of these various datasets or how they may be linked. The approach advocated in this paper utilizes statistical aspects of both modern rivers and ancient fluvial deposits such that the properties of analogue datasets allow their inclusion into reservoir modelling routines in an informed manner. This approach allows estimates of reservoir geometry at a variety of scales utilizing limited datasets such as are typically available in the subsurface. The scale invariant characteristics observed in fluvial systems and their deposits suggest that empirical predictions of key variables such as braid bar width and length can be made if additional data are known, or can be reasonably estimated (such as channel depth). In addition, confident observations of one level of reservoir hierarchy (e.g. bedsets) can allow the prediction of attributes of other levels of hierarchy (e.g. storeys) that may be more difficult to observe directly but are potentially significant for flow behaviour and reservoir performance.

Keywords Fluvial, braided, bar, channel, outcrop, scale invariance, reservoir.

INTRODUCTION

Ideally, both modern and ancient analogues should be used in the interpretation of fluvial sandbodies in the subsurface, particularly if information is limited to well data. In some aspects modern analogues are superior aids because it is possible to describe fluvial morphology and deposits directly and unambiguously, particularly the three-dimensional evolution of bedforms and bars (Bridge, 2001). While ancient analogues are potentially less reliable, because their origin must be interpreted (Bridge, 2001), they can be powerful as they allow detailed information on, for example, grain-size, sedimentary structure and facies to be collected, which is directly comparable to subsurface datasets from core.

Although most outcrops are not large enough to allow the full three-dimensional definition of the depositional architecture (e.g. channel-belt width), they contain a wealth of information on the architectural elements of fluvial reservoirs, that range in scale from beds to cosets, through bedsets to stratasets. Most published outcrop analogue datasets focus on the external geometry and dimensions of sandbodies with less emphasis on constituent architectural elements. However, it is possible that the geometry and statistical characteristics of smaller scale architectural elements (i.e. bedsets, cosets and stratasets) can provide the basis for estimating the dimensions of larger scale elements (Bridge & Tye, 2000).

Scale and self similarity

Scale is an important aspect of fluvial morphology. While superficial examination of braided fluvial systems across a wide range of channel widths (at least five orders of magnitude) often reveals a gross similarity in appearance (Bristow & Best, 1993), there have been few studies to investigate this similarity and its significance (Sapozhnikov & Foufoula-Georgiou, 1996, 1997, 1999). As stated by Bristow & Best (1993):

'The self-similarity of form across a range of scales or the scale independent nature of the geometry of braided rivers has several fundamental applications. First, self-similarity across scales of braiding may shed light upon the processes inherent in causing braiding, bar formation and growth. Second, when applying models of braided alluvial architecture deduced from one system to another of a completely different size (for instance, in braided alluvial reservoir heterogeneity models) it is essential to know which geometrical attributes are scale invariant or scale dependent.'

The objective of this paper is to link the scaling observed in modern rivers with the hierarchical scaling inherent in fluvial deposits in order to provide quantitative input for the modelling of braided fluvial reservoirs. The primary aims of this paper are to: (i) describe observations of hierarchy and scaling within modern river systems; (ii) describe the hierarchy and scaling aspects of outcrops of ancient fluvial deposits; (iii) provide a discussion of how these scaling relationships may be used in the reconstruction of fluvial deposits from limited information.

Fluvial reservoir models typically rely on routines that allow the translation of one-dimensional measurements from boreholes to three-dimensional estimates of architectural element geometry. For the objects that are to be modelled explicitly, these routines generally require quantitative constraints on size (thickness, length, width), shape and orientation (azimuth and dip), together with any significant trends between these variables (Geehan & Pearce, 1994; Lanzarini et al., 1997; Pinheiro et al., 2002). Of these, thickness is often the only measurement that can be taken directly from well data. Consequently, most of the modelling parameters have to be estimated. If such inferences can be made on quantitative grounds, based on reliable empirical analysis, the resultant models are more likely to be geologically realistic and therefore potentially result in a better match and prediction of reservoir performance. There is also compelling evidence (e.g. Hoimyr et al., 1993) that modelling different levels of detail and heterogeneity (and associated properties) within braided fluvial reservoirs results in significant variation in predicted reservoir performance (e.g. timing of water breakthrough). Given the vertical and lateral resolution constraints of most reservoir models it is important to: (i) define the critical level of reservoir heterogeneity; (ii) model it accurately, either directly or through the use of effective properties. This is best approached through an iterative process, modelling different levels of heterogeneity at various scales (degrees of upscaling) until an optimal compromise is found whereby critical heterogeneity is adequately retained but field-scale or sector models are able to be simulated with reasonable run times.

River planform and subsurface recognition

Although this paper focuses on braided rivers and their deposits, it is worthwhile briefly considering how specific river planform patterns are generated and the implications for preservation and recognition. There is evidence to suggest that, at least at a reach scale, rivers behave in a predictable manner regarding the development of their planform geometries, controlled by factors such as slope, grain size and discharge (Leopold & Wolman, 1957; Schumm et al., 1972). While the exact causes remain debated, it is commonly assumed that fluvial patterns arise as a complex response to a variety of internal and external mechanisms occurring on a variety of spatial and temporal scales (Hooke, 1997; Kirkby, 1999). Over periods of several days, weeks or months the same reach may switch between braided and meandering configurations in response to discharge variation (e.g. Graf, 1988). This may be important in interpreting the geological record as a fluvial sandbody or alluvial suite will often result from a palaeochannel system that changed its planform repeatedly over the period of alluvial deposition. Such systems may not be easily characterized as simply braided or meandering as they may have adopted both planforms at various periods. Furthermore, it is doubtful if information from boreholes, even closely spaced ones, will allow the reconstruction of palaeochannel pattern (Bridge, 1985).

To summarize, it appears that while a braided planform is a distinctive variant of many modern rivers, it is likely that: (i) channel patterns and planforms vary with time such that the fluvial deposits may reflect this variation; (ii) that in many cases it may be difficult to unambiguously distinguish the deposits of braided fluvial rivers from those of other types of river planform. For these reasons it is useful to not only evaluate the statistical character of modern braided rivers, but also other common planform geometries and associated fluvial forms, such as those that would typically be described as meandering. Furthermore, it is quite possible that specific palaeochannel patterns may not be very important in the subsurface due to the expected similarities between the channel belts of different channel patterns with similar sediment grain size (Bridge, 1985, 2001). This study focuses on sand and gravel bedload dominated rivers (with typically braided planforms) and the deposits of palaeochannels that are interpreted to have been of a broadly similar type. The premise made is that the two datasets (modern rivers and outcrops) are broadly analogous such that reasonable comparisons can be made concerning their scaling aspects.

Bedforms, hierarchy and superposition

Given that the scale of observation and associated terminology are important references when dealing with modern fluvial systems or braided fluvial deposits, it is worthwhile briefly summarizing the relevant terminology used in this study. Braided fluvial systems that carry sand or gravel bedloads can generate a variety of bedforms that exhibit several hierarchical attributes, including bedform size and the time span of existence of individual bedforms (Jackson, 1975). These two attributes allow the definition of fundamental groups of bedforms found in the majority of braided river systems.

Microforms

Jackson (1975) argued that a two-zone structural model of the turbulent boundary layers provides a genetic framework for the two smaller classes of bedforms, termed mesoforms and microforms. Microforms, such as ripples, are governed by flow structure in the inner part of the boundary zone. Microforms are not considered in this study.

Mesoforms

Mesoforms, such as dunes, respond to flow conditions in the outer zone of the turbulent boundary layer and often have lengths that scale with channel depth (Jackson, 1975; Yalin, 1977). Mesoforms such as dunes and bedload sheets are commonly superimposed on macroforms (Bridge, 1993b). Sorting and grain interaction play a significant role in coarse grained systems. Features described as transverse bars or linguoid bars are probably better referred to as mesoforms (Jackson, 1975).

Macroforms

The largest bedforms are macroforms, such as point bars, braid bars, unit bars or composite bars, that generally respond to the geomorphological regime of the environment, and are relatively insensitive to changes in fluid-dynamic regime during an individual event, for example a flood. Fluvial macroforms typically scale with active channel width and depth and are generally partially emergent at all but the highest flow stages. Much of the construction and modification of macroforms is achieved during bankfull flood discharge events. In meandering rivers the main type of macroform is the point bar. Braided rivers are characterized by large, periodic and geometrically similar macroforms or braid bars that scale with the main active channels (Jackson, 1975; Thorne, 1997; Bridge, 1993b).

Hierarchy and superposition

Braided rivers are typically characterized by a hierarchy of bars and associated channels (Bridge, 1993b). It is common for dunes or bars to be superimposed on larger macroforms in braided rivers (e.g. Coleman, 1969; Collinson, 1970). Sandy bedload braided rivers often have three to four levels of bedform hierarchy ranging from dunes through larger dunes to unit bars and composite braid bars (Jackson, 1975). Superposition may be either under equilibrium conditions when large bedforms effectively generate boundary layers in which smaller bedforms can exist (Rubin & McCulloch, 1980) or during non-equilibrium or fluctuating flow conditions with smaller dunes or bedforms developing on the stoss side of larger bedforms or macroforms during low flows, the larger bedforms or macroforms being formed or maintained during relatively high flow conditions (Jackson, 1975). In either case, it is possible to generate a complex hierarchy of bedforms and related macroforms, ultimately dominated by point bars in a meandering channel or braid bars in a braided river.

Observations of modern braided rivers strongly indicate scaling relationships between channel size (width, depth) and bar size (Jackson, 1975; Yalin, 1977). According to Jackson (1975) the depth of channel flow (i.e. thickness of the boundary layer) controls the ultimate size of macroforms and the occurrence and the size of dunes and bars. Yalin (1977) suggested that the depth of flow determines dune height and length, and also that channel width determines bar (macroform) length.

Scale invariance and universality of fluvial systems

Recent work (Sapozhnikov & Foufoula-Georgiou, 1996, 1997, 1999; Foufoula-Georgiou & Sapozhnikov, 1998, 2001; Nykanan et al., 1998; Sapozhnikov et al., 1998) has demonstrated that braided rivers of different spatial scales exhibit a form of scale invariance, i.e. the spatial scale of a braided fluvial system cannot be discerned by simply looking at its planform pattern.

The quantitative evidence for this scale invariance is derived from the analysis of aerial photographs and satellite imagery of modern braided rivers. Sapozhnikov & Foufoula-Georgiou (1996) concluded that braided rivers exhibit anisotropic spatial scaling (self-affinity) in their morphology. Stated simply, this implies that a smaller part of a river, stretched differently along the mainstream and perpendicular directions, is statistically identical to the larger part. Furthermore, the scaling exponents were found similar in braided rivers of diverse flow regimes, slopes, types of bed material and braidplain widths, indicating the presence of universal features in the underlying mechanisms responsible for the formation of their spatial structure.

The basis for the conclusions reached by Foufoula-Georgiou & Sapozhnikov (1998, 2001) and Sapozhnikov & Foufoula-Georgiou (1996, 1997, 1999) is most easily illustrated by the observation that common fluvial features such as braid bars are scale invariant in their shape. Sambrook Smith et al. (2005) have demonstrated this using a simple measure of bar shape, the width:length ratio of the bar, which in contrasting rivers, covering at least six orders of magnitude in spatial scale, is scale invariant.

MODERN RIVERS DATA

A similar dataset to that of Sambrook Smith et al. (2005) was also collected during this study. The data were collected from photographic images and drawings of 22 braided systems, including some flume studies, together with four meandering rivers for comparison (Table 1).

Bars (mesoforms and macroforms) were evaluated rather than individual channels. The outlines of all bars were manually digitized (e.g. Fig. 1) then measured automatically using image analysis software. The potential problems of using detailed surveys, aerial photographs or satellite images (and their interpretations) are listed below.

1 The datasets typically represent a single snapshot in time and do not reflect the dynamic morphology of rivers. The data were collected during a particular set of conditions including river discharge, sediment load, water clarity (related to turbulence and suspended load) and density of covering vegetation, all of which may affect the characteristics of the image or survey data. In order to assess temporal variation, where available, multiple images taken at separate time-steps were evaluated and the conclusion was reached that the results were essentially identical, the exception to this being images taken during extreme conditions of flood when bedforms and bars are typically fully submerged (e.g. comparison of fig. 2a of Thorne et al., 1993 with their fig. 2b, the latter being a satellite image taken during peak monsoon flood of 1987).

2 The type of data collected (aerial photograph or satellite image) inherently carries with it a certain resolution and quality which will potentially limit the detail one can interpret. The main effect of this is to place a lower limit of resolution in each data set,

Table 1 Summary of modern rivers and flume studies used in this study

Case	Location	Reference	Type*	Channel depth (m)	Channel width (m)	Length of reach evaluated
1	Allt Dubhaig	Wathen et al. (1997)	B	1.50	28.27	130.67
2	Arolla	Lane & Richards (1997)	B	0.55	24.09	53.65
3	Babbage	Miall (1996)	B	3.00	1268.50	3284.19
4	Brahmaputra	Coleman (1969)	B	45.00	17944.24	154380.34
5	Calamus	Bridge et al. (1986)	B	3.00	214.45	1029.69
6	Feshie	Ferguson & Werrity (1983)	B	1.25	38.96	164.53
7	Flume (G&S)	Germanoski & Schumm (1993)	B	0.15	1.78	15.00
8	Flume (L&W)	Leopold & Wolman (1957)	B	0.04	0.85	3.12
9	Ganges	LANDSAT	B	30.00	19251.87	51886.81
10	Gila	Graf (1988)	B	4.00	139.91	361.57
11	Horse Creek	Leopold & Wolman (1957)	B	1.56	82.26	229.71
12	Iceland Sandur 1	Rust (1978)	B	6.00	7363.04	22548.15
13	Iceland Sandur 2	Rust (1978)	B	3.00	1280.13	9727.58
14	Medano	Langford & Bracken (1987)	B	0.25	53.70	98.60
15	New Fork	Leopold & Wolman (1957)	B	1.68	151.67	940.19
16	Quatal Creek	Foley (1978)	B	1.12	68.67	276.38
17	Red Deer	Kellerhals et al. (1976)	B	4.00	1074.60	1901.11
18	Sandur	Bluck (1979)	B	2.50	210.17	269.92
19	Sarda	Inglis (1949)	B	6.00	2470.55	9291.53
20	Squamish	Brierley (1989) Brierley & Hicknin (1991)	B	4.50	683.96	2137.45
21	Sunwapta (large)	Ashmore (1993)	B	3.00	424.15	992.06
22	Sunwapta (small)	Ashmore (1993)	B	0.20	0.64	2.76
23	Endrick	Bluck (1971)	M	3.50	89.98	266.51
24	Lugg	Aerial Photo	M	2.00	195.96	514.44
25	Naver	Bluck (1976)	M	1.25	305.73	1033.46
26	Popo Agle	Leopold & Wolman (1957)	M	2.44	296.67	915.96

*B, braided; M, meandering.

i.e. bars or bedforms below a certain size will not be imaged. The data in this study were from a wide variety of sources and often the original image or photograph was unavailable to comment on this further. Additional investigation using a variety of images of varying sources and re-solution of the same target reach would be required to assess the impact of this potential source of error further.

3 When the images have been interpreted manually, there has been no attempt to classify the bars into type (e.g. dune, unit bar, compound bar), or their hierarchy as related to channels (cf. Bridge, 1993b). While this may be useful for future study, this evaluation avoided an additional interpretive step in the search for general relationships that could be applied to a wide variety of bedforms. Ideally, in order to correctly classify bedform and bar types, multiple time-step high-resolution images of the same reach would be required. With a few exceptions, such data were not generally available during the study. It is recommended that further studies utilize such data.

Results

A total of 1400 measurements of bar area (B_a), perimeter (B_p), minor (B_x) and major (B_y) axis and orientation were recorded (Fig. 2). Channel-belt width (C_w) of each measured reach was also

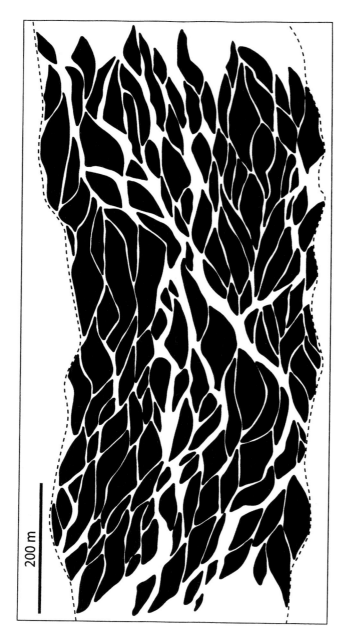

Fig. 1 Typical example of data from modern braided rivers analysed, Sunwapta River, Alberta, Canada (Ashmore, 1993).

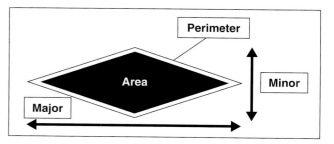

Fig. 2 Schematic diagram illustrating planform data collected from modern rivers, including bar area, perimeter length, minor and major axis length.

measured and maximum bankfull channel depth (C_d) derived from the literature. Geometric means for key variables were also calculated for each case, the aim being to remove the bias towards larger datasets with more measurements. The results are summarized in Table 2. All measurements are in metres. The results of this analysis were as follows.

1 Bar minor and major axes correlate strongly (R^2 = 0.96) over six orders of magnitude (Fig. 3) with the major axis typically being two to ten times the length of the minor axis. The regression relationship is $B_y = 4.95\ B_x^{0.96}$. Interestingly, bars from meandering rivers (typically point bars) lie in broadly the same trend but have a less robust regression, possibly reflecting the smaller dataset.

2 Bar area and perimeter correlate very strongly (R^2 = 0.99) over 10 orders of magnitude of bar area (Fig. 4). This is typical of objects that have the same basic shape character, particularly fractal or self-similar objects. The regression relationship is $B_p = 5.69\ B_a^{0.50}$.

3 The geometric mean of minor (B_x) and major (B_y) axis for each river study strongly correlates (R^2 = 0.99) (Fig. 5), again indicating an average ratio of 4.5–5.0. The regression relationship is $B_y = 4.62\ B_x^{0.96}$. It should be remembered that when enough data are available, log-normal distributions are characteristic for the bar area (B_a), perimeter (B_p), minor (B_x) and major (B_y) axis and as such this empirical relationship and those below describe only the geometric mean.

4 Channel-belt width (C_w) correlates strongly (R^2 = 0.95) with the geometric mean of the bar major axis (B_y) of the rivers studied (Fig. 6), suggesting that mean bar size is influenced by channel width. The regression relationship is $B_y = 1.04\ C_w^{0.82}$. The major axis length of the larger macroforms within a braided river can also be predicted by the expression $B_y = C_w/\pi$ (Yalin, 1977; Thorne et al., 1993).

5 Channel depth (C_d) correlates strongly (R^2 = 0.91) with bar minor axis (B_x) (Fig. 7). The regression relationship is $B_x = 7.39\ C_d^{1.46}$. This suggests that as channels become deeper, channel bars tend to become wider. This is most likely for mesoforms such as dunes that typically scale in size (height and length particularly) with flow depth (Coleman, 1969;

Table 2 Summary of results of analysis of datasets described in Table 1

Case	N	Geometric mean bar area (m^2)	Geometric mean bar perimeter (m)	Geometric mean bar major axis (m)	Geometric mean bar minor axis (m)
1	15	58.70	44.13	17.77	5.14
2	4	41.58	44.34	18.75	3.52
3	77	9557.76	615.31	263.55	54.94
4	37	13877013.16	19115.96	7636.53	2634.47
5	60	7932.55	484.69	193.04	63.28
6	12	265.24	97.09	42.60	9.17
7	96	0.15	2.10	0.92	0.23
8	4	0.06	1.56	0.70	0.16
9	14	7443925.19	13321.93	5061.50	1985.41
10	78	436.57	154.56	69.95	9.71
11	4	553.87	194.04	85.32	12.04
12	424	72511.85	1518.70	673.99	143.59
13	52	36697.04	963.94	405.28	112.94
14	31	19.96	25.29	10.82	2.77
15	17	577.39	118.91	49.09	20.06
16	3	859.50	202.90	78.68	18.52
17	27	14216.25	784.58	344.02	64.97
18	7	1176.49	192.69	73.17	24.57
19	79	19891.43	852.93	361.08	76.35
20	96	4574.85	373.38	160.29	40.73
21	153	834.88	159.52	70.87	15.89
22	33	0.13	2.11	0.93	0.22
23	31	1263.60	253.68	102.94	20.60
24	13	577.53	160.63	52.86	24.46
25	24	1327.06	222.76	86.01	24.59
26	7	636.66	189.70	83.75	12.16

Jackson, 1975). The correlation may also simply reflect the strong link between channel depth and channel-belt width in many natural braided rivers; channel width is possibly the main control on bar size, as indicated by the result above and earlier studies (Yalin, 1977).

6 In order to test the self-similarity of bar forms in braided (and meandering) rivers, the box dimension estimation method was used; results indicate that river bars are fractal with the fractal dimension (Db) in the range 1.66–1.85 (Fig. 8). This analysis confirms that bars in braided rivers are essentially self-similar in planform.

The results presented here clearly illustrate the tendency for bedforms and bars in braided rivers to scale with the channels that formed them. This result and that of Sambrook Smith et al. (2005) concur with the analysis of Sapozhnikov & Foufoula-Georgiou (1996, 1997) which also indicated that braided rivers are scale invariant (at least during equilibrium conditions). However, it remains unclear as to what sedimentological and hydrological processes within braided rivers lead to scale invariance.

FLUVIAL FACIES AND ALLUVIAL ARCHITECTURE

It is important to understand the implications of scale invariance to subsurface fluvial deposits (Bristow & Best, 1993). Sambrook Smith et al. (2005) recently addressed the issue of scale invariance and 'whether it is appropriate to apply data

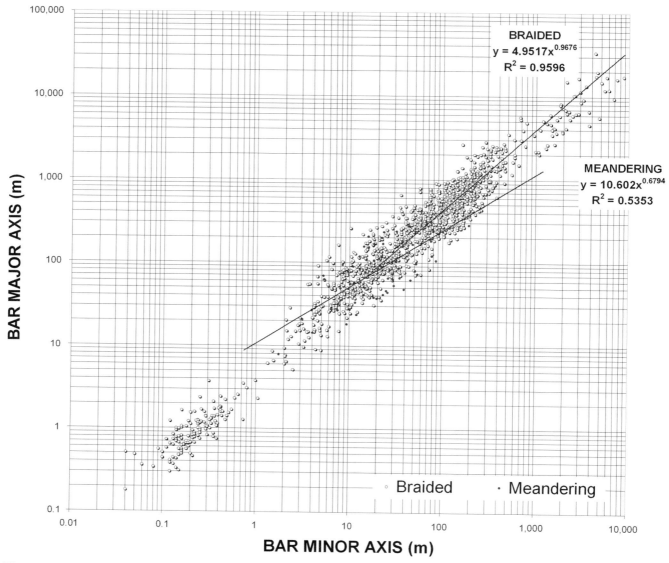

Fig. 3 Cross-plot of all bar minor and bar major axis length data. Data taken from braided (circles) and meandering (filled circles) planforms are differentiated.

from a single river or outcrop interpretation to others of either smaller or greater magnitude'. Sambrook Smith et al. (2005) considered scale invariance from the perspective of the subsurface sedimentary facies of sandy braided rivers by comparing ground-penetrating-radar datasets from three rivers spanning three orders of magnitude in spatial scale. From this comparison, Sambrook Smith et al. (2005) concluded that although sandy braided rivers do exhibit a degree of scale invariance, a single facies model cannot be applied to all rivers since other factors (e.g. the relative rate of bar migration) must be taken into account.

This conclusion of Sambrook Smith et al. (2005) indirectly supports the observation of Miall (1996) that fluvial facies models form a complete spectrum. The conclusion of Sambrook Smith et al. (2005) that similar surface morphologies need not necessarily be reflected in scale invariance in the subsurface facies is a valid one as the exact facies detail of any outcrop will reflect a large number of parameters (slope, discharge, flood frequency,

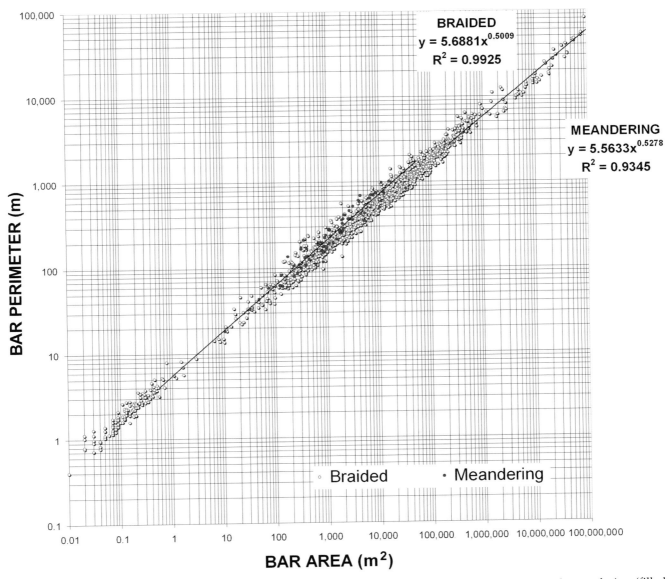

Fig. 4 Cross-plot of all bar area and bar perimeter length data. Data taken from braided (circles) and meandering (filled circles) planforms are differentiated.

bedload grain-size, bedform types, maximum scour depth, bank composition and stability). The non-uniqueness of outcrops reflects the fact that no two rivers and their associated depositional history are *exactly* the same. Locally developed facies models are essentially tools of convenience in describing where a particular example might sit in the continuum of potential facies models.

Although each river is different and its deposits could be described by a unique facies model, there are commonalities in the geometry of fluvial deposits which occur across a wide range of scales. Quantitative architectural studies of fluvial deposits are needed for establishing and evaluating any geometrical commonalities and investigating how they may be used for better understanding and modelling these sediments. This paper describes a database of geometries from modern and ancient fluvial systems and attempts to explore correlation within and between these two sources.

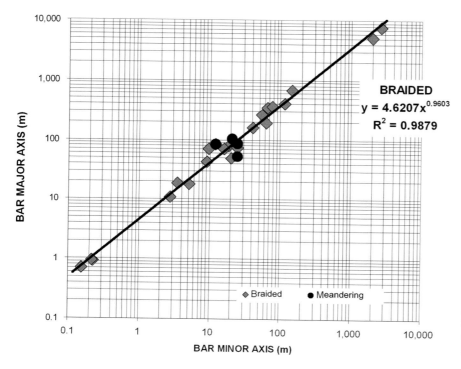

Fig. 5 Cross-plot of geometric means for bar minor and bar major axis length data. Each point represents a river. Data taken from braided (grey diamonds) and meandering (filled circles) planforms are differentiated.

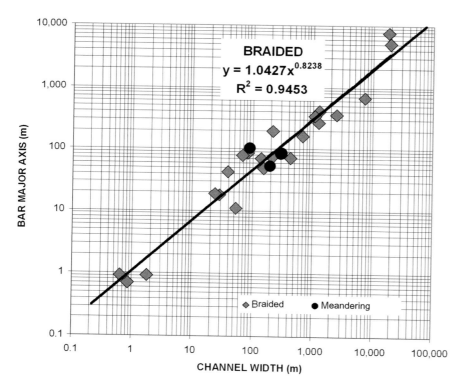

Fig. 6 Cross-plot of geometric means for channel width and bar major axis length data. Each point represents a river. Data taken from braided (grey diamonds) and meandering (filled circles) planforms are differentiated.

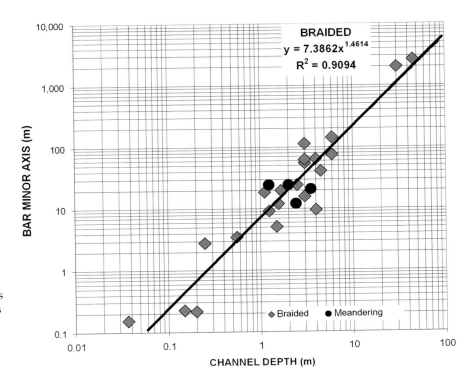

Fig. 7 Cross-plot of geometric means for channel depth and bar minor axis length data. Each point represents a river. Data taken from braided (grey diamonds) and meandering (filled circles) planforms are differentiated.

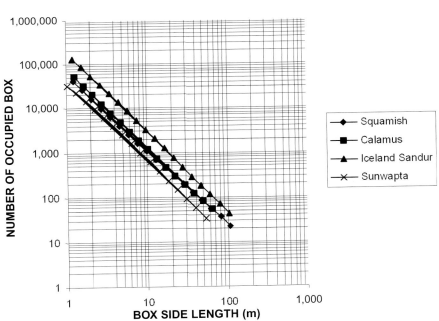

Fig. 8 Fractal dimension (Db) estimates of bars using box counting method. Note how different rivers have similar Db as indicated by the slope of the lines.

Nomenclature, architecture and hierarchy of fluvial sandbodies

A basic three-fold hierarchy is used here for describing the organization of fluvial sandstone bodies. The scheme utilized in this study is largely descriptive and depends on scales of sedimentary strata and relationships between these features and associated bounding surfaces. The simple descriptive hierarchy largely follows previous workers, such as Allen (1983) and Bridge & Diemer (1983). Despite the limitations of most

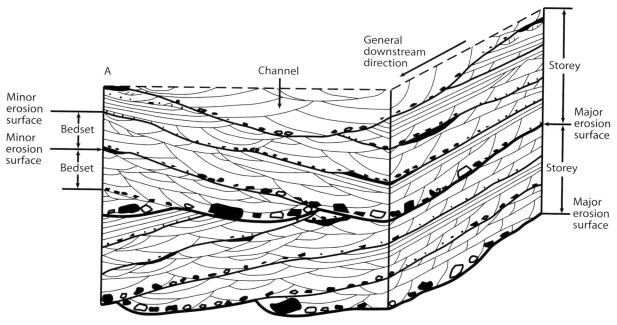

Fig. 9 Terminology used to describe fluvial deposits in outcrop. (From Bridge & Diemer, 1983.)

outcrops, useful data concerning sandstone body organization can be collected in the field (Friend, 1983). Although the spatial organization of texture and primary sedimentary structure in each sandstone body may be complex, it is often systematic (Bridge & Diemer, 1983). The description and internal subdivision of sandstone bodies utilizes sedimentary structures and the occurrence of distinctive erosional surfaces. Both of these features are generally recognizable in outcrops of limited lateral extent and are potentially identifiable in core or high-quality borehole images (BHIs). The hierarchy used follows previous workers such as Bridge & Diemer (1983) and recognizes: (i) beds, sets and cosets; (ii) bedsets and storeys; (iii) bodies or complexes.

Bedsets and storeys

Sedimentary beds or laminae within sandstone bodies are generally arranged into well defined sets which are in turn organized into cosets and larger units, the latter possibly containing a variety of primary sedimentary structures. These larger units or 'bedsets' (Campbell, 1967) are generally decimetre- to metre-scale and are separated by significant surfaces which often indicate some degree of erosion (Bridge & Diemer, 1983; Bridge, 1993a; Fig. 9).

The bedsets themselves are arranged into larger scale sets of inclined strata (Bridge, 1993a; Bridge & Tye, 2000) that may be termed 'storeys' (Friend et al., 1979) or 'stratasets'; in this study these terms are effectively used synonymously. Stratasets are typically composed of bedsets that are themselves composed of cosets of cross-strata or plane beds. The identification of individual storeys or stratasets is largely dependent on the recognition of significant erosional surfaces which typically show centimetre- to decimetre-scale relief, often marked by a change in grain size, and also possibly sedimentary structure. Storeys or stratasets may also be identified by significant changes in prevalent palaeocurrent azimuth and distinctive grain-size trends (e.g. fining-upward sequences) or abrupt or notable changes in the size or type of cross-strata or bedsets.

In addition to sandbody dimensions, the focus of this study has been bedsets and storeys as these are relatively easily determined in outcrop and they directly scale with channels and bars, information about which is important for estimating palaeochannel dimensions as part of a palaeohydrological reconstruction. With care, bedsets and storeys can also be identified in core and also potentially in wireline logs (Bridge & Tye, 2000).

Bodies and complexes

Bedsets and storeys comprise sandstone 'bodies'. Although no formal definition of a sandstone body is used in this study, the term is generally reserved for sandstone units that are thicker than 1 m, and are separated by mudrock intervals thicker than 0.5 m. The term 'complex' is a more general one and refers to closely related bodies, possibly deposited by the same palaeochannel system.

Statistical analysis of fluvial deposits

In order to evaluate the statistical properties of braided fluvial deposits, quantitative geometrical data were collected from 34 outcrops (20 published, 14 by the author; Table 3). A wide range of fluvial outcrops has been evaluated and although the precise planform of the formative river system can be debated, it is clear that all the deposits demonstrate good evidence of bar and macroform development and that the majority could safely be described as being deposited from sandy- and/or gravelly bedload rivers. For analogue data to be used effectively, the following points are important: (i) data are from a comparable depositional environment (as far as can be determined); (ii) the same category of architectural element is considered; (iii) elements should be of a comparable scale to that evaluated in the reservoir model (or appropriately rescaled).

Despite the significant amount of research on modern and ancient fluvial systems, there are relatively few outcrop studies that distinguish between genetic types of sandbody (e.g. Cuevas Gozalo & Martinius, 1993), and very few that describe the geometry of architectural elements (bedsets, storeys). The main difference between this study and previous studies (e.g. Collinson, 1978; Fielding & Crane, 1987; Hirst, 1991; Reynolds, 1999) is the focus on component architectural elements of sand bodies rather than the dimensions of the sand body.

The commonly quoted relationships between channel-belt width/thickness, channel pattern, grain size of sediment load, and bank stability are probably not valid as they are often built on very limited datasets and have large error bars (Bridge, 2001). Similarly, there have been several empirical attempts to reconstruct the palaeohydrology of ancient fluvial systems (e.g. Ethridge & Schumm 1978), but as pointed out by Miall (1996) these have largely been abandoned, presumably as a result of the inherent assumptions and inaccuracies of the methodology leading to imprecise results. This study attempts to utilize quantitative geometrical data to describe fluvial deposits rather than rely on factors that have a strong interpretative component (e.g. silt-content of channel banks) when dealing with ancient examples.

In total, over 1125 storeys and 3760 bedsets were measured using image analysis of manually digitized overlays of outcrops. An example of the outcrop data analysed is illustrated in Fig. 10. The key parameters measured include thickness (T), width (W), cross-sectional area (A), cross-sectional perimeter length (P) and dip (Z). In terms of hierarchy, three levels were measured: (i) bedset, (ii) storey and (iii) body or complex. Measurements are identified with subscripts b, s and c respectively. The results are summarized in Tables 4–7.

The majority of outcrops can be described as containing compound cross-bedding (Rubin & Hunter, 1983) of various types, the most common being the organization of cosets of cross-strata and plane-beds into bedsets which are in turn organized into stratasets or storeys. This hierarchical architecture is typical of sand-dominated braided fluvial deposits.

Limitations of two-dimensional outcrop

Before discussing the results of the statistical analysis of the outcrops, it is worthwhile describing the potential drawbacks of the data.

1 The data collected are two-dimensional only. Since the features exposed in two-dimensions arise from the intersection of the outcrop plane with the three-dimensional structure, it is logical to expect that measurements on the feature traces (e.g. bedset or storey boundaries) that are observed in outcrop can be utilized to obtain information about the same features in three-dimensions. However, there are several problems of sampling bias: (i) a two-dimensional outcrop is more likely to intercept larger features than smaller ones of the same type; (ii) the intersections

Table 3 Summary of outcrops utilized in this study

Outcrop	Name	Location	Unit	Age	Reference	Approximate size of outcrop height × width (m)
1	Whitby	UK	Saltwick	Jurassic	Kelly (this study)	27 × 125
2	La Serreta KP	Spain	Huesca	Oligo-Miocene	Hirst (1991)	20 × 890
3	La Serreta AJ	Spain	Huesca	Oligo-Miocene	Hirst (1991)	33 × 604
4	Willow Canyon	USA	Mesaverde	Cretaceous	Olsen et al. (1995)	380 × 1030
5	Price Canyon	USA	Mesaverde	Cretaceous	Olsen et al. (1995)	380 × 1700
6	Whitby	UK	Saltwick	Jurassic	Alexander & Gawthorpe (1993)	21 × 580
7	Profile 4	UK	Brownstones ORS	Devonian	Allen (1983)	3 × 31
8	Profile 5	UK	Brownstones ORS	Devonian	Allen (1983)	3 × 37
9	Profile 8	UK	Brownstones ORS	Devonian	Allen (1983)	5 × 23
10	Profile 9	UK	Brownstones ORS	Devonian	Allen (1983)	5 × 15
11	Profile 10	UK	Brownstones ORS	Devonian	Allen (1983)	5 × 41
12	Profile 11	UK	Brownstones ORS	Devonian	Allen (1983)	4 × 55
13	Chinji	Pakistan	Siwalik	Miocene	Willis (1993)	60 × 3970
14	Chinji	Pakistan	Siwalik	Miocene	Willis (1993)	80 × 3990
15	Chinji	Pakistan	Siwalik	Miocene	Willis (1993)	180 × 3055
16	Kerry Head	Ireland	ORS	Devonian	Kelly (this study)	13 × 86
17	Truscleve	Ireland	Namurian	Carboniferous	Kelly (this study)	30 × 116
18	Castlegate	USA	Castlegate	Cretaceous	Kelly (this study)	55 × 278
19	Dolores	USA	Kayenta	Jurassic	Stephens (1994)	48 × 536
20	Moab	USA	Kayenta	Jurassic	Kelly (this study)	40 × 265
21	Island 1	USA	Kayenta	Jurassic	Kelly (this study)	49 × 125
22	Island 2	USA	Kayenta	Jurassic	Kelly (this study)	29 × 171
23	Escalente	USA	Kayenta	Jurassic	Kelly (this study)	24 × 265
24	Shafer 1	USA	Kayenta	Jurassic	Kelly (this study)	71 × 329
25	Shafer 2	USA	Kayenta	Jurassic	Kelly (this study)	35 × 240
26	Shafer 3	USA	Kayenta	Jurassic	Kelly (this study)	16 × 74
27	Shafer 4	USA	Kayenta	Jurassic	Kelly (this study)	26 × 159
28	Shafer 5	USA	Kayenta	Jurassic	Kelly (this study)	31 × 160
29	Shafer 6	USA	Kayenta	Jurassic	Kelly (this study)	77 × 288
30	Blue Cove	USA	Straight Cliffs	Cretaceous	Dalrymple (2001)	294 × 2500
31	Chinji	Pakistan	Siwalik	Miocene	Friend et al. (2001)	282 × 2595
32	Flumen	Spain	Huesca	Oligo-Miocene	Hirst (1991)	146 × 892
33	Monte Aragon	Spain	Huesca	Oligo-Miocene	Hirst (1991)	95 × 1512
34	La Serreta	Spain	Huesca	Oligo-Miocene	Hirst (1991)	40 × 405

ORS, Old Red Sandstone.

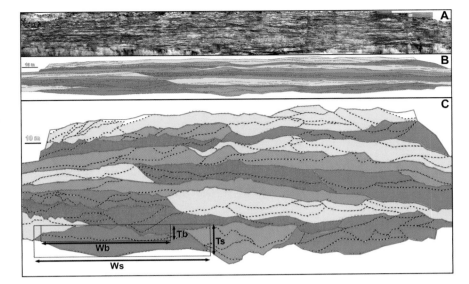

Fig. 10 Typical outcrop data utilized in this study (Kayenta Formation, Escalente, Utah). (A) Original photomosaic with line interpretation. (B) Final interpretation with different colours corresponding to individual storeys. (C) As B but with 4× vertical exaggeration. The width and thickness measurements taken are schematically illustrated; bar area and perimeter length follow the exact shape of the digitized form.

that are visible do not give the real size of the outcrop features (which are not often cut at their maximum diameter). Although these two characteristics cancel each other out to an extent, the sampling effects inherent in a two-dimensional outcrop are acknowledged but are not corrected for in any way.

2 Outcrops are finite (they have edges). Features intersecting the edge of an outcrop will be partial (Visser & Chessa, 2000). Consequently, relatively large outcrops were selected, typically such that the outcrop was larger than the storey scale (only five of the 34 outcrops have storeys that are typically larger than the outcrop, i.e. all storey measurements are partial). The main hierarchy level to suffer from limited outcrop is the body or complex level and therefore statistics of body/complex related parameters will always carry more uncertainty than those from bedsets or storeys. No attempt has been made to correct for partial lengths. However, data are reported with and without partial data included, thus allowing an assessment of their impact (Tables 4–7).

3 Although the general palaeocurrent or palaeo-transport direction may be known for an outcrop, it is rare to get outcrops either exactly parallel or orthogonal to this direction. It is possible to apply a simple trigonometric adjustment to measurements for this effect (not applied here) but this ignores the fact that local palaeocurrent or accretion direction will often be at some angle to the average slope or flow direction. During the evaluation of modern braided rivers it was noted that it is common to observe bar major axes oriented at angles up to 50° to the main channel orientation, although most are within ±30° of this azimuth. Consequently, some of the variation observed in width measurements may be related to variation in outcrop orientation relative to palaeoflow. All of the outcrops selected are generally oblique to palaeoflow direction but may be simultaneously subparallel to the accretion/migration direction of one bedform or bar and oblique or orthogonal to another in the same sandbody or outcrop.

Results

The main data discussed here are the thickness (T) and width (W) measurements from the outcrops studied. As can be seen from Fig. 11, which contains all data, storey dimensions were typically $T_s = 1–30$ m and $W_s = 10–2000$ m; bedsets were typically in the range $T_b = 0.1–15$ m and $W_b = 0.1–15$ m. All data plotted (Figs 11–19) include partial measurements.

The main results of this analysis of the outcrops are:

1 For the global database, the Thickness–Width relationships for both bedsets and storeys plot in a very similar way with reasonable correlation coefficients for both: $W_b = 17.15\ T_b^{1.04}$ ($R^2 = 0.64$); $W_s = 12.23\ T_s^{1.31}$ ($R^2 = 0.59$) (Fig. 11). As expected for self-similar shapes, the Area–Perimeter relationships correlate very strongly: $P_b = 9.29\ A_b^{0.57}$ ($R^2 = 0.92$); $P_s = 5.34\ A_s^{0.64}$ ($R^2 = 0.94$) (Fig. 12). The manner in which bedset data and storey data plot with strongly coincident fields supports a common underlying control and

Table 4 Summary of analysis of outcrop bedsets listed in Table 3, including partial measurements (i.e. bedsets that were not completely exposed due to the edges of the outcrop)

Data including partials		Arithmetic average (m)				Geometric mean (m)				Standard deviation (m)			
Outcrop	N	Area	Perimeter	Width	Thickness	Area	Perimeter	Width	Thickness	Area	Perimeter	Width	Thickness
1	100	13.14	39.78	14.48	1.10	10.03	34.66	12.52	1.02	9.28	21.74	8.23	0.40
2	91	70.15	96.91	40.69	2.03	44.15	76.48	31.64	1.78	70.07	72.39	32.70	0.98
3	154	1021.57	360.08	147.65	7.56	563.82	271.43	110.98	6.47	1192.87	275.42	118.07	4.14
4	397	215.57	141.79	59.01	4.05	137.65	114.03	46.55	3.76	244.00	108.15	47.12	1.50
5	798	160.16	123.79	50.44	3.48	102.47	100.20	40.42	3.23	203.47	95.52	39.99	1.41
6	12	509.88	312.46	131.77	4.64	390.94	285.56	114.30	4.35	318.09	129.99	64.02	1.80
7	26	1.55	9.73	3.80	0.36	0.55	6.33	2.55	0.27	1.89	8.62	3.17	0.24
8	63	1.10	8.77	3.60	0.31	0.59	6.70	2.76	0.27	1.23	6.58	2.70	0.16
9	21	3.45	15.77	6.51	0.57	2.08	12.75	5.30	0.50	3.70	10.51	4.06	0.33
10	15	3.07	12.82	5.03	0.65	1.85	10.58	4.12	0.57	3.42	7.65	3.12	0.36
11	22	4.26	24.21	9.24	0.53	2.82	19.96	7.85	0.46	3.66	14.87	4.87	0.28
12	28	4.80	20.40	8.24	0.58	2.41	15.13	6.06	0.51	8.64	20.15	8.15	0.31
13	138	563.30	418.07	177.32	2.98	258.17	304.96	130.80	2.51	769.06	335.32	141.87	1.78
14	101	916.03	506.99	222.21	3.84	440.21	390.52	174.11	3.22	1735.72	419.43	178.15	2.61
15	93	2207.99	1115.90	469.99	4.75	1070.40	763.12	325.65	4.19	2759.90	1015.42	423.23	2.34
16	41	16.55	44.94	17.95	1.04	11.80	39.70	15.90	0.94	16.23	22.94	9.23	0.46
17	348	4.36	25.42	9.27	0.51	2.61	19.41	7.41	0.45	4.55	19.26	6.43	0.26
18	251	37.98	73.76	25.06	1.57	20.52	56.59	19.75	1.32	38.26	49.72	16.36	0.78
19	262	61.44	121.91	44.34	1.36	26.63	78.98	30.29	1.12	78.45	113.41	36.90	0.78
20	88	101.42	117.47	49.26	2.43	75.43	104.17	43.14	2.23	72.89	55.62	25.12	0.85
21	77	58.40	104.78	39.05	1.71	39.14	87.23	33.34	1.49	45.57	55.95	19.11	0.82
22	92	41.05	84.38	34.52	1.45	33.28	75.50	30.68	1.38	28.87	39.68	16.69	0.48
23	114	42.79	83.16	33.28	1.43	28.60	68.31	27.31	1.33	41.80	58.04	22.40	0.57
24	38	541.07	272.21	110.78	5.29	346.76	225.44	91.19	4.84	559.21	167.37	68.27	2.37
25	81	83.93	116.16	46.67	1.98	43.98	83.92	32.64	1.74	85.97	82.92	34.42	0.86
26	41	21.77	55.80	21.36	1.15	13.97	45.41	17.79	1.00	20.11	34.32	12.01	0.57
27	68	45.67	91.27	36.46	1.39	26.03	69.88	27.72	1.20	41.11	58.26	24.28	0.64
28	44	96.25	132.11	52.41	2.15	63.32	105.36	42.18	1.91	65.61	74.49	27.61	0.83
29	156	72.60	120.57	42.92	1.86	42.18	94.21	34.65	1.55	74.32	77.70	25.59	1.17

Table 5 Summary of analysis of outcrop bedsets listed in Table 3, excluding partial measurements

Data excluding partials		Arithmetic average (m)				Geometric mean (m)				Standard deviation (m)			
Outcrop	N	Area	Perimeter	Width	Thickness	Area	Perimeter	Width	Thickness	Area	Perimeter	Width	Thickness
1	87	12.75	39.59	14.25	1.06	9.59	34.46	12.30	0.99	9.33	21.74	8.23	0.36
2	86	67.23	93.53	38.76	2.05	44.12	75.76	31.07	1.81	58.79	62.63	27.10	0.98
3	148	1000.02	355.19	145.35	7.47	543.80	266.21	108.71	6.37	1206.88	277.46	118.68	4.16
4	335	214.21	144.16	59.75	3.98	137.49	115.98	47.09	3.72	240.15	108.53	47.25	1.46
5	738	152.55	121.29	48.93	3.44	99.74	98.98	39.73	3.20	180.92	90.73	36.84	1.39
6	11	547.72	328.80	139.83	4.81	445.17	306.17	124.87	4.54	303.98	122.72	60.43	1.78
7	22	1.53	9.71	3.75	0.33	0.46	5.94	2.39	0.25	2.03	9.23	3.38	0.24
8	55	0.96	8.04	3.33	0.29	0.53	6.22	2.57	0.26	1.07	6.06	2.52	0.15
9	12	1.83	10.71	4.56	0.44	1.20	9.26	3.89	0.39	2.20	5.56	2.40	0.28
10	12	1.95	11.01	4.29	0.51	1.36	9.25	3.63	0.48	1.39	5.94	2.22	0.18
11	14	4.22	21.70	8.34	0.55	2.44	17.29	6.79	0.46	4.15	14.71	5.29	0.33
12	23	3.21	16.59	6.59	0.56	2.18	13.90	5.45	0.51	2.86	10.58	4.51	0.26
13	124	541.33	408.68	172.70	2.93	247.44	300.05	128.50	2.45	755.51	325.22	135.28	1.81
14	92	706.06	481.41	209.28	3.49	404.00	382.92	170.53	3.02	806.37	339.44	131.32	2.11
15	74	1790.94	1020.67	427.77	4.25	867.21	691.36	293.38	3.76	2379.69	989.04	412.73	2.06
16	24	15.26	44.69	17.66	0.95	10.51	40.46	15.94	0.84	17.45	20.75	8.47	0.48
17	289	3.93	24.11	8.69	0.49	2.34	18.26	6.98	0.43	4.26	19.08	5.94	0.25
18	203	33.75	70.24	23.76	1.46	17.60	53.34	18.50	1.21	35.39	47.84	15.91	0.75
19	225	50.36	109.93	40.06	1.25	22.48	72.09	27.54	1.04	63.58	101.34	34.12	0.68
20	63	91.75	116.60	47.96	2.21	65.94	102.65	41.77	2.01	71.25	56.75	25.27	0.79
21	33	43.63	97.41	34.01	1.41	26.93	76.75	28.07	1.22	40.32	58.72	17.50	0.70
22	57	34.61	79.02	32.35	1.34	29.64	71.97	29.21	1.29	19.17	33.25	14.43	0.36
23	97	42.53	83.11	33.15	1.42	28.10	67.91	27.20	1.32	41.55	58.57	22.03	0.60
24	11	356.63	218.50	86.76	4.59	242.92	186.32	72.98	4.24	274.88	105.46	44.30	1.86
25	64	63.08	96.74	38.53	1.82	33.58	71.22	27.43	1.58	53.52	56.26	22.87	0.79
26	24	20.41	50.66	19.84	1.18	13.05	41.60	16.34	1.02	15.93	28.53	11.16	0.58
27	35	28.66	69.46	25.65	1.20	14.88	51.84	19.47	0.97	24.35	45.11	16.54	0.59
28	18	74.56	105.86	43.34	1.98	40.22	77.95	31.41	1.63	50.77	58.91	25.80	0.80
29	105	65.99	117.35	41.53	1.66	35.92	91.43	33.53	1.36	74.69	76.14	24.25	1.13

Table 6 Summary of analysis of outcrop storeys listed in Table 3, including partial measurements (i.e. storeys that were not completely exposed due to the edges of the outcrop)

Data including partials		Arithmetic average (m)				Geometric mean (m)				Standard deviation (m)			
Outcrop	N	Area	Perimeter	Width	Thickness	Area	Perimeter	Width	Thickness	Area	Perimeter	Width	Thickness
1	21	69.78	82.87	32.36	2.45	46.36	70.50	27.24	2.17	52.45	42.11	17.48	0.87
2	17	416.00	278.08	121.05	3.88	279.42	234.08	99.65	3.57	378.15	171.84	78.64	1.48
3	27	6975.11	1101.04	461.87	17.25	4361.11	885.52	368.33	15.08	5974.78	690.88	308.72	8.42
4	87	1020.86	362.79	161.59	7.53	706.04	289.04	125.82	7.15	955.06	256.35	120.42	2.43
5	123	1035.16	375.85	165.23	7.31	717.14	306.55	133.54	6.84	906.62	245.25	109.89	2.64
6	3	2030.43	872.43	372.85	6.64	1808.43	841.94	354.53	6.49	1030.33	265.17	139.29	1.63
7	3	14.27	35.69	15.20	1.17	13.02	33.72	14.24	1.16	6.85	13.21	6.13	0.12
8	1	74.53	77.80	38.79	2.45	74.53	77.80	38.79	2.45	*	*	*	*
9	2	38.57	42.49	17.93	2.55	32.93	41.67	17.54	2.39	28.40	11.76	5.27	1.27
10	2	24.10	30.85	13.66	2.24	24.00	30.80	13.62	2.24	3.06	2.49	1.53	0.03
11	4	24.57	60.21	21.35	1.47	23.71	56.96	20.72	1.46	7.42	21.57	5.60	0.17
12	3	47.17	76.61	34.41	1.57	36.26	69.62	30.32	1.52	40.98	38.80	20.06	0.50
13	28	2848.86	962.31	430.65	7.29	1762.92	752.00	333.46	6.73	3068.02	633.11	294.66	2.75
14	15	4156.16	1090.84	503.80	7.66	1936.22	805.01	365.81	6.74	6371.42	924.46	442.37	3.96
15	15	12486.68	2123.15	1024.21	14.45	6839.35	1372.35	618.97	14.07	12040.27	1815.49	918.62	3.39
16	6	128.65	113.23	50.85	3.04	100.36	105.01	46.09	2.77	77.88	43.28	22.08	1.26
17	19	122.62	112.56	45.30	3.30	103.04	103.59	41.29	3.18	69.04	47.99	20.10	0.85
18	25	425.85	257.60	87.13	5.42	278.05	203.88	73.32	4.83	382.59	175.59	48.84	3.02
19	42	473.52	350.33	132.68	4.08	281.73	257.62	99.39	3.61	392.10	235.42	85.90	2.04
20	23	401.68	240.09	105.19	4.97	356.69	221.28	95.74	4.74	192.34	96.50	44.84	1.65
21	12	401.73	217.48	92.17	5.06	319.95	210.06	87.45	4.66	252.39	53.98	29.07	1.98
22	22	176.60	169.81	72.38	2.84	132.49	148.41	62.81	2.69	139.14	89.99	39.44	0.93
23	17	210.22	277.82	78.05	3.47	175.04	251.26	72.30	3.08	96.88	126.81	24.55	1.64
24	11	1943.69	502.41	198.87	11.70	1393.43	423.71	168.61	10.52	1414.43	230.64	92.64	6.61
25	24	286.43	205.59	80.39	4.35	156.52	150.39	60.09	3.32	216.88	119.45	45.54	3.29
26	8	117.24	113.58	49.18	2.87	94.98	99.61	43.43	2.78	58.89	51.12	21.55	0.68
27	13	261.99	201.09	91.07	3.52	209.56	182.77	80.49	3.31	173.31	89.05	46.35	1.28
28	8	554.36	294.87	127.89	5.17	469.92	286.88	121.99	4.90	311.01	68.39	38.65	1.83
29	23	580.33	277.64	111.65	6.34	413.47	232.49	91.78	5.74	473.44	146.75	62.98	3.22

*Not determined.

Table 7 Summary of analysis of outcrop storeys listed in Table 3, excluding partial measurements

Data excluding partials		Arithmetic average (m)				Geometric mean (m)				Standard deviation (m)			
Outcrop	N	Area	Perimeter	Width	Thickness	Area	Perimeter	Width	Thickness	Area	Perimeter	Width	Thickness
1	15	53.87	75.88	28.16	2.18	34.94	62.65	23.63	1.88	42.95	43.98	15.97	0.83
2	14	386.69	247.82	105.68	3.93	241.52	206.68	86.62	3.55	412.90	174.91	77.87	1.63
3	22	6871.25	1139.47	478.04	16.71	4540.73	966.73	397.10	14.56	5632.19	660.74	308.23	8.29
4	54	841.38	337.74	146.12	6.93	604.06	273.44	116.78	6.59	728.62	231.39	103.20	2.17
5	96	906.26	348.79	152.60	7.00	642.47	289.10	125.55	6.52	742.53	215.05	96.34	2.69
6	2	1874.39	824.44	371.81	5.98	1588.97	785.04	344.74	5.87	1406.09	356.11	196.96	1.65
7	1	7.11	20.44	8.45	1.07	7.11	20.44	8.45	1.07	*	*	*	*
8	0	*	*	*	*	*	*	*	*	*	*	*	*
9	0	*	*	*	*	*	*	*	*	*	*	*	*
10	0	*	*	*	*	*	*	*	*	*	*	*	*
11	1	25.81	54.31	21.11	1.56	25.81	54.31	21.11	1.56	*	*	*	*
12	0	*	*	*	*	*	*	*	*	*	*	*	*
13	22	2373.61	932.67	411.22	6.83	1654.81	751.69	329.55	6.39	1633.67	517.98	236.85	2.26
14	12	4343.29	1065.86	502.12	7.39	1681.26	731.11	337.60	6.34	7174.41	1038.89	497.17	4.37
15	6	3250.16	828.71	347.44	12.37	2452.83	644.93	258.33	12.09	2847.87	751.31	334.21	2.95
16	0	*	*	*	*	*	*	*	*	*	*	*	*
17	6	53.21	68.11	25.59	2.62	50.14	67.63	25.43	2.51	16.26	8.59	3.19	0.70
18	12	433.50	263.96	85.25	5.71	274.95	206.30	71.34	4.91	418.74	175.90	47.81	3.89
19	28	450.86	351.85	129.22	3.85	238.74	247.04	92.66	3.28	423.26	252.56	90.39	2.31
20	11	338.36	237.09	101.18	4.18	302.16	224.39	94.48	4.07	166.44	84.88	39.40	0.99
21	2	233.08	156.68	66.10	4.46	228.95	156.42	65.39	4.46	61.79	12.63	13.66	0.27
22	7	94.51	120.05	50.80	2.45	84.18	111.18	46.10	2.32	49.44	52.28	25.13	0.75
23	6	241.70	212.14	82.08	3.79	228.31	202.80	81.00	3.59	79.01	77.04	13.85	1.27
24	3	1965.80	360.47	129.75	14.55	994.12	289.81	111.97	11.30	2571.00	281.90	79.05	13.13
25	10	204.27	168.01	70.37	3.15	68.90	99.07	41.86	2.10	164.50	106.49	45.51	1.80
26	1	31.42	43.51	18.45	2.17	31.42	43.51	18.45	2.17	*	*	*	*
27	2	118.31	143.75	62.71	2.41	117.65	140.94	62.26	2.41	17.61	40.06	10.59	0.05
28	1	398.44	266.21	114.28	4.44	398.44	266.21	114.28	4.44	*	*	*	*
29	11	769.76	364.19	149.68	6.67	648.59	338.42	137.02	6.03	517.17	134.67	60.41	3.75

*Not determined.

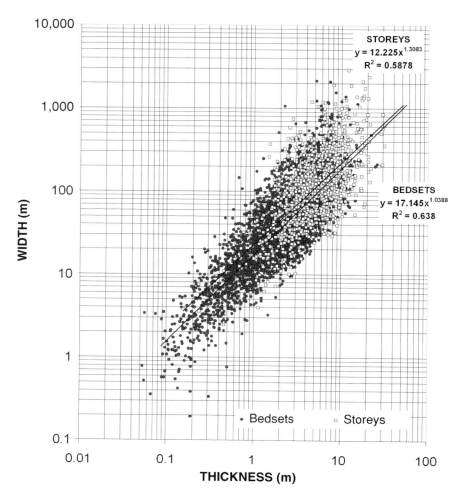

Fig. 11 Cross-plot of thickness and width data for all bedset (filled circles) and storey (open squares) data measured from two-dimensional outcrops.

formative mechanism. This accords with the observation that by definition storeys are simply bundles or stratasets of depositionally related bedsets. Consequently, the gross geometry of stratasets is likely to be similar to bedsets.

2 If individual outcrops are separated (Figs 13 & 14) it becomes clear that Thickness–Width fields for each example plots in a specific region and, although they may overlap, each formation clearly has its own range of geometrical properties and these presumably scale with the size of the formative river system. This clearly demonstrates that one cannot take measurements from one outcrop and assume that they will fit a subsurface example unless they are rescaled or calibrated using some common measure (e.g. channel/storey depth). The subtle but distinct differences between some formations relative to others in the cross-plots require further explanation but presumably relate to either: (i) primary factors such as channel style, rate of deposition/preservation potential; or (ii) secondary effects such as compaction and cementation which influence the stratal thicknesses observed.

3 If a single formation is evaluated in detail, in this case the Kayenta Formation (Lower Jurassic) of the Colorado Plateau region, USA (Miall, 1988; Bromley, 1991; Stephens, 1994; North & Taylor, 1996; Sanabria, 2001), then variations between locations can also be seen that may be significant (Fig. 15a & b). These differences could be apparent due to: (i) orientation differences of the outcrop to the dominant palaeotransport direction; (ii) differences in the resolution of field data (photographs, logs) which result in differences in the resolution of the data. All of the Kayenta data were collected by the author except for that from the Dolores location (Stephens, 1994), which is located some distance away from the other outcrops and this possibly explains the slight difference in the Dolores data, particularly in the bedset measurements. The majority of variations in the Kayenta data are thought to reflect differences

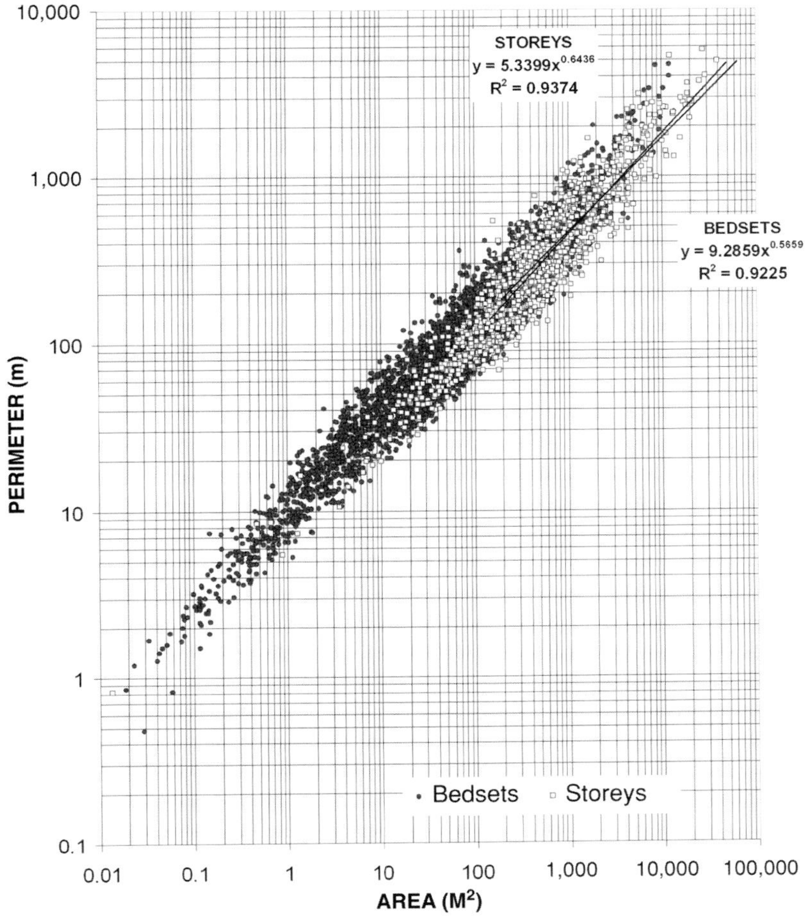

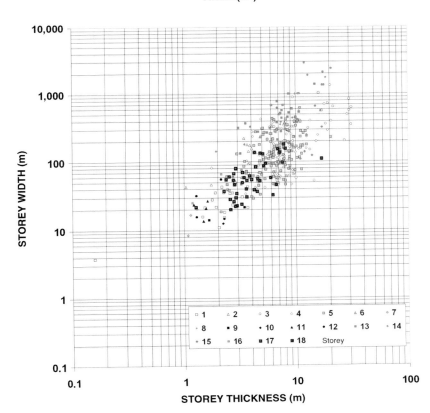

Fig. 12 Cross-plot of area and perimeter length data for all bedset (filled circles) and storey (open squares) data measured from two-dimensional outcrops.

Fig. 13 Cross-plot of thickness and width data for all storey data measured from two-dimensional outcrops. Individual outcrops are indicated.

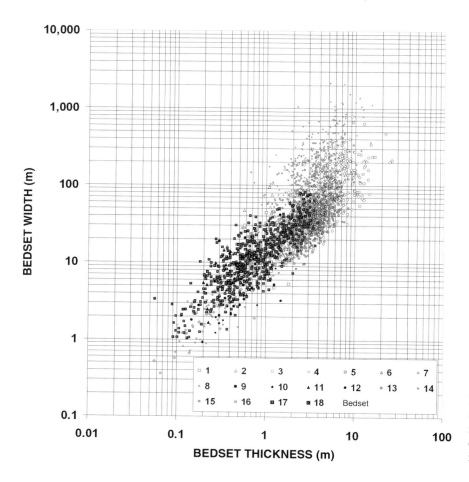

Fig. 14 Cross-plot of thickness and width data for all bedset data measured from two-dimensional outcrops. Individual outcrops are indicated.

in the orientation of the outcrop to the dominant palaeocurrent direction. One subset (outcrop 24, Shafer 1) stands out due to the significantly larger means for bedset and storey dimensions; this appears to reflect the presence of some very large scale bedsets within this outcrop, presumably representing deposition within relatively deep channels. There remains the possibility that these thick bedsets are aeolian deposits as there are aeolian intervals locally developed in the uppermost Kayenta (North & Taylor, 1996; Sanabria, 2001). However, the form and style of bedding in this outcrop are thought to indicate fluvial deposits rather than aeolian dunes. Notwithstanding these differences it can be seen that for a given depositional unit the scaling and distribution of geometrical parameters remain persistent across a number of effectively random samples (assuming the palaeogeography and gross depositional conditions do not change significantly). Furthermore, statistical evaluation of geometries can highlight variants (e.g. the Shafer 1 outcrop) that may require further investigation or explanation.

4 In order to derive empirical relationships that describe the relationships between T_b, T_s, W_b, W_s the geometric mean of these variables was taken for each outcrop and then the data for all outcrops were pooled. By taking the geometric mean, the bias towards larger outcrops with more measurements should be removed. Several important relationships can be described: (i) the mean thickness of bedsets (T_b) is a good predictor of mean bedset width (W_b), with $W_b = 15.95\ T_b^{1.37}$ ($R^2 = 0.88$) (Fig. 16); (ii) the mean thickness of storeys (T_s) is a good predictor of mean storey width (W_s), with $W_s = 10.06\ T_s^{1.52}$ ($R^2 = 0.78$) (Fig. 17); (iii) mean storey thickness (T_s) can be reasonably predicted from mean bedset thickness (T_b), with $T_s = 3.31\ T_b^{0.66}$ ($R^2 = 0.83$) (Fig. 18); (iv) mean storey width (W_s) can be reasonably predicted from mean bedset thickness (T_b), with $W_s = 58.33\ T_b^{1.09}$ ($R^2 = 0.81$) (Fig. 19).

This dataset suggests that in two-dimensional outcrops the deposits of rivers have characteristic geometries that are essentially self-similar across an appropriate range of scales. The scaling rela-

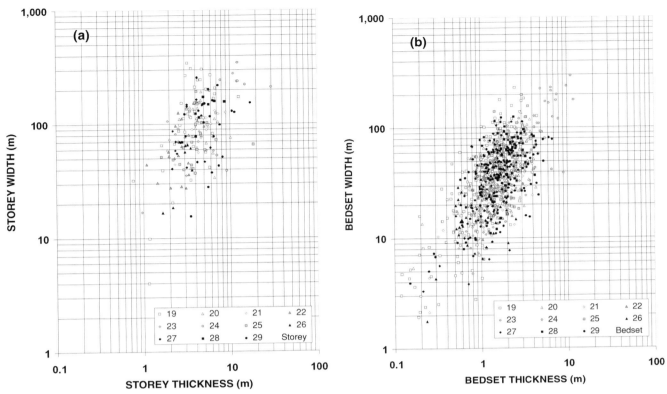

Fig. 15 Cross-plot of thickness and width data for (a) storey and (b) bedset data measured from Kayenta Formation two-dimensional outcrops. Individual outcrop locations are indicated.

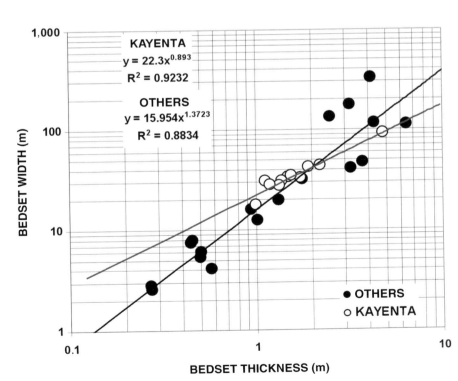

Fig. 16 Cross-plot of geometric means of thickness and width data for bedset data measured from all two-dimensional outcrops. Individual points represent outcrops. Outcrop data from the Kayenta Formation are differentiated. Compare with Fig. 11.

KAYENTA
$y = 22.3x^{0.893}$
$R^2 = 0.9232$

OTHERS
$y = 15.954x^{1.3723}$
$R^2 = 0.8834$

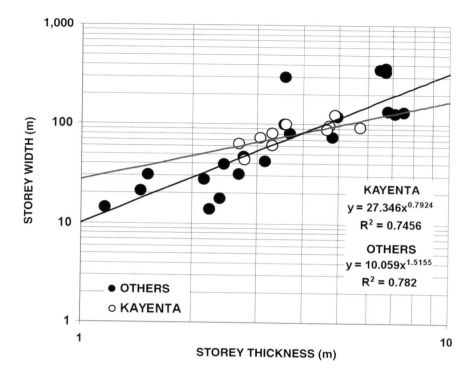

Fig. 17 Cross-plot of geometric means of thickness and width data for storey data measured from all two-dimensional outcrops. Individual points represent outcrops. Outcrop data from the Kayenta Formation are differentiated. Compare with Fig. 11.

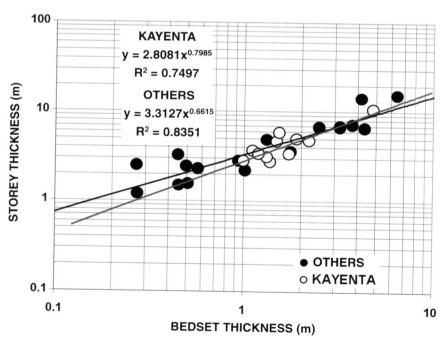

Fig. 18 Cross-plot of geometric means of bedset thickness and storey thickness data measured from all two-dimensional outcrops. Individual points represent outcrops. Outcrop data from the Kayenta Formation are differentiated.

tionships between different hierarchies of elements (e.g. bedsets and storeys) opens up the possibility of predicting the geometries and frequency distributions of larger scale features (e.g. storeys) from smaller ones (e.g. bedsets), and vice versa. The scaling characteristics of larger elements such as bodies or complexes are discussed below.

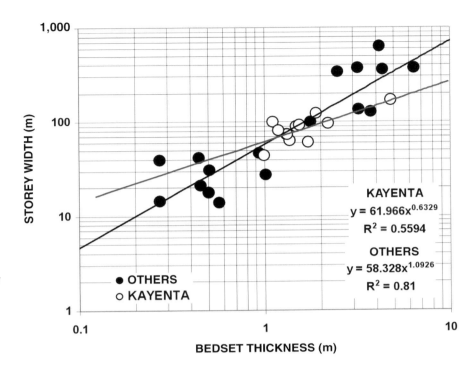

Fig. 19 Cross-plot of geometric means of bedset thickness and storey width data measured from all two-dimensional outcrops. Individual points represent outcrops. Outcrop data from the Kayenta Formation are differentiated.

COMBINING DATA FROM MODERN RIVERS WITH OUTCROP DATA

The key challenge ahead is to effectively and meaningfully combine the observations from modern rivers (mainly planform data) with cross-sectional data from a wide range of outcrops in order to derive suitable empirical estimates of architectural element geometries that can be used as input to object-based models of braided fluvial reservoirs. While this study does not provide a completely satisfactory solution to this problem, it is worthwhile exploring potential links between these datasets.

The depth of a fluvial channel (C_d) can be relatively easily measured and, given that the larger bars and macroforms scale with flow depth (Jackson, 1975), it is possible to use T_s as a proxy for channel depth (cf. Ethridge & Schumm, 1978). However, in addition to the problems of compaction, there remains the inherent variability of stratasets, which can vary in thickness by a factor of two or more within a channel belt sandbody (Bridge & Tye, 2001). As the maximum thickness for each storey/strataset has been recorded in the outcrop study presented here (Fig. 10) this problem is reduced with the thickness more likely to correspond to the channel depth, recognizing that 'channel depth as routinely measured and inferred from sedimentary features (i.e. point bars) will be less than total scour depth' (Salter, 1993). It should also be noted that maximum scour and aggradation will tend to produce sandbodies that are thicker than channel depth (Fielding & Crane, 1987; Salter, 1993). Despite these caveats, it is suggested that reliable depth estimates from stratasets or storeys can be used to estimate palaeochannel depth, and this estimate can then be used in estimating bar size and distribution from the empirical relationships derived for modern braided rivers. As a secondary cross-check the depth of formative flows can be estimated using a careful evaluation of cross-strata formed by dunes (Leclair et al., 1997; Bridge & Tye, 2000; Leclair & Bridge, 2001). Channel-belt width can also be predicted from mean bankfull depth estimates (Bridge & Mackey 1993; Bridge & Tye, 2000).

In order to investigate the potential link between channel depth/bar dimensions and storey thickness/storey width, the means for the collected datasets are plotted in Fig. 20. The observation that the two pairs of variables appear to show a similar relationship to each other tentatively supports using one as proxy for the other.

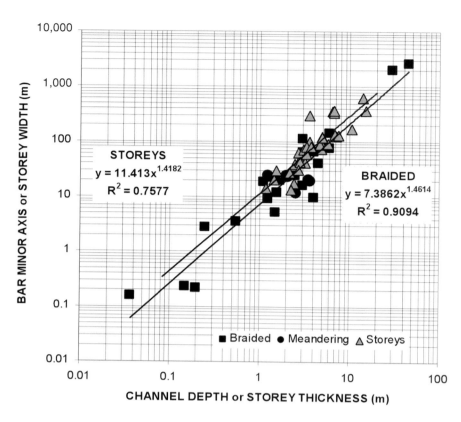

Fig. 20 Cross-plot of geometric means of storey thickness and storey width data measured from all two-dimensional outcrops, combined with a cross-plot of geometric means for channel depth and bar minor axis length data. A combination of Figs 7 & 17.

IMPLICATIONS FOR RESERVOIR MODELLING OF BRAIDED FLUVIAL RESERVOIRS

An empirical route is of potential value in the investigation of sandbody geometries in the subsurface (Miall, 1996; Bridge & Tye, 2000) when few other data are available (e.g. high-resolution seismic). The logic of this approach is based on two key assumptions (Bridge & Tye, 2000; Bridge, 2001) that: (i) fluvial sandbodies will scale with the rivers that deposited them; (ii) estimates of the size of the formative flows based on smaller scale features such as preserved bedforms can allow the size of the channels (and the resultant sandbodies) to be inferred.

The most comprehensive attempt to date to describe the reconstruction of sandbody dimensions in the subsurface is that of Bridge & Tye (2000). The approach of Bridge & Tye involves: (i) models for the lateral and vertical variation of lithofacies and petrophysical-log response of river-channel deposits, with explicit recognition of the different superimposed scales of strata; (ii) distinction between single and superimposed channel bars, channels and channel belts; (iii) interpretation of maximum palaeochannel depth from the thickness of channel bars and the thickness of sets of cross-strata formed by dunes; (iv) evaluation of various methods for estimation of widths of sandbodies that represent either single or connected channel belts using empirical equations relating channel depth, channel width and channel-belt width.

Although it is difficult to unambiguously interpret fluvial depositional forms from one-dimensional data, an interpretation step is necessary as deposits associated with particular depositional forms will have particular geometries and related reservoir quality or associated heterogeneity (Bridge, 2001). The detail and level of interpretation that will provide input to the reconstruction of sandbody architecture and dimensions will generally reflect the quality and quantity of data available. One major potential problem is the confident identification of storeys or stratasets (macroforms) in the subsurface, particularly in multistorey sandbodies (Miall, 1996). The differentiation of erosional surfaces that define bedsets, from those which define storeys, is somewhat subjective and may not be possible even in core or outcrop exposures of limited lateral extent. Bridge & Tye (2000) provided useful recognition criteria for the identification of storeys or stratasets from wireline logs. Additional data, such as high quality dipmeter or borehole image data can potentially resolve shale-draped bar (macroform) surfaces as well as dune or bar cross-strata (Herweijer *et al.*, 1990; Bridge & Tye, 2000).

It is possible to estimate sandbody dimensions (width) from the width of component elements (bedset or storey width). Although this is not without pitfalls, this method is offered as an alternative to the channel-belt width predictions based on channel depth estimates (Bridge & Mackey, 1993; Bridge & Tye, 2000). The method simply uses the relationships between storey or bedset width and body width, based on the assumption that, on average, larger rivers will deposit larger bars and also larger sandbodies than those left by smaller rivers. The observed relationships from a limited database are: $W_c = 6.75\, T_c^{1.33}$ ($R^2 = 0.80$) (Fig. 21), $W_c = 4.28\, W_b^{1.07}$ ($R^2 = 0.81$) (Fig. 22) and $W_c = 0.82\, W_s^{1.22}$ ($R^2 = 0.87$) (Fig. 23). One of the principal drawbacks is that a number of the sandbodies evaluated are

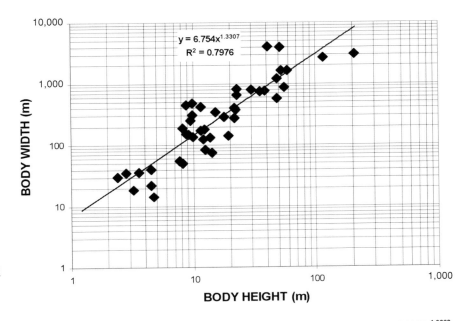

Fig. 21 Cross-plot of geometric means of body height (thickness) and body width data. Individual points represent outcrops.

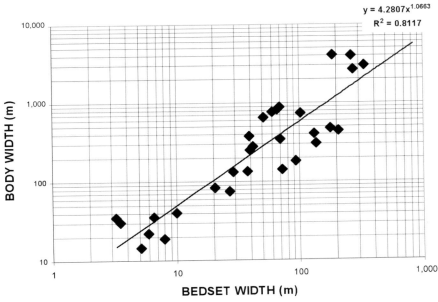

Fig. 22 Cross-plot of geometric means of bedset width and body width data. Individual points represent outcrops.

Fig. 23 Cross-plot of geometric means of storey width and body width data. Individual points represent outcrops.

limited by extent of outcrop (a geometric mean has been used when more than one sandbody occurs within an outcrop), which would tend to result in an underestimation of sandbody dimensions. Although the dataset is limited by these conditions, the regression coefficients suggest this approach may be of some value in estimating sandbody dimensions (width), if only as a cross-check of other methods.

In contrast to other types of fluvial reservoirs, sandbody dimensions are often not used as input to braided-fluvial reservoir models, with more attention being placed on modelling the non-reservoir heterogeneities (e.g. shales or mudrocks) within mapped or correlated sandbodies (e.g. Robinson & McCabe, 1997). Emphasis is commonly placed on detailed heterogeneities such as fine-grained bar/bedform drapes, abandoned channel fills, low permeability mudclast lags or other depositional features that strongly influence fluid flow, e.g. high permeability gravel facies (Martin, 1993). These types of feature often have specific problems relating to identification and modelling, including the resolution and non-uniqueness of the log response of different shale/mudrock types (Martin, 1993). While the modelling of such detailed heterogeneity within braided fluvial reservoirs is beyond the scope of this study, it is anticipated that any attempts to model such features would benefit from an appreciation of their scaling characteristics relative to the more easily quantified architectural elements (e.g. bedsets and storeys). Case-specific evaluation is probably required to establish the significance and scaling relationship between such features (e.g. shale drapes) with the sand-prone reservoir facies elements such as those which have been evaluated here. The detailed architectural approach outlined will assist in the building of detailed reservoir models that potentially allow the inclusion of such heterogeneity once these relationships have been established.

SUMMARY

Braided rivers display a wide range of bedform and bar sizes and are often characterized by hierarchy of channels, bedforms and associated superposition. As concluded by earlier studies, the planform geometries of modern rivers are typically scale invariant (Sapozhnikov & Foufoula-Georgiou, 1996; Sambrook Smith *et al.*, 2005) and suggest that these rivers have reached a dynamic equilibrium (Sapozhnikov & Foufoula-Georgiou, 1997). The scaling exponents derived from an extensive database covering five orders of channel width indicate that it is possible to predict bar size (minor and major axis length, area, perimeter length) when channel-belt width or channel depth is known or vice versa. Hierarchy and superposition appear to be a ubiquitous feature of braided fluvial systems and are thought to be related directly to the scale invariant character observed.

By analogy with modern rivers we should anticipate an equivalent wide range of bar/bedform sizes in the outcrop and subsurface deposits of braided rivers. A large dataset of measurements from outcrops indicates that bodies/storeys/bedsets scale to each other. From this dataset it is possible to develop an additional set of empirical relationships that allow the size/geometry and distribution of one parameter (e.g. bedsets) to be predicted from another (e.g. channel storey size).

The geometry (length, thickness) of preserved braided river deposits is likely to be similar to that of the associated bedforms. This link together with the established relationship of bedform (dunes, bars) wavelength and height to channel depth and width (Jackson, 1975) allows the comparison of modern braided rivers and ancient fluvial outcrops. Channel depth and storey thickness are thought to provide the best scaling link between fluvial outcrops and modern systems. Estimates of depth using storey or strataset thickness are relatively easily made from outcrop, and with care can be obtained from borehole data (Bridge & Tye, 2000).

Most fluvial outcrop databases simply record the width and thickness of sandbodies rather than data on surfaces, lithofacies or architectural elements within the sandbodies. Although these data may be useful, it is often difficult to relate directly to subsurface datasets, especially when sandbodies are multistorey in character. The common practice of using sandbody width:thickness relationships is fraught with difficulties, typically reflecting limited or inappropriate data, or simply indicating that there is no strong relationship between these parameters (Bridge & Tye, 2000; Bridge, 2001). The quantitative description of bedsets and storeys provides an alternative method of estimating sandbody width, although this may underestimate maximum width due to the common limitations of outcrops which provide database input.

To conclude, the issue of scale is fundamental to an understanding of braided rivers and their deposits. Although scale invariance appears to characterize many aspects of these rivers and their depositional record, each river and each outcrop are unique and hence a unifying facies model for all braided rivers cannot be developed (Sambrook Smith *et al.*, 2005). However, it is possible to derive generic process-based models (Bridge, 1993) that act as guides and predictors for facies or reservoir architecture. In a similar way, scale invariance can be used to provide quantitative predictions of the geometries of braided river deposits. Simply assuming that data collected from any outcrop can be applied to subsurface data without recognizing critical scaling aspects will likely lead to serious error and does not take into account the unique characteristics of braided rivers and their sedimentary record.

ACKNOWLEDGEMENTS

The continued enthusiasm and encouragement of Shaun Sadler has been invaluable over the years. The support of Shell Exploration and Production UK Ltd for conference attendance and publication is gratefully acknowledged. All views expressed are solely those of the author and do not necessarily reflect the opinion of Shell UK. Diego Sanabria is thanked for helpful discussions in the field on the Kayenta. Ian Lunt and Dave Moreton are thanked for their constructive reviews of the original manuscript. Special thanks to Greg Sambrook Smith for his editorial advice and patience. Finally, big thanks to my family Rosie, Roisin and Iona for their support and tolerance.

REFERENCES

Allen, J.R.L. (1983) Studies in fluviatile sedimentation:bars, bar complexes and sandstone sheets (low-sinuosity braided streams) in the Brownstones (L. Devonian), Welsh Borders. *Sediment. Geol.*, **33**, 237–293.

Alexander, J. and Gawthorpe, R.L. (1993) The complex nature of a Jurassic multistorey, alluvial sandstone body, Whitby, North Yorkshire. In: *Characterization of Fluvial and Aeolian Reservoirs* (Eds C.P. North and D.J. Prosser), pp. 123–142. Special Publication 73, Geological Society Publishing House, Bath.

Ashmore, P.E. (1993) Anabrach confluence kinetics and sedimentation processes in gravel-bed streams. In: *Braided Rivers* (Eds J.L. Best and C.S. Bristow), pp. 129–146. Special Publication 75, Geological Society Publishing House, Bath.

Bluck, B.J. (1971) Sedimentation in the meandering River Endrick. *Scot. J. Geol.*, **7**, 93–138.

Bluck, B.J. (1976) Sedimentation in some Scottish rivers of low sinuosity. *Trans. Roy. Soc. Edinb.*, **69**, 425–455.

Bluck, B.J. (1979) Structure of coarse grained braided alluvium. *Trans. Roy. Soc. Edinb.*, **70**, 181–221.

Bridge, J.S. (1985) Paleochannel patterns inferred from alluvial deposits: a critical evaluation. *J. Sediment. Petrol.*, **55**, 579–589.

Bridge, J.S. (1993a) Description and interpretation of fluvial deposits. *Sedimentology*, **40**, 801–810.

Bridge, J.S. (1993b) The interaction between channel geometry, water flow, sediment transport, erosion and deposition in braided rivers. In: *Braided Rivers* (Eds J.L. Best and C.S. Bristow), pp. 13–71. Special Publication 75, Geological Society Publishing House, Bath.

Bridge, J.S. (2001) Characterization of fluvial hydrocarbon reservoirs and aquifers: problems and solutions. *Asoc. Argentina Sedimentol. AAS Rev.*, **8**, 87–114.

Bridge, J.S. and Diemer, J.A. (1983) Quantitative interpretation of evolving ancient river system. *Sedimentology*, **30**, 599–623.

Bridge, J.S. and Mackey, S.D. (1993) A theoretical study of fluvial sandstone body dimensions. In: *The Geological Modelling of Hydrocarbon Reservoirs and Outcrop Analogues* (Eds S. Flint and I.D. Bryant), pp. 213–236. Special Publication 15, International Association of Sedimentologists. Blackwell Scientific, Oxford.

Bridge, J.S. and Tye, R.S. (2000) Interpreting the dimensions of ancient fluvial channel bars, channels, and channel belts from wireline-logs and cores. *Am. Assoc. Petrol. Geol. Bull.*, **84**, 1205–1228.

Bridge, J.S., Smith, N.D., Trent, F., Gabel, S.L. and Bernstein, P. (1986) Sedimentology and morphology of the low-sinuosity Calamus River, Nebraska Sandhills. *Sedimentology*, **36**, 851–870.

Brierley, G.J. (1989) River planform facies models: the sedimentology of braided, wandering and meandering reaches of the Squamish River, British Columbia. *Sediment. Geol.*, **61**, 17–35.

Brierley, G.J. and Hickin, E.J. (1991) Channel planform as a non-controlling factor in fluvial sedimentology: the case of the Squamish River floodplain, British Columbia. *Sediment. Geol.*, **75**, 67–83.

Bristow, C.S. and Best, J.L. (1993) Braided rivers: perspectives and problems. In: *Braided Rivers* (Eds J.L. Best and C.S. Bristow), pp. 1–9. Special Publication 75, Geological Society Publishing House, Bath.

Bromley, M. (1991) Variations in fluvial style as revealed by architectural elements, Kayenta Formation, Mesa Creek, Colorado, U.S.A. In: *The Three-dimensional Facies Architecture of Terrigenous Clastic Sediments and its Implications for Hydrocarbon Discovery and Recovery* (Eds A.D. Miall and N. Tyler), pp. 94–103. Concepts in Sedimentology and Palaeontology 3, Society of Economic Paleontologists and Mineralogists, Tulsa, OK.

Campbell, C.V. (1967) Lamina, laminaset, bed and bedset. *Sedimentology*, **8**, 7–26.

Coleman, J.M. (1969) Brahmaputra River: channel processes and sedimentation: *Sediment. Geol.*, **13**, 129–239.

Collinson, J.D. (1970) Bedforms of the Tana River, Norway. *Geogr. Ann.*, **52A**, 31–56.

Collinson, J.D. (1978) Vertical sequence and sand body shape in alluvial sequences. In: *Fluvial Sedimentology* (Ed. A.D. Miall), pp. 577–586. Memoir 5, Canadian Society of Petroleum Geologists, Calgary.

Cuevas Gozalo, M.C. and Martinius, A.W. (1993) Outcrop data-base for the geological characterization of fluvial reservoirs: an example from distal fluvial fan deposits in the Loranca Basin, Spain. In: *Characterization of Fluvial and Aeolian Reservoirs* (Eds C.P. North and D.J. Prosser), pp. 79–94. Special Publication 73, Geological Society Publishing House, Bath.

Dalrymple, M. (2001) Fluvial reservoir architecture in the Statfjord Formation (northern North Sea) augmented by outcrop analogue statistics. *Petroleum Geoscience*, **7**, 115–122.

Ethridge, F.G. and Schumm, S.A. (1978) Reconstructing palaeochannel morphologic and flow characteristics: methodology, limitations and assessment. In: *Fluvial Sedimentology* (Ed. A.D. Miall), pp. 703–721. Memoir 5, Canadian Society of Petroleum Geologists, Calgary.

Ferguson, R.I. and Werritty, A. (1983) Bar development and channel changes in the gravelly River Feshie, Scotland. In: *Modern and Ancient Fluvial Systems* (Ed. J.D. Collinson and J. Lewin), pp. 181–193. Special Publication 6, International Associated of Sedimentologists. Blackwell Scientific, Oxford.

Fielding, C.R. and Crane, R.C. (1987) An application of statistical modelling to the prediction of hydrocarbon recovery factors in fluvial reservoir sequences. In: *Recent Developments in Fluvial Sedimentology* (Eds F.G. Ethridge, R.M. Flores and M.D. Harvey), pp. 321–327. Special Publication 39, Society of Economic Paleontologists and Mineralogists, Tulsa, OK.

Foley, M.G. (1978) Scour and fill in steep, sand-bed ephemeral streams. *Geol. Soc. Am. Bull.*, **89**, 559–570.

Foufoula-Georgiou, E. and Sapozhnikov, E. (1998) Anisotropic scaling in braided rivers: an integrated theoretical framework and results from application to an experimental river. *Water Resour. Res.*, **34**, 863–867.

Foufoula-Georgiou, E. and Sapozhnikov, V. (2001) Scale invariances in the morphology and evolution of braided rivers. *Math. Geol.*, **33**, 273–291.

Friend, P.F. (1983) Towards the field classification of alluvial architecture or sequence. In: *Modern and Ancient Fluvial Systems* (Ed. J.D. Collinson and J. Lewin), pp. 345–354. Special Publication 6, International Associated of Sedimentologists. Blackwell Scientific, Oxford.

Friend, P.F., Slater, M.J. and Williams, R.C. (1979) Vertical and lateral building of river sandstone bodies. *J. Geol. Soc. Lond.*, **136**, 39–46.

Friend, P.F. Raza, S.M., Geehan, G. and Sheikh, K.A. (2001) Intermediate-scale architectural features of the fluvial Chinji (Miocene), Siwalik Group, northern Pakistan. *J. Geol. Soc. Lond.*, **158**, 163–177.

Geehan, G.W. and Pearce, A.J. (1994) Geological Reservoir Heterogeneity Databases and Their Application to Integrated Reservoir Description. In: *North Sea Oil and Gas Reservoirs—III* (Eds J.O. Aasen et al.), pp. 131–140. Kluwer, Dordrecht.

Germanoski, D. and Schumm, S.A. (1993) Changes in braided river morphology resulting from aggradation and degradation. *J. Geol.*, **101**, 451–466.

Graf, W.L. (1988) *Fluvial Processes in Dryland Rivers.* Springer-Verlag, Berlin.

Herweijer, J.C., Hocker, C.F.W., Williams, H. and Eastwood, K.M. (1990) The relevance of dip profiles from outcrops as reference for the interpretation of SHDT dips. In: *Geological Applications of Wireline Logs* (Eds A. Hurst, M.A. Lovell, and A.C. Morton), pp. 39–43. Special Publication 48, Geological Society Publishing House, Bath.

Hirst, J.P.P. (1991) Variations in alluvial architecture across the Oligo-Miocene Huesca Fluvial System, Ebro Basin, Spain. In: *The Three-dimensional Facies Architecture of Terrigenous Clastic Sediments and its Implications for Hydrocarbon Discovery and Recovery* (Eds A.D. Miall and N. Tyler), pp. 111–121. Concepts in Sedimentology and Palaeontology 3, Society of Economic Paleontologists and Mineralogists, Tulsa, OK.

Hoimyr, O., Kleppe, A. and Nystuen, J.P. (1993) Effects of heterogeneities in a braided stream channel sandbody on the simulation of oil recovery: a case study from the Lower Jurassic Statfjord Formation, Snorre Field, North Sea. In: *Advances in Reservoir Geology* (Ed. M. Ashton), pp. 105–134. Special Publication 69, Geological Society Publishing House, Bath.

Hooke, J.M. (1997) Styles of channel change. In: *Applied Fluvial Geomorphology for River Engineering and Management* (Eds C.R. Thorne, R.D. Hey and M.D. Newson), pp. 237–268. Wiley, Chichester.

Inglis, C.C. (1949) *The Behaviour and Control of Rivers and Canals (with the Aid of Models), Meandering of Rivers.* Research Publication No. 13, Central Water-Power, Irrigation and Navigation Research Station, Poona, 143 pp.

Jackson, R.G. II (1975) Hierarchical attributes and a unifying model of bed forms composed of cohesionless material and produced by shearing flow. *Geol. Soc. Am. Bull.*, **86**, 1523–1533.

Kellerhals, R., Church, M. and Bray, D.I. (1976) Classification and Analysis of River Processes. *J. Hydrol. Div. Am. Soc. Civ. Eng.*, **102**, 813–829.

Kirkby, M.J. (1999) Towards an understanding of varieties of fluvial form. In: *Varieties of Fluvial Form* (Eds A.J. Miller and A. Gupta), pp. 507–514. Wiley, Chichester.

Lane, S.N. and Richards, K.S. (1997) Linking river channel form and process: time, space and causality revisited. *Earth Surf. Proc. Landf.*, **22**, 249–260.

Langford, R. and Bracken, B. (1987) Medano Creek, Colorado, a model for upper-flow-regime fluvial deposition. *J. Sediment. Petrol.*, **57**, 863–870.

Lanzarini, W.L., Poletto, C.A., Tavares, G. and Pesco, S. (1997) Stochastic modelling of geometric objects and reservoir heterogeneities. *Fifth Latin-American and Caribbean Petroleum Engineering Conference, Rio de Janeiro*, Society of Petroleum Engineering, Paper 38953.

Leclair, S.F. and Bridge, J.S. (2001) Quantitative interpretation of sedimentary structures formed by river dunes. *J. Sediment. Res.*, **71**, 714–717.

Leclair, S.F., Bridge, J.S. and Wang, F. (1997) Preservation of cross-strata due to migration of subaqueous dunes over aggrading and non-aggrading beds: comparison of experimental data with theory. *Geosci. Can.*, **24**, 55–66.

Leopold, L.B. and Wolman, M.G. (1957) River channel patterns: braided, meandering and straight. *US Geol. Surv. Prof. Pap.*, **282-B**, 39–85.

Martin, J.H. (1993) A review of braided fluvial hydrocarbon reservoirs: the petroleum engineer's perspective. In: *Braided Rivers* (Eds J.L. Best and C.S. Bristow), pp. 333–367. Special Publication 75, Geological Society Publishing House, Bath.

Miall, A.D. (1988) Architectural elements and bounding surfaces in fluvial deposits: anatomy of the Kayenta Formation (Lower Jurassic), Southwest Colorado. *Sediment. Geol.*, **55**, 233–262.

Miall, A.D. (1996) *The Geology of Fluvial Deposits.* Springer-Verlag Berlin Heidelberg.

North, C.P. and Taylor, K.S. (1996) Ephemeral-fluvial deposits: integrated outcrop and simulation studies reveal complexity. *Am. Assoc. Petrol. Geol. Bull.*, **80**, 811–830.

Nykanan, D., Foufoula-Georgiou, E. and Sapozhnikov, V. (1998) Study of spatial scaling in braided river patterns using synthetic aperture radar imagery. *Water Resour. Res.*, **34**, 1795–1807.

Olsen, T., Steel, R. Høgseth, Skar, T. and Røe, S.L. (1995) Sequential architecture in a fluvial succession: sequence stratigraphy in the Upper Cretaceous Mesaverde Group, Price Canyon, Utah. *J. Sediment. Res.*, **B65**, 265–280.

Pinheiro, F., Menezes, L., Poletto, C.A., Tavares, G., Lopes, H. and Pesco, S. (2002) Fluvial outcrops parametrization applied to object based geological modelling for reservoirs of Potiguar basin. *Terra Nostra*, **3**, 173–179.

Reynolds, A.D. (1999) Dimensions of paralic sandstone bodies. *Am. Assoc. Petrol. Geol. Bull.*, **83**, 211–229.

Robinson, J.W. and McCabe, P.J. (1997) Sandstone-body and shale-body dimensions in a braided fluvial system: Salt Wash Sandstone Member (Morrison Formation). *Am. Assoc. Petrol. Geol. Bull.*, **81**, 1267–1291.

Rubin, D.M. and McCulloch, D.S. (1980) Single and superimposed bedforms: a synthesis of San Francisco Bay and flume observations. *Sediment. Geol.*, **26**, 207–231.

Rubin, D.M. and Hunter, R.E. (1983) Reconstructing bedform assemblages from compound crossbedding. In: *Eolian Sediments and Processes* (Eds M.E. Brookfield and T.S. Ahlbrandt), pp. 407–427. Elsevier, Amsterdam.

Rust, B.R. (1978) A classification of alluvial channel systems. In: *Fluvial Sedimentology* (Ed. A.D. Miall), pp. 187–198. Memoir 5, Canadian Society of Petroleum Geologists, Calgary.

Salter, T. (1993) Fluvial scour and incision: models for the influence on the development of realistic reservoir geometries. In: *Characterization of Fluvial and Aeolian Reservoirs* (Eds C.P. North and D.J. Prosser), pp. 33–51. Special Publication 73, Geological Society Publishing House, Bath.

Sambrook Smith, G.H., Ashworth, P.J., Best, J.L., Woodward, J. and Simpson, C.J. (2005) The morphology and facies of sandy braided rivers: some considerations of spatial and temporal scale invariance. In: *Fluvial Sedimentology VII* (Eds M.D. Blum, S.B. Marriott and S.F. Leclair), pp. 145–158. Special Publication 35, International Association of Sedimentologists. Blackwell Publishing, Oxford.

Sanabria, D.I. (2001) *Sedimentology and stratigraphy of the Lower Jurassic Kayenta Formation, Colorado Plateau, USA*. Unpublished PhD thesis, Department of Geology and Geophysics, Rice University, Houston, Texas.

Sapozhnikov, V. and Foufoula-Georgiou, E. (1996) Self-affinity in braided rivers. *Water Resour. Res.*, **32**, 1429–1439.

Sapozhnikov, V. and Foufoula-Georgiou, E. (1997) Experimental evidence of dynamic scaling and self-organized criticality in braided rivers. *Water Resour. Res.*, **33**, 1983–1991.

Sapozhnikov, V. and Foufoula-Georgiou, E. (1999) Horizontal and vertical self-organization of braided rivers towards a critical state, *Water Resour. Res.*, **35**, 843–851.

Sapozhnikov, V., Murray, B., Paola, C. and Foufoula-Georgiou, E. (1998) Validation of braided-stream models: spatial state-space plots, self-affine scaling and island shapes. *Water Resour. Res.*, **34**, 2353–2364.

Schumm, S.A., Khan, H.R., Winkley, B.R. and Robbins, W.G. (1972) Variability of river patterns. *Nature*, **237**, 75–76.

Stephens, M. (1994) Architectural element analysis within the Kayenta Formation (Lower Jurassic) using ground penetrating radar and sedimentological profiling, southwestern Colorado. *Sediment. Geol.*, **90**, 179–211.

Thorne, C.R. (1997) Channel types and morphological classification. In: *Applied Fluvial Geomorphology for River Engineering and Management* (Eds C.R. Thorne, R.D. Hey and M.D. Newson), pp. 175–222. Wiley, Chichester.

Thorne, C.R., Russell, A.P.G. and Alam, M.K. (1993) Planform pattern and channel evolution of the Brahmaputra River, Bangladesh. In: *Braided Rivers* (Eds J.L. Best and C.S. Bristow), pp. 257–276. Special Publication 75, Geological Society Publishing House, Bath.

Visser, C.A. and Chessa, A.G. (2000) A new method for estimating lengths for partially exposed features. *Math. Geol.*, **32**, 109–126.

Wathen, S.J., Hoey, T.B. and Werritty, A. (1997) Quantitative determination of the activity of within-reach sediment storage in a small gravel-bed river using transit time and response time. *Geomorphology*, **20**, 113–114.

Willis, B.J. (1993) Ancient river systems in the Himalayan foredeep, Chinji Village area, northern Pakistan. *Sediment. Geol.*, **88**, 1–76.

Yalin, M.S. (1977) *Mechanics of Sediment Transport*. Pergamon, Oxford, 290 pp.

Approaching the system-scale understanding of braided river behaviour

STUART N. LANE

Department of Geography, University of Durham, Science Laboratories, South Road, Durham DH1 3LE, UK
(Email: s.n.lane@durham.ac.uk)

ABSTRACT

This paper reviews three major areas of development in the system-scale study of braided river behaviour: (i) computational modelling; (ii) spatial and dynamic scaling analyses; and (iii) remote sensing of bed morphology and morphological change. With each, the paper explains the principles behind them and addresses the importance of their contribution to the system-scale understanding of braided river form. It is argued that reduced complexity approaches to understanding braiding behaviour have resulted in a step change in the kinds of questions that can be asked about the braiding process. In particular, the work has shown that braiding appears to be the fundamental instability that arises when water moves over non-cohesive material in the absence of lateral confinement. The classic continuum of channel pattern is recast, with braided channels being the default channel pattern that a river reverts to in the absence of lateral confinement or channel resistance effects due to geomorphological (i.e. valley width), sedimentological (presence of cohesive sediment) or vegetation effects (root binding and sediment resistance to entrainment). Whilst it is possible to assess the conceptual basis of these reduced complexity models critically, as well as empirically, this step change in understanding is likely to be robust. The more complex models involving fewer process descriptions and proper numerical solution offer additional prospects. However, both reduced and increased complexity modelling approaches require different ways of approaching model validation: it is not possible to compare a time-series of predictions of braided river morphology with measured data. Severe non-linearities and boundary conditions mean that all models will diverge rapidly from the system they are designed to represent. For this reason, synthetic descriptors of river braiding have been developed as a means of comparing models, as well as increasing our fundamental understanding of the braiding process. Commonly, these are based upon spatial and dynamic scaling analyses and they have provided considerable insight into the fundamental nature of the braiding process. The key finding has been that braided rivers may well be self-organized, and this has been shown for a number of rivers, but that this state may be perturbed by the geological, geomorphological and probably ecological setting through which the river passes. The work has also shown that the evolution of channel pattern in smaller areas is forced by the evolution of channel pattern in larger areas. Research questions remain, notably in relation to the role of lateral constraints in causing a switch from self-organization to self-affinity, as well as over the extent to which these methods can be extended to remotely sensed measurements of braided river channel change. In turn, answering the latter question is being facilitated by developments in remote sensing, and notably laser altimetry and photogrammetry. These developments have allowed us to produce high-resolution digital elevation models of braided river morphology for areas with considerable spatial extent (rivers > 1 km wide and 3 km in length). This morphological evidence shows that:

1 braiding, as is conventionally produced in laboratory scaled models, appears to occur at all flows, regardless of magnitude, but is confined to a weakly meandering braid bar belt, whose location is fixed in the short term by the location of larger scales of active braidplain morphology;

2 the braid belt, at its margins, is responsible for the longer term modification of these larger scales of braidplain morphology, through both erosional and depositional processes;

3 larger storm-related events appear to accelerate these modifications by providing the conditions necessary for a wider range of system responses, including very rapid migration of individual braid bar anabranches and major avulsions.

Tests suggest that deposition is no different to erosion and, whilst stage change is not necessary for braiding to occur in the braid belt, stage change does leave an important morphological legacy in terms of causing lobes of migrating sediment to stall in individual anabranches. These kinds of data allow us to increase significantly our understanding of the nature of the braiding process as well as providing key datasets for driving and assessing models and for undertaking advanced scaling analyses.

Keywords Fluvial, braided river behaviour, system scale, modelling, bed morphology.

INTRODUCTION

This paper provides a review of progress in developing a system-scale understanding of braided river behaviour. By the early 1990s, most progress in braided river research had been achieved through laboratory-scaled modelling of the braiding process (Ashmore, 1982, 1991, 1993; Young & Davies, 1991) and the study of relatively small reaches of valley-confined systems (e.g. Ashworth & Ferguson, 1986; Ferguson et al., 1992; Ferguson, 1993). A few studies had attempted understanding of larger scale systems (e.g. Carson & Griffiths, 1989), assisted by synoptic data such as aerial photography. However, by the mid-1990s, there had been no real studies of either the continuous variation of morphology or morphological change in large braided rivers in the field. This is surprising as braided rivers represent the only type of river where behaviour is continual and rapid enough for significant change to be measured over short time periods. As such, they may be the only channel pattern amenable to morphodynamic study over relatively short time-periods (Ashmore, personal communication, 2005). With this observation in mind, the past 10 yr has seen major progress in three main areas of study of braided rivers.

1 Numerical models (e.g. Murray & Paola, 1994) have been developed that can reproduce aspects of the braiding process over reach lengths that are many multiples of braidplain width.

2 Synoptic remote sensing (e.g. Nykanen et al., 1998; Walsh & Hicks, 2002) has proved the fundamental data necessary to provide alternative quantifications of emergent braided river properties (e.g. Sapozhnikov & Foufoula-Georgiou, 1996, 1997). These are 'alternative' in that they have recognized explicitly the scale dependence of braided river pattern as opposed to conventional descriptors of braidplain pattern.

3 There has been progress in the quantification of braided river morphology and morphological change using digital elevation modelling methods based upon remote sensing techniques such as digital photogrammetry and laser altimetry (Westaway et al., 2000, 2001, 2003; Lane et al., 2003). For the first time this has allowed us to visualize and to analyse the morphology of a large braided river (active braidplain >1 km wide).

This paper reviews the development of these approaches, illustrates the contribution that they are making to our understanding of braided river behaviour, and provides a critical assessment of the research questions that they raise.

The paper begins with a review of numerical modelling of braided rivers, with a particular emphasis upon the key model developed since the 1990s: the Murray & Paola (1994, 1997) model of braided river behaviour, based upon a reduced complexity approach. This section argues that to understand the debate over the Murray & Paola model it is necessary to think more broadly and philosophically about the role that models can play in increasing our understanding of fluvial systems. This is because the Murray & Paola approach is contested (Paola, 2001), and that without thinking about what makes a valid model it is difficult to appreciate the need to undertake a conceptual assessment of the physical basis of the model, and also to understand the needs: (i) to generate synthetic descriptors of braiding, reviewed in the middle section of the paper; and (ii) to try to measure braided river topography in large river systems, reviewed in the last section of the paper. The first section contains a full assessment of the physical concepts underlying the Murray & Paola (1994, 1997) model. It concludes by contrasting the Murray & Paola reduced complexity approach with more complex models that, for example, try to treat the dispersion terms associated with secondary circulation and sediment sorting effects more explicitly.

It is highly unlikely that any model, however well-grounded, will reproduce the exact time-dependent evolution of a braided system. This is because of non-linearities in the coupled flow–morphology–sediment transport linkage but also because of ongoing uncertainties in representing model boundary conditions. The solution to this problem has been to generate synthetic descriptors of river braiding, and both static and dynamic scaling have proved extremely valuable in this respect (Paola & Foufoula-Georgiou, 2001). In turn, these have shed light on the nature of river braiding, and this is described in the second part of the paper. All of the scaling work to date has been undertaken for maps of planform inundation. Thus, and recognizing the need to measure braided river morphology to initialize models of braided river behaviour, the final section reviews progress in the measurement of braided river morphology over very large spatial scales. As with both modelling and scaling analyses, this also reveals important understanding of how braided rivers work.

NUMERICAL MODELS OF BRAIDED RIVERS AT THE SYSTEM-SCALE

Perhaps the major development over the last 10 years in terms of system-scale analyses is in terms of braided river models. This development has two aspects:

1 reduced complexity approaches;
2 depth-averaged, coupled, flow and sediment transport models.

This section: (i) describes the basis of reduced complexity approaches; (ii) introduces the need to think about how these approaches, which are hotly contested, should be assessed; (iii) undertakes a conceptual assessment of the model, which seeks to link the model back to fundamental governing equations; and (iv) contrasts reduced complexity approaches with more traditional increased complexity approaches, based upon depth-averaged, coupled, flow and sediment transport models.

Braided river models I: reduced complexity approaches

Murray & Paola (1994, 1997) presented the first numerical representation of the spatially distributed, time-dependent evolution of river braiding. It is probably the most important achievement in braided river science over the last decade. A full review of this development was provided by Paola (2001). The model is based on: (i) strict enforcement of mass conservation of both water and sediment; coupled to (ii) a flow routing approximation (Eq. 1), (iii) a downstream sediment transport law which can have a non-linear dependence upon routed flow (Eq. 2), and (iv) a lateral sediment transport law (Eq. 3). The flow routing law is given as:

$$Q_{-ij} = \frac{Q_o s_{-ij}^k}{s_{-ij}^k + s_i^k + s_{ij}^k} \quad Q_i = \frac{Q_o s_i^k}{s_{-ij}^k + s_i^k + s_{ij}^k}$$

$$Q_{ij} = \frac{Q_o s_{ij}^k}{s_{-ij}^k + s_i^k + s_{ij}^k} \quad \text{for} \quad s_i > 0 \tag{1}$$

where Q_o is the flow discharge through cell o, Q_* are the flow discharges in the three grid cells downstream from cell o (i.e. two diagonal (* = $-ij$, * = $+ij$), one downstream (* = i)) and s_* are the corresponding slopes into those three cells. If $s_i^k \leq 0$, then $Q_i = 0$. If $s_* \leq 0$ for all *, then Q_o is routed analogously with Eq. 1 according to the least negative slopes. The sediment transport law is based on a stream power treatment where:

$$Q_{s*} = K_*[Q_*(s_* + C_s) - Th]^m \tag{2}$$

where Q_{s*} is the downstream sediment flux through from cell o to cell * (for * = $-ij, i, +ij$), K is a scaling constant, C_s is a constant used to allow sediment transport up negative slope gradients, Th is a measure of resistance to erosion, and m is an exponent that controls the degree of non-linearity in the relationship between sediment transport and excess stream power. The lateral (j) sediment transport was driven by lateral slope (s_j):

$$Q_{sj} = K_j s_j Q_{so} \qquad (3)$$

This approach has resulted in a number of fundamental changes to the way that we think about braided rivers. These have been well-reviewed (e.g. Paola, 2001). In summary, the model suggests that (i) braiding is a consequence of flow contraction and expansion coupled to (ii) a non-linear sediment transport law, which leads to (iii) sediment flux over-reacting to variation in flow strength and never managing to optimize the bed topography (Paola, 2001). Braiding appears to be the fundamental instability that arises when water moves over non-cohesive material in the absence of lateral confinement (Paola, 2001). These two points led Paola (2001) to suggest that the classic continuum of channel pattern (after Leopold & Wolman, 1957) be recast, with braided channels being the default channel pattern that a river reverts to in the absence of lateral confinement or channel resistance effects due to geomorphological (i.e. valley width), sedimentological (presence of cohesive sediment) or vegetation effects (root binding and sediment resistance to entrainment). Indeed, Murray & Paola (2003) were able to incorporate vegetation into the same model framework to assess this process.

Prior to a critical assessment of the model, it is important to emphasize that this is the first model to have produced the spatially distributed time–space feedbacks that are characteristic of river braiding and it does so with relatively low process complexity. The fundamental changes to the way we think about braided rivers almost certainly hold regardless of the criticisms made of the model. However, embarking upon a critical analysis is important as it helps to identify the future prospects for modelling the braiding process. This includes realization of the probable impossibility of a quantitative predictive tool (cf. Haff, 1996) that can predict the exact behaviour of a braided channel system through time. In turn, this has major implications for the requirements of field and laboratory data collection campaigns and the development of synthetic descriptors of braiding as the focus of alternative approaches to assessing model performance.

Braided river models 2: what makes a valid model?

The Murray & Paola (1994, 1997) model is contentious: it produces what looks like braided river behaviour but without much of the hydraulic complexity that has been thought to be necessary in previous attempts at modelling braided river processes (e.g. Lane & Richards, 1998; Nicholas & Sambrook Smith, 1999). Paola (2001) saw the debate about the Murray & Paola model as being about synthesist versus reductionist approaches to river braiding: is the model:

1 '... a creative new approach that bypasses much of the gritty detail that underpins traditional model schemes...' (Paola, 2001, p. 13)? or;
2 a model where '... decades of work aimed at building a stable foundation one brick at a time [are] discarded in favour of unsubstantiated rules invented and changed at the whim of the modeller ...' (Paola, 2001, p. 13)?

The synthesists would answer '1' and the reductionists '2'. However, as Paola (2001) emphasized, all models that might attempt to represent the braiding process, even those based upon explicit solution of the full form of the shallow water equations (see below), must necessarily involve simplifications or assumptions. Even when modelling only a very small component of a braided river model (e.g. three-dimensional flow processes in a confluence with a stable bed; Lane et al., 1999), major simplifications have to be made. If the 'human social dynamics' (Paola, 2001, p. 13) that lead to debate between 1 and 2 are to be addressed, it is necessary to consider what constitutes a valid model.

Validation is currently the subject of much debate in the environmental and earth sciences (e.g. see syntheses in Anderson & Bates; 2001; Wilcock & Iverson, 2003). Traditionally, validation has been seen as an empirical activity in which the theoretical underpinnings are subsumed (and numerical models are little more than computationally embedded theoretical statements) to observations and, more specifically, measurements. Braided river topography can now be measured rapidly and over a large spatial extent (see below) and hence repeat digital elevation models (DEMs) can be generated that allow the comparison of successive sets of model predictions through time. However, process simplification, boundary condition uncertainty (e.g.

distribution of inflow fluxes of water and sediment, spatial distribution of grain sizes at a range of depths) and the possible divergence of model predictions in highly non-linear systems (Haff, 1996) all mean that it is highly unlikely that braided river models will ever reproduce the exact detail of the braiding process. Indeed, to require them to do so might be unnecessary (i.e. reductionist) if they still capture the *essence* of the braiding process and allow us to make statements about aspects of braiding that are otherwise hidden in experimental or field-based observations. As Paola (2001, pp. 12–13) noted, '... the standard for evaluation of a high level model is how it performs in tests at the level for which it was designed ... it is not acceptable to judge a model according to how one feels about the method by which it is constructed'.

What should the nature of those tests be? There are two main approaches. The first recognizes that the model should reproduce those emergent properties of the braiding process that the study sets out to reproduce. Thus, the Murray & Paola approach has been assessed through exploration of geometric scaling properties (e.g. Murray & Paola, 1996; Doeschl *et al.*, this volume, pp. 177–197) as well as conventional properties that have been used to describe empirical aspects of braided river geometry (see Doeschl *et al.*, this volume, pp. 177–197). 'High level' validation assumes that if a particular process representation is solved in a numerical model and applied to representative initial conditions, and this reproduces observed high-level characteristics, then the model may have captured the essence of the braiding process at that level. For this to hold, there are a number of requirements.

1 It is necessary to demonstrate that the descriptors of the high-level characteristics can distinguish between erroneous formulations of the model and the ones that are used to infer properties from the model. In essence, the validation statistics must themselves be 'valid' (i.e. precise, accurate, sensitive) descriptors of the prototype system.
2 In the absence of other tests, an assumption of symmetry must be made with caution.

'High level' or 'reduced complexity' models allow us to understand properties of systems that are emergent at a particular level of enquiry. If they are validated at that level of enquiry, under the principle of symmetry, it may be argued that this means that the fundamental details of the model must also be correct. However, whilst the model may appear to reproduce high-level characteristics of the braiding process, and is valid in this sense, this does not mean that lower level characteristics of the process are properly represented, unless those lower level characteristics have also been properly assessed. Thus, a high-level model should be used with care when making statements about lower level attributes as this would mean overinterpretation. This is well-established in applications of computational fluid dynamics to natural river channels: Lane *et al.* (2005) noted that a model validated on streamwise velocity components will not necessarily predict non-streamwise velocity components accurately; similarly, a model validated on velocity components will not necessarily reproduce turbulence quantities adequately. In relation to their model, Murray & Paola (1997) compared model simulations with and without lateral sediment transport to show that lateral sediment transport is important for reproducing the richness of braided river behaviour. The precise form of the lateral sediment transport rule was simplified from a relationship in Parker (1984), and this was not validated by the model. However, the dependence of geometrical scaling properties upon some form of lateral transport allows the general statement of its importance for reproducing those properties to be made. Measuring actual lateral sediment transport might suggest that model estimates of lateral transport are imprecise, but this does not undermine the general conclusion that lateral transport matters at the high level. This also implies that internal 'validation' of the Murray & Paola model (e.g. comparison with field measurements of sediment transport and/or channel change) needs to be undertaken with caution, as the model was not designed to reproduce the detailed internal behaviour of the system being considered.

Unfortunately, one problem emerges with high-level validation: the possibility that the right model results (e.g. braiding) have been obtained for the wrong reasons. Thus, a second group of activities are sometimes required. Lane & Richards (2001) labelled these as 'post-positivist' in that they place less emphasis upon agreement with field or laboratory measured data because there will always be good reasons as to why a model does not reproduce reality. Rather, it uses a range of techniques to understand and to justify model formulation. The best analogy here is with a cake: a cake that had been mistakenly baked using salt might look just as appealing as one baked using sugar. Indeed, the salty cake would share some identical emergent properties with the sugary cake (it should have risen, for instance). There are two ways to respond to this problem: (i) to subject the cake to probing empirical testing (i.e. to evaluate it in a number of ways: visual, taste, smell, etc.), where those tests are carefully honed on the nature of model predictions; or (ii) to go back to the ingredients and to gain confidence about what the cake is, by discovering what has been put into it (i.e. by assessing the underlying concepts) and what is produced when the cake is put together in different ways. This second approach is 'post-positivist' in that it requires no further observation or measurement. Neither of the approaches are 'reductionist'. Rather, they are both part of the natural process of scientific enquiry in which given hypotheses (in this case, a set of hypotheses combined together into a numerical model) are subject to careful testing and development. 'Post-positivist' analysis will therefore involve a range of activities (see Lane & Richards, 2001), including assessment of the conceptual basis of the model, checks on mass balance, assessment of model sensitivity and visualization. The Murray & Paola model has been subject to many of these (Murray & Paola, 1997). However, Murray & Paola (1997) raised conceptual issues that need reinterpretation and this is undertaken in the next section.

Braided river models 3: a conceptual assessment

The philosophical approach to the Murray & Paola model was well-stated by Paola (2001, p. 13): 'Models of complex, spatially extended systems may become simpler because passage up the hierarchy of scales culls important processes...'. This 'culling' of processes is the route into a conceptual evaluation of the Murray & Paola approach. This culling can take one of two forms: (i) the removal of processes that are not relevant; and (ii) the simplification of processes to make them more amenable to numerical solution.

In relation to the flow routing, the main argument made is that the flow distribution in Eq. 1 is a crude representation of momentum conservation (Murray & Paola, 1997). This is used to justify the 0.5 exponent in Eq. 1, but not the discretization of Eq. 1 explicitly. Murray & Paola (1997) derived the justification from the equation of motion for uniform flow. Indeed, it is possible to show that a slope exponent of 0.5 (and the equation of motion for uniform flow) can be derived explicitly from simplification of the two-dimensional, depth-averaged form of the shallow water equations, ignoring dispersion, Coriolis, wind stress and turbulence terms, and using the Manning friction law to account for bottom stresses (see the Appendix; Bradbrook et al., 2004). However, the problem emerges in the discretization of Eq. A5 in the Murray & Paola model. Using the notation shown in Fig. 1c, and discretizing the slope as the line of steepest descent, Eq. A5 gives:

$$Q_i = \frac{d^{5/3}\left(\frac{h_{i,j} - h_{i\pm1,j}}{w}\right)}{n\left[\left(\frac{h_{i,j} - h_{i\pm1,j}}{w}\right)^2 + \left(\frac{h_{i,j} - h_{i,j\pm1}}{w}\right)^2\right]^{1/4}} \quad (4a)$$

$$Q_j = \frac{d^{5/3}\left(\frac{h_{i,j} - h_{i,j\pm1}}{w}\right)}{n\left[\left(\frac{h_{i,j} - h_{i\pm1,j}}{w}\right)^2 + \left(\frac{h_{i,j} - h_{i,j\pm1}}{w}\right)^2\right]^{1/4}} \quad (4b)$$

There is a substantial difference between the discretization described in Eq. 4a and Eq. 4b and that shown in Fig. 1a. First, the routing in Eq. 4a and b includes water surface elevation explicitly. This prevents the determination of unrealistically high flux values. Indeed, application of the basic Murray & Paola model to real braided river topography resulted in lateral water surface slopes equivalent to as much as 45°, as a result of the overconcentration of flux. Obtaining braiding in the Murray & Paola model required a non-linear sediment transport relationship (Murray & Paola, 1997), with an exponent greater than unity as well as this overconcentration of flux. Second, if it is assumed that $Q_1 = Q_3 = 0$ and $Q_j = 0$, then Eq. 1 will simply pass Q_o downstream whereas Eq. 4a will pass flux downstream in relation to flow depth and water surface slope. Third, the maximum flow divergence that comes from Eq. 1 is ±45° (Murray & Paola, 2003), and this will restrict the maximum anabranch sinuosity that the model can predict. The flow angle defined by Eq. 4a and b is given by:

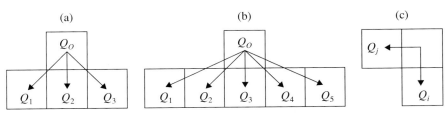

Fig. 1 Three flow routing discretizations: (a) the original Murray & Paola (1994) representation; (b) the Thomas & Nicholas (2002) representation; and (c) the representation associated with direct discretization of the simplified form of the depth-averaged shallow water equations.

$$\alpha = \tan^{-1}\left(\frac{s_j}{s_i}\right) \quad (5)$$

That given by the Murray & Paola model is:

$$\alpha_{MP} = \tan^{-1}\left(\frac{s_{ij}^{0.5} - s_{-ij}^{0.5}}{s_{-ij}^{0.5} + s_{ij}^{0.5} + \sqrt{2}s_i^{0.5}}\right) \quad (6)$$

Thus, the maximum flow deviation is achieved only when the slopes into both cell $-ij$ and cell i are either zero or negative. Finally, Eq. 4a can be simplified to the conventional Manning equation with s_i raised to the power of 0.5 if, and only if, $s_j = 0$. Thus, the justification of $k = 0.5$ in Murray & Paola (1997) must be treated with some caution.

This analysis is based upon comparison of the Murray & Paola discretization with a cellular discretization of the shallow water equations. Cellular discretizations can be problematic as they tend to provide only a poor representation of diffusion processes, something that has been well-recognized in hydrological routing studies (e.g. Tarboton, 1997). One alternative is a hexagonal mesh, with three possible transport pathways rather than a square mesh with only two. Thus, whilst the Murray & Paola discretization restricts the maximum flow divergence possible, it has advantages in allowing routing to more than two cells. However, for it to be correct, a hexagonal treatment is required and this would simultaneously allow much greater flow turning in response to topographic variation.

The sediment transport relations have a strong range of justifications. However, the same problems of discretization identified for the flow model also apply to the sediment transport relations. The Appendix shows the discretized form of Eq. 2, ignoring the C_s and Th terms, but applied to Eq. 4a and b. This leads to the following expressions for Q_{si} and Q_{sj}:

$$Q_{si} = K\left(\frac{d^{5/3}\left(\frac{h_{i,j} - h_{i\pm1,j}}{w}\right)^2}{n\left[\left(\frac{h_{i,j} - h_{i\pm1,j}}{w}\right)^2 + \left(\frac{h_{i,j} - h_{i,j\pm1}}{w}\right)^2\right]^{1/4}}\right)^m$$

$$= K(q_i \underline{\nabla}(h))^m \quad (7a)$$

$$Q_{sj} = K\left(\frac{d^{5/3}\left(\frac{h_{i,j} - h_{i,j\pm1}}{w}\right)^2}{n\left[\left(\frac{h_{i,j} - h_{i\pm1,j}}{w}\right)^2 + \left(\frac{h_{i,j} - h_{i,j\pm1}}{w}\right)^2\right]^{1/4}}\right)^m$$

$$= K(q_j \underline{\nabla}(h))^m \quad (7b)$$

If the directions of flow (α) and sediment transport (α_s) from Eqs 4 and 7 respectively are compared then, for $s_i > 0$:

$$\alpha = \tan^{-1}\left(\frac{s_j}{s_i}\right) \quad (8a)$$

and:

$$\alpha_s = \tan^{-1}\left(\left[\frac{s_j}{s_i}\right]^{2m}\right) \quad (8b)$$

This results in an interesting property of a cellular discretization: application of Eq. 2 to the cell unit discharge produces divergence between the flow and sediment transport vectors for all $m \neq 0.5$. For $m > 0.5$, the sediment transport vector deviates from the line of maximum topographic slope to a greater degree than the flow vector. For $m < 0.5$, the opposite occurs. Thus, in the above formulation, m is implicitly a control on the sensitivity of sediment transport routing to topographic forcing, as well as an explicit control on the non-linearity of the dependence of sediment transport upon discharge. At higher values of m, the magnitude of sediment transport will be greater but the sediment will be less readily steered by the bed topography. The presence of gravitational forcing of routed sediment will serve to counter this, and hence the m parameter can be seen as trading off topographic steering of the sediment due to gravitational effect (lower values of m) against inertial effects (higher values of m). The extent to which this occurs depends upon the sensitivity of sediment transport to flow as given by the m parameter, and it is logical that as the non-linearity increases, inertial effects should also increase.

If a similar analysis is applied to the Murray & Paola model (still ignoring the C_s and Th terms), then

$$\alpha_{\mathrm{MP}} = \tan^{-1}\left(\frac{s_{ij}^{0.5} - s_{-ij}^{0.5}}{s_{-ij}^{0.5} + s_{ij}^{0.5} + \sqrt{2} s_i^{0.5}} \right) \quad (9)$$

and

$$\alpha_{\mathrm{sMP}} = \tan^{-1}\left(\frac{K_i \sqrt{2}(s_{ij}^{1.5})^m - K_i \sqrt{2}(s_{-ij}^{1.5})^m + K_j s_j}{K_i[(s_i^{1.5})^m + \sqrt{2}(s_{-ij}^{1.5})^m + \sqrt{2}(s_{ij}^{1.5})^m]} \right) \quad (10)$$

Comparison of Eqs 9 and 10 shows that one of the odd elements of the process treatment that comes from the combination of Eqs 2 and 3 is the possibility that sediment can diverge more rapidly than flow. For instance, if $s_{-ij} \leq 0$ and $s_i \leq 0$, Eq. 10 becomes:

$$\alpha_{\mathrm{sMP}} = \tan^{-1}\left(1 + \frac{K_j s_j}{K_i \sqrt{2}(s_{ij}^{1.5})^m} \right) \quad (11)$$

Provided $s_j > 0$, this is unconditionally greater than 45°, and sediment will diverge more than flow. The extent to which this is the case depends on K_i, K_j and m and emphasizes that, as with the cellular discretization above, parameters that control the magnitude of sediment transport also control the magnitude of flow–sediment divergence. The tendency for sediment to diverge more than flow is countered by the C_s term in Eq. 2, which allows uphill sediment transport in the presence of negative slopes. Murray & Paola (1997) noted that this can be set as a product of upstream (i.e. inherited) stream power, which represents the inertial effects that will result in uphill transport. Murray & Paola observed that neither explicit specification of C_s nor a stream power based determination is especially well-constrained. Even with specification, there remains the possibility that sediment diverges more than flow, depending upon the magnitude of s_j as compared with the inherited inertia, as the flow routing is restricted in the extent to which it can be forced laterally. The issue here is not the principle of a sediment transport law that is driven by lateral slope. Rather, it is the way in which the law is discretized in relation to the flow relationship that may be problematic. Ironically, the discretization of the simplified shallow water equations in Eq. 4 coupled to the sediment routing rule shown in Eq. 7 produces a much simpler modelling framework for braided rivers. Not only does it have fewer adjustable parameters, but it has parameters that have implicit process representation. However, whether or not such a formulation produces braiding and the nature of the braiding that is produced are issues that merit further research.

The above points have one implicit theme: the discretization of continuous equations onto a numerical grid (i.e. into finite difference form). To understand the nature of the discretization, it is necessary to consider the nature of the model coupling, as well as the way in which time and space are resolved in the model. The nature of the coupling implied in Eq. 2 is such that the sediment flux to row i is dependent upon the flow flux to the cells in row i, not a simultaneous solution of the flow and transport equations (i.e. the model is loosely coupled rather than full embedded). In other words, the model is solved for flow to produce a new flow field, and then solved for sediment flux to produce a new topography, and then solved for flow, etc. The main issue here is that this type of iterative solution has a tendency to be highly diffusive and will be strongly dependent upon mesh resolution and model time-step (i.e. spatial and temporal discretization).

The time-step treatment used in the Murray & Paola model involves an explicit enforcement of mass conservation: all flow is moved into the downstream row of cells during each model iteration (Murray & Paola, 1997). Murray & Paola (1997) showed that the duration of each iteration is determined by K in Eq. 2, and this is not surprising as K determines the sediment transport volume for a given stream power and m combination: larger values of K increase the sediment transport volume and hence should simultaneously be associated with larger time-steps. Strictly speaking, for a given mesh resolution, there should be some form of time-step control based on the Courant number (i.e. the model time-step should scale with mesh resolution), such that material cannot pass through a mesh cell in a time-step without being influenced by the properties of that cell. If the model is independent of its discretization, as the mesh resolution (and time-step) are fined further, the magnitudes of change in both flow and topography should scale with the fining. This fining will also lead to a closer approximation of a tightly coupled model (i.e. the feedback between flow, sediment transport and topography becomes stronger). As the time-step control is implicit in K, it ought to be possible to obtain the same model behaviour on different mesh resolutions through changing the value of K.

Unfortunately, changing mesh resolution not only changes model discretization but also the resolution of topography (i.e. the discretization of slopes) (Doeschl-Wilson & Ashmore, 2005), which can have a major effect as a result of the non-linearity of some of the slope terms. Indeed, Doeschl-Wilson & Ashmore (2005) found that the nature of braiding, including whether or not it is observed at all, appears to depend upon mesh resolution, with only the coarser mesh resolutions appearing to mimic the behaviour observed in the original Murray & Paola model. It follows that there are two views of the Murray & Paola model. The extreme interpretation is that the model has artificial numerical diffusion associated with a coarse discretization of a loosely coupled model. The coarse resolutions, with correspondingly long time-steps, will reduce the precision of estimates of change at any one instance. As there is no correction of these estimates, as might be expected in a higher order scheme, there is the possibility of propagation of error, or numerical diffusion. The more probable interpretation is that there are parameter sets, given the processes represented in the model, that yield just the right process dynamical behaviour to produce the exaggerated response of sediment flow to water flux (Paola, 2001) that seems to be the essence of braiding. This is the 'high level' aspect of model performance.

The detail of the model, notably its discretization, can be easily criticized, but it may not matter for reproducing simplified but generic braided model behaviour. Greater confidence in this conclusion might be gained if the process representation had a stronger conceptual grounding. As it is, the acceptability of the model rests upon the extent to which the model validation based upon scaling properties is deemed to be sufficient. Doeschl et al. (this volume, pp. 177–197) present alternative forms of validation of the Murray & Paola (1997) scheme that show that the model does not reproduce real braided river topography. However, by starting with a conceptual assessment of the model, it is clear that there is no reason why the model should reproduce the detail of braided river behaviour at this lower level of enquiry as the basic responses of flow and sediment to topographic forcing are not correctly represented.

Braided river models 4: increased complexity approaches

One way of approaching the Murray & Paola model is to recognize that if the model is correct it ought to be able to reproduce the full range of river behaviour—braiding, meandering and straight—with the exact type of channel and its dynamics, dependent upon the initial conditions applied to the model. Murray & Paola (2003) viewed this as an additional question, as the algorithm does not allow flow to deviate by more than 45° from the downstream direction and because curvature-inducing effects develop over downstream length-scales that are commonly many channel widths. This immediately implies a practical problem for any increase in process complexity: the spatial resolution and model time-step required for more complex process models, as well as the increase in computational complexity, will necessarily restrict possible spatial scales of enquiry. One way of viewing a braided river is as a filter of upstream influences (i.e. flow, sediment supply). As the length-scale of the simulation is increased, so the sensitivity of model development to the imposed initial conditions should be reduced. This provides a further challenge for modelling reaches of braided rivers that are shorter, where initially imposed conditions may have a dramatic effect upon model predictions.

The question then arises as to whether or not a fuller process description is capable of reproducing aspects of river braiding. Initial modelling of flow processes in a braiding river by Lane et al. (1995a) and Lane & Richards (1998) was not encouraging, for although it was possible to obtain a reasonable level of agreement between model predictions and field observations (Lane et al., 1995a) and to show that model behaviour in response to parameter perturbation was robust (Lane & Richards, 1998), it was very difficult to obtain reliable estimates of the bed shear stresses necessary to estimate sediment entrainment, transport and deposition (Lane et al., 1999). Indeed, the response to this was somewhat tangential to the goal of enhancing understanding of river braiding at the system-scale, in that it involved application and development of three-dimensional computational fluid dynamics, initially for river confluences (Lane et al., 1999; Bradbrook et al., 2000) and eventually at even smaller spatial scales in relation to flow around clusters of gravel grains (Lane & Hardy, 2002; Lane et al., 2004). Although there were other motivations for these research activities, this seems to be a classic reductionist approach (in this case in terms of process representation as well as spatial scale) to only one challenge in modelling the braiding process. However, McArdell & Faeh (2001) have demonstrated that it is possible to reproduce some aspects, if not the richness, of the river braiding process by coupling a full solution of the depth-averaged sediment transport equations to a sediment entrainment and transport model, including sediment sorting.

The starting point for the McArdell & Faeh (2001) approach is the depth-averaged conservation equations for flow and sediment mass and flow momentum. By comparison with Eq. A1, the flow component of the model reintroduces the acceleration terms. It also separates out turbulent shear stresses from the friction law and, in theory, allows for specification of any one of a number of friction laws. As compared with other depth-averaged models of flow in curved channels (e.g. Lane & Richards, 1998), it does not represent explicitly the dispersion terms that arise during depth-averaging and these are, essentially, the secondary circulation terms. If dispersion terms are not treated, depth-averaged models will tend to constrain the largest velocities to lie near the inside of channel bends (Bernard & Schneider, 1992) instead of predicting gradual velocity migration to the outside, as observed in the bends of curved channels (Dietrich & Whiting, 1989). Indeed, whilst there have been measurements of secondary circulation in braided river channels (e.g. Ashmore et al., 1992; Bridge & Gabel, 1992), there has been no definitive demonstration of the extent to which it affects characteristics of the river braiding process. The sediment transport model is based upon the two-dimensional form of the Exner equation, with an explicit representation of individual size fractions. McArdell & Faeh (2001) used empirical relationships for both sediment transport rate (based upon the Meyer–Peter Müller relationship) and bank collapse.

One of the very real difficulties of modelling river channels is when there is a change in the spatial extent of the inundated area. This will not be an issue in the original formulations of the reduced complexity approaches described above, as there is no determination of water surface elevation, and a cell will automatically become dry when there is a negative slope into it but positive slopes into its neighbours. However, with a water surface elevation treatment, water depths can become vanishingly small during both wetting and drying and this can lead to severe computational instabilities. This will be compounded by erosion and deposition in a mobile bed model, both in terms of bed-level-change effects and also bank erosion effects. There are now some clever methods for dealing with these types of problems, although they have been restricted to single channel flows (e.g. Nagata et al., 2000; Duan et al., 2001) and are unlikely to be suited to braided rivers. McArdell & Faeh (2001) used a cellular grid and a simple method for managing the wet–dry and dry–wet transition: in modelling a river with a mean flow depth of 0.820 m, they forced a cell to become dry for depths <0.0010 m and allowed it to become wet for depths >0.0011 m. This does raise issues of mass conservation, but the lost depths of flow are likely to be so small that they will have a negligible impact upon the final solution.

McArdell & Faeh (2001) tested their model on laboratory experimental data in which they compared the predicted evolution of the system to that measured by Fujita (1989). The results were reasonably successful: the model predicted a similar reduction in bar mode through time, although the mode value was slightly higher for the numerical simulation than was observed by Fujita. Whilst the initial relief was much greater in the laboratory experiments (different by an order of magnitude), at the end of the run, both the model and the laboratory data had similar relief. McArdell & Faeh (2001) were able to conclude that the model reproduced basic braiding processes, including bar emergence, upstream bar migration, confluence scour and evolution of braiding intensity (as given by bar mode) through time. It was less successful at reproducing the large dry areas often found in natural braided river channels, which the reduced complexity approaches have been more successful at reproducing. However, as with the reduced complexity approaches described above, there remains the issue of how to validate these types of models. As this is a model that aims to reproduce higher level behaviour from a lower level process representation it would seem logical to expect that some validation of the lower level representation is required. Yet, the system-scale measurements of either flow or sediment transport in shallow braided river channels that might form the basis of such an analysis remain unavailable.

Braided river models 5: models in comparative perspective

Sufficient uncertainties remain surrounding existing modelling approaches to warrant considerable research effort into system-scale models of the braiding process. This probably needs a two-pronged attack. Further development of the Murray & Paola approach (e.g. Thomas & Nicholas, 2002) has already been shown to have merit. However, approaching the same problem from a model with a more sophisticated process representation will be valuable, as it can provide a base for exploring how model predictions change as processes are progressively simplified. Choosing between different realizations of the same model, and different models, rests upon a reliable validation framework. Accepting the argument that prediction of the detail of the braiding process is futile as a result of the challenges of boundary condition determination and system divergence in the presence of strong non-linearity, it follows that reliable system-scale generalizations of braided rivers will be required in order to generate the validation data for model assessment. The next two sections approach this issue from two different perspectives. The first explores the significant progress in spatial scaling analyses that have provided a new and exciting way of describing braided river behaviour through explicit consideration of scale dependence. The second addresses developments in the measurement of braided river morphology, and morphological change, that provide new data for the population of these descriptions.

THE SCALING CHARACTERISTICS OF BRAIDED RIVER SYSTEMS

Approach and findings

Paola & Foufoula-Georgiou (2001) noted that one of the fundamental characteristics of braided rivers is their visual similarity, both as the same system is viewed over different spatial scales and when different systems, even with different perimeter sedimentologies and slopes, are compared. This has resulted in the consideration of braided rivers as an example of geometrical, statistical self-similarity (where scaling is isotropic) and self-affinity (where scaling is anisotropic or, in essence, it depends upon direction). Consider a region X by Y. If the system shows anisotropic spatial scaling (Foufoula-Georgiou & Sapozhnikov, 1998) then if the region were rescaled from i to j according to Eq. 12 there should be no statistical difference between the original and the rescaled region (Sapozhnikov & Foufoula-Georgiou, 1996).

$$\left(\frac{X_i}{X_j}\right)^{1/v_x} = \left(\frac{Y_i}{Y_j}\right)^{1/v_y} \quad (12)$$

In terms of spatial scaling, the majority of research to date (e.g. Sapozhnikov & Foufoula-Georgiou, 1996) has focused upon the readily derived channel pattern revealed from aerial photography and satellite imagery. This has shown that braided channel geometry is self-affine over a restricted range of spatial scales, with a ratio of v_x to v_y of 1.4 to 1.5 indicating the degree of anisotropy, and similar values of v_x and v_y when three different rivers were compared. The anisotropy is not surprising, as braided rivers have a preferred downstream direction (Paola & Foufoula-Georgiou,

2001), but these findings suggest an element of universality in the braiding process (Sapozhnikov & Foufoula-Georgiou, 1996) that mirrors the success with which the Murray & Paola (1994) model was able to produce braiding for the first time. The presence of spatial scaling of this kind implies that braided rivers may be self-organized. This appears to be found in a range of rivers provided that they are not influenced by a change in external controls (Paola & Foufoula-Georgiou, 2001), such as channel steering by a constrained valley, or a progressive change in channel slope. It appears (Nykanen *et al.*, 1998) that braiding can be viewed as a general case of a self-organizing system, but one that may be perturbed by the geological, the geomorphological and probably the ecological setting through which it passes. This has opened up a much more developed spatial and temporal consideration of the dynamics of braided rivers which, when taken with the Murray & Paola (1994) modelling approach, is confirming that what looks like an extremely complex (statically and dynamically) system has some simple underlying physics which create some fundamental emergent properties.

Sapozhnikov & Foufoula-Georgiou (1997) and Foufoula-Georgiou & Sapozhnikov (1998) extended consideration to dynamic scaling by using video imagery acquired from a laboratory study of braiding, allowing for possible self-affinity:

$$\left(\frac{X_i}{X_j}\right)^{z_x} = \left(\frac{Y_i}{Y_j}\right)^{z_y} = \left(\frac{t_i}{t_j}\right) \qquad (13)$$

and $z_x = 0.32$ and $z_y = 0.35$ (Foufoula-Georgiou & Sapozhnikov, 1998). Sapozhnikov & Foufoula-Georgiou (1997) concluded that values of z close to 2.0 would imply a diffusive process (i.e. one driven by the diffusion of relatively small changes across space), and a value close to zero would indicate a system where small area dynamics are driven by the behaviour of larger areas. Whether or not their area-based analysis extends to individual channels (or other units such as the confluence–diffluence unit) is an interesting question, and one that has yet to be resolved (Paola, personal communication, 2005). As the dynamic scaling parameters are closer to 0, this implies that the evolution of smaller areas is forced by the evolution of larger areas (Foufoula-Georgiou & Sapozhnikov, 1998).

Interestingly, Foufoula-Georgiou & Sapozhnikov (1998) found that the ratio of v_x to v_y was around 1.1, implying a greater degree of self-similarity in the laboratory case, but recognizing that different methods were being used to determine the scaling parameters. The question remains as to whether or not these differences reflect laboratory versus real river differences, methodological differences, or substantive differences. For instance, the spatial scaling methods reflect a snap shot in time, and are not indicative of the full range of the braid plain width that may be occupied over long time periods. It would be logical to expect the degree of self-similarity, as well as the degree to which spatial scaling is found (Nykanen *et al.*, 1998), to depend upon the extent to which braid plain width is laterally constrained. Whilst the Foufoula-Georgiou & Sapozhnikov (1998) experiment used a laboratory flume, and hence was laterally constrained, the level of scaling is such that it was much less constrained than some of the field examples considered. For instance, Nykanen *et al.* (1998) considered flow rates of between 3 and 6×10^4 cfs (800–1500 m^3 s^{-1}), which is about six orders of magnitude greater than the 20 g s^{-1} (0.0010 m^3 s^{-1}) used in the flume studies of Foufoula-Georgiou & Sapozhnikov (1998). Applying this scaling to the 0.75 m flume width used by Foufoula-Georgiou & Sapozhnikov, the effective braid plain width in the flume is around 750 km, significantly less constrained than in the field case of Nykanen *et al.* (1998) where self-affinity was clear.

The future role of spatial scaling analyses

Spatial scaling work is important in that it ties consideration of braiding river behaviour into much wider issues regarding dynamics in systems. The fact that the system appears to be 'self-organizing' in the absence of exogenic influences suggests that braiding is a generic process that manifests itself in the kind of conditions identified by the Murray & Paola (1994) model. Spatial scaling analyses, in combination with traditional braiding indices, will be useful in assessing the next generation of system-scale braided river models, especially in relation to dynamic scaling behaviour. However, their use may go much further than this. Of particular interest is whether or not there is spatial scaling in morphological change, as opposed to planform inundation, or even braided river morphology.

For a braided river in an unconfined state, consider a region of X by Y. As the area of this region increases then the probability of encountering an elevation change greater than some critical value (z_c) should increase. It might be expected that this probability increases more rapidly in the cross-stream direction than in the downstream direction: running water has a preferred direction (controlled by valley slope) and, even with the excess response of sediment to flow forcing, this appears to be central to the braiding process (Paola, 2001). This leads to significant morphological complexity in the downstream direction, but the down-valley slope direction is likely to be relatively smoother, and this is reflected in the anisotropic spatial scaling reviewed above. Thus, statistically on average, spatial variation in elevation change in the downstream direction should be less than in the cross-stream direction. As the maximum depths of erosion and deposition will be partly determined by the minimum and maximum elevations in an area (i.e. more scoured areas of a river are more fillable; less scoured areas are more erodible), and because elevation variability is greater in the cross-stream direction, the probability of greater elevation changes occurring as Y is increased should be greater than larger changes occurring as X is increased:

$$n_z(X_i, Y_i, z > z_c) = n_z(X_j, Y_j, z > z_c)\left(\frac{X_i}{X_j}\right)^{1/v_x} \\ = n_z(X_j, Y_j, z > z_c)\left(\frac{Y_i}{Y_j}\right)^{1/v_y} \quad (14)$$

Thus, it would be expected that:

$$\frac{1}{v_y} > \frac{1}{v_x} \quad (15)$$

and hence anisotropy in the scaling with:

$$v_x > v_y \quad (16)$$

This is the same result as that found for the static analysis of channel pattern by Sapozhnikov & Foufoula-Georgiou (1996). This considers the result for elevation changes where $z > z_c$. It can be extended to the case of all z_c. If the area of change increased, it might also be expected that the maximum elevation change might increase and that this might also show some sort of characteristic scaling:

$$\left(\frac{X_i}{X_j}\right)^{1/c_x} = \left(\frac{Y_i}{Y_j}\right)^{1/c_y} = \left(\frac{z_i}{z_j}\right) \quad (17)$$

i.e.:

$$\left(\frac{z_i}{X_i^{1/c_x}}\right) = \text{constant and} \left(\frac{z_i}{Y_i^{1/c_y}}\right) = \text{constant} \quad (18)$$

Note that there is a direct analogy here with the dynamic scaling of Foufoula-Georgiou & Sapozhnikov (1998). They considered that the scale at which an area is considered (i.e. the size of X by Y) would be related to the rate at which that area changed (slower for larger scale). In Eq. 17 it is argued that the magnitude of change expected should scale with the area considered (bigger change with larger scale). If such a relationship does exist, it is of potential importance because it would allow a statistical determination of the range of depths of change expected associated with much more readily determined information on planform extent of change.

The interesting question that then follows from the above is the role that should be played by time in the analysis. The dynamic scaling analysis of Sapozhnikov & Foufoula-Georgiou (1997) and Foufoula-Georgiou & Sapozhnikov (1998) implies that there should be a time-scale dependence. This is not considered in the framework for morphological change outlined above. One of the striking differences in the DEMs of difference presented in Lane et al. (2003) is that by the time an annual change is considered there is a breakdown in the coherence of erosion and deposition patterns, as a result of multiple superimposed braiding events. As explained below, this research has identified a number of time-scales of channel change (Hicks et al., 2002), ranging from continual braiding to flood event changes. It follows that the coherence of the erosion and deposition patterns may vary spatially as a function of the time-scale being considered, and this may break down the detectability of the kind of scaling relationships hypothesized above.

At present, this analysis is cast in terms of scaling in relation to a changing region of interest. To exploit a possible relationship between the area of change and the magnitude of change, it is necessary to move to individual areas of erosion and deposition of different spatial extents. Again, analogy can be made with the dynamic scaling treatment of Foufoula-Georgiou & Sapozhnikov (1998) to do this. However, before it can be done, DEMs of braided rivers are needed that can be used to determine distributed data on patterns of erosion and deposition, and that have a temporal resolution sufficient for patterns of erosion and deposition to remain coherent.

BRAIDED RIVER MORPHOLOGY AND MORPHOLOGICAL CHANGE

The analyses of scaling in braided river systems have been almost entirely based upon estimations of inundation as an indication of channel pattern. The stage dependence of channel pattern has been widely recognized. Similarly, whilst it is unlikely that braided river models will ever produce the exact detail of channel morphology and morphological change, a more robust approach to model validation would involve exploring the distribution of elevations and elevation changes predicted when the model is initialized on real braided river topography, as well as the extent to which the model reproduces established styles of channel change (e.g. Ashmore, 1991; Ferguson, 1993). This is now possible.

The third major development in system-scale analysis has been in the approach to measurement of braided river morphology, which in turn is causing new questions to be asked about system-scale braiding processes. Up until the late 1990s, the prime focus of field measurement of braided river morphology was cross-section survey. In general, this involved narrow, valley-confined braid bar systems (Ashworth & Ferguson, 1986; Goff & Ashmore, 1994), although there are examples of long-term survey of very wide braided rivers (greater than 1 km width) using cross-sections (e.g. Carson & Griffiths, 1989), normally where braided river management issues required operational data collection. These analyses were supported by quantification of braided river planform measured using various forms of aerial survey (e.g. Laronne & Duncan, 1992).

The above situation began to change in the 1990s with the application of oblique terrestrial analytical photogrammetry, coupled with ground survey of inundated areas, in order to obtain three-dimensional surface models (e.g. Lane et al., 1994). The focus remained relatively small-scale, but was increasingly founded upon representation of the continuous variation of topography using DEMs. However, the most dramatic developments have come with: (i) automated digital photogrammetry, primarily using airborne platforms (Westaway et al., 2000, 2001, 2003; Brasington et al., 2003; Lane et al., 2003), but also using terrestrial oblique imagery (Pittaluga et al., 2003; Chandler et al., 2004); (ii) laser altimetry (Charlton et al., 2003; Lane et al., 2003); and (iii) global positioning systems (GPS, Brasington et al., 2000). The result of these technologies is visualization of braided river topography at a resolution and scale that has never been seen before (Hicks et al., 2002; Fig. 2).

This section identifies some of the issues associated with these emerging technologies and identifies ways in which system-scale measurement of braided rivers might be used to address some of the research issues being raised by both the numerical modelling and the scaling analyses described above.

A brief introduction to the main techniques

Only digital photogrammetry and laser altimetry are capable of generating truly system-scale data. Digital photogrammetry is based upon well-established photogrammetric principles (see Lane et al. (1993) for review) in which stereo-images, generated by sensors of known geometry (including sensor focal length and principal point position), position and orientation, are subject to automated data extraction to identify the image positions of the same data point. Reliance upon automation means that significant levels of post-processing may be required in order to identify erroneous data points associated with incorrect matching (Lane, 2000; Lane et al., 2004). Laser altimetry (also known as LIght Detection And Range, or LIDAR) is a more recent technique in which an airborne (or now spaceborne) laser scanner is positioned and oriented in real-time using GPS technologies. As the laser scans, each scanning time (the time for

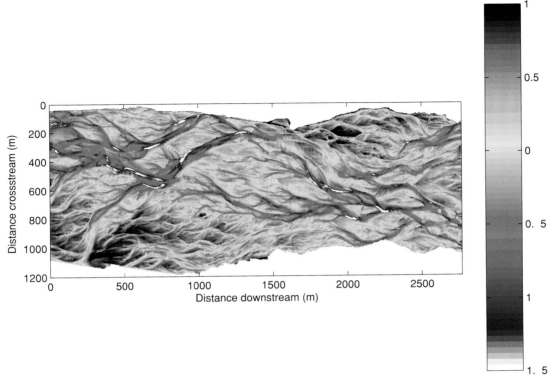

Fig. 2 Detrended digital elevation model of the Waimakariri River, New Zealand. Scale bar gives deviation from mean elevation in metres.

an emitted pulse to reach the ground surface and be reflected back to the sensor) can be converted to a distance to ground surface. From knowledge of the beam orientation in combination with the GPS data it is possible to determine ground elevations. As with photogrammetry, post-processing routines are required to remove erroneous data points.

Issues surrounding their application

System-scale measurement of braided river morphology represents a serious methodological challenge for a number of reasons:

1 the spatial extent of the field of interest is large;
2 the rates of change of elevation across space can be locally high (e.g. at channel margins);
3 the system relief is low, commonly a very small fraction of sensor heights;
4 there can be substantial areas of inundation that can hide regions of the river that may be of major interest;
5 the braidplain may be vegetated to a varying degree, obscuring the surface morphology that is of interest;
6 braiding appears to occur at all flow stages (Hicks et al., 2002), such that capturing change may require repeat data collection at a reasonably high frequency (possibly even daily).

These characteristics create a number of problems. First, the spatial extent of the system-scale means that remote sensing remains the only feasible means of measuring continuous elevation variability in large braided rivers (e.g. with widths of greater than 0.5 km over lengths of greater than 1 km). Remote sensing involves a fundamental trade-off between the scale of survey and the resolution of data that can be achieved. Coverage of larger areas normally requires a higher flying height (whether the imagery is spaceborne or airborne). A higher flying height will degrade possible data resolution and, in the case of photogrammetry in particular, degrade data quality.

This is a particular problem, as the relief is low; flow and sediment transport processes may be driven by quite subtle changes in topography that can be readily swamped by noise in the data. In all applications, either lowering flying height or repeating flight lines can be used to densify the surface and, in the case of photogrammetry, may be crucial for introducing sufficient texture into the imagery in order to assist the stereo-matching process, especially for low-relief bar surfaces where the grain size is in the gravel or sand size range (Westaway et al., 2003). Lowering flying height raises the costs (money and time) of the survey: (i) greater levels of ground control may be required if the platform does not have onboard GPS; (ii) the time to acquire imagery is longer; and (iii) more data analysis and post-processing will be required. It follows that resolving this trade-off is not straightforward.

The second problem is what to do with inundated areas. These are commonly the regions of most active change and hence most scientific interest. The problem can be reduced by constraining data collection to the lower flow periods, either seasonally or in relation to storm events. This does not necessarily yield data when it is required, and most rivers retain some inundation even at low flows. Two solutions have been adopted. Both require image data. This is automatically generated as part of photogrammetric data collection. However, for altimetric applications, it is necessary to undertake simultaneous laser scanning and image collection. The first solution applies to situations where the water is sufficiently clear for the bed to be visible. Provided the flow depth is less than the optical depth (which it commonly is in shallow rivers) and the flow is subcritical (so the water surface is not 'broken'), then stereo-matching may successfully estimate subaqueous points (e.g. Westaway et al., 2000), and these can be subject to a two-media refraction correction (Westaway et al., 2001). If the water is turbid, the bed textural signature will commonly be lost. However, there may still be a depth signature. In theory, this could be theoretically derived from knowledge of the bed substrate colour and composition combined with sun angle and information on atmospheric characteristics.

In practice, sufficient knowledge of the required boundary conditions for physically based modelling is rarely available and so semi-empirical approaches have been adopted (e.g. Westaway et al., 2003; after Lyzenga (1981) and Winterbottom & Gilvear (1997)). These require calibration data to establish a relationship between measured water depths and the spectral signature found in the imagery at the location of each depth at the time of data acquisition. These relationships are assumed to hold through space, despite possible variations in bottom reflectance. They will rarely hold through time due to variation in factors such as suspended sediment concentration, water depth and sun angle. Both the clear-water and the turbidity-based approaches involve a set of image processing with estimation of water surface elevations and water depths in order to estimate subaqueous point elevations. The methods are detailed in full for the clear-water case in Westaway et al. (2000) and the turbid-water case in Westaway et al. (2003).

The third problem is how to identify and to correct possible errors. The DEM shown in Fig. 2 was collected to a 1 m resolution, covering an area of 1.2 km by 3.5 km. This involves 4,200,000 data points. Individual point correction, using stereo-vision for instance, is simply not possible with this many data points. Indeed, prior to using the DEMs derived using the digital photogrammetric methods, it was necessary to embark upon a lengthy process of data handling and management in order to determine and to improve the associated data quality (Lane et al., 2003). The quality of DEM data is all too often overlooked (Cooper, 1998; Lane, 2000). Thus, the methods developed had a number of characteristics.

1 It was necessary to identify possibly erroneous data points. Braided rivers are topographically complex, but this complexity commonly involves relatively smooth surfaces separated by breaklines within which slope is locally much greater. This provides basic a priori estimates of local elevation variance (essentially defined by an estimate of local grain size) and allows for development of locally-intelligent point based filters. Removal of those data points associated with these errors results in a significant improvement in the quality of surface representation (Lane et al., 2004). The presence of vegetation in DEM data is also a form of error, and simple filtering routines based on image processing have been used to identify and to remove vegetated locations (Westaway et al., 2003).

2 Following their removal it was necessary to replace removed data points. Fortunately, the richness of the

Table 1 Data quality of digital elevation models based upon comparison of independent check data (from Lane et al., 2003)

Digital elevation model	Theoretical precision (m)	Check data precision (m)	
		Dry bed	Wet bed
Photogrammetry: February 1999	±0.070	±0.261	±0.318
Photogrammetry: March 1999	±0.070	±0.257	±0.256
Photogrammetry: February 2000	±0.056	±0.131	±0.219
Laser altimetry: May 2000	n/a	±0.100	±0.250

associated datasets is such that removing many hundreds of data points does not reduce data quality significantly (Lane et al., 2004). However, in the vicinity of the locally more rapid topographic change (i.e. close to river banks), it has been found that more intelligent interpolation of removed data results in better surface representation (Lane et al., 2004).

3 Even after identification and removal of possibly erroneous points, residual error remained. This imparts uncertainty into estimates of properties from the DEMs, including both slope and change between subsequent DEMs. Thus, Lane et al. (2003) used propagation of error techniques to develop a minimum level of detection (e.g. Brasington et al., 2003) based upon a statistical analysis.

Table 1 shows the error associated with data obtained for each of the four DEMs collected by Lane et al. (2003). In this case, the wet-bed data were obtained using image processing based upon spectral signatures. Two points follow: (i) at least for the photogrammetry, the theoretical data precision, as defined by sensor geometry, position and location, is significantly better than the actual precision; and (ii) the dry-bed precision is better than the wet-bed precision. Degradation from theoretical precision is largely linked to the stereo-matching process in general and a significant loss of surface texture when 1:5000 scale imagery was used (in 1999). For February 2000, data collection was redesigned to use 1:4000 scale imagery, and this resulted in a substantial improvement in matching success and hence in surface precision, approaching that of the laser altimetry survey. The magnitude of the wet-bed precision emphasizes that the spatial extent of wet-bed inundation needs to be minimized in applications of this type. In this case study, this meant that data collection was restricted to either side of a flood event, and collection of within-flood data was not attempted.

The real interest to the braided rivers scientist is not the magnitude of surface errors but the way in which they propagate through into estimates of patterns of erosion and deposition and also into volumes of surface change. Figure 3 used the statistical basis of the method to weight point estimates of erosion and deposition by the probability that they were reliable (see Lane et al. (2003) for a full explanation of the method). Table 2 propagates point uncertainties into system-scale estimates of volumes of erosion and deposition. Two points arise (Lane et al., 2003).

1 Measured changes are generally sound. The dry–wet zones are all associated with net erosion and wet–dry zones are associated with net deposition (Table 2). Similarly, the spatially averaged change, with the averaging undertaken with respect to the area of a particular type of change, shows that the average change for dry–wet and wet–dry areas is greater than for both dry–dry and wet–wet areas (Table 2). Again, this is as expected.

2 The volumetric uncertainties in Table 2 demonstrate the extent to which these survey methods yield highly reliable volume estimates. This may appear somewhat surprising given the values of precision in Table 1. However, only a proportion of the elevation changes are close to the threshold of detection and, as this analysis assumes that the error in change does not scale with the magnitude of change, the reach-scale volumes of change are high as compared with their estimated uncertainty. Also as expected,

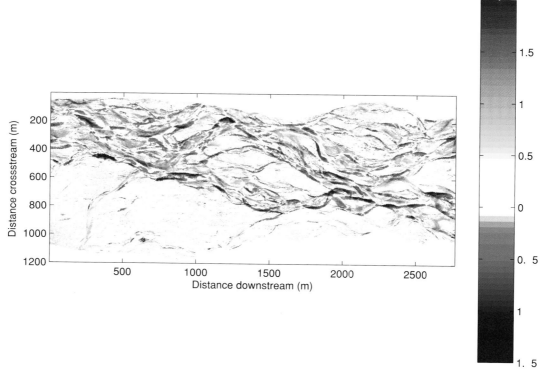

Fig. 3 An example digital elevation model of difference (based upon change between February 1999 and February 2000 on the Waimakariri River, New Zealand) where changes are weighted according to their statistical reliability using propagation of error techniques (Lane et al., 2003).

the relative uncertainty in net volumes of change is greater than in either cuts or fills alone. This is because the magnitudes of net change are smaller than cuts or fills and a much larger number of cells contribute to the total volumetric uncertainty. The relative precision of estimates of volume change is good in relation to the changes observed, and certainly enough to suggest that volume of change estimates derived from DEMs may be used for process studies at the system scale.

Potential applications of these new data

The richness of these datasets is confirmed when the information generated is compared with traditional methods of morphological measurement. One of the means of doing this is to take the full topography and to evaluate the morphological estimates that would have been made using more traditional methods based upon repeat survey of cross-sections, such as the Waimakariri River case case study of Lane et al. (1994, 2003). In this application, cross-sections were extracted from each DEM at a given spacing, to calculate a mean bed level for that DEM for that cross-section spacing. The minimum cross-section spacing was defined by the DEM point spacing (i.e. 1 m). The maximum cross-section spacing was set as half of the DEM of difference downstream length: 1507 m for the Waimakariri River. For each cross-section spacing, bed-level change was calculated for a window equal in length to the maximum cross-section spacing. The procedure was repeated multiple times for each cross-section spacing in order to quantify the effects of sampling error. The number of repeats possible increases as the cross-section spacing is reduced. Thus, 150 repetitions were used (set as around ten times less than the Waimakariri maximum cross-section spacing).

Figure 4 shows the mean and standard deviation of relative change in mean bed level plotted against cross-section spacing for the Waimakariri River. Relative values are based upon comparing the estimated change in mean bed level with

Table 2 Results of propagating the errors shown in Table 1 into volume of change estimates (from Lane et al., 2003). Data are shown for three periods (W1, W2 and W3), for both the full area and for individual zones classified into a contingency table based upon start and end, and inundated and not inundated

Zone	Digital elevation model of difference	Cut Area (m²)	Cut Volume (m³)	Cut Zone-averaged (m³ m⁻²)	Fill Area (m²)	Fill Volume (m³)	Fill Zone-averaged (m³ m⁻²)	Net change Volume (m³)	Net change Zone-averaged (m³ m⁻²)
Dry–dry	W1 (Feb 99–Mar 99)	575,081	56,724 ±278	0.027 ±0.0005	1,513,401	262,081 ±451	0.125 ±0.0003	+205,356 ±529	+0.098 ±0.0004
	W2 (Feb 99–Feb 00)	752,717	189,902 ±253	0.095 ±0.0003	1,232,778	304,062 ±324	0.152 ±0.0003	+114,160 ±412	+0.057 ±0.0003
	W3 (Feb 00–May 00)	786,090	113,406 ±146	0.060 ±0.0002	1,069,760	185,136 ±171	0.099 ±0.0002	+71,731 ±225	+0.038 ±0.0002
Dry–wet	W1 (Feb 99–Mar 99)	288,890	110,291 ±197	0.304 ±0.0007	73,812	12,110 ±99	0.033 ±0.0014	−98,181 ±220	−0.271 ±0.0030
	W2 (Feb 99–Feb 00)	412,345	240,161 ±219	0.537 ±0.0005	34,730	6549 ±64	0.015 ±0.0018	−233,612 ±228	−0.523 ±0.0066
	W3 (Feb 00–May 00)	372,987	192,234 ±172	0.444 ±0.0005	59,287	11,129 ±69	0.026 ±0.0012	−181,105 ±186	−0.418 ±0.0031
Wet–dry	W1 (Feb 99–Mar 99)	11,974	1769 ±45	0.012 ±0.0037	131,749	59,271 ±148	0.412 ±0.0011	+57,502 ±155	+0.400 ±0.0012
	W2 (Feb 99–Feb 00)	24,391	3672 ±54	0.012 ±0.0022	278,252	159,785 ±181	0.525 ±0.0007	+156,113 ±189	+0.513 ±0.0007
	W3 (Feb 00–May 00)	11,848	1797 ±26	0.006 ±0.0022	287,823	170,473 ±129	0.569 ±0.0004	+168,676 ±132	+0.563 ±0.0005
Wet–wet	W1 (Feb 99–Mar 99)	151,087	42,357 ±151	0.133 ±0.0011	166,868	48,009 ±174	0.151 ±0.0010	+5652 ±230	+0.018 ±0.0013
	W2 (Feb 99–Feb 00)	77,134	24,885 ±107	0.158 ±0.0014	80,026	25,889 ±109	0.165 ±0.0014	+1004 ±153	+0.006 ±0.0019
	W3 (Feb 00–May 00)	114,401	27,866 ±112	0.091 ±0.0010	190,184	63,391 ±145	0.208 ±0.0008	+35,525 ±184	+0.117 ±0.0010
All areas	W1 (Feb 99–Mar 99)	1,012,352	211,141 ±375	0.209 ±0.0004	1,900,277	381,471 ±515	0.201 ±0.0003	+170,330 ±637	+0.058 ±0.0002
	W2 (Feb 99–Feb 00)	1,266,587	458,620 ±356	0.362 ±0.0003	1,625,786	496,285 ±392	0.305 ±0.0002	+37,665 ±530	+0.013 ±0.0002
	W3 (Feb 00–May 00)	1,285,326	335,303 ±254	0.261 ±0.0002	1,607,054	430,129 ±267	0.268 ±0.0002	+94,826 ±369	+0.033 ±0.0001

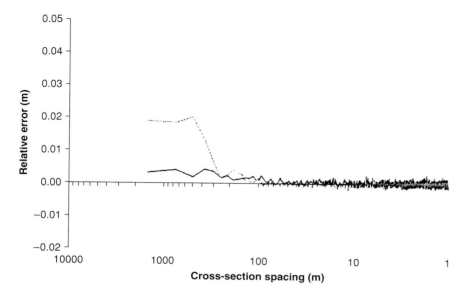

Fig. 4 Plots of mean (black) and standard deviation (grey) of the relative error in mean bed-level estimates against cross-section spacing for the digital elevation model of difference of February to March 1999 (from Lane et al., 2003).

respect to estimates with a 1.0 m cross-section spacing. This shows how the error generally increases as the cross-section spacing is increased. The pattern is consistent in all three comparison DEMs: mean errors are low below 100 m section spacings, increasingly rapidly thereafter. Most notable is increasing sensitivity to the choice of cross-section location at spacings of 100 m or more. Thus, for the typical cross-section spacing used for long-term river monitoring (500–800 m; Griffiths, 1993), significant differences are found between the detected pattern of bed-level change and that estimated from the DEM of difference. In this case, the required section spacing is of the same order of magnitude as the low-flow braidplain width, and much narrower than the active channel or braidplain width (cf. Ashmore & Church, 1998).

Similar results were obtained when the analysis was repeated for channel-change estimates. Table 3 shows the effects upon volume of change estimates relative to a 1.0 m cross-section spacing. This mirrors the patterns shown in Fig. 4. Up to cross-section spacings of around 100 m, the percentage error in volume change is generally less than 10%, except for W1. Higher relative errors might be expected for W1 given the relatively small change in reach-aggregated volume change during this period, and this reflects the important point that the ease of change detection in any surveying system increases with the magnitude of change. At cross-section spacings greater than 100 m, there is a marked increase in the volume error in all cases, but especially for the W2 case. Section spacings greater than 500 m are probably useless for estimating erosion and deposition volumes.

However, scaling cross-section spacing by width is something that needs to be approached with caution. Scaling by width implies that larger rivers contain the same topographical variability as smaller rivers, but over length-scales that are a function of river width. There are two problems with this.

1 The scaling appears to be self-affine: it scales at different rates in the cross-stream and downstream directions. Thus, scaling a river by its width will not reduce downstream derived estimates (e.g. topographical variability) in the same way for rivers of different size.
2 The scaling analysis shows that the scaling is over reduced length scales, and this also implies that there is a restricted scale range over which a width-based scaling might work.

There are also good reasons for questioning this in terms of what is known about sediment transfer in these braided rivers, especially where they comprise multiple interacting sinuous channels (see Fig. 3), which tends to be from outer bank to downstream point bar (Carson & Griffiths, 1989), and which will require a much closer cross-section spacing where there are interacting individual anabranches than is implied by scaling by total active width.

Given the much improved surface representation, what are the potential applications of these data?

Table 3 The effect of cross-section spacing upon the error in volume estimates expressed with reference to the 1.0 m estimate, from Lane et al. (2003). Data are shown for three periods (W1, W2 and W3) for the full area

Approximate downstream spacing (m)	W1 February 1999–March 1999		W2 February 1999–February 2000		W3 February 2000–May 2000	
	Reach-aggregated volume change (m³)	Error in volume as a % of 1.0 m volume change	Reach-aggregated volume change (m³)	Error in volume as a % of 1.0 m volume change	Reach-aggregated volume change (m³)	Error in volume as a % of 1.0 m volume change
1	+9710		+25,638	–	–37,583	–
2	+9090	–6.3	+26,066	1.6	–37,787	–0.5
5	+11,907	–22.6	+25,192	–1.7	–33,126	+11.9
10	+7129	–26.5	+28,997	+13.0	–37,845	–0.7
50	+9536	–1.7	+21,501	–16.1	–36,185	+3.7
100	+10,022	+3.0	+23,074	+10.0	–34,670	+7.8
200	+7691	–20.8	+17,888	–30.2	–30,649	+18.4
500	+17,364	+78.8	+45,196	+76.3	–25,609	+31.9
750	–6934	–171.4	–8332	–132.5	–28,173	+25.0
1500	+12,236	+26.0	–10,751	–141.9	–3991	+89.3

First, they allow the first visualization of the morphology and morphological change of a braided river at the system-scale. Figure 2 shows that there is dramatically more topography in this braided river than is suggested by an analysis of inundated area alone. The low-flow inundated areas take the form of an active braiding belt, and time-lapse video imagery (Hicks et al., 2002) suggests that braiding within this zone is near-continuous, confirming flume observations (e.g. Ashmore, 1982) that flow variability is not a necessary requirement of the braiding process. However, the low-flow braid belt as revealed by patterns of inundation sites is found within a much richer active braidplain topography (Fig. 2). The visual characteristics of this non-inundated topography appear to be different to those in the low-flow braid belt: (i) a network of channels that, in places, appears to be dendritic and is superimposed upon (ii) larger point bars between which the low-flow braid belt actively meanders. There is at least some evidence that migration of the low-flow braid belt leads to eventual wash out of these larger scale point bars, and their eventual realignment and reformation in different locations.

The volume of change data confirm the observation of Griffiths (1979) that this section of the Waimakariri River is generally aggradational. It is unlikely that this additional topography is simply relict topography, formed as the low-flow braid belt migrates around the active braidplain, leading to this aggradation, because (i) it is visually different to the low-flow braid belt topography and (ii) it is superimposed upon higher elevation topography, which is inundated for only short periods of time during relatively high flows. Murray & Paola (1997) found that the reduced complexity model resulted in dendritic network development if deposition was not allowed, and it is interesting to find a dendritic network developing in parts of a braided river that comprise non-cohesive sediment. These networks imply that the point bars are predominantly erosional, temporary (modified by both the surface erosion and lateral migration of the active braiding belt), and possibly related to storm-related inundation of flow but not sediment from upstream (i.e. the water can get onto them, but the sediment cannot). It leads to the kind of dissection of high points observed by Carson & Griffiths (1987) in the Waimakariri River.

It is also noteworthy (e.g. Fig. 3) that there is at least some sense in which the erosion and deposition patterns record the progressive migration of narrow and sinuous meandering streams. As noted above, this was hypothesized to be the case on the basis of planform inundation patterns (Carson & Griffiths, 1989), and the DEM of difference confirms the extent to which this is the case. It has also been seen in laboratory flume experiments: braiding involves a history of channel migration more than simultaneously active channels (Ashmore, 2001). This has some important implications. First, analysis of the scaling properties of the morphological data shown in Fig. 2 would be interesting, but would involve simultaneous analysis of a topographic legacy created over time rather than at a point in time. The emphasis on planform inundation of the scaling analysis reported above may well be preferable to the use of morphological data. Second, as Bridge (1993) emphasized in his development of quantitative models of braid bar stratigraphy, the flow and sediment transport processes in a braided river are no different to those found in straight or meandering rivers, or confluences in dendritic drainage networks. What does differ is the time-scale over which change happens, with much greater process rates, and hence more dynamic system behaviour. As Paola (2001) discussed, elements of the braiding process may slow, or even be lost, in rivers where either the sedimentology (cohesive material) or ecology (bank stabilization) allow and it is this that creates straight or meandering patterns.

Given the two scales of morphology measured in the river, there is an emerging issue in relation to flow variability. Much has been made of the fact that flow variability is not a necessary precursor for the braiding process. These data emphasize that there are secondary questions to be answered: for example, how do braided systems with pronounced flow variability differ from those expected in the absence of flow variability? If such differences exist, how does the nature (e.g. rate of change, falling limb form) of the stage change impact on the differences? Both laboratory scaled models and field situations have braiding, but is the braiding the same? Initial inspection of the Waimakariri River DEMs (Hicks et al., 2002) suggested that there is a range of scales of morphological response:

1 braiding, as is conventionally produced in laboratory scaled models, appears to occur at all flows, regardless of magnitude, but is confined to a weakly meandering braid bar belt, the location of which is fixed in the short term by the location of larger scales of active braidplain morphology;

2 this braid belt, at its margins, is responsible for the longer term modification of these larger scales of braidplain morphology, through both erosional and depositional processes;

3 larger storm-related events appear to accelerate these modifications by providing the conditions necessary for a wider range of system responses, including very rapid migration of individual braid bar anabranches and major avulsions.

There is no evidence to suggest that, at least in a statistical sense if not in quantitative detail, these more rapid and larger-scale changes would not occur through the progressive and continual operation of braiding processes under a steady flow regime. However, the flow variability may be important in increasing the rate of larger-scale system change and in maintaining the active braidplain width, especially through reducing the rate at which colonization by vegetation occurs. Whether or not the superimposition of braiding and flood events leads only to such acceleration, or whether it changes fundamentally the quantitative distribution of mechanisms (as opposed to the mechanisms themselves) of channel change, has yet to be established. This issue may need to be addressed as it has implications for the interpretation of the sedimentary record in particular.

In relation to depositional processes and the sedimentary record, an assessment of the datasets generated suggests that deposition is no more difficult to detect than erosion. This is not surprising as a substantial part of the depositional signal must be the filling of locations of scour (pools in the outer parts of meander bends, confluences) in order to maintain the braiding process. Qualitative observations from continuously measuring video cameras (Hicks et al., 2002) have confirmed that within the periods delimited by the DEMs change was restricted to an active braiding belt, with relatively high areas inundated during extreme flow events, but experiencing relatively little topographic change. Deposition was observed to occur at diffluence bars and particularly as gravel lobes stalled, either due to local avulsion processes or simply on flood recessions. The lobes were much more sheet-like: wider and thinner and sometimes only 20 cm or so thick. In theory, this is close to the associated levels of detection (Lane et al., 2003). In practice, the detectability of these lobes will depend upon the duration between surveys and the extent to which multiple lobe deposition provides a detectable depositional signature. However, substantial in-channel deposition was also observed, associated with the falling limb of storm hydrographs, when channel-confined lobes stalled within individual anabranches. These may be a major part of the depositional process in this type of environment.

The generated DEMs are suitable for both evaluating and developing the numerical modelling strategies described above. Doeschl-Wilson & Ashmore (2005) and Doeschl et al. (this volume, pp. 177–197) presented an evaluation of the Murray & Paola (1994) modelling approach based upon topographic characteristics of the simulated braidplain morphology. Their evaluation was not based upon a detailed comparison of the time-dependent evolution of modelled and measured topographies. As noted above, this is unlikely to be a useful activity because of system divergence in response to detail within the model simulation. However, reflecting the higher (or simplified) level of process representation there are a range of topographic attributes that can be extracted from both DEMs and DEMs of difference that might provide a useful foundation for validation. Doeschl-Wilson & Ashmore (2005) and Doeschl et al. (this volume, pp. 177–197) use basic measures of relief. To these it might be possible to add: (i) a wider range of measures of relief (e.g. distributions of slope, planform convergence and divergence); (ii) summative statistics of channel change (e.g. maximum scour and fill, skewness and kurtosis in the distribution), although this will require careful scaling of time-scales of the model with the field or laboratory prototype as the distributions of change will depend upon the duration between DEMs; and (iii) summative characteristics of different styles of channel change, and their relative importance, to assess which occur within the model and at what frequency. In a similar vein, it will be interesting to apply the DEMs and DEMs of difference to the scaling analyses described above. This will allow a test of the extent to which analysis of braiding behaviour based upon inundated area produces similar scaling results to that obtained from consideration of the full channel topography. It may also allow a linkage between areas of change and magnitude of change to be developed, which in turn allows for morphological estimation of transport rates from information on planform inundation.

The above data will allow the performance of other means of understanding braided river behaviour to be assessed. This will include determination of the range of braided river types that are represented by laboratory, Froude-scaled, braided river models. The key issue here is grain-size scaling, and preliminary work using the same methods reported above, but applied to a sand-bedded braided river (the South Saskatchewan), confirms that both the bed morphology and the nature of channel change appear to be very different to those observed in the Waimakariri. This is not surprising given results from quantitative assessments of river channel pattern. Robertson-Rintoul & Richards (1993) confirmed that differences in braided river sinuosity within the braided river class are controlled by similar variables to those that discriminate between meandering and braided rivers: a broadly hydraulic control (essentially stream power) and perimeter sedimentology explained 74% of the variability in sinuosity. Hydraulic controls dominated over sedimentological controls in the sand-bed case, but are of similar importance in the gravel-bed case (Robertson-Rintoul & Richards, 1993). Again, these kinds of differences are not due to differences in braiding mechanisms, rather in the intensity of sediment transfer processes, which in turn create differences in channel pattern.

Finally, there may be a level of analysis that falls between interpretation of the DEMs and DEMs of difference and system-scale modelling approaches, in which coupling of flow (and sediment) routing algorithms to the DEMs, and interpretation of the results in relation to channel change, will allow some of the basic form–flow–change interactions that characterize the braiding process to be understood. This may allow a system-scale estimation of coarse sediment transport rates in braided rivers to be obtained (e.g. Lane et al., 1995b; Martin & Church, 1996; Ashmore & Church, 1998; McLean & Church, 1999), where direct measurement of transport rate can be difficult (Carson & Griffiths, 1989). It should also allow quantification of the visual interpretations of much of the behaviour of systems such as the Waimakariri. Thus, the contribution that these new DEMs will make is likely to divide itself into:

1 the generation of new questions and new datasets to answer those questions, perhaps stimulating new modelling, field or laboratory analyses;
2 confirming the qualitative or visual understanding that has come from both scaled laboratory analyses and field interpretations.

As an example, one of the striking aspects of the collected DEM data is the extent to which pools and scour at confluences fill (i.e. they are ephemeral). With notable exceptions in relation to sediment transport at confluences (Mosley, 1976; Best, 1987, 1988), observations of styles of confluence sedimentation in scaled braided rivers (Ashmore, 1993) and interpretations of sedimentary fill (Siegenthaler & Huggenberger, 1993; Beres et al., 1999; Lunt & Bridge, 2004; Lunt et al., 2004), relatively little is known about the situations that lead to confluence fill. This is a good example where coupling morphological monitoring of filling confluences with ground-penetrating radar surveys of those confluences after they have filled could lead to a much enhanced understanding of confluence dynamics.

Technological developments

It is probable that the next 10 yr will see technological developments that will cement remote sensing as *the* technology that allows us to obtain system-scale measurements of gravel-bed rivers. For instance, LIDAR systems are now capable of delivering similar precisions to those obtained for the dry areas of the Waimakariri River reported above (see Table 1), but simultaneously below and above the water surface (e.g. Guenther et al., 2000). They have yet to be applied to braided river systems, but they should significantly improve the quality of the data that are obtained in underwater zones. Similarly, Carbonneau et al. (2004, 2005) report on the use of optical imagery for grain-size estimation for a clear water stream, both under water and on exposed bars. They were able to generate highly precise estimates of mean grain size on a 1 m resolution for an 80 km long reach of gravel-bed river. These new technological developments will undoubtedly expand the range of questions that can be explored in relation to system-scale braided river behaviour.

CONCLUSIONS: PERSPECTIVES FOR THE NEXT 10 YEARS

The past 10 yr has seen a step change in our ability to understand braided rivers at the system scale. This has been achieved through a series of thought-provoking developments in numerical modelling, as well as the application of scaling analyses and a range of remote sensing techniques. The modelling has provided a fundamentally different view of the nature of braided rivers: they are the default river channel state, and it takes geomorphological, geological and/or vegetation influences to cause deviation from this state. In other words, rivers in the absence of these controls are fundamentally unstable. The scaling analyses are important because they confirm that a unifying description of the braiding process is possible, regardless of the system being studied. In turn, both the modelling and scaling approaches rely on new ways of measuring braided rivers at the system scale, and this is being delivered by the rapid development of remote sensing technology. When taken together, the revolution in braided river understanding that followed from the laboratory-based studies of the 1980s (see Ashmore (2001) for a review) has now been extended through system-scale analyses, and our view of braided rivers today is fundamentally different to that of 10 yr ago.

These methods will be refined (e.g. through a developing understanding of the role of bifurcations in controlling flow and sediment transport partitioning) and applied in exciting and novel ways such as through extension to interactions with ecological processes, which may be important in explaining differences in channel pattern between different braided river systems (e.g. Murray & Paola, 2003). Over the next 10 yr, research will need to progress through two related directions. On the one hand, there remains no widely accepted methodology for modelling the braiding process, nor even agreement as to whether or not a single unifying methodology can be developed that is suitable for all possible scales or levels of enquiry. The debate between 'synthesists' and 'reductionists' is still there, but if different models are suitable for different levels/scales, and one of the distinguishing features of those different models is their level of process complexity, then the middle ground between synthesism and reductionism will require acceptance. As Paola (2001) argued, the physics in those models should be sufficient for the target level of enquiry. What is probably central to this is: (i) demonstration that the models are valid, in the broadest sense; and (ii) interpretation of the field datasets, coupled to more traditionally reductionist modelling approaches, to understand what the sufficient level of physics should be, where and when. On the other hand, the much more exciting aspect of this research will involve seeing just where these models and measurements can take us in terms of braided river understanding. There is already emerging progress in terms of ecological linkage. Similar progress may be possible in terms of modelling the spatially distributed evolution of the record of sedimentary fill: it is now possible to map grain-size parameters to a 1.0 m resolution from airborne digital imagery (e.g. Carbonneau et al., 2004). When coupled with repeat data on morphological change, we will not only have the kinds of data that can be used to generalize validation information for modelling studies, but also the data needed to understand the relationship between different scales of behaviour in braided river systems.

ACKNOWLEDGEMENTS

This research was made possible by the involvement of the author in a study that was partially funded by the New Zealand Foundation for Research, Science and Technology under contracts CO1818 and CO1X0014 awarded to Dr D.M. Hicks (NIWA, Christchurch, New Zealand), a NERC studentship held by Dr R.M. Westaway, and a Royal Society grant awarded to SNL. Discussions with Murray Hicks and Richard Westaway were particularly important for the ideas in this paper.

A very large number of people supported the NIWA research, including Maurice Duncan (NIWA), B. Fraser, D. Pettigrew and W. Mecchia (Environment Canterbury), Air Logistics, Auckland, AAM Geodan, Brisbane, and G. Chisholm (Trimble Navigation N.Z. Ltd.). Peter Ashmore, Jim Best, Greg Sambrook Smith and Chris Paola provided very valuable reviews of an earlier draft of this paper.

APPENDIX

Simplification of the depth-averaged shallow water equations to a diffusion wave equation (from Bradbrook et al., 2004) and coupling to sediment transport

River and floodplain flows are governed by gravitational, inertial and frictional forces. The effects of gravity are that the flow at any point will tend to be in the direction of the steepest water surface slope. Frictional effects are proportionally smaller for deeper flows where inertial effects become more important. Well known hydraulic equations such as Manning's equation and the St Venant equations (and various derivatives thereof, such as the diffusive wave equation) represent some or all of these factors. Consider the depth-averaged form of the Navier–Stokes conservation equation for momentum with a Manning type friction law to represent the associated friction sink term:

$$\frac{DV}{Dt} + (V \cdot \nabla)V + g\nabla(z_o + d) + \frac{n^2 g V |V|}{d^{4/3}} = 0 \quad (A1)$$

where V is the depth-averaged velocity vector; t is time; z_o is the bed elevation; d is the flow depth; g is the gravity constant; and n is Manning's n. The first step is to make a diffusion wave approximation by ignoring the acceleration terms: the first two terms on the left of Eq. 4. This is assuming that the temporal acceleration within a time step is negligible and that the spatial accelerations or inertial terms are also negligible. In other words, the flow is being driven by the balance between water surface slope and bottom resistance. If $z_o + d = h$, where d is water depth, and dividing through by g, Eq. A1 becomes:

$$V|V| = -\frac{d^{4/3}}{n^2}\mathbf{V}(h) \tag{A2}$$

Equation A2 can be rearranged to solve for the modulus of V (i.e. velocity magnitude) through:

$$|V| = -\frac{d^{2/3}}{n}|\mathbf{V}(h)|^{1/2} \tag{A3}$$

Substitution of Eq. A3 into Eq. A2 and rearranging gives:

$$V = \frac{d^{2/3}}{n}\frac{\mathbf{V}(h)}{|\mathbf{V}(h)|^{1/2}} \tag{A4}$$

Given that $Q = wdV$, where Q is vector discharge and w is the grid spacing:

$$Q = \frac{wd^{5/3}}{n}\frac{\mathbf{V}(h)}{|\mathbf{V}(h)|^{1/2}} \tag{A5}$$

Taking the absolute value of Eq. A5 and dividing through by w gives Q per unit width:

$$Q = \frac{d^{5/3}}{n}\frac{(s_x^2 + s_y^2)^{1/2}}{(s_x^2 + s_y^2)^{1/4}} = \frac{d^{5/3}s^{1/2}}{n} \tag{A6}$$

where s is the absolute value of the vector slope.

Taking the following form of the sediment transport relation:

$$Q_s = K(Q_s)^m \tag{A7}$$

and then combining with Eq. A6 gives:

$$Q_s = K\left(\frac{d^{5/3}s^{3/2}}{n}\right)^m \tag{A8}$$

which when discretized into vector form becomes:

$$\mathbf{Q_s} = K\left(\frac{d^{5/3}}{n}\frac{\mathbf{V}(h)^2}{|\mathbf{V}(h)|^{1/2}}\right)^m \tag{A9}$$

REFERENCES

Anderson, M.G. and Bates, P.D. (2001) *Model Validation in the Hydrological Sciences*, Wiley, Chichester, 500 pp.

Ashmore, P.E. (1982) Laboratory modelling of gravel braided stream morphology. *Earth Surf. Process. Landf.*, **7**, 201–225.

Ashmore, P.E. (1991) How do gravel bed streams braid? *Can. J. Earth Sci.*, **28**, 326–41.

Ashmore, P.E. (1993) Anabranch confluence kinetics and sedimentation processes in gravel-bed streams. In: *Braided Rivers* (Eds J.L. Best and C.S. Bristow), pp. 129–46. Special Publication 75, Geological Society Publishing House, Bath.

Ashmore, P.E. (2001) Braiding phenomena: statics and kinetics. In: *Gravel Bed Rivers V* (Ed. M.P. Mosley), pp. 95–114. New Zealand Hydrological Society, Wellington.

Ashmore, P.E. and Church, M.A. (1998) Sediment transport and river morphology: a paradigm for study. In: *Gravel Bed Rivers in the Environment* (Eds P.C. Klingeman, R.L. Beschta, P.D. Komar and J.B. Bradley), pp. 115–148. Water Resources Publications, Highlands Ranch, Colorado.

Ashmore, P.E., Ferguson, R.I., Prestegaard, K.L., Ashworth, P.J. and Paola, C. (1992) Secondary flow in anabranch confluences of a braided gravel-bed stream. *Earth Surf. Process. Landf.*, **17**, 299–311.

Ashworth, P.J. and Ferguson, R.I. (1986) Interrelationships of channel processes, changes and sediments in a proglacial braided river. *Geogr. Ann.*, **68**(4), 361–371.

Beres, M., Huggenberger, P., Green, A.G. and Horstmeyer, H. (1999) Using two- and three-dimensional georadar methods to characterize glaciofluvial architecture. *Sediment. Geol.*, **129**, 1–24.

Bernard, R.S. and Schneider, M.L. (1992) *Depth-averaged Numerical Modelling for Curved Channels*. Technical Report HL-92-9, US Army Corps Engineers, Waterways Experiment Research Station, Vicksburg, Mississippi.

Best, J.L. (1987) Flow dynamics at river channel confluences: implications for sediment transport and river morphology. In: *Recent Developments in Fluvial Sedimentology* (Eds F.G. Ethridge, R.M. Flores and M.D. Harvey), pp. 27–35. Special Publication 39, Society of Economic Paleontologists and Mineralogists, Tulsa, OK.

Best, J.L. (1988) Sediment transport and bed morphology at river channel confluences. *Sedimentology*, **35**, 481–98.

Bradbrook, K.F., Lane, S.N. and Richards, K.S. (2000) Numerical simulation of time-averaged flow structure at river channel confluences. *Water Resour. Res.*, **36**, 2731–46.

Bradbrook, K.F., Lane, S.N., Waller, S.G. and Bates, P.D. (2004) Two dimensional diffusion wave modelling of flood inundation using a simplified channel representation. *Int. J. River Basin Manage.*, **3**, 1–13.

Brasington, J., Rumsby, B.T. and McVey, R.A. (2000) Monitoring and modelling morphological change in a braided gravel-bed river using high resolution GPS-based survey. *Earth Surf. Process. Landf.*, **25**, 973–990.

Brasington, J., Langham, J. and Rumsby, B. (2003) Methodological sensitivity of morphometric estimates of coarse fluvial sediment transport. *Geomorphology*, **53**, 299–316.

Bridge, J.S. (1993) The interaction between channel geometry, water flow, sediment transport and deposition in braided rivers. In: *Braided Rivers* (Eds J.L. Best and C.S. Bristow), pp. 13–71. Special Publication 75, Geological Society Publishing House, Bath.

Bridge, J.S. and Gabel, S.L. (1992) Flow and sediment dynamics in a low sinuosity braided river: Calamus River, Nebraska Sandhills. *Sedimentology*, **39**, 125–42.

Carbonneau, P.E., Lane, S.N. and Bergeron, N.E. (2004) Catchment-scale mapping of surface grain size in gravel-bed rivers using airborne digital imagery. *Water Resour. Res.*, **40**, W07202 JUL 22 2004.

Carbonneau, P.E., Bergeron, N.E. and Lane, S.N. (2005) Automated grain size measurements from airborne remote sensing for long profile measurements of fluvial grain sizes. *Water Resour. Res.*, **41**, W11426, doi:ID.1029/2005WR003994.

Carson, M.A. and Griffiths, G.A. (1987) Bedload transport in gravel channels. *N. Z. J. Hydrol.*, **26**, 1–151.

Carson, M.A. and Griffiths, G.A. (1989) Gravel transport in the braided Waimakariri River: mechanisms, measurements and predictions. *J. Hydrol.*, **109**, 201–20.

Chandler, J., Ashmore, P., Paola, C., Gooch, M. and Varkaris, F. (2004) Monitoring river-channel change using terrestrial oblique digital imagery and automated digital photogrammetry. *Ann. Assoc. Am. Geogr.*, **92**, 631–44.

Charlton, M.E., Large, A.R.G. and Fuller, I.C. (2003) Application of airborne LiDAR in river environments: the River Coquet, Northumberland, UK. *Earth Surf. Process. Landf.*, **28**, 299–306.

Cooper, M.A.R. (1998) Datums, coordinates and differences'. In: *Landform Monitoring, Modelling and Analysis* (Eds S.N. Lane, K.S. Richards and J.H. Chandler), pp. 21–36. Wiley, Chichester.

Dietrich, W.E. and Whiting, P.J. (1989) Boundary shear stress and sediment transport in river meanders of sand and gravel. In: *River Meandering* (Eds S. Ikeda and G. Parker), pp. 1–50. Water Resources Monograph 12, American Geophysical Union, Washington, DC.

Doeschl, A.B., Ashmore, P.E. and Davison, M., et al. (2006) Methods for assessing exploratory computational models of braided rivers. In: *Braided Rivers: Process, Deposits, Ecology and Management* (Eds G.H. Sambrook Smith, J.L. Best, C.S. Bristow and G.E. Petts), pp. 177–197. Special Publication 36, International Association of Sedimentologists. Blackwell, Oxford.

Doeschl-Wilson, A.B. and Ashmore, P.E. (2005) Assessing a numerical cellular braided-stream model with a physical model. *Earth Surf. Process. Landf.*, **30**, 519–40.

Duan, J.G., Wang, S.S.Y. and Jia, Y. (2001) The application of the enhanced CCHE2D model to study the alluvial channel migration process. *J. Hydraul. Res.*, **39**, 469–480.

Ferguson, R.I. (1993) Understanding braiding processes in gravel-bed rivers: Progress and unresolved problems. In: *Braided Rivers* (Eds J.L. Best and C.S. Bristow), pp. 13–71. Special Publication 75, Geological Society Publishing House, Bath.

Ferguson, R.I., Ashmore, P.E., Ashworth, P.J., Paola, C. and Prestegaard, K.L. (1992) Measurements in a braided river chute and lobe: I. Flow pattern, sediment transport and channel change. *Water Resour. Res.*, **28**, 1877–1886.

Foufoula-Georgiou, E. and Sapozhnikov, V.B. (1998) Anisotropic scaling in braided rivers: an integrated theoretical framework and results from application to an experimental river. *Water Resour. Res.*, **34**, 863–7.

Fujita, Y. (1989) Bar and channel formation in braided streams. In: *River Meandering* (Eds S. Ikeda and G. Parker), pp. 417–462. Water Resources Monograph 12, American Geophysical Union, Washington, DC.

Goff, J.R. and Ashmore, P.E. (1994) Gravel transport and morphological change in braided Sunwapta River, Alberta, Canada. *Earth Surf. Process. Landf.*, **19**, 195–212.

Griffiths, G.A. (1979) Recent sedimentation history of the Waimakariri River, New Zealand. *N. Z. J. Hydrol.*, **18**, 6–28.

Griffiths, G.A. (1993) Sediment translation waves in braided gravel-bed rivers. *J. Hydraul. Eng.*, **119**, 924–937.

Guenther, G.C., Brooks, M.W., LaRocque, P.E. (2000) New capabilities of the 'SHOALS' airborne lidar bathymeter. *Remote Sensing of the Environment*, **73**, 247–55.

Haff, P.K. (1996) Limitations on predictive modelling in geomorphology. In: *The Scientific Nature of Geomorphology* (Eds B.L. Rhoads, and C.E. Thorn), pp. 337–358. Wiley, Chichester.

Hicks, D.M., Duncan, M.J., Walsh, J.M., Westaway, R.M. and Lane, S.N. (2002) New views of the morphodynamics of large braided rivers from high-resolution topographic surveys and time-lapse video. In: *The Structure, Function and Management Implications of Fluvial Sedimentary Systems* (Eds F.J. Dyer, M.C. Thomas and J.M. Olley). Publication 276, International Association of Hydrological Sciences, Wallingford.

Lane, S.N. (2000) The measurement of river channel morphology using digital photogrammetry. *Photogramm. Rec.*, **16**, 937–957.

Lane, S.N. and Hardy, R.J. (2002) Porous rivers: a new way of conceptualising and modelling river and floodplain flows? In: *Transport Phenomena in Porous Media*, Vol. 2 (Eds D.B. Ingham and I. Pop), pp. 425–449. Pergamon, Oxford.

Lane, S.N. and Richards, K.S. (1998) Two-dimensional modelling of flow processes in a multi-thread channel. *Hydrol. Process.*, **12**, 1279–98.

Lane, S.N. and Richards, K.S. (2001) The 'validation' of hydrodynamic models: some critical perspectives. In: *Model Validation for Hydrological and Hydraulic Research* (Eds P.D. Bates and M.G. Anderson), pp. 413–38. Wiley, Chichester.

Lane, S.N., Richards, K.S. and Chandler, J.H. (1993) Developments in photogrammetry: the geomorphological potential. *Progr. Phys. Geogr.*, **17**, 306–328.

Lane, S.N., Chandler, J.H. and Richards, K.S. (1994) Developments in monitoring and terrain modelling small-scale river-bed topography. *Earth Surf. Process. Landf.*, **19**, 349–368.

Lane, S.N., Richards, K.S. and Chandler, J.H. (1995a) Within reach spatial patterns of process and channel adjustment. In: *Rivers* (Ed. E.J. Hickin), pp. 105–130. Wiley, Chichester.

Lane, S.N., Richards, K.S. and Chandler, J.H. (1995b) Morphological estimation of the time-integrated bedload transport rate. *Water Resour. Res.*, **31**, 761–772.

Lane, S.N., Bradbrook, K.F., Richards, K.S., Biron, P.M. and Roy, A.G. (1999) The application of computational fluid dynamics to natural river channels: three-dimensional versus two-dimensional approaches. *Geomorphology*, **29**, 1–20.

Lane, S.N., Westaway, R.M. and Hicks, D.M. (2003) Estimation of erosion and deposition volumes in a large gravel-bed, braided river using synoptic remote sensing. *Earth Surf. Process. Landf.*, **28**, 249–71.

Lane, S.N., Reid, S.C., Westaway, R.M. and Hicks, D.M. (2004) Remotely sensed topographic data for river channel research: the identification, explanation and management of error. In: *Spatial Modelling of the Terrestrial Environment* (Eds R.E.J. Kelly, N.A. Drake and S.L. Barr), pp. 157–74. Wiley, Chichester.

Lane, S.N., Hardy, R.J., Ferguson, R.I. and Parsons, D.R. (2005) A framework for model verification and validation of CFD schemes in natural open channel flows. In: *Computational Fluid Dynamics: Applications in Environmental Hydraulics* (Eds P.D. Bates, S.N. Lane and R.I. Ferguson), pp. 169–191. Wiley, Chichester.

Laronne, J.B. and Duncan, M.J. (1992) Bedload transport paths and gravel bar formation. In: *Dynamics of Gravel-bed Rivers* (Eds P. Billi, C.R. Hey, C.R. Thorne and P. Tacconi), pp. 177–200. Wiley, Chichester.

Leopold, L.B. and Wolman, M.G. (1957) River channel patterns—braided, meandering and straight. *U.S. Geol. Surv. Prof. Pap.*, **282B**.

Lunt, I.A. and Bridge, J.S. (2004) Evolution and deposits of a gravely braid bar, Sagavanirktok River, Alaska. *Sedimentology*, **51**, 415–32.

Lunt, I.A., Bridge, J.S. and Tye, R.S. (2004) A quantitative, three-dimensional depositional model of gravely braided rivers. *Sedimentology*, **51**, 377–414.

Lyzenga, D.R. (1981) Remote sensing of bottom reflectance and water attenuation parameters in shallow water using aircraft and Landsat data. *Int. J. Remote Sens.*, **2**, 71–82.

Martin, Y. and Church, M. (1996) Bed-material transport estimated from channel surveys—Vedder River, British Columbia, *Earth Surf. Process. Landf.*, **20**, 347–61.

McArdell, B.W. and Feah, R. (2001) A computational investigation of river braiding. In: *Gravel Bed Rivers V* (Ed. M.P. Mosley), pp. 11–46. New Zealand Hydrological Society, Wellington.

McLean, D.G. and Church, M. (1999) Sediment transport along lower Fraser River. 2. Estimates based on long-term gravel budget. *Water Resour. Res.*, **35**, 2549–2559.

Mosley, M.P. (1976) An experimental study of channel confluences. *Journal of Geology*, **84**, 535–62.

Murray, A.B. and Paola, C. (1994) A cellular model of braided rivers. *Nature*, **371**, 54–57.

Murray, A.B. and Paola, C. (1996) A new quantitative test of geomorphic models, applied to a model of braided streams. *Water Resour. Res.*, **32**, 2579–87.

Murray, A.B. and Paola, C. (1997) Properties of a cellular braided stream model. *Earth Surf. Process. Landf.*, **22**, 1001–1025.

Murray, A.B. and Paola, C. (2003) Modelling the effect of vegetation on channel pattern in bedload rivers. *Earth Surf. Process. Landf.*, **28**, 131–43.

Nagata, N., Hosoda, T. and Muramoto, Y. (2000) Numerical analysis of river channel processes with bank erosion. *J. Hydraul. Eng.*, **126**, 243–252.

Nicholas, A.P. and Sambrook Smith, G.H. (1999) Numerical simulation of three-dimensional flow hydraulics in a braided channel. *Hydrol. Process.*, **13**, 913–929.

Nykanen, D.K., Foufoula-Georgiou, E. and Sapozhnikov, V.B. (1998) Study of spatial scaling in braided river patterns using synthetic aperture radar imagery. *Water Resour. Res.*, **34**, 1795–807.

Paola, C. (2001) Modelling stream braiding over a range of scales. In: *Gravel Bed Rivers V* (Ed. M.P. Mosley), pp. 11–46. New Zealand Hydrological Society, Wellington.

Paola, C. and Foufoula-Georgiou, E. (2001) Statistical geometry and dynamics of braided rivers. In: *Gravel Bed Rivers V* (Ed. M.P. Mosley), pp. 47–69. New Zealand Hydrological Society, Wellington.

Parker, G. (1984) Lateral bed load transport on side slopes. *J. Hydraul. Eng.*, **110**, 197–199.

Pittaluga, M.B., Repetto, R. and Tubino, M. (2003) Channel bifurcation in braided rivers: Equilibrium configurations and stability. *Water Resour. Res.*, **39**, 1323 NOV 25 2003.

Robertson-Rintoul, M.S.E. and Richards, K.S. (1993) Braided-channel pattern and palaeohydrology using an index of total sinuosity. In: *Braided Rivers* (Eds J.L. Best and C.S. Bristow), pp. 113–118. Special Publication 75, Geological Society Publishing House, Bath.

Sapozhnikov, V.B. and Foufoula-Georgiou, E. (1996) Do the current landscape evolution models show self-organized criticality?. *Water Resour. Res.*, **33**, 1983–1991.

Sapozhnikov, V.B., Foufoula-Georgiou, E. (1997) Experimental evidence of dynamic scaling and indications of self-organized criticality in braided rivers. *Water Resour. Res.*, **32**, 1109–1112.

Siegenthaler, C. and Huggenberger, P. (1993) Pleistocene river gravel: deposits of a braided rievr system with dominant pool preservation. In: *Braided Rivers* (Eds J.L. Best and C.S. Bristow), pp. 147–62. Special Publication 75, Geological Society Publishing House, Bath.

Tarboton, D.G. (1997) A new method for the determination of flow directions and upslope areas in grid digital elevation models. *Water Resour. Res.*, **33**, 309–19.

Thomas, R. and Nicholas, A.P. (2002) Simulation of braided river flow using a new cellular routing scheme. *Geomorphology*, **43**, 179–95.

Walsh, J. and Hicks, D.M. (2002) Braided channels: self-similar or self-affine? *Water Resour. Res.*, **38**, 1082 JUN 2002.

Westaway, R.M., Lane, S.N. and Hicks, D.M. (2000) Development of an automated correction procedure for digital photogrammetry for the study of wide, shallow gravel-bed rivers. *Earth Surf. Process. Landf.*, **25**, 200–26.

Westaway, R.M., Lane, S.N. and Hicks, D.M. (2001) Airborne remote sensing of clear water, shallow, gravel-bed rivers using digital photogrammetry and image analysis. *Photogr. Eng. Remote Sens.*, **67**, 1271–81.

Westaway, R.M., Lane, S.N. and Hicks, D.M. (2003) Remote survey of large-scale braided rivers using digital photogrammetry and image analysis. *Int. J. Remote Sens.*, **24**, 795–816.

Wilcock, P.R. and Iverson, R.M. (2003) *Prediction in Geomorphology*. Monograph 135, American Geophysical Union, Washington, DC, 256 pp.

Winterbottom, S.J. and Gilvear, D.J. (1997) Quantification of channel bed morphology in gravel-bed rivers using airborne multispectral imagery and aerial photography. *Regul. River. Res. Manage.*, **13**, 489–499.

Young, W.J. and Davies, T.R.H. (1991) Bedload transport in a braided gravel-bed river model. *Earth Surf. Process. Landf.*, **16**, 499–511.

Cellular modelling of braided river form and process

A.P. NICHOLAS, R. THOMAS *and* T.A. QUINE

Department of Geography, University of Exeter, Exeter EX4 4RJ, UK (Email: A.P.Nicholas@exeter.ac.uk)

ABSTRACT

Cellular approaches to modelling fluvial processes typically implement simplified rules describing water and sediment routing and have, in the past, been used to provide qualitative insight into fundamental aspects of braided river behaviour. A new cellular model of flow, sediment transport and vegetation effects is presented here and used to simulate the formation and evolution of braided rivers over periods of up to 200 yr. Through comparison with field data and the results of a more sophisticated hydraulic model, which solves the shallow-water equations, the cellular model is shown to generate accurate predictions of inundation extent, spatial patterns of unit discharge and flow routing pathways. The coupled flow–sediment–vegetation model is also shown to replicate many of the static characteristics of braided river morphology and dynamic aspects of channel behaviour. These results illustrate that cellular approaches may have considerable potential as tools for simulating river behaviour over periods of 100 yr or longer. However, in order to fulfil this potential it is necessary to develop approaches that emphasize both simplicity and quantitative accuracy, at least where the two are not mutually exclusive.

Keywords Cellular model, flow hydraulics, channel morphology, dynamic behaviour.

INTRODUCTION

In reviewing progress in braided river research a decade ago Ferguson (1993) observed that relatively little numerical modelling of multi-thread rivers had been conducted at that time. Over the past 10 years this situation has changed dramatically and significant advances in modelling have been made over a range of scales. For example, two-dimensional and three-dimensional computational fluid dynamics models have been used to simulate flow hydraulics both at the scale of the bar–pool unit (Lane *et al.*, 1999; Nicholas & Sambrook Smith, 1999; Bradbrook *et al.*, 2001; Nicholas, 2005) and the larger reach-scale (Nicholas, 2003). Recently, such approaches have also been coupled with bedload transport models to simulate the initiation and evolution of braided channel morphology (McArdell & Faeh, 2001; Kurabayashi *et al.*, 2002; Shimizu, 2002). These approaches offer the prospect of providing an improved physically based understanding of process–form interactions within braided rivers, albeit over relatively short temporal and spatial scales.

Cellular approaches that use simplified rules representing flow and sediment routing have also emerged as a distinct class of model capable of simulating braided river dynamics at larger scales (Murray & Paola, 1994, 1997; Thomas & Nicholas, 2002; Thomas *et al.*, 2002). The model of Murray & Paola (1994) was developed, in part, to identify the simplest possible process rules required to generate braiding, and was not intended to make quantitative predictions of flow characteristics, process rates or morphological variables. Subsequent reviews by Paola (2001) and Murray (2003) have highlighted these objectives and emphasized this as an important difference between reductionist approaches and cellular (synthesist) models (Paola's terminology). Despite this distinction, the basis and framework of cellular models are very similar to those of two-dimensional reductionist approaches in many respects; for example, both are underpinned by the principle of mass conservation. Furthermore, the more simplistic process representation inherent in cellular models may be considered less significant if one accepts that many natural processes (e.g. bedload transport) are poorly understood at present, and that many elements of relatively sophisticated geomorphological 'theory' are essentially empirical in nature (Church, 1996). For this and other reasons (Richards, 2001), it can be argued that cellular models have much in common with reductionist approaches and that, in principle, it may be possible to generate quantitatively meaningful predictions of form and process in braided channels using the former model type. This is a tantalizing prospect, since the greater computational efficiency of cellular approaches affords the potential to simulate braided river evolution over periods of decades to centuries. This paper examines these issues using a new cellular model that represents flow, sediment routing, morphological change and braidplain vegetation.

THE CELLULAR MODEL

The cellular model that is applied here uses simple rules to route water and sediment throughout a regular grid of cells that represent the channel topography. Calculations commence when water and sediment are introduced into some, or all, of the upstream boundary cells. Routing is carried out from each cell in a row to the five immediate neighbours in the row downstream (Fig. 1a). This allows channels to form at angles of up to approximately 60° to the downstream direction. The discharge received by each of the five downstream cells, q_i, from the upstream cell that water is leaving is given by:

$$q_i = \frac{P_i}{\sum_{i=1}^{5} P_i} q_0 \quad (1)$$

where q_0 is the discharge leaving the upstream cell and P_i is a routing potential, which determines the distribution of discharge between the five downstream cells. The routing potential at each cell is determined by first estimating the depth of flow at the cell from which water is being routed (h_0) using:

$$h_0 = \left(\frac{q_0}{f}\right)^D \quad (2)$$

where appropriate values for f and D can be determined by equating Eq. 2 to a rearranged form of the Manning equation. On this basis D is assigned a value of 0.6 and f is estimated to be:

$$f = \frac{\sqrt{S_0}}{n} \quad (3)$$

where S_0 and n are reach-averaged values of the longitudinal bed slope and the Manning roughness coefficient, respectively. Flow depths (h_i) at each of the five downstream cells are then approximated from:

$$h_i = h_0 + z_0 - z_i - S_0 \Delta x \quad \text{for} \quad h_i > 0 \quad (4)$$

where z_0 and z_i are, respectively, the bed level at the cells that water is being routed from and to, and Δx is the downstream distance between these cells. Implicit in this is the assumption that the drop in water level between the two cells can be estimated using the mean channel gradient. Where Eq. 4 yields negative flow depths these are set equal to zero. If at least one of the five cells in question has a positive depth the downstream routing potential for each cell is then calculated using:

$$P_i = h_i^{1.67} S_i^m \quad (5)$$

where S_i is the local bed slope between the cells that water is being routed from and to, and m is a

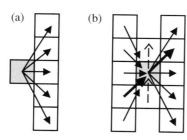

Fig. 1 (a) Water routing pathways from a single central cell (shaded grey) to its five immediate downstream neighbours. (b) Sediment routing pathways into and out of a single central cell (shaded grey). Dashed vectors with open heads indicate lateral sediment transport (determined using Eq. 14). Solid vectors indicate actual sediment transport rates determined using Eq. 12. Bold vectors provide an example of routing pathways upstream and downstream of the central cell that lie parallel to one another (see explanation of Eq. 13).

parameter that must be provided as an input to the model, and which might be expected to take a value of 0.5 based on the Manning equation (the value of the depth exponent used here is also derived from the Manning equation). Since bed slope values used in Eq. 5 provide only a first order estimate of the local energy slope and may be negative they are constrained so that they are not less than a threshold value, set equal to the reach-averaged bed slope. This approach will, on average, tend to over-estimate local slopes, hence one might expect m to take a value lower than 0.5. Here m is set to 0.25 based on a previous assessment of model performance (Thomas & Nicholas, 2002).

In the situation where all five downstream cells are calculated to have zero or negative flow depths (using Eq. 4) the flow is assumed to be critical (i.e. the Froude number = 1). This approach is based on the observation that water is able to flow over negative slopes for short distances (e.g. at riffles) and that in such situations Froude numbers are typically high. Accordingly, the water level across the five cells is determined by estimating a maximum flow depth at these cells, h_{max}, using:

$$h_{max} = \left(\frac{q_0}{\sqrt{g}}\right)^{0.67} \quad (6)$$

where g is the acceleration due to gravity. Flow depths, h_i, at each of the five downstream cells are then approximated using:

$$h_i = h_{max} + z_{min} - z_i \quad \text{for} \quad h_i > 0 \quad (7)$$

where z_{min} is the lowest elevation of the five downstream cells. The downstream routing potential for each wet cell is then calculated from:

$$P_i = h_i^{1.5} \quad (8)$$

where the exponent value of 1.5 is based on the definition of the Froude number.

Bedload is routed from each cell to its five downstream neighbours by first calculating the sediment transport capacity per unit width, T_i^{CAP}, along each flow pathway leaving the upstream cell using:

$$T_i^{CAP} = a(\omega_i - \omega_{cr})^b \quad (9)$$

where a and b are empirical constants (assigned values of 1 and 1.5, respectively, in model runs reported here) and ω_i and ω_{cr} represent the unit stream power and the critical unit stream power for entrainment of sediment. Unit stream power is calculated as:

$$\omega_i = q_i <S_i> \quad (10)$$

where the slope $<S_i>$ is defined as a function of the local bedslope between the two adjacent cells, S_i, and a spatially averaged bedslope, S^*:

$$<S_i> = \lambda S_i + (1 - \lambda)S^* \quad (11)$$

This definition is adopted here since it provides a simple means of accounting for the fact that stream power is a function of the local energy slope, which is positively correlated with bed slope, but typically exhibits a damped response to fluctuations in bed topography. The parameter λ is set equal to 0.1 here. This parameter influences the scaling of the channel morphology generated by the model (see below). The value of S^* is determined by calculating the average downstream bed slope at each point along the model reach (i.e. the average of the downstream slopes for all wet cells in a single row of the model grid). These average values are then smoothed by averaging the slope at each row with those at the rows upstream and downstream. This smoothing procedure is repeated a number of times to obtain a value of S^* that represents the weighted sum of the row-averaged slopes over a specified distance either side of the row that contains cell i. In the simulations reported here this distance was set equal to 40 m (approximately half the valley floor width).

The actual rate of sediment transport, T_i^{ACT}, along each of the five flow pathways leaving a cell is calculated as:

$$T_i^{ACT} = kT_i^{CAP} + (1 - k)T_i^{SUP} \quad (12)$$

The parameter k varies between 0 and 1, and is used to define the extent to which actual sediment transport rates are influenced by T_i^{CAP}, the sediment transport capacity along the pathway, and T_i^{SUP}, the potential rate of sediment supply to the pathway from the upstream cell in the absence of erosion and deposition in that cell. Values of T_i^{SUP} are determined by distributing the sediment that enters the upstream cell between the five possible routing pathways downstream in proportion to a transport potential, TP_i, calculated using:

$$TP_i = T_i^{CAP} + \Omega T_0^{ACT} \quad (13)$$

where T_0^{ACT} is defined as the actual rate of sediment transport entering the upstream cell in the direction parallel to the downstream flow routing pathway under consideration (see Fig. 1b), and Ω is a parameter that represents the effects of inertia on sediment transport.

Lateral sediment transport (i.e. perpendicular to the downstream direction) in a downslope direction is also incorporated in the cellular model using the approach of Murray & Paola (1994):

$$T^{LAT} = ES_L T_i^{ACT} \quad (14)$$

where T^{LAT} is the lateral sediment transport rate per unit width, E is a constant that represents the bank erodibility and S_L is the lateral bed slope. The parameters k, Ω and E were assigned values of 0.3, 0.12 and 0.2, respectively, in the model runs reported here. The effects on simulated channel morphology of changes in these parameter values are discussed below.

Changes in the age of topographic surfaces and associated effects on sediment entrainment (e.g. as a result of the development of vegetation) have been incorporated recently in cellular models (e.g. Murray & Paola, 2003) and are represented here using a similar simplified approach. Based on the classification scheme devised by Reinfelds & Nanson (1993) for the Waimakariri River, vegetation colonization is assumed to begin after a cell has been 'inactive' for 3 yr (including periods of low flow). After this time the age of vegetation in each cell is monitored continuously. Vegetation age is reset to zero if the surface experiences a depth of scour or burial greater than a defined threshold, or if the cell is inundated during low flows. Sediment entrainment thresholds (e.g. ω_{cr} in Eq. 9) are varied according to vegetation age, which controls the type and density of vegetation present. These relationships are defined for a range of vegetation classes within age brackets based on the scheme of Reinfelds & Nanson (1993). Entrainment thresholds for bare gravel and vegetated surfaces >150 yr old were defined as 0.003 m^{-2} s^{-1} and 0.1 m^{-2} s^{-1} respectively, for the model runs reported here. These values represent the upper and lower limits of the entrainment thresholds used here.

The model described above is used to simulate a series of flood events with specified peak discharges. Flood events are subdivided into a series of time-steps, for each of which the total discharge entering the model reach is specified. The discharge within each cell at the reach inlet can be determined using a rearranged version of Eq. 2, assuming a uniform water level across the section. Rates of sediment transport at the upstream boundary are calculated from Eq. 9. Stream power is determined using the specified unit discharge values at each cell in conjunction with a section-averaged bed-slope. If the slope at the upstream boundary is set equal to the reach-averaged channel gradient then one would expect the channel within the model reach to transport the sediment supplied at the upstream boundary without substantial net erosion or deposition occurring, thus resulting in approximate equilibrium conditions. Consequently, the upstream boundary slope can be varied to represent fluctuations in sediment supply and enforce periods of aggradation or degradation. Changes in bed elevation at each cell are determined by integrating sediment transport rates (calculated using Eqs 12 & 14) into and out of that cell over the model time-step using a discretized form of the Exner equation (see also Fig. 1b):

$$\Delta z = \left(\sum T_{IN}^{ACT} + \sum T_{IN}^{LAT} - \sum T_{OUT}^{ACT} - \sum T_{OUT}^{LAT} \right) \frac{\Delta t}{\Delta x} \quad (15)$$

where the time-step, Δt, is defined to limit the maximum change in elevation in the model reach to a predefined value, and porosity effects are assumed to be incorporated within the sediment transport constant, a, in Eq. 9.

VALIDATION OF THE WATER ROUTING SCHEME

The water routing component of the cellular model was applied to a 470 m long braided reach of the Avoca River, South Island, New Zealand. This was carried out using a digital elevation model

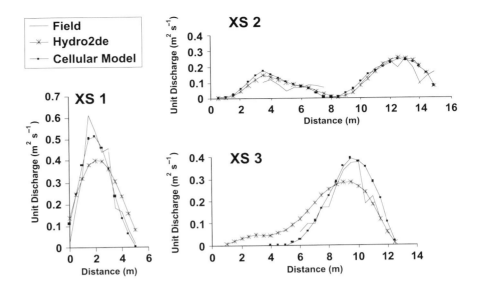

Fig. 2 Distributed patterns of unit discharge (m² s⁻¹) at three cross-sections (XS1–3) for field data, HYDRO2DE and the cellular model output.

(DEM) with a horizontal resolution of 1 m constructed from topographic data obtained using a total station. Model validation was conducted using field measurements of unit discharge obtained at 19 channel cross-sections during a period of low flow. As a further test of the cellular water routing scheme unit discharge predictions were also compared with the output from a more sophisticated two-dimensional hydraulic model (HYDRO2DE) that solves the depth-averaged shallow water equations (cf. Nicholas, 2003). This was carried out within the same reach for discharges of 5, 10, 20 and 50 m³ s⁻¹. These results are summarized in greater detail by Thomas & Nicholas (2002).

Coefficients of determination derived by comparison of field data with the predictions of the cellular model and HYDRO2DE demonstrate that both models explain approximately 50% of the observed variation in unit discharge for the dataset of 256 measurements. A substantial proportion of the unexplained variance is associated with grain-scale flow structures that are not represented by either model due to the 1 m resolution of the DEM. The high level of subgrid-scale variability in flow conditions within the field reflects the coarse nature of the bed material and shallow flow depths within the study reach during the measurement period. Figure 2 shows model predictions and field measurements of unit discharge at three typical cross-sections within the reach. Both models capture the broad lateral changes in unit discharge observed in the field. There is some tendency for the cellular model to concentrate the flow within the deeper parts of the channel to a greater extent than the more sophisticated model. This is also seen in Fig. 3 which shows unit discharge patterns and flow vectors generated by the two models for a discharge of 20 m³ s⁻¹.

Vector patterns for the cellular model have been converted from diagonal water fluxes into two perpendicular components for ease of comparison with HYDRO2DE results (Thomas & Nicholas, 2002). Concentration of flow in the channel thalweg is more pronounced in the field than for HYDRO2DE predictions at many locations (e.g. cross-section 1). This may be due to higher relative roughness at channel margins in the field that is not represented by either model. The cellular model is more successful at replicating the tendency for flow to be concentrated in this way. However, this is probably a product of the fact that the five possible discharge contributions received by each cell from upstream are calculated independently of one another. As a consequence of this, discharge routing to low points on the bed is not moderated by the fact that these areas may receive water from several upstream cells.

The good agreement that is evident in Fig. 3 between results generated using the two modelling approaches is typical of simulations conducted over a range of discharges. Predicted inundation extents for the two models at all four discharges were within 4% of each other. In addition, correlation coefficients calculated between unit discharge predictions for the two models were all in the range

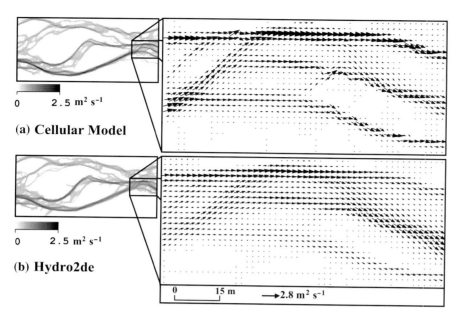

Fig. 3 Patterns of unit discharge predicted by (a) the cellular model and (b) HYDRO2DE, for a discharge of 20 m³ s⁻¹. Small boxes show the total model reach (470 m by 230 m). Large boxes show flow vectors for a small portion of the reach.

0.8–0.9. Slight differences are apparent in the flow routing patterns and there is further evidence of the tendency for the cellular model to concentrate the flow over a narrower region of the channel bed than HYDRO2DE. However, overall both approaches predict very similar patterns of flow routing within channels and around bars. These results illustrate that the water routing component of the cellular model is able to generate quantitatively realistic flow predictions when compared with both field data and the results of a more sophisticated model that solves the full shallow-water equations. This is encouraging given the simplicity of the cellular water routing scheme and the fact that this approach does not attempt to conserve fluid momentum. The comparative success of the cellular model reflects two factors. First, the inclusion of a relationship between unit discharge and flow depth (Eq. 2). This was neglected by Murray & Paola (1994) and is essential in order to provide a realistic representation of the relationship between total discharge and inundation extent. Second, the fact that flow routing is controlled strongly by channel topography. Given the latter it is, perhaps, not surprising that the water routing scheme performs well where topography is specified and where cross-stream relief is substantial. A more rigorous assessment of the cellular model must be based upon predictions derived for channel morphology generated by the model itself.

MODELLED CHANNEL MORPHOLOGY

The combined water and sediment routing scheme can be used to simulate channel evolution over a wide range of temporal and spatial scales. Here results are presented from model runs using a 1 m resolution grid with 450 cells in the downstream direction and 85 in the cross-stream direction. A period of 200 yr containing 5000 flood events was simulated. Peak discharges were sampled from an 80 yr flow record available for the Harper-Avoca River (scaled by the ratio of grid width to the Harper valley floor width). Initial conditions for these simulations consisted of a flat surface with a uniform downstream slope of 0.01 m m⁻¹. Random perturbations in the bed topography with a maximum amplitude of ±0.1 m were imposed on this surface. This slope value was also used at the upstream boundary to calculate rates of sediment supply to the first row of cells in the grid.

Figure 4a & b shows simulated patterns of unit discharge and surface vegetation age at seven points in time over the course of a model run. Figure 5a & b shows simulated volumes of erosion and deposition and their timing in relation to the flood series applied at the upstream grid boundary. During the first few years of a simulation, alternating peaks and depressions form as a series of low amplitude bedforms, which promote a relatively regular network of flow divergences and

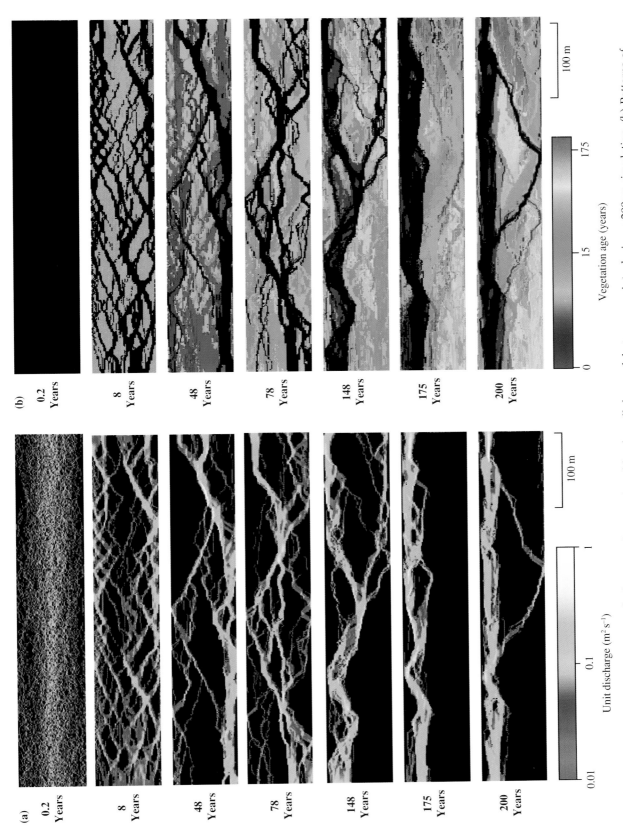

Fig. 4 (a) Patterns of unit discharge (m² s⁻¹) at low flow produced by the cellular model at seven points during a 200 yr simulation. (b) Patterns of surface vegetation age (years) produced by the cellular model at seven points during a 200 yr simulation.

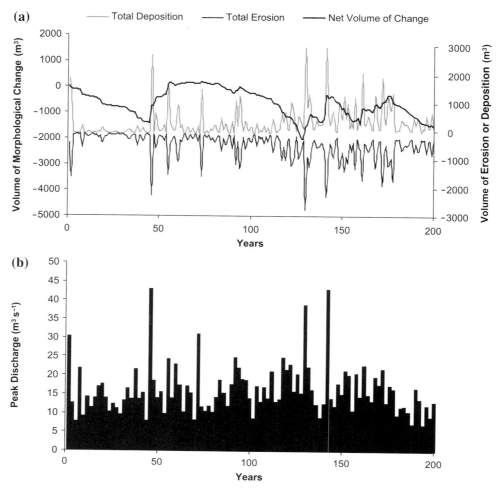

Fig. 5 Temporal changes over the course of a 200 yr simulation in: (a) the total volume of erosion, deposition and net morphological change within the model reach; and (b) maximum discharge in each 5 yr time interval.

convergences. Sediment transport rates are relatively low during this phase, but as the simulation proceeds net removal of sediment from the reach is driven by increasing transport rates as a channel system develops and flow becomes concentrated. Broadly speaking, the channel adopts a braided form over the first 150 yr of the simulation, after which there is some tendency for flow to occupy a single dominant channel, particularly at low discharges. Substantial increases in topographic relief are associated with high magnitude floods (e.g. after 48 yr). These events typically initiate phases of aggradation that may last several years, but also promote localized erosion and the formation of deep channels that grow by headward incision. This leads to unsteadiness in channel behaviour, with phases of aggradation and degradation that last for between 10 and 75 yr.

The planform characteristics of the braided channels simulated by the cellular model (e.g. in Fig. 4b) appear visually similar to those of natural braided rivers. To provide a more quantitative assessment of the similarity between modelled and natural channels a number of static indices of channel morphology were examined. Figure 6 shows the relationship between bar area and perimeter derived from an aerial photograph of the Rakaia River, New Zealand, and cellular model results obtained at an equivalent inundation extent. Trend lines fitted to these data are described by power laws with exponents that are not significantly different for the cellular model (0.62) and Rakaia

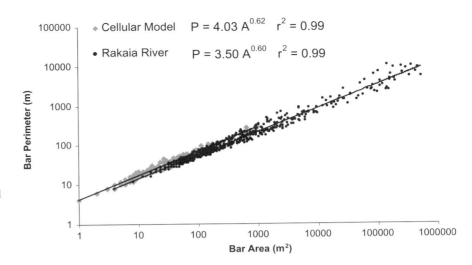

Fig. 6 Comparison of morphological characteristics of braid bars for cellular model output (grey diamonds) and Rakaia River data (black circles).

Table 1 Downstream trend in width between three adjacent cross-sections (A, B and C) for Avoca River topography and topography simulated by the cellular model after 0.2 yr (pre-channel development), 78 yr (braided channel) and 200 yr (single channel). W+, W– and W= indicate increasing, decreasing and constant channel width between successive cross-sections

	Avoca River DEM			Cellular model (0.2 yr)			Cellular model (78 yr)			Cellular model (200 yr)		
	$W+^{AB}$	$W=^{AB}$	$W-^{AB}$	$W+^{AB}$	$W=^{AB}$	$W-^{AB}$	$W+^{AB}$	$W=^{AB}$	$W-^{AB}$	$W+^{AB}$	$W=^{AB}$	$W-^{AB}$
$W+^{BC}$	17.2	9.4	14.6	8.0	5.6	23.7	17.9	5.6	17.2	11.6	8.5	16.1
$W=^{BC}$	6.9	2.5	8.7	8.3	11.4	6.9	7.1	5.8	8.0	8.0	11.0	9.2
$W-^{BC}$	17.0	6.3	17.4	20.7	9.8	5.6	15.4	9.8	13.2	16.5	8.7	10.4

(0.60). The form of these relationships reflects systematic changes in bar shape as bar size increases. Exponents >0.5 are indicative of a tendency for bars to become more elongate as their area increases.

Spatial changes in channel form (e.g. total width) can also be examined using dynamical systems analysis (e.g. Murray & Paola, 1996; Sapozhnikov et al., 1998). To do this total water surface widths for sections throughout the Avoca River DEM were calculated based on the predictions of the water routing component of the cellular model. These data were compared with water surface width changes for channel morphologies generated by the cellular model. Comparisons were made for a discharge equivalent to the peak of the mean annual flood. State space plots of total channel width plotted against width at the next section downstream were found here to provide a limited measure of model performance in some respects (channels with a range of morphologies produced similar attractors that could not be discriminated between easily using available techniques). As an alternative approach the sequence of changes in total water surface width between three adjacent sections was examined. Widths were classified either to increase, decrease or remain constant between each pair of sections. When considering three adjacent sections this leads to nine possible sequences of width adjustment. The frequency of each of these nine sequences is shown in Table 1 for the Avoca River and for the topography generated by the cellular model. Cellular model results are shown for three of the channel morphologies in Fig. 4a & b. These results are characterized by significant differences between width adjustments in the three channels generated by the cellular model. Furthermore, the

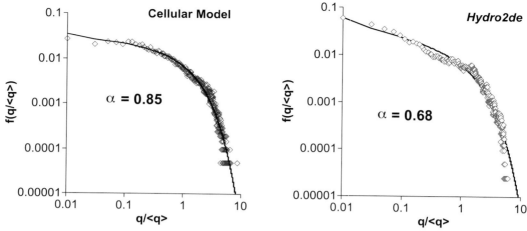

Fig. 7 Gamma distributions fitted to cellular model and HYDRO2DE unit discharge predictions at a discharge equivalent to the peak of the mean annual flood.

similarity between the Avoca River and cellular model results is clearly greatest for the braided channel pattern generated using the model (after 78 yr). Sequences of width adjustment for the Avoca and cellular braided channel are indicative of a tendency for channels to widen and narrow in a downstream direction at similar rates. This is consistent with the observation that when viewed in planform as a simple wet/dry image it can often be difficult to determine flow direction in braided channels. This is true of both natural rivers and those generated by the cellular model.

Several previous studies have shown that a range of braided river hydraulic variables can be described by gamma distributions (e.g. Paola, 1996; Nicholas, 2000, 2003). Figure 7 shows gamma distributions of dimensionless unit discharge fitted to results from the cellular model (for the channel morphology shown in Fig. 4a after 78 yr) and HYDRO2DE predictions for the Avoca River DEM. Data are shown for a discharge equivalent to the peak of the mean annual flood. Once again, there is good agreement between the two model results. Both distributions are described by a low value of the shape parameter α, which indicates that unit discharge values are described approximately by an exponential distribution (i.e. large areas of low unit discharge and increasingly smaller areas of higher unit discharge). In conjunction with the comparison of HYDRO2DE and cellular model results carried out on the Avoca River (see above, and Thomas & Nicholas, 2002), these results provide further evidence that the cellular model is replicating previously observed characteristics of braided channels and is, in this respect, generating quantitatively realistic results.

MODELLED CHANNEL DYNAMICS

In order to investigate dynamic aspects of model behaviour the cellular model was used to simulate the response of a braided channel to changes in upstream sediment supply. This was carried out using a 1 m resolution grid 900 m long by 110 m wide. The model was run for a period of 60 yr with the slope used to calculate sediment transport rates at the upstream boundary set equal to the reach-averaged bed slope (0.01 m m^{-1}). This led to the formation of a braided channel in a state of approximate equilibrium (i.e. no net long-term trend towards either aggradation or degradation). The resulting 'equilibrium' channel topography was then used as the starting point for two additional simulations, each run for a further 50 yr. Slopes of 0.015 m m^{-1} and 0.005 m m^{-1} were imposed at the upstream boundary of the model reach in these simulations. The effect of these changes in boundary slope is to promote abrupt changes in sediment supply to the model reach. River responses, in the form of changes in bed elevation and planform morphology, were observed to be initiated at the upstream boundary and to propagate downstream over time.

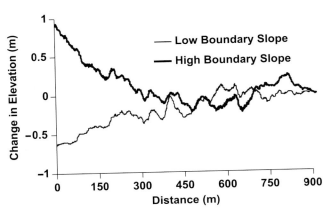

Fig. 8 Changes in mean long-profile elevation after 110 yr (relative to the long profile after 60 yr) for simulations of low and high boundary slopes.

Figure 8 shows the simulated changes in section-mean elevation in the model reach (after 110 yr in total) relative to the river long-profile after 60 yr. The region experiencing substantial channel change extends for a distance of approximately 400 m downstream of the inlet to the reach. Cross-sections illustrating the topography at various locations along the reach (e.g. 50, 100, 200 and 300 m) are shown in Fig. 9. No obvious differences in cross-section shape or total relief are evident between equilibrium and aggrading channel topography. In contrast, degrading channel cross-sections at 50 m and 100 m are characterized by the concentration of incision over a restricted part of the section, leading to the formation of paired and unpaired terraces. Figure 10 shows the channel morphology in planform and the surface vegetation age for the three simulations. Clear differences in morphology and surface age are evident in the upstream half of the model reach, where greater sediment supply (Fig. 10c) has led to an increase in braid intensity and the deposition of large areas of fresh sediment and vegetated surfaces <5 yr old. Reduced sediment supply (Fig. 10b) has driven a decrease in braid intensity and the creation of terraces with distinct ages associated with well-defined periods of incision and channel abandonment. There is clear evidence of the effects of the downstream base level on both the rate and extent of simulated channel adjustment. However, despite this, the changes in channel morphology and surface age characteristics

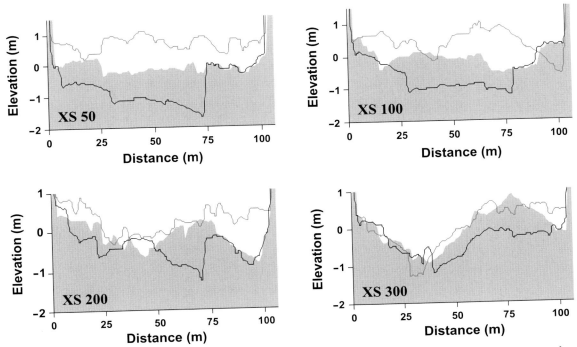

Fig. 9 Cross-section (XS) topography at locations 50, 100, 200 and 300 m downstream of the upstream boundary of the model reach: (i) after 60 yr (shaded); (ii) after a further 50 yr using a low boundary slope (black line); and (iii) after a further 50 yr using a high boundary slope (grey line).

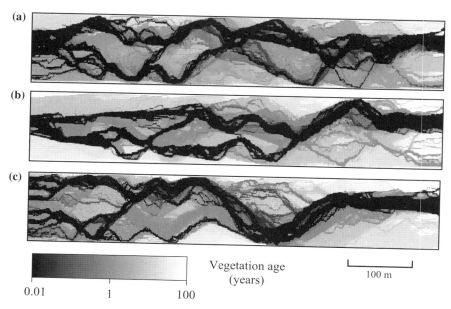

Fig. 10 Patterns of surface vegetation age: (a) after 60 yr; (b) after a further 50 yr using a low boundary slope; and (c) after a further 50 yr using a high boundary slope. Note that the cross-stream scale is exaggerated for ease of viewing. Scale bar relates to the cross-stream direction. Reach dimensions are 900 m by 110 m.

described above are consistent with the observations of previous studies of braided river response to changes in sediment supply (e.g. Germanoski & Schumm, 1993), and provide a further indication that the cellular model is generating physically realistic predictions of process–form interactions.

DISCUSSION

The cellular model presented here appears to be capable of replicating many of the static and dynamic characteristics of braided channels. The agreement between the results of the water routing component of the cellular model and the physically based hydraulic model for the Avoca study reach is particularly good, although this is, perhaps, not surprising given that routing patterns are controlled strongly by topography. It is also encouraging that when coupled with the simple sediment routing scheme the cellular model generates channels that are morphologically similar to natural braided streams, both in terms of their planform shape and cross-sectional relief.

Analysis of model sensitivity to variations in parameter values indicates that, broadly speaking, braiding is encouraged by higher rates of lateral sediment transport (higher values of E) and lower values of the critical unit stream power required for entrainment (ω_{cr}). However, the exact value of these parameters is not critical to the occurrence of braiding (channels with realistic braided planform morphology are generated for a wide range of parameter values). In this sense the predictions of the cellular model can be considered robust (Murray, 2003). Furthermore, these results indicate that braiding will be predicted to occur for a wide range of bed and bank material types (E and ω_{cr} represent the main properties of such material) which is consistent with field observations. In contrast, some parameter combinations appear either to promote channels with planform morphology that is unrealistic, or to encourage the formation of a straight single-thread river.

The development of braided channels with planform morphology that is both visually and statistically similar to that of natural streams appears to be strongly dependent on the parameters λ and k. These control the local slope values used in sediment transport calculations and the extent to which transport rates are supply or capacity limited. In both cases, braiding is encouraged by moderate parameter values, while lower and higher values lead to single channel formation and/or unrealistic morphology. It is interesting to note that both parameters play a similar role in influencing model behaviour, since both are, in effect, controlling the relative importance of local and upstream form and process. They act to regulate the feedbacks associated with flow convergence and divergence that initiate and then maintain topographic highs and lows, and effectively control the scaling of bar and channel features. In this sense, they may perform a similar role to channel-scale coherent flow structures that regulate meander wavelength and confluence–diffluence morphology in natural rivers (Yalin, 1971; Rhoads & Welford, 1991; McLelland et al., 1996), in that they drive the growth and decay of sediment transport perturbations. Of course, although these components of the cellular model may mimic the effect of the physical processes that control the scaling of natural channel morphology, they do not seek to represent these processes explicitly. Furthermore, it must be recognized that the parameterization of these physical processes is dependent upon the resolution of the grid used to represent channel topography.

This raises important questions concerning the role of cellular models in fluvial geomorphology and their ability to generate results that are either qualitatively or quantitatively meaningful. Recent reviews (e.g. Paola, 2001; Murray, 2003) have emphasized the distinction between reductionist approaches (typically based on fundamental principles of mass and momentum conservation) and cellular (synthesist or exploratory) models. Murray (2003) argued that exploratory models (such as the Murray–Paola cellular braided stream model) seek to maximize the simplicity of process representation so as to identify the mechanisms that are essential in producing fundamental behaviour. Such simplified models are generally considered to be unsuitable for deriving quantitative predictions of form or process. The latter task is considered to be performed better by reductionist models that attempt to represent a larger number of the governing processes more accurately. In reality, two-dimensional cellular models and reductionist approaches that couple the shallow-water equations with bedload transport and bank erosion laws (e.g. McArdell & Faeh, 2001) often share common structures and governing principles, and may generate very similar results, as is the case of the flow routing models examined here. Furthermore, many of the equations used in reductionist approaches are little more sophisticated than the rules used by cellular models, and may contain significant empirical components (e.g. in the case of sediment transport). Consequently, the ability of both ap-proaches to generate accurate predictions of sediment transport and channel change is uncertain.

In the case of exploratory models it has been argued (Murray, 2003) that one need not strive to obtain quantitative accuracy in order to provide plausible explanations for observed phenomena. While this is a valid viewpoint, the degree to which any model is able to provide meaningful insight into system behaviour must be influenced, to some extent, by the reliability of model predictions in quantitative terms. The Murray & Paola (1994) model illustrates that flow convergence and divergence, coupled with the non-linear dependence of sediment transport on flow strength, represents a fundamental control on braiding. This view is consistent with the results of a large body of field and laboratory research into flow–sediment–morphology interactions at the scale of individual braid bars or channel confluences (e.g. Davoren & Mosley, 1986; Ashmore, 1991; McLelland et al., 1996; Bradbrook et al., 2001).

While approaches based on maximum simplicity may allow the identification of necessary conditions for braiding, they may neglect process mechanisms that are fundamental to braided river behaviour. For example, experiments conducted using the cellular model presented here are characterized by at least two styles of channel behaviour at different discharges. At moderate flows in excess of the threshold for sediment motion, morphological change is dominated by bar formation, migration and reworking, and channel switching and migration over relatively short distances. These patterns largely reflect feedbacks associated with flow convergence and divergence at the scale of the bar–pool unit. During more extreme flows (floods with a recurrence interval of several decades) major reorganization of the braided network occurs as a result of avulsion processes and floodplain dissection. In this sense, the cellular model is displaying multiple stage-dependent mechanisms of channel pattern formation (cf. Lewin & Brewer, 2001). While both styles of channel change may be evident in the results of models that employ fairly simple flow routing schemes (e.g. the Murray & Paola model), capturing the relative importance of these mechanisms and their dependence on flood magnitude requires at least a basic treatment of the relationship between discharge, depth and inundation extent, since this controls the frequency of floodplain inundation and dissection. The cellular model presented here incorporates flow routing rules that address these requirements, resulting in quantitatively accurate predictions of braided river flow characteristics and an improved physical basis for the simulation of sediment transport and morphological change. Given the limitations inherent in studying braided rivers over large time- and space-scales using either field techniques or computationally expensive, coupled hydraulic–sediment-transport models, it is in this area that cellular models may have the greatest potential as tools for investigating braided river behaviour. However, in order to fulfil this potential it is necessary to develop approaches that emphasize both simplicity and quantitative accuracy, at least where the two are not mutually exclusive.

SUMMARY

The past decade has seen a significant increase in the use of numerical models to investigate braided river form and process over a wide range of scales. Cellular approaches in this field have become increasingly popular following the pioneering work of Murray and Paola (1994) and have been shown to have the potential to elucidate fundamental aspects of braided river behaviour. The cellular model presented here differs from previous approaches in a number of ways. Most notably, by incorporating explicit representation of the relationship between flow depth and unit discharge, and including a simple parameterization of the influence of sediment supply on bedload transport. The main conclusions of this study are:

1 Cellular flow routing schemes are capable of generating accurate predictions of inundation extent and distributed patterns of unit discharge in channels with complex bed topography. However, while these predictions are not affected significantly by the simple treatment of the relationship between depth, roughness and slope used here, this does preclude the disaggregation of unit discharge into flow depth and velocity.

2 The combined flow and sediment routing model developed here produces channels with planform morphology and cross-sectional topography that are both visually and statistically similar to that of natural braided rivers. However, many aspects of model performance remain untested (e.g. simulated three-dimensional bed topography and its dynamic evolution) and process parameterization requires further attention, particularly with respect to the physical basis of key parameters and their dependence on grid resolution.

3 Dynamic channel behaviour simulated by the cellular model in response to changing upstream sediment supply is consistent with previous laboratory observations. This suggests that, following further development and validation, cellular models may provide suitable tools for simulating medium-term (circa hundreds of years) braided river evolution. Overall, these results illustrate that cellular models need not be considered to be purely 'maximum simplicity' approaches, but rather 'optimal' approaches, which seek to balance the necessity for quantitative accuracy in the description of critical processes against the computational demands of competing modelling strategies.

ACKNOWLEDGEMENTS

This research was funded by the UK Natural Environment Research Council (studentship GT04/1999/FS/0092) and the Royal Society. We are grateful to Bob and Val Brown for their hospitality and for allowing access to the Harper-Avoca study site, and to Les Basher at Landcare Research for assisting with field work and for introducing us to the Harper-Avoca. Thanks also to Rob Ferguson and Brad Murray for their valuable reviews.

REFERENCES

Ashmore, P.E. (1991) How do gravel-bed rivers braid? *Can. J. Earth Sci.*, **28**, 326–341.

Bradbrook, K.F., Lane, S.N., Richards, K.S., Biron, P.M. and Roy, A.G. (2001) Role of bed discordance at asymmetrical river confluences. *J. Hydraul. Eng.*, **127**, 351–368.

Church, M. (1996) Space, time and the mountain—how do we order what we see? In: *The Scientific Nature of Geomorphology*. (Eds B.L. Rhoads and C.E. Thorn), pp. 147–170. Wiley, Chichester.

Davoren, A. and Mosley, M.P. (1986) Observations of bedload movement, bar development and sediment supply in braided Ohau River. *Earth Surf. Process. Landf.*, **11**, 643–652.

Ferguson, R.I. (1993) Understanding braiding processes in gravel-bed rivers: progress and unsolved problems. In: *Braided Rivers* (Eds J.L. Best and C.S. Bristow), pp. 73–87. Special Publication 75, Geological Society Publishing House, Bath.

Germanoski, D. and Schumm, S.A. (1993) Changes in braided river morphology resulting from aggradation and degradation. *J. Geol.*, **101**, 451–466.

Kurabayashi, H., Shimizu, Y. and Hoshi, K. (2002) Numerical analysis on bed configuration in braided stream with emerged mid-channel bars. In: *River Flow 2002: Proceedings of the International Conference on Fluvial Hydraulics*, Vol. 2, Louvain la Neuve, Belgium, 4–6 September (Eds D. Bousmar and Y. Zech), pp. 803–808. A.A. Balkema, Lisse.

Lane, S.N., Bradbrook, K.F., Richards, K.S., Biron, P.A. and Roy, A.G. (1999) The application of computational fluid dynamics to natural river channels: three-dimensional versus two-dimensional approaches. *Geomorphology*, **29**, 1–20.

Lewin, J. and Brewer, P. (2001) Predicting channel patterns. *Geomorphology*, **40**, 329–339.

McArdell, B.W. and Faeh, R. (2001) A computational investigation of river braiding. In: *Gravel Bed Rivers V* (Ed. M.P. Mosley), pp. 73–86. New Zealand Hydrological Society, Wellington.

McLelland, S.J., Ashworth, P.J. and Best, J.L. (1996) The origin and development of coherent flow structures at channel junctions. In: *Coherent Flow Structures in Open Channels* (Eds P.J. Ashworth, S.J. Bennett, J.L. Best and S.J. McLelland), pp. 705–723. Wiley, Chichester.

Murray, A.B. (2003) Contrasting the goal, strategies, and predictions associated with simplified numerical models and detailed simulations. In: *Prediction in Geomorphology* (Eds D. Iverson and P. Wilcock), pp. 151–165. Monograph 135, American Geophysical Union, Washington, DC.

Murray, A.B. and Paola, C. (1994) A cellular model of braided rivers. *Nature*, **371**, 54–57.

Murray, A.B. and Paola, C. (1996) A new quantitative test of geomorphic models, applied to a model of braided streams. *Water Resour. Res.*, **32**, 2579–2587.

Murray, A.B. and Paola, C. (1997) Properties of a cellular braided stream model *Earth Surf. Process. Landf.*, **22**, 1001–1025.

Murray, A.B. and Paola, C. (2003) Modeling the effects of vegetation on channel pattern in bedload rivers, *Earth Surf. Process. Landf.*, **28**, 131–143.

Nicholas, A.P. (2000) Modelling bedload yield in braided gravel bed rivers, *Geomorphology*, **36**, 89–106.

Nicholas, A.P. (2003) Investigation of spatially distributed braided river flows using a two-dimensional hydraulic model. *Earth Surf. Process. Landf.*, **28**, 655–674.

Nicholas, A.P. (2005) Roughness parameterization in CFD modelling of gravel-bed rivers. In: *Computational Fluid Dynamics: Applications in Environmental Hydraulics* (Eds P.D. Bates, S.N. Lane and R.I. Ferguson), pp. 329–355. Wiley, Chichester.

Nicholas, A.P. and Sambrook Smith, G.H. (1999) Numerical simulation of three-dimensional flow hydraulics in a braided channel. *Hydrol. Process.*, **13**, 913–929.

Paola, C. (1996) Incoherent structure: Turbulence as a metaphor for stream braiding. In: *Coherent Flow Structures in Open Channels* (Eds P.J. Ashworth, S.J. Bennett, J.L. Best and S.J. McLelland), pp. 705–723. Wiley, Chichester.

Paola, C. (2001) Modelling stream braiding over a range of scales. In: *Gravel Bed Rivers V* (Ed. M.P. Mosley), pp. 11–38. New Zealand Hydrological Society, Wellington.

Reinfelds, I. and Nanson, G. (1993) Formation of braided river floodplains, Waimakariri River, New Zealand. *Sedimentology*, **40**, 1113–1127.

Rhoads, B.L. and Welford, M.R. (1991) Initiation of river meandering. *Progr. Phys. Geogr.*, **15**, 127–56.

Richards, K.S. (2001) Discussion of: Paola, C. (2001) Modelling stream braiding over a range of scales. In: *Gravel Bed Rivers V* (Ed. M.P. Mosley), p. 40. New Zealand Hydrological Society, Wellington.

Sapozhnikov, V.B., Murray, A.B., Paola, C. and Foufoula-Georgiou, E. (1998) validation of braided-stream models: Spatial state-space plots, self-affine scaling, and island shapes. *Water Resour. Res.*, **34**, 2353–2364.

Shimizu, Y. (2002) A method for simultaneous computation of bed and beank deformation of a river. In: *River Flow 2002: Proceedings of the International Conference on Fluvial Hydraulics*, Vol. 2, Louvain la Neuve, Belgium, 4–6 September (Eds D. Bousmar and Y. Zech), pp. 793–801. A.A. Balkema, Lisse.

Thomas, R. and Nicholas, A.P. (2002) Simulation of braided river flow using a new cellular routing scheme. *Geomorphology*, **43**, 179–195.

Thomas, R., Nicholas, A.P. and Quine, T.A. (2002) Development and application of a cellular model to simulate braided river process-form interactions and morphological change. In: *River Flow 2002: Proceedings of the International Conference on Fluvial Hydraulics*, Vol. 2, Louvain la Neuve, Belgium, 4–6 September (Eds D. Bousmar and Y. Zech), pp. 783–791. A.A. Balkema, Lisse.

Yalin, M.S. (1971) On the formation of dunes and meanders. *Proceedings of the 14th International Congress of the International Association for Hydraulic Research*, Paris, Vol. 3, Paper C13, pp. 1–8.

Numerical modelling of alternate bars in shallow channels

A. BERNINI, V. CALEFFI *and* A. VALIANI

Dipartimento di Ingegneria, Università degli Studi di Ferrara, via G. Saragat 1, 44100 Ferrara, Italy
(Email: abernini@ing.unife.it; vcaleffi@ing.unife.it; avaliani@ing.unife.it)

ABSTRACT

A code for the numerical solution of two-dimensional shallow-water equations over a movable bed is developed in order to reproduce the generation, growth and migration of alternate bars in a straight, rectangular channel. Many studies suggest that alternate bar formation may be related to hydrodynamic instability of a cohesionless bed. Such a phenomenon is reproduced by numerical simulations, which are carried out with both supercritical and subcritical uniform flow as the initial reference condition. The instability of the bed is generated by a localized bed disturbance. The evolution of the bed from this initial disturbance is modelled until a dynamic equilibrium of bed topography is achieved, characterized by alternate bars with a defined space–time evolution. Particular attention is devoted to the analysis of gravity effects due to the transverse bed slope on the equilibrium values of the geometric and kinematic bar characteristics. The simulated equilibrium values of bar height and bar length are compared with semi-empirical, experimental and numerical results from the literature.

Keywords River morphodynamics, alternate bars, numerical modelling, fluvial bar evolution, straight channel, transverse slope.

INTRODUCTION

According to the widely adopted bar evolution model of Bridge (1993), bed forms may initially be organized in single or multiple rows of bars. Flow around these bars will then cause non-uniform erosion of banks resulting in increases of sinuosity and channel width, as reported in many laboratory experiments and field observations (Karcz, 1971; Parker, 1976; Ashmore, 1982, 1991; Ferguson & Werritty, 1983; Bridge *et al.*, 1986; Fujita, 1989; Bridge, 1993; Bertoldi *et al.*, 2002). The alternate bar crests then grow and develop into unit bars before subsequently evolving into compound point and braid bars. Thus even the most complex bar forms have a common origin from alternate bars. This simple conceptual model for bar evolution belies a great deal of complexity inherent within braided rivers that has led to a variety of approaches being used to further understanding of these complex systems. For example, there have been studies on theoretical analysis of bar stability (Colombini *et al.*, 1987; Seminara & Tubino, 1989; Seminara, 1998; Federici & Seminara, 2002), the dynamics of confluences (Best, 1987, 1988), the development of quantitative and qualitative models of braided river deposits (Bridge, 1985, 1993), numerical simulation of bar evolution (Takebayashi *et al.*, 2001; Defina & Lanzoni, 2002; Federici & Colombini, 2002; Defina, 2003) in addition to a range of more general laboratory experiments (Ashmore, 1982, 1991; Lanzoni, 2000) and field observations (Bridge *et al.*, 1986; Bristow, 1987; Thorne *et al.*, 1993; Ashworth *et al.*, 2000). Despite these advances, a physically based numerical model that adequately simulates braid bar development does not exist, although some progress has been made with simpler cellular models (e.g. Murray & Paola, 1994).

In the present study the attention is focused on the first part of braided river evolution, with the aim to numerically simulate the physical mechanisms that determine the generation and development of alternate bars in a straight channel. Some recent progress has been made in this area that will be briefly reviewed. Federici & Seminara (2002) demonstrated that perturbations in the river bed behave as wave groups, increasing their amplitude as they migrate downstream. Federici & Colombini (2002) further found that such wave groups are sensitive to the amplitude of the initial perturbations during the growth phase, but are independent of the perturbation amplitude at the asymptotic stage, after an appropriate time and distance from the initial perturbation. These simulations were qualitatively similar to the experimental results of Fujita & Muramoto (1985).

Defina & Lanzoni (2002) also investigated bar train stability with imposed periodic boundary conditions. They found that an asymptotic equilibrium configuration results as an intermediate equilibrium state, although this is not stable over time. They suggested that the variance of bar amplitude over time is important and provides a good indication of the state (equilibrium or not) of the system. Their model was also able to manage partially dry areas, using a subgrid conceptual model for ground irregularities, previously conceived by Defina (2000). This is particularly relevant when flow stage decreases, and the higher portions of bars emerge, causing significant parts of the computational domain to be dry.

Defina (2003) assessed the influence of the initial bed-level perturbation on the flow field and bar form development by varying the initial disturbance. This was either a single bump in the up-stream part of the channel, a periodic (in space) disturbance, a 'mixed' type of perturbation (a certain number of bumps distributed in space along a sidewall), or a periodic in time but localized in space bump (at the upstream inlet of the channel). In the last case, the bar celerity was shown to depend on the bump frequency generation, at least for sufficiently high values of such a frequency. However, generally, bar characteristics were shown to be dependent on the history of their generation, and influenced by the space/time distribution of the generating disturbance. On the contrary, equilibrium wavelength, height and celerity are related to one another, no matter what the type of generating disturbance. In addition to studies of the response of alternate bar evolution to the initial perturbation, Takebayashi & Egashira (2001) analysed the influence of non-uniform sediment size. Bar wave height for a non-uniform sediment bed was smaller than for a uniform bed. Additionally, longitudinal and transverse sediment sorting significantly affects the sediment discharge, and also influences the wavelength (which decreases for non-uniform sediments) and the migration speed (which, on the contrary, increases).

The present work builds on these previous studies in two key ways. First, consideration is given to modelling bar growth under supercritical and subcritical flow conditions as well as transitions between these states that may arise during bar evolution. Takebayashi et al. (1999) suggested such a capability of a numerical model to be important. Second, much greater consideration is given to modelling the influence of gravity on sediment transport across bars and how this influences bar evolution and growth. A first contribution on these topics is given in a related paper (Bernini et al., 2003). In order to validate the model, the numerical results are compared with laboratory data instead of field data. This is due to the greater reliability and completeness of laboratory data, and also to avoid additional difficulties related to the presence of vegetation, cohesive sediments and sediment heterogeneity. In this work simulations characterized by natural field parameters were not performed, however it is reasonable to suggest that the simulated bars could be comparable with those observed in natural rivers. In fact, no scale effects are observed in computing the flow field and the bar shape passing from small channels (Takebayashi et al., 1998, 1999, 2001) to large channels (Lanzoni, 2000).

BASIC EQUATIONS

Hydrodynamic modelling

The classic two-dimensional shallow-water scheme is used to describe the hydrodynamics of the system. This is justified by the small values of the ratio between vertical and horizontal bar dimensions, whose heights are comparable with the flow depth, and whose lengths scale with the channel width.

In the present model, no attempt is made for simulating the vertical velocity distribution. This simplification is justified based on the low sinuosity of the modelled stream; depth-averaged velocity vectors are nearly parallel to the channel axis and consequently the stream curvature value is always very low. Thus the effects of secondary currents and sinuous flow in general are negligible. It is known that the numerical simulation of typical mesoscale bed-form (i.e. dunes) generation cannot avoid a detailed turbulence modelling (Shimizu et al., 1999), which only is able to take into account the recirculation mechanism in the lee side of the dune. On the contrary, the large spatial scale of bars leads to a limited local flow separation, which is confined in very small regions near the front, and thus the overall flow configuration is weakly affected by turbulence phenomena. Such a consideration is supported by three-dimensional model results (Vignoli & Tubino, 2002). As a direct consequence, detailed representation of turbulence effects was not performed.

The mass and momentum balance equations, written in conservative form, are:

$$\frac{\partial U}{\partial t} + \nabla \cdot F + S = 0 \tag{1}$$

where t is time, U is the vector of conservative variables, F is the flux term matrix and S is the source term. Introducing a Cartesian reference system (x, y, z) in which the z axis is vertical and the x–y plane is horizontal, Eq. 1 may be written in the following form as:

$$\frac{\partial U}{\partial t} + \frac{\partial E}{\partial x} + \frac{\partial G}{\partial y} + S = 0 \tag{2}$$

where E and G are Cartesian components of the flux term. Making explicit the role of physical variables, the expression of these vectors is:

$$U = \begin{bmatrix} h \\ Uh \\ Vh \end{bmatrix}; \quad E = \begin{bmatrix} Uh \\ U^2h + g\frac{h^2}{2} \\ UVh \end{bmatrix};$$

$$G = \begin{bmatrix} Vh \\ UVh \\ V^2h + g\frac{h^2}{2} \end{bmatrix}; \quad S = \begin{bmatrix} 0 \\ gh(S_{fx} - S_{0x}) \\ gh(S_{fy} - S_{0y}) \end{bmatrix} \tag{3}$$

where g is the acceleration due to gravity, h is the water depth and U and V are the vertically depth-averaged velocities in the x and y directions. Variables S_{0x} and S_{0y} correspond to the Cartesian components of the bed slope. Using z_b as the bed elevation then it follows that:

$$S_{0x} = -\frac{\partial z_b}{\partial x}; \quad S_{0y} = -\frac{\partial z_b}{\partial y} \tag{4}$$

Finally, S_{fx} and S_{fy} are the Cartesian components of the friction slope, evaluated by using empirical resistance relationships in the Chézy form:

$$S_{fx} = \frac{U\sqrt{U^2 + V^2}}{C^2 gh}; \quad S_{fy} = \frac{V\sqrt{U^2 + V^2}}{C^2 gh} \tag{5}$$

The non-dimensional Chézy coefficient C is evaluated by a simple power law: $C = C_0(h/h_0)^{1/6}$, where the 0 subscript refers to uniform reference flow, whilst current flow status is written without indexes (Chow, 1959).

Sediment transport and bed evolution modelling

Bed evolution is modelled using the standard equation for the sediment balance derived by Exner (1925). Bed load transport is assumed to be dominant with respect to suspended load transport and a uniform grain size is used. Under these assumptions, the Exner relationship may be written as:

$$\frac{\partial z_b}{\partial t} = -\frac{1}{1-p} \nabla \cdot q_b \tag{6}$$

where p is bed porosity (assumed to be equal to 0.4 for each simulation) and q_b is the vectorial volumetric bed load discharge. Introducing the magnitude, q_b, and the direction with respect to the x axis, γ, of the bed load transport, the Cartesian components of q_b are defined by:

$$q_{bx} = q_b \cos \gamma; \quad q_{by} = q_b \sin \gamma \tag{7}$$

and thus Eq. 6 becomes:

$$\frac{\partial z_b}{\partial t} = -\frac{1}{1-p}\left(\frac{\partial q_{bx}}{\partial x} + \frac{\partial q_{by}}{\partial y}\right) \tag{8}$$

q_b is calculated using the Meyer-Peter & Muller (1948) formula:

$$q_b = 8\sqrt{g(S_g - 1)d^3}\,(\theta - \theta_{cr})^{3/2} \qquad (9)$$

where d is the particle diameter, S_g is the specific density of sediment (assumed to be equal to 2.65 for each simulation) and θ and θ_{cr} are the non-dimensional shear stress and its critical value respectively. According to the Meyer-Peter & Muller (1948) approach a value of $\theta_{cr} = 0.047$ is used.

Studies have shown (e.g. Hasegawa, 1981; Ikeda, 1982; Parker, 1984; Kovacs & Parker, 1994; Talmon et al., 1995) that when a finite local bed slope is present the component of the gravitational force tangential to the bed affects the local direction of sediment transport such that it deviates from the flow direction. Each of the above mentioned approaches leads to a relationship between the deviation angle, the local bed slope and the local shear stress. As reported by Sekine & Parker (1992), assuming a Cartesian reference system s–n, with s and n axes parallel and transverse to the average flow direction respectively, these different relations can be reduced to the following form:

$$\tan\beta = \frac{q_{bn}}{q_{bs}} = \frac{u_{bn}}{u_{bs}} - k\frac{\partial z_b}{\partial n} \qquad (10)$$

where q_{bn} and q_{bs} are the transverse and downstream sediment transport components respectively; β is the angle between the local direction of sediment transport and the average flow direction (Fig. 1); u_{bn} and u_{bs} are the corresponding near-bed velocities linked to secondary flows; k is a dimensionless parameter given by:

$$k = k^*(\theta_{cr}/\theta)^m \qquad (11)$$

with k^* and m coefficients depending on the selected approach. Neglecting secondary flows ($u_{bn} = 0$) and selecting the approach proposed by Ikeda (1982), subsequently modified by Parker (1984) and Tubino et al. (1999), Eq. 10 becomes:

$$\tan\beta = \frac{q_{bn}}{q_{bs}} = -\frac{r}{\sqrt{\theta}}\frac{\partial z_b}{\partial n} \qquad (12)$$

in which the magnitude of the deviation angle increases with increasing local transverse bed slope. The parameter r varies with the critical shear stress and has values that range between 0.3 and 0.6 (Sekine & Parker, 1992). The direction of bed load transport is given by:

$$\gamma = \alpha + \beta \qquad (13)$$

where $\alpha = \tan(V/U)$ is the angle between the average flow direction and the x axis.

NUMERICAL MODEL

Given that the time-scale of bed evolution is much greater than the time-scale of the fluid flow, the use of a decoupled scheme is appropriate in which the bed is assumed to be fixed during each time step of the hydraulic computations. Numerical integration of shallow water equations is performed adopting an alternating-direction implicit (ADI), Beam–Warming, three-point backward, finite difference method that is second-order accurate in space and time (Beam & Warming, 1976; Hirsch, 1990; Chaudhry, 1993; Valiani, 1998). The classic Beam–Warming implicit scheme has been modified in order to reduce (to a negligible quantity) the mass conservation error related to the critical transition, as suggested by Jha et al. (1994; see also Jha et al., 1996).

Basic formulation

The structure of Beam–Warming schemes may be described starting from a time-difference approximation of the form:

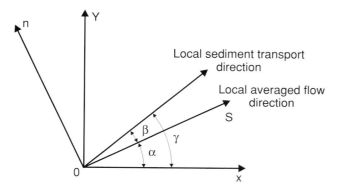

Fig. 1 Reference systems and angles of deviations of averaged flow and sediment transport rates.

$$U^{k+1} = U^k + \Delta t \left[\frac{\varsigma}{1+\xi} \left(\frac{\partial U}{\partial t} \right)^{k+1} + \frac{1-\varsigma}{1+\xi} \left(\frac{\partial U}{\partial t} \right)^k + \frac{\xi}{1+\xi} \left(\frac{\partial U}{\partial t} \right)^{k-1} \right] \quad (14)$$

where superscripts denote time step. The term $k + 1$ refers to the advanced time-step of unknown variables, while k and $k - 1$ refer to the time step of current variables and previously calculated variables, and ς and ξ are the Beam-Warming weights. Different values of ς and ξ lead to different formulations of the scheme. In this work a three-point backward formulation ($\varsigma = 1$, $\xi = 1/2$) is adopted. Using Eq. 1, Eq. 14 becomes:

$$U^{k+1} = U^k - \Delta t \left[\frac{\varsigma}{1+\xi} \left(\frac{\partial E}{\partial x} + \frac{\partial G}{\partial y} + S \right)^{k+1} + \frac{1-\varsigma}{1+\xi} \left(\frac{\partial E}{\partial x} + \frac{\partial G}{\partial y} + S \right)^k \right] + \frac{\xi \Delta t}{1+\xi} \left(\frac{\partial U}{\partial t} \right)^{k-1} \quad (15)$$

The treatment of the spatial derivatives of E and G at the advanced time-step is performed introducing a local series expansion of the first order of flux terms themselves, for instance:

$$E^{k+1} = E^k + \Delta t \frac{\partial E^k}{\partial t} \quad (16)$$

Using the chain rule, introducing the Jacobian matrix $\mathbf{A} = \mathbf{A}(U) = \partial E / \partial U$, and a proper discretized form for the time derivative of U, Eq. 16 becomes:

$$E^{k+1} = E^k + \mathbf{A}^k (U^{k+1} - U^k) \quad (17)$$

Similar expansions are obtained for G^{k+1} and S^{k+1}:

$$G^{k+1} = G^k + \mathbf{B}^k (U^{k+1} - U^k)$$
$$S^{k+1} = S^k + \mathbf{Q}^k (U^{k+1} - U^k) \quad (18)$$

where $\mathbf{B}$ and $\mathbf{Q}$ are the Jacobian matrices of G and S, respectively. Inserting Eqs 17 & 18 into Eq. 15, after some manipulation, the following relation is obtained:

$$U^{k+1} + \Delta t \frac{\varsigma}{1+\xi} \left[\frac{\partial}{\partial x} (\mathbf{A}^k U^{k+1}) + \frac{\partial}{\partial y} (\mathbf{B}^k U^{k+1}) + \mathbf{Q}^k U^{k+1} \right]$$
$$= U^k + \Delta t \frac{\varsigma}{1+\xi} \left[\frac{\partial}{\partial x} (\mathbf{A}^k U^k) + \frac{\partial}{\partial y} (\mathbf{B}^k U^k) + \mathbf{Q}^k U^k \right] +$$
$$- \Delta t \frac{1}{1+\xi} \left(\frac{\partial E}{\partial x} + \frac{\partial G}{\partial x} + S \right)^k + \Delta t \frac{\xi}{1+\xi} \left(\frac{\partial U}{\partial t} \right)^{k-1} \quad (19)$$

which, in delta form, becomes:

$$\Delta U^k + \Delta t \frac{\varsigma}{1+\xi} \left[\frac{\partial}{\partial x} (\mathbf{A}^k \Delta U^k) + \frac{\partial}{\partial y} (\mathbf{B}^k \Delta U^k) + \mathbf{Q}^k \Delta U^k \right]$$
$$= RHS^k \quad (20)$$

with:

$$RHS^k = -\Delta t \frac{1}{1+\xi} \left(\frac{\partial E}{\partial x} + \frac{\partial G}{\partial x} + S \right)^k + \frac{\xi}{1+\xi} \Delta U^{k-1} \quad (21)$$

and with $\Delta U^k = U^{k+1} - U^k$.

Alternating-direction implicit factorization

The resultant linear system, represented by Eqs 20 & 21, is characterized by a matrix coefficient with a large bandwidth: its numerical solution, even if possible, requires a large amount of computational resources. However, this system may be solved more efficiently by applying factoring techniques to the left-hand side (Beam & Warming, 1976). The selected approach is a two-step alternating-direction implicit (ADI) procedure, in which a two-dimensional problem is reduced to a pair of one-dimensional problems, with a corresponding coefficient matrix with a small bandwidth. The ADI procedure is described by the following expressions:

Step 1

$$\Delta \tilde{U}^k + \Delta t \frac{\varsigma}{1+\xi} \left[\frac{\partial}{\partial x} (\mathbf{A}^k \Delta \tilde{U}^k) + \mathbf{Q}_x^k \Delta \tilde{U}^k \right]$$
$$= RHS^k \quad (22)$$

Step 2

$$\Delta U^k + \Delta t \frac{\varsigma}{1+\xi}\left[\frac{\partial}{\partial y}(\mathbf{B}^k \Delta U^k) + \mathbf{Q}_y^k \Delta U^k\right] = \Delta \tilde{U}^k \quad (23)$$

where $\Delta \tilde{U}$ is the time increment of the vector of conservative variables after the first step and the Jacobian matrix $\mathbf{Q}$ is split into $\mathbf{Q}_x$ and $\mathbf{Q}_y$, which are defined as:

$$\mathbf{Q}_x = \frac{\partial S_x}{\partial U}; \quad \mathbf{Q}_y = \frac{\partial S_y}{\partial U} \quad (24)$$

with:

$$S_x = \begin{bmatrix} 0 \\ gh(S_{fx} - S_{0x}) \\ 0 \end{bmatrix}; \quad S_y = \begin{bmatrix} 0 \\ 0 \\ gh(S_{fy} - S_{0y}) \end{bmatrix} \quad (25)$$

In the classic formulation of the scheme (Chaudhry, 1993), the splitting of $\mathbf{Q}$ is not performed and this element appears only in the second step of the described approach (Eq. 23). In this work only the decomposition described by Eq. 24, based on the physical meaning of different terms, is tested (Valiani, 1998).

Jacobian decomposition and derivatives discretization

In order to achieve a numerical model able to simulate flows in which subcritical and supercritical states may coexist, an upwind discretization of the spatial derivatives, based on a proper decomposition of Jacobian matrixes, is performed. The Jacobian matrices $\mathbf{A}$ and $\mathbf{B}$ may be written in diagonalized form as follows:

$$\mathbf{A} = \mathbf{A}_R \cdot \mathbf{D}_A \cdot \mathbf{A}_L; \quad \mathbf{B} = \mathbf{B}_R \cdot \mathbf{D}_B \cdot \mathbf{B}_L \quad (26)$$

with $\mathbf{A}_R$ and $\mathbf{A}_L$, $\mathbf{B}_R$ and $\mathbf{B}_L$ being the right and left eigenvectors matrixes of $\mathbf{A}$ and $\mathbf{B}$, respectively, and $\mathbf{D}_A$ and $\mathbf{D}_B$ the diagonal eigenvalues matrix of $\mathbf{A}$ and $\mathbf{B}$. The following quantities have to be defined (no summation over repeated indexes):

$$\begin{aligned} D_{Aii}^+ &= \max(D_{Aii}, 0); \quad D_{Aii}^- = \min(D_{Aii}, 0) \\ D_{Bii}^+ &= \max(D_{Bii}, 0); \quad D_{Bii}^- = \min(D_{Bii}, 0) \end{aligned} \quad (27)$$

where D_{Aii} and D_{Bii} are the eigenvalues of $\mathbf{A}$ and $\mathbf{B}$, respectively. Introducing:

$$\begin{aligned} \mathbf{A}^+ &= \mathbf{A}_R \cdot \mathbf{D}_A^+ \cdot \mathbf{A}_L; \quad \mathbf{A}^- = \mathbf{A}_R \cdot \mathbf{D}_A^- \cdot \mathbf{A}_L \\ \mathbf{B}^+ &= \mathbf{B}_R \cdot \mathbf{D}_B^+ \cdot \mathbf{B}_L; \quad \mathbf{B}^- = \mathbf{B}_R \cdot \mathbf{D}_B^- \cdot \mathbf{B}_L \end{aligned} \quad (28)$$

the following decompositions are possible:

$$\begin{aligned} \frac{\partial}{\partial x}(\mathbf{A}^k \Delta \tilde{U}^k) &= \frac{\partial}{\partial x}(\mathbf{A}^{+k} \Delta \tilde{U}^k) + \frac{\partial}{\partial x}(\mathbf{A}^{-k} \Delta \tilde{U}^k) \\ \frac{\partial}{\partial y}(\mathbf{B}^k \Delta \tilde{U}^k) &= \frac{\partial}{\partial y}(\mathbf{B}^{+k} \Delta \tilde{U}^k) + \frac{\partial}{\partial y}(\mathbf{B}^{-k} \Delta \tilde{U}^k) \end{aligned} \quad (29)$$

Spatial derivatives have to be discretized taking into account the sign of the eigenvalues of Jacobian matrices: using backward finite differences associated to positive eigenvalues (i.e. associated to matrices $\mathbf{A}^+$ and $\mathbf{B}^+$) and forward finite differences associated to negative eigenvalues (i.e. associated to matrices $\mathbf{A}^-$ and $\mathbf{B}^-$), the discrete form of Eqs 22 & 23 becomes:

$$\begin{aligned} \Delta \tilde{U}_{i,j}^k &+ \frac{\varsigma}{1+\xi}\frac{\Delta t}{\Delta x}(\mathbf{A}_{i,j}^{+k}\Delta \tilde{U}_{i,j}^k - \mathbf{A}_{i-1,j}^{+k}\Delta \tilde{U}_{i-1,j}^k) \\ &+ \frac{\varsigma}{1+\xi}\frac{\Delta t}{\Delta x}(\mathbf{A}_{i+1,j}^{-k}\Delta \tilde{U}_{i+1,j}^k - \mathbf{A}_{i,j}^{-k}\Delta \tilde{U}_{i,j}^k) \\ &+ \frac{\varsigma}{1+\xi}\Delta t \mathbf{Q}_{xi,j}^k \Delta \tilde{U}_{i,j}^k = RHS_{i,j}^k) \end{aligned} \quad (30)$$

and

$$\begin{aligned} \Delta U_{i,j}^k &+ \frac{\varsigma}{1+\xi}\frac{\Delta t}{\Delta y}(\mathbf{B}_{i,j}^{+k}\Delta U_{i,j}^k - \mathbf{B}_{i,j-1}^{+k}\Delta U_{i,j-1}^k) \\ &+ \frac{\varsigma}{1+\xi}\frac{\Delta t}{\Delta y}(\mathbf{B}_{i,j+1}^{-k}\Delta U_{i,j+1}^k - \mathbf{B}_{i,j}^{-k}\Delta U_{i,j}^k) \\ &+ \frac{\varsigma}{1+\xi}\Delta t \mathbf{Q}_{yi,j}^k \Delta U_{i,j}^k = \Delta \tilde{U}_{i,j}^k \end{aligned} \quad (31)$$

Regarding the discretization of $RHS_{i,j}^k$, two different formulations are presented in the following, but only the latter is implemented in the final version of the numerical model. In fact, preliminary tests (not reported here) suggest a much better performance of such a model, as specified in the following.

Standard formulation for RHS discretization

In the original formulation of the Beam–Warming scheme, $RHS_{i,j}^k$ is obtained by discretizing Eq. 21 using the approach previously applied to the left-hand side. After the linearization $E = \mathbf{A}U$ and $G = \mathbf{B}U$, using the Jacobian matrices factorization, spatial derivatives in Eq. 21 may be written as:

$$\frac{\partial E}{\partial x} = \mathbf{A}\frac{\partial U}{\partial x} = \mathbf{A}^+\frac{\partial U}{\partial x} + \mathbf{A}^-\frac{\partial U}{\partial x}$$
$$\frac{\partial G}{\partial y} = \mathbf{B}\frac{\partial U}{\partial y} = \mathbf{B}^+\frac{\partial U}{\partial y} + \mathbf{B}^-\frac{\partial U}{\partial y} \quad (32)$$

Discretizing Eq. 32 according to the sign of the eigenvalues, the expression for $RHS_{i,j}^k$ becomes:

$$RHS_{i,j}^k = -\frac{1}{1+\xi}\frac{\Delta t}{\Delta x}[\mathbf{A}_{i,j}^{+k}(U_{i,j}^k - U_{i-1,j}^k)] +$$
$$-\frac{1}{1+\xi}\frac{\Delta t}{\Delta x}[\mathbf{A}_{i,j}^{-k}(U_{i+1,j}^k - U_{i,j}^k)] +$$
$$-\frac{1}{1+\xi}\frac{\Delta t}{\Delta y}[\mathbf{B}_{i,j}^{+k}(U_{i,j}^k - U_{i,j-1}^k)] + \quad (33)$$
$$-\frac{1}{1+\xi}\frac{\Delta t}{\Delta y}[\mathbf{B}_{i,j}^{-k}(U_{i,j+1}^k - U_{i,j}^k)] +$$
$$-\frac{\Delta t}{1+\xi}S_{i,j}^k + \frac{\xi}{1+\xi}\Delta U_{i,j}^{k-1}$$

The main limitation of this formulation is the linearization of fluxes E and G, which are strongly non-linear functions of U. The resulting numerical scheme, after derivatives discretization, is non-conservative. While this formulation gives satisfactory results when the expected solution is continuous, a mass-conservation error is found when the expected solution is discontinuous. In order to reduce such an error, a conservative linearization must be performed.

Modified formulation for RHS discretization

The modification, proposed by Jha et al. (1994, 1996) for one-dimensional open channel flow, is based on a conservative evaluation of the space derivative of flux terms E and G. Such a modification is extended here to a two-dimensional framework.

This approach is based on Roe's theory on approximate Riemann solvers for Euler equations (Roe, 1981). The key elements of this approach are appropriate approximate Jacobian matrices ($\hat{\mathbf{A}}$ and $\hat{\mathbf{B}}$) that can be used in a linear conservative approximation of flux terms. Such approximate Jacobian matrices are constructed for each pair of adjacent computational nodes in the x and y directions, separately. For example, considering the x direction, the matrix $\hat{\mathbf{A}}$ and the two adjacent nodes (x_i, y_j) and (x_{i+1}, y_j), the general form of $\hat{\mathbf{A}}$ is:

$$\hat{\mathbf{A}}_{i+1/2,j} = \hat{\mathbf{A}}(U_{i,j}, U_{i+1,j}) \quad (34)$$

The explicit expression of such a function must satisfy the following properties (Roe, 1981; Toro, 1999).

1 Consistency with the exact Jacobian:

$$\hat{\mathbf{A}}(U, U) = \mathbf{A}(U) \quad (35)$$

2 Conservative behaviour across discontinuities:

$$E(U_{i+1,j}) - E(U_{i,j}) = \hat{\mathbf{A}}_{i+1/2,j}(U_{i+1,j} - U_{i,j}) \quad (36)$$

3 Hyperbolicity of the system: $\hat{\mathbf{A}}$ must have real eigenvalues and a complete set of linearly independent right eigenvectors.

Following the approach proposed by Roe & Pike (1985) and introducing the Roe averages:

$$h_{i+1/2,j} = \sqrt{h_{i,j}h_{i+1,j}};$$
$$U_{i+1/2,j} = \frac{U_{i,j}\sqrt{h_{i,j}} + U_{i+1,j}\sqrt{h_{i+1,j}}}{\sqrt{h_{i,j}} + \sqrt{h_{i+1,j}}}; \quad (37)$$
$$V_{i+1/2,j} = \frac{V_{i,j}\sqrt{h_{i,j}} + V_{i+1,j}\sqrt{h_{i+1,j}}}{\sqrt{h_{i,j}} + \sqrt{h_{i+1,j}}}$$

and the corresponding vector of conserved variables $U_{i+1/2,j}$, approximate Jacobian matrices satisfying the previously introduced requirements are obtained using:

$$\hat{\mathbf{A}}_{i+1/2,j} = \hat{\mathbf{A}}(U_{i,j}, U_{i+1,j}) = \mathbf{A}(U_{i+1/2,j}) \quad (38)$$

Similar expressions are valid for $\hat{\mathbf{A}}_{i-1/2,j}$, $\hat{\mathbf{B}}_{i,j+1/2}$ and $\hat{\mathbf{B}}_{i,j-1/2}$.

The scheme of the standard approach may now be modified using the substitutions: $\hat{\mathbf{A}} \to \mathbf{A}$ and $\hat{\mathbf{B}} \to \mathbf{B}$. After the now conservative linearization, $E = \hat{\mathbf{A}}U$ and $G = \hat{\mathbf{B}}U$, space derivatives of fluxes inside the right-hand side of Eq. 21 may be factorized as follows:

$$\frac{\partial E}{\partial x} = \hat{\mathbf{A}}\frac{\partial U}{\partial x} = \hat{\mathbf{A}}^+\frac{\partial U}{\partial x} + \hat{\mathbf{A}}^-\frac{\partial U}{\partial x}$$
$$\frac{\partial G}{\partial y} = \hat{\mathbf{B}}\frac{\partial U}{\partial y} = \hat{\mathbf{B}}^+\frac{\partial U}{\partial y} + \hat{\mathbf{B}}^-\frac{\partial U}{\partial y}$$
(39)

Matrixes $\hat{\mathbf{A}}^+$, $\hat{\mathbf{A}}^-$, $\hat{\mathbf{B}}^+$ and $\hat{\mathbf{B}}^-$ are computed using the procedure described by Eqs 26, 27 and 28 with the substitutions $\hat{\mathbf{A}} \to \mathbf{A}$ and $\hat{\mathbf{B}} \to \mathbf{B}$.

Finally, the discretized expression for $RHS_{i,j}^k$ is:

$$RHS_{i,j}^k = -\frac{1}{1+\xi}\frac{\Delta t}{\Delta x}[\hat{\mathbf{A}}_{i-1/2,j}^{+k}(U_{i,j}^k - U_{i-1,j}^k)] +$$
$$-\frac{1}{1+\xi}\frac{\Delta t}{\Delta x}[\hat{\mathbf{A}}_{i+1/2,j}^{-k}(U_{i+1,j}^k - U_{i,j}^k)] +$$
$$-\frac{1}{1+\xi}\frac{\Delta t}{\Delta y}[\hat{\mathbf{B}}_{i,j-1/2}^{+k}(U_{i,j}^k - U_{i,j-1}^k)] + \quad (40)$$
$$-\frac{1}{1+\xi}\frac{\Delta t}{\Delta y}[\hat{\mathbf{B}}_{i,j+1/2}^{-k}(U_{i,j+1}^k - U_{i,j}^k)] +$$
$$-\frac{\Delta t}{1+\xi}S_{i,j}^k + \frac{\xi}{1+\xi}\Delta U_{i,j}^{k-1}$$

Numerical discretization of Exner equation

Bed updating is performed by discretizing the Exner (1925) equation, decoupled from the governing equations of the liquid phase:

$$\Delta z_{b\,i,j}^k = -\frac{\varsigma}{(1+\xi)}\frac{\Delta t}{(1-p)}[\nabla \cdot q_b]_{i,j}^{k+1} +$$
$$-\frac{(1-\varsigma)}{(1+\xi)}\frac{\Delta t}{(1-p)}[\nabla \cdot q_b]_{i,j}^k + \frac{\xi}{(1+\xi)}\Delta z_{b\,i,j}^{k-1} \quad (41)$$

$$z_{b\,i,j}^{k+1} = z_{b\,i,j}^k + \Delta z_{b\,i,j}^k \quad (42)$$

The divergence of the sediment discharge must be computed before and after the updating of hydrodynamics:

$$[\nabla \cdot q_b]_{i,j}^k = \frac{q_{bx\,i+1/2,j}^k - q_{bx\,i-1/2,j}^k}{\Delta x} + \frac{q_{by\,i,j+1/2}^k - q_{by\,i,j-1/2}^k}{\Delta y}$$

$$[\nabla \cdot q_b]_{i,j}^{k+1} = \frac{q_{bx\,i+1/2,j}^{k+1} - q_{bx\,i-1/2,j}^{k+1}}{\Delta x} + \frac{q_{by\,i,j+1/2}^{k+1} - q_{by\,i,j-1/2}^{k+1}}{\Delta y}$$
(43)

NUMERICAL SIMULATIONS

Two numerical simulations (Case 1 and Case 2) were carried out in order to evaluate the influence of gravity effects, due to a transverse bed slope, on the development of alternate bars. The effect of gravitational force on bedload particle movement was calculated using Eq. 12 (Tubino et al., 1999) in both simulations. Case 1 and Case 2 differ only in the chosen value of the r parameter in Eq. 12; r is assumed to be 0.6 and 0.3 in Case 1 and Case 2 respectively. The former value was proposed by Hasegawa (1981; as reported in Sekine & Parker, 1992), the latter is from Parker (1984). In Case 1 and Case 2 flow is initially supercritical. A further numerical simulation, Case 3, was carried out with subcritical flow in order to test the capability of the code to manage different flow conditions.

General framework

Hydraulic and sedimentological parameters and the geometry of the simulated channel used in Case 1 and Case 2 are similar to those used by Takebayashi et al. (1999) in their numerical analysis. A 60×0.3 m^2 channel, with fixed and impermeable lateral walls and characterized by a constant initial slope, was used for the numerical domain in these simulations. With insufficient grid resolution bed perturbations can readily disappear, even for those cases in which experimental observations confirm bar growth. Thus some test cases were performed in order to detect an appropriate maximum grid size. Such an analysis suggests the use of 0.03×0.01 m^2 grid elements. The use of an implicit scheme allows the choice of a Courant number above unity. The developed model maintains a stable behaviour up to a value of 20 for such a parameter. However, in order to achieve a high accuracy and to avoid one-dimensionalization effects a Courant number equal to 3 was assumed.

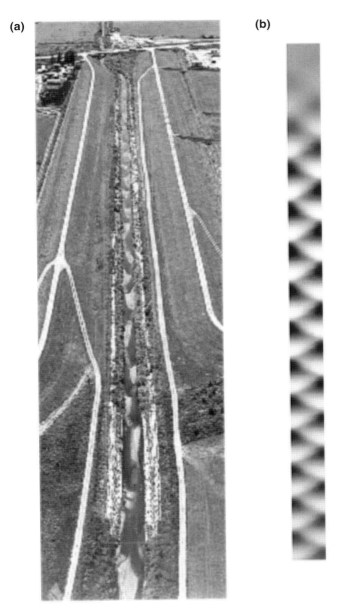

Fig. 7 Comparison between (a) natural alternate bars in Reisakubetsu River, Japan and (b) simulated alternate bars after 2 h (7200 s) from the beginning of the run. (Reisakubetsu River photograph courtesy of the Civil Engineering Research Institute of Hokkaido.)

reduction of bar heights. The amplification of bar equilibrium height due to the reduction of the gravity term is in agreement with results obtained by Takebayashi *et al.* (1999).

For Case 2, a stable configuration is reached faster than in Case 1 (Fig. 9a & b). Additionally, equilibrium height is reached by bars generated at the early stage of the simulation in contrast to Case 1. Bars generated under subcritical conditions (Case 3, Fig. 9c) are characterized by a slower development. Bars formed during the first stage of simulation are longer lived with respect to the corresponding bed forms generated under supercritical flow conditions. The asymptotic state is not fully achieved because the channel is not sufficiently long to allow a dynamic equilibrium stable configuration to be reached.

A comparison between the equilibrium values of bar height obtained by the numerical analysis of Case 1, Case 2 and Case 3, the numerical results of Takebayashi *et al.* (1998) and experimental data obtained by Hasegawa & Yamaoka (1982; as reported by Takebayashi *et al.*, 1999) and by Shimizu (1991; as reported in Takebayashi *et al.*, 1998) is shown in Fig. 9d, in which dimensionless bar height is plotted against width to depth ratio. Good agreement exists between the equilibrium values of bar height, obtained considering $r = 0.3$, and the experimental results of Hasegawa & Yamaoka (1982). The increasing gravity effects, related to local transverse slope, act to suppress the growth of bed forms. Given that a value of 0.6 is considered as the upper limit of variability for the r parameter, the corresponding equilibrium height can be considered the lowest value for the bar heights to have any real physical significance. This consideration is also supported by the comparison of this result with experimental data, which appear to be underestimated if a 0.6 value is adopted.

In Fig. 9d equilibrium bar heights relative to the corresponding numerical simulations carried out by Takebayashi *et al.* (1999) are also plotted. Such simulations, named (a) and (b) by the authors, differ from each other in that the effect of gravity in (a) is double that of simulation (b). A comparison between bar heights of Case 1 with Case (a) and Case 2 with Case (b) suggests a good agreement between the two approaches, even though they have quite different numerical formulations and sediment transport formula. Several curves are plotted in order to present the results in a more general context. Such curves represent the functional relationships between the dimensionless bar height and the width to depth ratio, expressed by three different semi-empirical formulae from Ikeda (1984, 1990) and Yalin (1992). In this analysis attention is focused on the comparison between numerical and semi-empirical results for a fixed width to

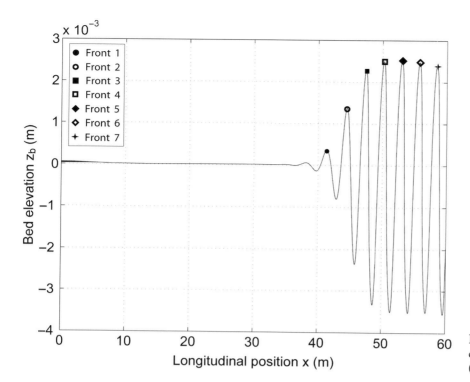

Fig. 8 Longitudinal position of the crests of the seven selected bars (Case 1) after t = 3 h (10,800 s).

depth ratio. A detailed study of the role of B/h_0 on numerical results has not been performed. These formulae are characterized by the following analytical expressions:

$$\frac{H_b}{h_0} = 9.34 \left(\frac{B}{d}\right)^{-0.45} \exp\left[f\left(\frac{B}{h_0}\right)\right] \quad (44)$$

with:

$$f\left(\frac{B}{h_0}\right) = 2.53 \text{ erf}\left(\frac{\log_{10}(B/h_0) - 1.22}{0.594}\right) \quad (45)$$

from Ikeda (1990);

$$\frac{H_b}{h_0} = 0.044 \left(\frac{B}{h_0}\right)^{1.45} \left(\frac{B}{d}\right)^{-0.45} \quad (46)$$

from Ikeda (1984); and

$$\frac{H_b}{h_0} = 0.18 \left(\frac{B}{h_0}\right) \left(\frac{B}{d}\right)^{-0.45} \quad (47)$$

from Yalin (1992).

In Fig. 10, a time sequence of bed elevation with respect to the initial flat configuration is plotted from Case 1. Three longitudinal sections corresponding to the left and right side and the centreline of the channel are used for the analysis. Figure 10a shows the bed perturbation after 1 h (3600 s) from the beginning of the simulation. Bed forms located upstream of the bar with maximum height disappear without achieving the equilibrium value shown in Fig. 9a, while bed forms downstream of the maximum amplitude bar continue to evolve. After 2 h (7200 s), bed forms located downstream of the highest bar increase in height, with dimensions that tend towards asymptotic values, as suggested in Fig. 10b. The bar train plotted after 3 h (10,800 s) is characterized by a stable configuration in which each single bar reaches the equilibrium height during its development. After 4 h (14,400 s), the bed perturbation exits the computational domain, leaving an unperturbed bed configuration in the upstream portion of the channel. Maximum deposition height and maximum bed scour occur along the lateral sides of the domain, as suggested by all the four plots shown in Fig. 10.

Bar length

Bar length is calculated as the longitudinal distance between two subsequent zero-downcrossing points,

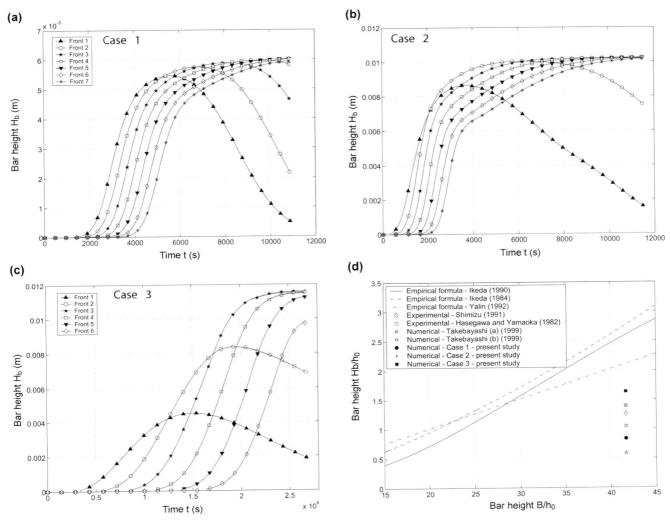

Fig. 9 Bar evolution with time. (a)–(c) Cases 1–3. (d) Comparison between numerical results, empirical formulae and experimental data.

located around the position of the bar crest. The evolution of bar lengths is shown in Fig. 11a–c for Cases 1–3, respectively. In order to reduce the influence of the numerical resolution on the computation of bar length, a 21-point moving average filter was adopted. The general trend of the plotted curves shows that bar length increases its value until an asymptotic equilibrium value is reached (Colombini *et al.*, 1987). Each curve is plotted from the time in which the selected bar height is higher than a threshold value of about 0.9×10^{-4} m, which is less than the grain size. Bed perturbations characterized by heights lower than the selected threshold are not considered as physically significant.

During the last stages of development the bars deviate from the asymptotic trend as length increases; they also become less steep when compared with the early stage of their evolution. Bar length is strongly controlled by the type of conditions prescribed at the open boundaries of the computational domain, as suggested by Defina (2003). Periodic conditions limit the range of values that bar lengths can have (Defina & Lanzoni, 2002), because the number of bars contained within the domain must be an integer. Non-periodic boundary conditions are only used in order to influence as little as possible the main characteristics of simulated bars. The equilibrium bar length tends to a final value of 2.8 m in Case 1 (Fig. 11a) and 3.5 m in Case 2

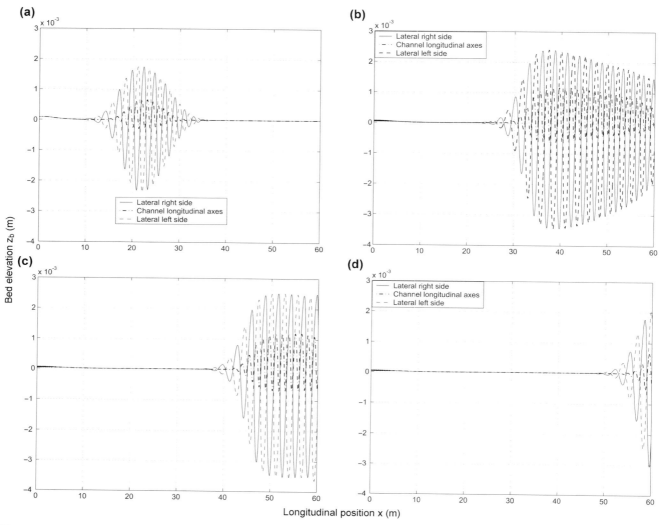

Fig. 10 Bed elevation with respect to initial flat configuration after: (a) t = 1 h (3600 s); (b) t = 2 h (7200 s); (c) t = 3 h (10,800 s); (d) t = 4 h (14,400 s). Left and right sidewalls and channel centreline are considered, Case 1.

(Fig. 11b). This difference is due to the influence of gravity effects that are responsible for a reduction in bed form slope. The gravity force acts on the incoherent material of a movable bed to reduce the bed forms to a flat bed, thus also leading to a corresponding shortening of bar lengths. By increasing the sensitivity of the numerically simulated sediment transport direction to the effects of gravity (i.e. increasing r value), bar height decreases, bar steepness decreases and bar length decreases.

Analysis of the curves plotted in Fig. 11a & b again shows how bars tend to a stable configuration faster in Case 2 as compared with Case 1. In Case 2 (Fig. 11b), the asymptotic state is reached, while in Case 1 (Fig. 11a) the approach to the asymptotic value is not fully completed. The equilibrium length achieved in Case 3 is ~2.7 m, although the relatively slow rate of bed evolution has not allowed the complete development of bars. Such an equilibrium value is quite similar to the asymptotic lengths obtained in the previous cases. Thus the flow regime seems to only weakly influence bar length.

In Fig. 11d, dimensionless bar length is plotted against width to depth ratio. The plot suggests that equilibrium bar lengths obtained in all three cases are overestimated with respect to the experimental data, and also with respect to the numerical results of Takebayashi *et al.* (1999). Conversely, a good

U	depth-averaged velocity along x axis (m s^{-1})
V	depth-averaged velocity along y axis (m s^{-1})
x	longitudinal coordinate in a Cartesian reference system (m)
y	transversal coordinate in a Cartesian reference system (m)
z	vertical coordinate in a Cartesian reference system (m)
z_b	bottom elevation (m)
α	counterclockwise angle between x axis and depth averaged velocity vector (0)
β	counterclockwise deviation angle between the flow direction and the direction of vectorial bedload discharge (0)
γ	counterclockwise angle between x axis and vectorial bedload discharge (0)
Δt	time step in numerical integration (s)
Δx	spatial step in numerical integration, x direction (m)
Δy	spatial step in numerical integration, y direction (m)
Δz_b	time increment of the bottom elevation (m)
θ	non-dimensional shear stress on sediment grains (0)
θ_{cr}	critical value of non-dimensional shear stress on sediment grains (0)
λ_b	bar length (m)
ξ	weight (memory degree) in the Beam–Warming method (0)
ς	weight (implicity degree) in the Beam–Warming method (0)
$\mathbf{A}$	Jacobian matrix of vector E
$\mathbf{A}_R$	right eigenvector matrix, relative to $\mathbf{A}$
$\mathbf{A}_L$	left eigenvector matrix, relative to $\mathbf{A}$
$\mathbf{A}^+$	positive part of $\mathbf{A}$ matrix as defined by Eq. 28
$\mathbf{A}^-$	negative part of $\mathbf{A}$ matrix as defined by Eq. 28
$\hat{\mathbf{A}}$	$\mathbf{A}$ matrix, evaluated at an appropriate intermediate point $(I + 1/2, j)$ between (i, j) and $(i + 1, j)$
$\mathbf{B}$	Jacobian matrix of vector G
$\mathbf{B}_R$	right eigenvector matrix, relative to $\mathbf{B}$
$\mathbf{B}_L$	left eigenvector matrix, relative to $\mathbf{B}$
$\mathbf{B}^+$	positive part of $\mathbf{B}$ matrix as defined by Eq. 28
$\mathbf{B}^-$	negative part of $\mathbf{B}$ matrix as defined by Eq. 28
$\hat{\mathbf{B}}$	$\mathbf{B}$ matrix, evaluated at an appropriate intermediate point $(i, j + 1/2)$ between (i, j) and $(i, j + 1)$
$\mathbf{D}_A$	diagonal matrix of eigenvalues of $\mathbf{A}$ matrix
$\mathbf{D}_A^+$	diagonal matrix of non-negative eigenvalues of $\mathbf{A}$ matrix
$\mathbf{D}_A^-$	diagonal matrix of non-positive eigenvalues of $\mathbf{A}$ matrix
$\mathbf{D}_B$	diagonal matrix of eigenvalues of $\mathbf{B}$ matrix
$\mathbf{D}_B^+$	diagonal matrix of non-negative eigenvalues of $\mathbf{B}$ matrix
$\mathbf{D}_B^-$	diagonal matrix of non-positive eigenvalues of $\mathbf{B}$ matrix
$E = (Uh\ U^2h + gh^2/2\ UVh)^T$	flux term (first column in $\mathbf{F}$ matrix) vector in conservation equation
$F = (E, G)$	flux term matrix in conservation equation
$G = (Vh\ Uvh\ V^2h + gh^2/2)^T$	flux term (second column in $\mathbf{F}$ matrix) vector in conservation equation
q_b	vectorial volumetric bedload discharge per unit width (m^2 s^{-1})
$\mathbf{Q}$	Jacobian matrix of vector S
$\mathbf{Q}_x$	Jacobian matrix of vector S, linked to x direction
$\mathbf{Q}_y$	Jacobian matrix of vector S, linked to y direction
RHS	right-hand side in Eq. 20, defined as in Eq. 21
$S = [0\ gh(S_{fx} - S_{0x}) gh(S_{fy} - S_{0y})]^T$	source term in conservation equation
$U = (h\ Uh\ Vh)^T$	vector of conservative variables
ΔU	time increment of the vector of conservative variables after the final step
$\Delta \tilde{U}$	time increment of the vector of conservative variables after the first step

REFERENCES

Ashmore, P.E. (1982) Laboratory modeling of gravel braided stream morphology. *Earth Surf. Process.*, **7**, 201–225.

Ashmore, P.E. (1991) How do gravel-bed rivers braid?. *Can. J. Earth Sci.*, **28**, 326–341.

Ashworth, P.J., Best, J.L., Roden, J.E., Bristow, C.S. and Klaassen G.J. (2000) Morphological evolution and dynamics of a large, sand braid-bar, Jamuna River, Bangladesh. *Sedimentology*, **47**, 533–555.

Beam, R.M. and Warming, R.F. (1976) An implicit finite difference algorithm for hyperbolic systems in conservation law form. *J. Comput. Phys.*, **22**, 87–110.

Bernini, A., Caleffi, V. and Valiani, A. (2003) Generation and development of alternate bars: numerical modelling. *XVI AIMETA Congress of Theoretical and Applied Mechanics*, 9–12 September, Ferrara, Italy.

Bertoldi, W., Tubino, M. and Zolezzi, G. (2002) Experimental observation of river bifurcation with uniform and graded sediments. In: *River Flow 2002: Proceedings of the International Conference on Fluvial Hydraulics*, Vol. 2, Louvain la Neuve, Belgium, 4–6 September (Eds D. Bousmar and Y. Zech), pp. 751–759. Balkema, Lisse.

Best, J.L. (1987) Flow dynamics at river channel confluences: implications for sediment transport and bed morphology. In: *Recent Developments in Fluvial Sedimentology* (Eds F.G. Ethridge, R.M. Flores and M.D. Harvey), pp. 27–35. Special Publication 39, Society of Economic Paleontologists and Mineralogists, Tulsa, OK.

Best, J.L. (1988) Sediment transport and bed morphology at river channel confluences. *Sedimentology*, **35**, 481–498.

Bridge, J.S. (1985) Palaeochannel patterns inferred from alluvial deposits: a critical evaluation. *J. Sediment. Petrol.*, **55**, 579–589.

Bridge, J.S. (1993) The interaction between channel geometry, water flow, sediment transport and deposition in braided rivers. In: *Braided Rivers* (Eds J.L. Best and C.S. Bristow), pp. 13–71. Special Publication 75, Geological Society Publishing House, Bath.

Bridge, J.S., Smith, N.D., Trent, F., Gabel, S.L. and Bernstein P. (1986) Sedimentology and morphology of a low-sinuosity river: Calamus River, Nebraska Sand Hills. *Sedimentolology*, **33**, 851–870.

Bristow, C.S. (1987) Brahmaputra River: channel migration and deposition. In: *Recent Developments in Fluvial Sedimentology* (Eds F.G. Ethridge, R.M. Flores and M.D. Harvey), pp. 63–74. Special Publication 39, Society of Economic Paleontologists and Mineralogists, Tulsa, OK.

Chaudhry, M.H. (1993) *Open-Channel Flow*. Prentice-Hall, New Jersey.

Chow, V.T. (1959) *Open-Channel Hydraulics*. McGraw-Hill, New Jork.

Colombini, M., Seminara, G. and Tubino, M. (1987) Finite-amplitude alternate bars. *J. Fluid Mech.*, **181**, 213–232.

Defina, A. (2000) Two-dimensional shallow flow equation for partially dry areas. *Water Resour. Res.*, **36**, 3251–3264.

Defina, A. (2003) Numerical experiments on bar growth. *Water Resour. Res.*, **39**.

Defina, A. and Lanzoni, S. (2002) Simulazione numerica dell'evoluzione di barre alternate. *28° Convegno di Idraulica e Costruzioni Idrauliche*, **3**, 103–110, Potenza, Italy.

Exner, F.M. (1925) Uber die Wechselwirkung zwischen Wasser und Geschiebe. *Flussen, Sitzber Akad. Wiss*, **134**, 165–180.

Federici, B. and Colombini, M. (2002) Barre alternate di ampiezza finita: analisi spaziale. *28° Convegno di Idraulica e Costruzioni Idrauliche*, **3**, 95–102, Potenza, Italy.

Federici, B. and Seminara, G. (2002) Sulla natura convettiva dell'instabilità delle barre fluviali. *28° Convegno di Idraulica e Costruzioni Idrauliche*, **3**, 87–94, Potenza, Italy.

Ferguson, R.I. and Werritty, A. (1983) Bar development and channel changes in the gravelly River Feshie, Scotland. In: *Modern and Ancient Fluvial Systems* (Eds J.D. Collinson and J. Lewin), pp. 181–193. Special Publication 6, International Association of Sedimentologists. Blackwell Scientific, Oxford.

Fujita, Y. (1989) Bar and channel formation in braided streams. In: *River Meandering* (Eds S. Ikeda and G. Parker), pp. 417–462. Water Resources Monograph 12, American Geophysical Union, Washington, DC.

Fujita, Y. and Muramoto, Y. (1985) Studies on the process of development of alternate bars. *Bull. Disaster Prevent. Res. Inst.*, **35**, 55–86.

Hasegawa, K. (1981) Bank erosion discharge based on a non equilibrium theory. *Trans. Jpn Soc. Eng.*, **316**, 37–52.

Hasegawa, K. and Yamaoka, H. (1982) Experiments and analysis on the characteristic of developed alternating bars. *Ann. J. Hydraul. Eng., Jpn Soc. Civ. Eng*, **26**, 31–38.

Hirsch, C. (1990) *Numerical Computation of Internal and External Flows*, Vols I and II. Wiley, Chicester.

Ikeda, S. (1982) Lateral bedload transport on side slopes. *J. Hydrol. Div. Am. Soc. Civ. Eng.*, **108**, 1369–1373.

Ikeda, S. (1984) Prediction of alternate bar wavelength and height. *J. Hydraul. Eng. Am. Soc. Civ. Eng.*, **110**, 371–386.

Ikeda, S. (1990) Experiments by Engels on alternate bars. *Summer School on Stability of River and Coastal Forms*, La Colombella, Perugia, Italy, 3–14 September.

Jha, A.K., Akiyama, J. and Ura, M. (1994) The modeling of unsteady open-channel flows—modification to the Beam and Warming scheme. *J. Hydraul. Eng., Am. Soc. Civ. Eng.*, **120**, 461–476.

Jha, A.K., Akiyama, J. and Ura, M. (1996) A fully conservative Beam and Warming scheme for transient open channel flows. *J. Hydraul. Res.*, **34**, 605–621.

Karcz, I. (1971) Development of a meandering thalweg in a straight, erodible laboratory channel. *J. Geol.*, **79**, 234–240.

Kovacs, A. and Parker, G. (1994) A new vectorial bedload formulation and its application to the time evolution of straight river channels. *J. Fluid Mech.*, **267**, 153–183.

Kuroki, M. and Kishi, T. (1984) Regime criteria on bars and braids in alluvional straight channels. *Proc. Hydraul. Eng., Jpn Soc. Civ. Eng.*, **342**, 87–96.

Lanzoni, S. (2000) Experiments on bar formation in a straight channel flume. *Water Resour. Res.*, **36**, 3337–3349.

Meyer-Peter, E. and Muller, R. (1948) Formulas for bed-load transport. *Proceedings of the 2nd International Association for Hydraulic Research Meeting*, Stockholm, Sweden, pp. 39–64.

Murray, A.B. and Paola, C. (1994) A cellular model of braided rivers. *Nature*, **371**, 54–57.

Parker, G. (1976) On the cause and characteristic scales of meandering and braiding in rivers. *J Fluid Mec*, **76**, 457–480.

Parker, G. (1984) Discussion of: Lateral bed load transport on side slopes, by S. Ikeda. *J. Hydraul. Eng. Am. Soc. Civ. Eng.*, **110**, 197–199.

Roe, P.L. (1981) Approximate Riemann solvers, parameter vectors and difference schemes. *J. Comput. Phys.*, **43**, 357–372.

Roe, P.L. and Pike, J. (1985) Efficient construction and utilisation of approximate Riemann solution. In: *Proceedings of the Sixth International Symposium on Computing Methods in Applied Science and Engineering* (Eds R. Glowinski and J. Liions), pp. 499–518. North-Holland Publishing, Amsterdam.

Sanders, B.F. (2001) Non-reflecting boundary flux function for finite volume shallow-water models. *Adv. Water Res.*, **25**, 195–202.

Sekine, M. and Parker, G. (1992) Bed-load transport on transverse slope. *J. Hydr. Eng. Am. Soc. Civ. Eng.*, **118**, 513–535.

Seminara, G. (1998) Stability and Morphodynamics. *Meccanica*, **33**, 59–99.

Seminara, G. and Tubino, M. (1989) Alternate bars and meandering: free, forced and mixed interactions. In: *River Meandering* (Eds S. Ikeda and G. Parker), pp. 267–320. Water Resources Monograph 12, American Geophysical Union, Washington, DC.

Shimizu, Y. (1991) *A study on prediction of flows and bed deformations in alluvial streams*. Unpublished PhD thesis, Hokkaido University, Hokkaido, Japan.

Shimizu, Y., Schmeeckle, M. W., Hoshi, K. and Tateya, K. (1999) Numerical simulation of turbulence over two-dimensional dunes. *International Association for Hydraulic Research Symposium on River, Coastal and Estuarine Morphodynamics*, Genoa, Italy, 6–10 September, Vol. 1, pp. 251–260.

Takebayashi, H., Egashira, S. and Jin, H.S. (1998) Numerical simulation of alternate bar formation. *Proceedings of the 7th International Symposium on River Sedimentation and 2nd International Symposium on Environmental Hydraulics*, pp. 733–738.

Takebayashi, H., Egashira, S. and Kuroki, M. (1999) Prediction of geometric characteristics of alternate bars. *Proceedings of XXVIII International Association for Hydraulic Research Congress*, Graz, Austria.

Takebayashi, H. and Egashira, S. (2001) Instability of developed alternate bar on non-uniform sediment bed. *Proceedings of the 29th International Association of Hydraulic Engineering and Research* (Eds Wang Zhao Yin et al.). Tsinghua University Press, Beijing.

Takebayashi, H., Egashira, S. and Okabe, T. (2001) Stream formation process between confinig banks of straight wide channels. *Proceedings of the 2nd International Association for Hydraulic Research Symposium on River Coastal and Estuarine Morphodynamics*, 10–14 September, Obihiro, Japan.

Talmon, A.M., Mierlo, M.C.L.M. van and Struiksma, N. (1995) Laboratory measurements of the direction of sediment transport on transverse alluvial bed slopes. *J. Hydraul. Res.*, **33**, 519–534.

Thorne, C.R., Russel, A.P.G. and Alam, M.K. (1993) Planform pattern and channel evolution of the Brahmaputra River, Bangladesh. In: *Braided Rivers* (Eds J.L. Best and C.S. Bristow), pp. 257–276. Special Publication 75, Geological Society Publishing House, Bath.

Toro, E. (1999) *Riemann Solvers and Numerical Methods for Fluid Dynamics*. Springer-Verlag, Berlin.

Tubino, M., Repetto, R. and Zolezzi, G. (1999) Free bars in rivers. *J. Hydraul. Res.*, **37**, 759–775.

Valiani, A. (1998) Propagazione di un'onda di sommersione in una canale con brusca variazione di direzione. *26° Convegno di Idraulica e Costruzioni Idrauliche*, **1**, 223–236, Catania, Italy.

Vignoli, G. and Tubino, M. (2002) A numerical model for sand bar stability. In: *River Flow 2002: Proceedings of the International Conference on Fluvial Hydraulics*, Vol. 2, Louvain la Neuve, Belgium, 4–6 September (Eds D. Bousmar and Y. Zech), pp. 833–841. Balkema, Lisse.

Yalin, M.S. (1992) *River Mechanics*. Pergamon Press, Oxford.

Methods for assessing exploratory computational models of braided rivers

ANDREA B. DOESCHL*,[1], PETER E. ASHMORE[†] and MATT DAVISON*

*Department of Applied Mathematics, The University of Western Ontario, London, Ontario, Canada N6A-5B7
[†]Department of Geography, The University of Western Ontario, London, Ontario, Canada N6A-5C2

ABSTRACT

Numerical models of braided rivers range from detailed computational fluid dynamics studies designed to predict the evolution of a particular braided river, to the simple cellular automata-based model of Murray & Paola (hereafter the MP model), designed to identify in a broad-brush way the most important conditions responsible for river braiding. Evaluating exploratory models, such as Murray & Paola's, is challenging as the evaluation cannot be reduced to a direct comparison between model output and a particular natural or laboratory river, but must involve the identification of characteristic properties common to all braided rivers. The MP model has provided a valuable service to the braided river community by stimulating research aimed at identifying such properties. Existing research suggests that the planform characteristics of the MP model are similar to those of real braided rivers. Some new model evaluation techniques are proposed to address additional aspects of braided river morphology and these are applied to comparisons of a physical model and the MP numerical model with a real braided river, the Sunwapta River. The new techniques show that the flume river shares the statistical characteristics of the braided river. However, certain planform characteristics of the MP model seem different from that of a real river. The topographic characteristics of the MP river are completely different from that of the real river. This research is not intended to attack the work of Murray & Paola, which was not, after all, designed to bear this kind of analysis. However, to evaluate generic braided river models, a variety of statistical metrics including topographic ones must be utilized, and several possible metrics are proposed here for future evaluation.

Keywords Model evaluation, braided rivers, scale-invariant properties, cellular models.

INTRODUCTION

During the past few decades, the demand for reliable methods to analyse and predict braided river behaviour has grown considerably. Physical and numerical models reveal various aspects of braided rivers that cannot easily be assessed in natural braided rivers, such as the constant interaction between planform and topographic characteristics, and their collective response to variations in hydraulic and geometric conditions. As a consequence, there is increased awareness that braided river dynamics cannot be described by the development of the channel structure independently from the development of the topography (Murray & Paola, 1997; Ashmore, 2000; Furbish, 2003). Additionally, appropriate model evaluation tools that capture these multiple aspects of braided river morphodynamics are also required. Quantitative methods for model evaluation concentrate mainly on static flow or planform properties, with the evaluation of topography and dynamics primarily restricted to qualitative assessment (Howard et al., 1970; Murray & Paola, 1996, 1997; Sapozhnikov et al., 1998; Paola, 2000;

[1]Present address: Animal Health and Nutrition, Scottish Agricultural College, Bush Estate, Penicuik, EH26, UK (Email: Andrea.wilson@sac.ac.uk)

Thomas & Nicholas, 2002; Jagers, 2003). However, adequate model evaluation requires quantitative methods that cover multiple aspects of the modelled system, thus the problem is to find quantitative criteria that characterize braided rivers and allow modellers to assess the extent to which model output reproduces the morphology, dynamics and response to external forcing of 'real' braiding.

This paper presents quantitative methods to describe braided river characteristics, including previous work (focused primarily on planform) and new methods that incorporate topography and morphodynamics. These methods are also used to evaluate the cellular braided river model of Murray & Paola (1994; hereafter MP model). The MP model was intended as a simple tool to show that braided channel patterns arise spontaneously from flow over non-cohesive sediment, and to find the minimal set of ingredients needed to generate a realistic braided behaviour. The model reproduces many of the planform characteristics of natural braided rivers. However, little attention has been paid to whether it also reproduces the topographic characteristics of natural braided rivers and whether the modelled rivers respond correctly to given external forcing. The present study extends the model evaluation by simultaneously considering planform and topographic patterns and by using data from a laboratory flume for comparing the intrinsic spatial and temporal scales of both systems as a means to assess their response to external forcing.

The point of this analysis is not to criticize the MP model, but rather to use it as an example with which to demonstrate the need for a larger range of assessment tools, and the necessity to consider multiple aspects of braided river behaviour simultaneously in the model assessment. The latter imposes stricter criteria on a model's validity, but it is essential for fully identifying a model's capability and, as a consequence, for further progress on understanding fundamentals of braiding in the exploratory model, as described by Murray (2003).

The suggestions and methods introduced in this paper do not offer a complete solution for robust model evaluation, but should be considered a first step toward adequate approaches in modelling braided river morphodynamics. The methods of this paper have proved particularly useful for cellular automata based models. For traditional numerical models in computational fluid dynamics these assessment methods may need to be accompanied by methods that focus on numerical aspects such as robustness, accuracy and stability (Davison & Doeschl, 2004; Doeschl et al., 2004) or the coupling between flow and sediment flux.

EXISTING METHODS TO ANALYSE BRAIDED RIVER CHARACTERISTICS

Two main sets of characteristics identify and describe a braided river: the intrinsic spatial and temporal scales that distinguish a particular braided river from another, and the scale-invariant properties common to all braided rivers. The characteristic spatial and temporal scales of planform and topographic features are the river's response to the driving hydraulic and geometric conditions. A braided river model that aims to improve the understanding of braided river mechanics must therefore represent the true relationship between the hydraulic and geometric controls and the resulting planform, topography and morphodynamics. Although braided rivers are characterized by continually unstable and evolving morphology, the overall average morphology is statistically stable (Paola, 2000) and this provides a general basis for assessing the river morphology and response.

Whereas the definition of characteristic scales and their functional controls helps to distinguish quantitatively one braided river from another, scale-invariant properties distinguish braided rivers from other systems. These properties express statistical similarities between braided rivers of different scales which result from the same underlying mechanical processes inherent in all braided rivers. Examples of scale-invariant properties in braided rivers may range from a characteristic organization of pool–bar units to the sorting of sediment along a braid bar. A model that claims to represent the main mechanical processes of braided rivers must therefore reproduce these scale-invariant properties.

Characteristic spatial and temporal scales

Characteristics that describe the style or degree of braiding include the braiding index, the total or average width, the average confluence–confluence spacing and the total sinuosity (for a review of some of these characteristics see Mosley (1983) and Bridge (1993)). Topography can be described by scour depth or bar height, as well as measures that describe the structural variability in the topography (Zarn, 1997). Empirical studies that relate the scales of these features to flow and material properties via multiple regression analyses are invaluable for model testing. As Ashmore (2000) pointed out, few empirical formulae are available for braided rivers, although Mosley (1983) provided a comprehensive data set on channel dimensions and the distribution of flow at different stages that is equivalent to the 'at-a-section' hydraulic geometry for single thread channels.

The large spatial variability in characteristics, along with ambiguity or a lack of consensus on measurements and definitions, contributes to current uncertainty in defining braiding characteristics. For example, does braiding intensity refer to all channels, or only those transporting bed sediment at high stage, and are all channel segments included regardless of length or width? Seldom are these characteristics defined over a range of flow conditions or even at a well-defined single flow state. In some cases the complete frequency distributions rather than mean values may be more diagnostic (Ashmore, 2000), but these are difficult to obtain and therefore seldom known.

An even greater challenge than finding appropriate definitions of spatial scales is the definition of characteristic time-scales. Identification of characteristic temporal scales is important in assessing the rate at which models function and the relationship to static spatial characteristics. Without the definition of characteristic temporal scales, it is impossible to quantify the dynamics of braided rivers.

Morphology changes abruptly and locally and adequate characterization of process rates in the field may require several years of observation. While rates of morphological evolution are tied to rates of bed material transport, the relationship between them remains uncertain (Ashmore, 2000; Hoey et al., 2000), so that measurement of morphological change is required in addition to direct measurement of bed load transport rates. At present, the rates of morphological development and bed sediment transport are known to fluctuate widely at constant total discharge (Ashmore & Church, 1995), but average rates also increase with discharge in a given river or, experimentally, at higher total stream power for a given grain size (Ashmore, 2000). Large spatial features take longer to develop than small features, and the lifetime of larger channels must depend on the lifetime of their smaller tributaries. However, temporal characteristics of the morphodynamics have not been quantified and related to spatial characteristics or hydraulic conditions. Characteristic time-scales of braided rivers may be defined in various ways. Possibilities include: (i) the time that the system takes to establish its statistical equilibrium; (ii) the average time of existence of channel confluences and diffluences (Paola, 2000); (iii) the time in which individual anabranches are active sediment conveyers relative to their widths; or (iv) the time in which a channel with increasing sinuosity forms before different processes such as avulsion or cut-off take place.

Knowledge about the scales of characteristic static braided river features is invaluable for model evaluation. The present lack of understanding in the relationship between spatial and temporal scales may be overcome by analysing the dynamic behaviour in laboratory and computational river models, which successfully reproduce the static scales expected from empirical studies.

Scale-invariant properties

Various methods have been proposed to assess scale-invariant planform properties of braided rivers, ranging from statistical methods to state space plots in dynamical system theory or the theory of fractals (Nikora, 1991; Sapozhnikov & Foufoula-Georgio, 1995, 1996, 1997; Sapozhnikov et al., 1998). Each of these methods reveals one or more aspects of braided rivers that act as fingerprints of single or multiple characteristic mechanical processes.

Sequential organization of planform patterns (dynamical systems approach)

Sapazhnikov et al. (1998) used a dynamical systems approach to explore the presence of a sequential organization of total width in various braided rivers. They plotted sequences of the total channel widths per cross-section in state space plots, and compared the plots of various rivers by dividing the state space into distinct regions and by assigning a probability to each region according to the number of plots that fall into the region. The studies suggested that braided rivers of different average discharge and grain sizes have a similar sequence of total widths, scaled by the average total width. Rosatti (2002) presented similar results for a small-scale laboratory flume, suggesting that laboratory rivers have a similar sequential organization of standardized widths. The similarity in sequential width variations between different braided rivers of different scales indicates the presence of a characteristic distance between wide and shallow sedimentation zones and narrow, deep scour zones. A numerical model that incorporates sediment transport in braided rivers should therefore reproduce similar state space plots for standardized width variations. Assessment of the model according to state space plots requires the establishment of a measure for the similarity of state space plots of different systems as well as a tolerance level for the acceptance or rejection of a model. A metric to quantify the difference between the probability distributions corresponding to different rivers was defined by Moeckel & Murray (1997), but for the establishment of an adequate tolerance level, state space plots of a larger range of natural braided rivers are required.

Self-affinity and dynamic scaling anisotropy (fractal methods)

Various universal planform properties of braided rivers have been discovered using the theory of fractals. Sapozhnikov & Foufoula-Georgiou (1996) traced the flow patterns of various natural and laboratory braided rivers to separate wet from dry areas, and calculated mass dimensions, which are related to the number of black pixels (wet area) relative to the number of white pixels (dry area) within specified regions. This approach revealed that natural braided river channel network planform has fractal properties and exhibits an anisotropic scaling invariance. Smaller parts of the river look statistically similar to larger parts of the same river, but the scaling differs slightly between the downstream and cross-stream direction. Moreover, similar fractal exponents v_x and v_y for the anisotropic scaling invariance were obtained for braided rivers of different scales, pointing towards further geometric similarities.

The statistical invariance of the mechanisms generating scale invariance in the entire river network continues to hold for braided river islands. Sapozhnikov & Foufoula-Georgiou (1996) found anisotropic scaling in island sizes with a similar exponent for several braided rivers, which was always 10–20% lower than the scaling anisotropy of the rivers as a whole. Further scale-invariant properties referring to island areas have been reported by Barzini & Ball (1993) and Rosatti (2002) for various natural and laboratory braided rivers, respectively.

Caution is necessary when using fractal properties as sole evidence for model success. Fractal properties of a channel network can be attributed only to the self-organizing nature of flow and sediment flux, but do not specify the particular fundamental processes and their degree of influence on these properties. It is therefore difficult to state a priori to what degree these properties are specific for braided river systems and distinguish them from other self-organizing systems, such as branch structures in trees or dry and wet areas in coastal regions. This uncertainty has several implications for the evaluation of braided river models. If a braided river model does not reproduce these scale-invariant properties, the model must lack some of the fundamental processes of braiding. It is, however, difficult to identify from this method which model ingredients are missing, since fractal properties of braided rivers cannot be linked directly to their causes. If, on the other hand, a model does reproduce the specified fractal properties, it is not guaranteed that the model incorporates all of the essential processes contributing to a realistic braiding behaviour. The same properties may be generated by different processes, allowing self-organization of flow and sediment flux. In addition, establishing a tolerance level for the differences between the proposed fractal measures is important for the application of the above proposed methods for the assessment of braided river models.

Network link scaling

A distinctive property of braided rivers is that characteristic measures often differ greatly from their averages. Some mechanical processes in braided rivers may therefore show their fingerprints more clearly in frequency distributions of characteristic

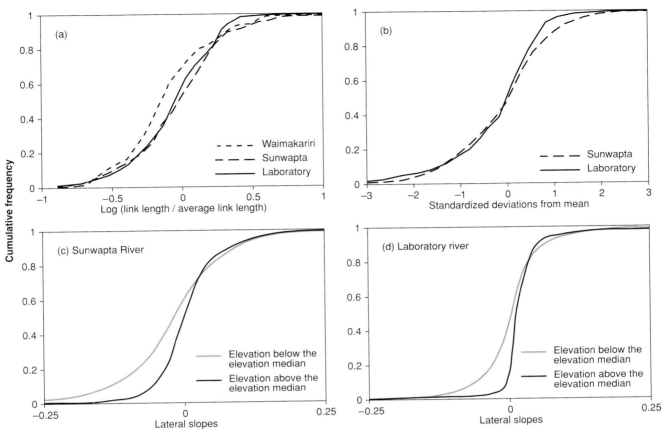

Fig. 1 Cumulative frequency distributions for natural and laboratory braided rivers. (a) Distribution of standardized link length. (From Ashmore, 2000.) (b) Distribution of standardized deviations of elevation from the median elevation. The deviations from the median in (b) are standardized by the standard deviation $\sigma = [1/n \times \Sigma^n_{i=1}(z_i - z_m)^2]^{0.5}$, where z_m is the elevation median for the cross-section and z_i are the measured elevations. (c & d) Cumulative distributions of lateral slopes for areas above the elevation median and for areas below the elevation median for (c) the Sunwapta River and (d) the laboratory river.

measures than in their standardized average values. Ashmore (2000) investigated the distributions of confluence–confluence and diffluence–diffluence spacings standardized by the mean link length in small-scale laboratory models and large braided rivers and found great similarities in the shapes and value ranges of the distributions for various rivers (Fig. 1a). The resulting link length distributions of all rivers are negatively skewed and the range of link lengths covers two orders of magnitude.

Like similarities in sequential width variations, similarities in the frequency distributions of standardized link lengths indicate the presence of a characteristic distance between erosion and redeposition of sediment, although the latter lack the sequential information provided by the state space plots. Standard statistical tests can be applied to test whether different braided rivers correspond to similar frequency distributions of standardized link lengths.

NEW METHODS TO ANALYSE BRAIDED RIVER CHARACTERISTICS

Whereas characteristic scales have been defined and analysed for both planform and topographic patterns, studies of scale-invariant properties in natural braided rivers have been completed only for planform patterns. The lack of methods using topographic or dynamic properties is because they require large amounts of topographic information at a high spatial and temporal resolution. Aerial photographs provide sufficient information about static planform properties, but analysis of the topography or the dynamic behaviour of braided rivers requires terrain data, which are much harder to acquire. Nevertheless, topographic properties cannot be neglected in the characterization of braiding, since planform and topography are strongly linked by bar formation and sediment erosion and redeposition. For example, sequential organization of total channel width and frequency distributions of network links point towards characteristic distances between erosion and redeposition of sediment. Topographic characteristics, such as scour depth and bar height, provide additional information about the relative quantities of transported sediment and thus add another dimension to the description of braiding.

Extension of existing methods to topographic patterns

Recent progress in the development of technologies for topographic data acquisition (Lane, this volume, pp. 107–135) suggest that many of the methods for identifying planform characteristics could be extended for describing topographic properties (Lane, this volume, pp. 107–135). For example, the state space approach could be readily extended to assess organization in the river topography, as total width sequences could be replaced by sequential scour depths at channel confluences or heights of successive bars. The fractal methods described above could prove useful for identifying scaling invariance of river topographies by using elevations below and above a certain reference level (e.g. water surface level) instead of wet and dry pixels as criteria. Further, similar distributions of standardized distances between successive confluences and diffluences in different braided rivers may be paralleled by similar patterns of variations in bed elevation along a channel, reflecting relevant characteristics in the bar–pool structure underlying the channel network pattern.

Statistical scale-invariant properties of topographic patterns using transect data

Not all aspects of the topography require the acquisition of a dense data set describing the river bed. Some useful topographic information can also be obtained from cross-section surveys. For example, in most of the cross-sections shown in Fig. 2, maximum scour depth below the water surface exceeds the maximum height of bars above the water surface. The areas of the rivers that are exposed at low discharge are also generally flatter than the areas that remain under water. Thus, the elevation median generally exceeds the elevation mean and the standardized frequency distribution of deviations from the elevation median is skewed (Fig. 1b).

Both prototype and model distributions have similar shapes and ranges of scaled deviations from the median. The two-sample Kolmogorov-Smirnov test with a 95% confidence level identifies the frequency distributions of the deviations from the median for the Sunwapta River and the laboratory river as samples from the same population. The strong similarity between the distributions is surprising, particularly as differences in the distributions are expected due to the different data acquisition techniques (point survey versus digital photogrammetry) and flow properties (discharge was held constant in the laboratory model). The boundaries of the transect elevations in the Sunwapta River frequently consist of high elevation values corresponding to steep banks and artificial rip-rapped embankments (Fig. 2a & b), which may account for the higher occurrence of deviations greater than one standard deviation from the median in the positive direction.

The degree of structure in the topography can be estimated from the frequency of elevations crossing a reference level, for example the median elevation, per river width (cross-section). In this example, to allow comparison between rivers of different scales, cross-sections were divided into the same number of subintervals, and the number of relative sign changes across the reference line per cross-section was defined as the ratio between the number of actual sign changes and the number of subintervals. Averaged over many cross-sections and various sampling times, the Sunwapta River and the flume have 0.13 and 0.12 relative sign changes per cross-section respectively. The temporal variations about these averages exhibited a random behaviour and were about 14% for the Sunwapta River and about 10% for the laboratory river.

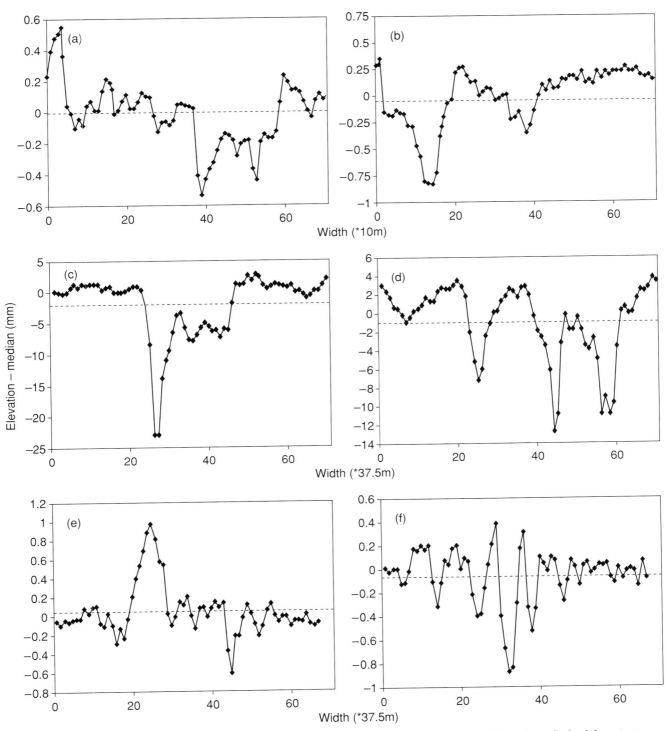

Fig. 2 Deviations from the elevation median for cross-sections of (a & b) the Sunwapta River, (c & d) the laboratory river and (e & f) the MP modelled river. The dotted line represents the deviation of the elevation mean from the elevation median.

The prevalence of relatively wide and flat areas above the water surface and relatively narrow, deep areas below the water surface can be captured by distributions of lateral bed slopes. Using the elevation median as a reference level instead of the water surface level, which is often unknown and often exhibits strong variations, lateral slope distributions below and above the elevation median have distinct shapes. The frequency distribution curves of lateral slopes corresponding to areas with elevations less than the elevation median (Fig. 1c & d, grey lines) are significantly wider than those corresponding to the areas with elevations above the median (Fig. 1c & d, black lines), reflecting the observed shape differences between exposed and submerged areas. Further, the ranges of lateral slopes corresponding to elevations below the median elevation are similar for both rivers. For elevations above the median, the variation of lateral slope values is, however, significantly higher in the Sunwapta River.

The two-sample Kolmogorov-Smirnov test with a 95% confidence level identifies the distributions of lateral slopes below the elevation median for the Sunwapta River and laboratory river as samples from the same population but the distributions corresponding to elevations above the elevation median as samples from different populations. In this case the distribution of lateral slopes could not be classified as a scale-invariant property of braided rivers. Mechanical properties such as shear strength of the sand or gravel, or the presence of vegetation, could contribute to this observed discrepancy (Gran & Paola, 2001, Murray & Paola, 2003). Alternatively, variable discharge in the natural river compared with the flume model could also cause the discrepancy.

The overall similarities between various frequency distributions derived from topographic transect data of different rivers point towards the presence of scale-invariant properties in the river topographies. The observed discrepancies between the distributions of lateral slopes suggest, however, that some distributions are more sensitive to variations in hydraulic conditions or material properties than others. Topographic data from a greater variety of braided rivers of different scales are needed to test the robustness and sensitivity of this method.

USING CHARACTERISTIC SCALES AND SCALE-INVARIANT PROPERTIES TO EVALUATE THE MP MODEL FOR BRAIDED RIVERS

The cellular model of Murray & Paola (1994) can be used to demonstrate the advantages of model evaluation using multiple criteria because:

1 It was designed to model the dynamic behaviour of a large braided network using equations that describe the interaction between flow and topography.
2 Its purpose is not prediction of future behaviour of a particular river, but rather aims to enhance the understanding of the fundamental processes taking place in all braided rivers (Murray & Paola, 1997; Paola, 2000). Model evaluation is therefore challenging as it cannot be restricted to a direct comparison between the modelled and a particular natural river, but must comprise tools that refer to properties that are characteristic for braided rivers in general. In fact, the model has provoked the development of various quantitative methods for the assessment of modelled braided river planforms (Murray & Paola, 1996, Sapozhnikov et al., 1998).
3 It has stimulated additional development and application because it reproduces many characteristic aspects of braided rivers from simple rules (Coulthard et al., 2000; Thomas & Nicholas, 2002; Murray & Paola, 2003; Doeschl & Ashmore, 2004).
4 Although extensive model evaluation has been undertaken (Murray & Paola, 1994, 1996, 1997, Sapozhnikov et al., 1998), the analysis of its performance is incomplete because it has concentrated only on planform characteristics at the scale of the braided network. Visual inspection of the topographic patterns suggests strong discrepancies between the topographies of the modelled and natural braided rivers. There remains uncertainty about the extent to which the model reproduces braided river morphodynamics. Its validity with respect to produc-ing both realistic planform and topographic characteristics, along with realistic dynamic behaviour of these properties, has not been tested.

The MP model and previous results

The MP model uses the concept of cellular automata theory to generate a continuously changing braided river network from random initial conditions. It operates on a two-dimensional lattice of square cells, along which water and sediment are routed into the downstream direction and lateral direction (the latter refers to sediment transport only) according to the equations:

$$Q = Q_0 S^n / \sum S_j^n \quad \text{if} \quad S > 0 \quad (1)$$

$$Q_s = K[QS + \varepsilon \sum (Q_j S_j)]^m \quad (2)$$

$$Q_{s,l} = K_l S_l Q_s^{out} \quad (3)$$

where Q is the fraction of local discharge delivered to a neighbouring cell from the total discharge Q_0 in the distributing cell, S is the local bed slope, Q_s and $Q_{s,l}$ are the amounts of sediment transferred between the downstream and lateral cells, respectively. The subscript j refers to the three downstream neighbours of a distributing cell, and n, m, K, ε and K_l are model parameters. The original model included some variations of the downstream sediment transport rule, but for the present study rule 2 was used, because it produced the most realistic results (Murray & Paola, 1997; Doeschl, 2002). Change in cell elevation as a result of sediment movement obeys the mass balance law. Equations (1) to (3) are representations of various physical laws, including the law of conservation of mass for water flux and Exner's equation for mass balance in the sediment flux, as well as crude representations of the bedload formula of Engelund & Hansen (1967) and the lateral sediment transport formula of Parker (1984). A detailed description of the model is provided in Murray & Paola (1997).

In Murray & Paola's computational experiments, the model started from a scale-free initial topography of constant downhill slope with superimposed random perturbations. Model parameters were not calibrated to simulate a particular physical setup, but were chosen freely, on the basis of how realistic the generated patterns looked. As a consequence, the generated patterns have an intrinsic scale, which is difficult to compare with that of natural braided rivers. Although Murray & Paola (1997) proposed a method to relate a time-scale with the scaling constant K in the sediment transport rule 2, this setup did not lend itself to a direct comparison of the characteristic scales of the generated modelled rivers with those of natural rivers. Model evaluation concentrated therefore on the assessment of scale-invariant properties (Murray & Paola, 1996, 1997; Sapozhnikov et al., 1998).

State space plots revealed strong similarities between the sequences of total cross-sectional widths of the model runs and several natural braided rivers. The transportation distances corresponding to the probability distributions of the width sequences of the modelled and various natural rivers were relatively small, although identified as statistically significantly different (Sapozhnikov et al., 1998). Like natural braided rivers, flow and sediment flux in the modelled rivers were found to exhibit a self-organizing nature, which is expressed by a self-affine behaviour of the planform patterns (Sapozhnikov et al., 1998). Moreover, the agreement between spatial scaling characteristics, such as the ranges of the fractal exponents, was generally high between natural, laboratory and modelled rivers. The scaling anisotropy of islands as opposed to the entire river network was found to be about 10–20% lower in natural than in the modelled braided rivers.

Reproduction of these scale-invariant properties in the modelled rivers could be considered as indicating that the model correctly represents the mechanisms that control the planform structure of braided rivers. However, subtle differences between the modelled and natural rivers, which were also observed by Sapozhnikov et al. (1998), may point to some missing ingredients in the MP model. For example, statistical tests not only indicated significant differences between the distribution of transportation distances between the modelled and natural braided rivers, but also found subtle discrepancies in the conditions for sudden width variations. In natural braided rivers, these occurred predominantly in wide sections, while they occurred equally in narrow and wide sections in the modelled rivers (Sapozhnikov et al., 1998). Further, modelled rivers were found on average to exhibit a higher scaling anisotropy with higher variance than natural rivers, which reflects the presence of fewer channels and the absence of a long-scale sinuosity in the modelled rivers (Sapozhnikov et al., 1998).

Without estimates of error bars associated with the variations of these scale-invariant properties in natural braided rivers, or a thorough understanding of the relationship between the observed symptoms and the underlying controls, it is difficult to determine to what extent the observed discrepancies point to missing essential components in the representation of the system. The answer to this problem requires the establishment of a tolerance level within which a model is accepted or additional methods for describing planform characteristics are implemented.

The MP model also produces detailed information about the topographies of the modelled rivers which, after visual inspection, raise some doubt about their resemblance to those of natural braided rivers. This is difficult to describe, but while real braided rivers show clear channels and bars with gradual transitions between areas of low and high elevation (e.g. Westaway et al., 2003), the topography generated by the MP model appears less organized, with abrupt transitions in elevation. However, except for the observed association between the formation of scour holes and bars, and flow contraction and expansion, respectively (Murray & Paola, 1997), no detailed validation of the modelled topographies has been performed.

Assessment of the characteristic scales of the modelled rivers

The dimension of the computational grid representing the braid plain, the introduced water and sediment fluxes, and the values of the model input parameters, control the structure and scales of the characteristic features of the modelled rivers. Under the assumption that the model incorporates the main mechanisms that lead to a realistic braided behaviour, it is hypothesized that the model would also reproduce the characteristic scales of a physical braided river, if the necessary information for initial values and parameter estimates corresponding to those of the natural river were available. To test this hypothesis, the MP model was implemented to simulate the initial setup of a laboratory flume experiment, for which the required information was available.

This approach provides the possibility of assigning a spatial or temporal scale to the modelled river networks, and enables a direct comparison between the characteristic spatial and temporal scales of the modelled and the physical river planforms and topographies. Since many empirical equations that relate hydraulic and geometric conditions to scales of planform and topographic patterns have not been tested for small-scale laboratory braided rivers, it cannot be assumed a priori that the flume model reproduces the expected spatial and temporal scales. For this reason, the scales of both the laboratory and the computationally modelled rivers are compared with those derived from natural braided rivers under equivalent external forcing.

The flume experiment

The laboratory river used for this study represented approximately a 1:40 generic scale hydraulic model of a gravel-bed braided stream. The flume experiment consisted of time periods (typically 10–15 min) during which water and sediment were introduced at a constant rate at the upstream end and the river morphology evolved. At the end of each period water was drained from the flume and digital photographs were taken of the bed, from which a sequential set of digital elevation models of the flume-bed was derived (Stojic et al., 1998; Fig. 3). In the digital elevation models, the river bed is divided into a grid of cells with (x,y,z) co-ordinates, which has the same structure as the computational lattice in the MP model. Measures describing the flume dimensions and the physical setup are listed in Table 1.

The experimental river evolved from an initially straight channel to meandering and then braided patterns in approximately 3 h. Once the braided stage was developed, a statistical stability was reached, which is characterized by small temporal fluctuations but overall constant values of characteristic measures such as the braiding index, average channel widths and average scour depth/bar height. Despite this robustness in the statistical measures, the river constantly reconfigured itself and exhibited a vast range of characteristic dynamical processes, such as avulsion, channel migrations and chute cut-off (Ashmore et al., 2001). The consequences of these dynamic processes on the bed topography are shown in Fig. 3.

Adoption of the setup of the flume experiment

The dimension of the computational lattice, the initial values for the model variables and the model input parameters in the MP model were calculated from the dimensions of the flume, the median grain

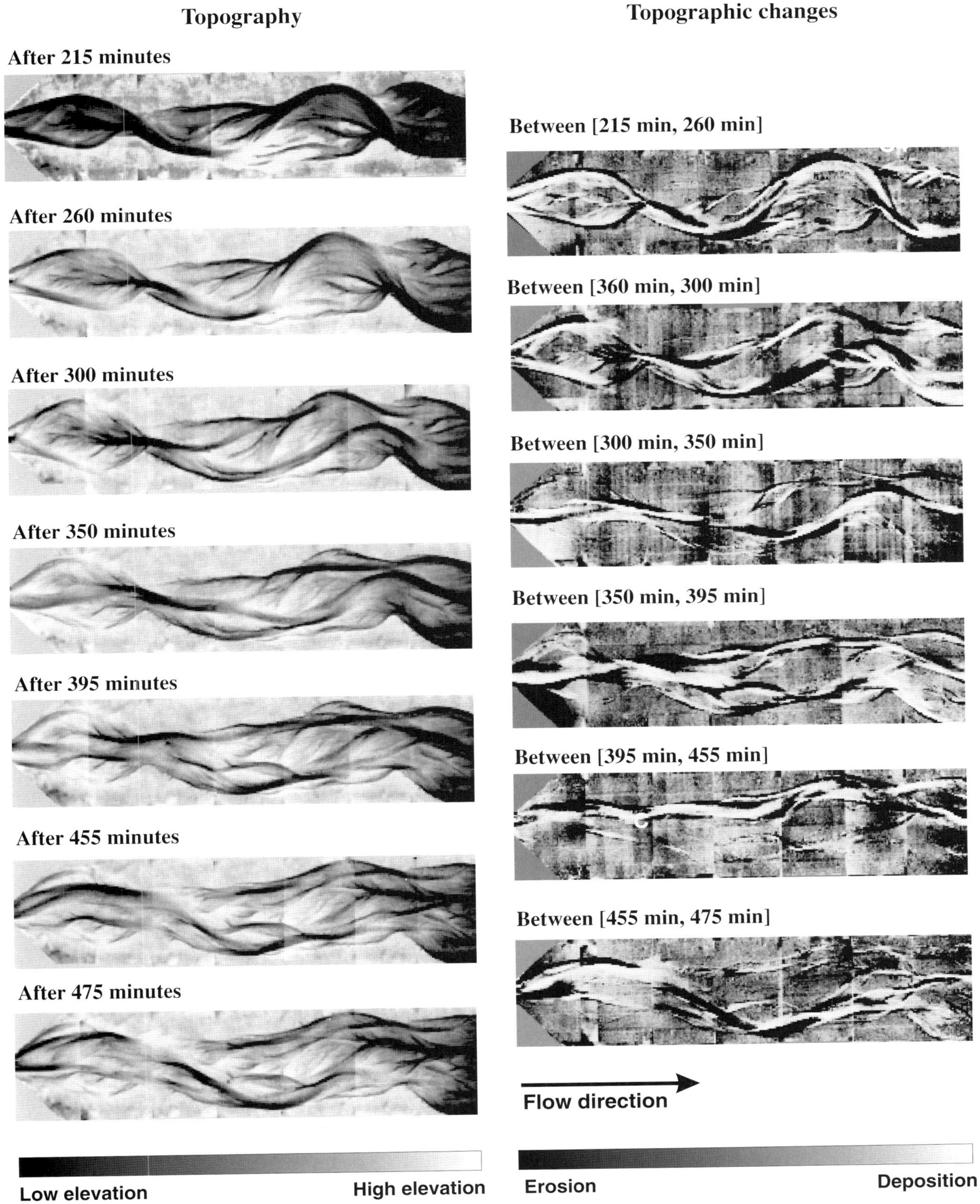

Topography

After 215 minutes

After 260 minutes

After 300 minutes

After 350 minutes

After 395 minutes

After 455 minutes

After 475 minutes

Low elevation — **High elevation**

Topographic changes

Between [215 min, 260 min]

Between [360 min, 300 min]

Between [300 min, 350 min]

Between [350 min, 395 min]

Between [395 min, 455 min]

Between [455 min, 475 min]

→ **Flow direction**

Erosion — **Deposition**

Fig. 3 Digital elevation models of the topography and topographic changes of the laboratory river. The overall slope was removed before plotting.

Table 1 Dimensions of the physical flume experiment

Flume dimensions	11.5 m × 2.9 m
Average bed slope	0.012
Introduced discharge	0.0015 m³ s⁻¹
Introduced sediment discharge	0.25 kg min⁻¹ (average)
Median grain size	7 mm
Cell length in raster digital elevation model	37.5 mm

size and the hydraulic conditions for the flume experiment (Table 1).

The flume's initial topography was represented by a smooth surface of the flume's constant downhill slope and superimposed white noise. The maximal amplitudes of the white noise perturbations were chosen to be 10 times the valley slope, multiplied by the cell length. However, computational experiments with different amplitudes of perturbation (3 to 20 times the valley slope times cell length) showed that the amplitudes have little impact on the planform and topographic scales of the fully developed modelled braided rivers. Dimen-sion and resolution of the computational lattice were chosen to equal the raster structure of the digital elevation models. The time step, corresponding to one iteration of the computational model, was determined by the initial volumetric discharge and the grid spacing. This setup allowed a 1:1 comparison between the spatial and temporal scales of the MP modelled rivers and the laboratory river.

Due to their heuristic origin, calibration of the five model parameters n, m, K, ε and K_l from the physical experiment was not straightforward. Following Murray & Paola's outline (1997), the parameters K and m could be derived from Engelund & Hansen's (1967) sediment transport formula. Using uniform flow formulae, the latter formula can be rewritten in the form of Eq. (2) with $\varepsilon = 0$, $m = 5/3$ and $K = 0.4 \times [5.66 g^{1/3} f^{1/6} D_{50} (s-1)]^{-1}$, where f is the Darcy Weisbach friction coefficient, D_{50} is the average grain size and s is the specific gravity. Substituting appropriate values into the equation for K leads to $K = 2.4 \times 10^{-5}$. Similarly, the lateral sediment transport Eq. (3) could be considered as a crude representation of Parker's (1984) lateral sediment transport formula, which yielded $K_l = 0.875$, after substituting appropriate values of the average water depth in the flume experiment, grain diameter and bed slope and an estimate of the critical shear stress on the level bed (Doeschl, 2002). The parameters n and ε were unconstrained and were chosen to obtain optimal results.

The modelled river was not expected to exhibit the same evolution as the laboratory river. The question is whether the MP model represents the physics of the laboratory river sufficiently to generate braided patterns and topographic features of similar scales to those in the physical braided river, and whether the modelled rivers exhibit a similarly rich dynamic behaviour with similar intrinsic time-scales to the laboratory river.

Comparison of the evolution of the laboratory and the modelled river

The above parameter estimates together with $n > 1$ and $0.3 \leq \varepsilon \leq 0.5$ led to braiding, but with unrealistic channel structures and poor structural changes over time. More realistic braided patterns, which constantly reconfigured, could however be generated by increasing the lateral scaling factor to $K_l = 20$ and the sediment transport exponent m to 2.5.

In contrast to the laboratory river, which evolved from straight to meandering and finally braided patterns in approximately 3 h, the MP modelled river had a large number of frequently intertwining channels during the first few hours flume time, with small and rapidly altering wet and dry areas (Fig. 4a & b). Eventually several narrow channels combined to fewer wider channels and the wet and dry areas increased in size. Only after approximately 30 h simulated flume time is a statistically stable state reached, at which the river constantly reconfigured without changing the average channel width or the average number of channels per cross-section (Fig. 4c–f). However, in contrast to the statistical equilibrium in the planform structure, bars and pools kept increasing in size in the MP modelled rivers (Fig. 4b, d, f & h). This increasing trend in topographic features eventually affects the statistical balance in the planform pattern. After approximately 280 h flume time, the braiding index

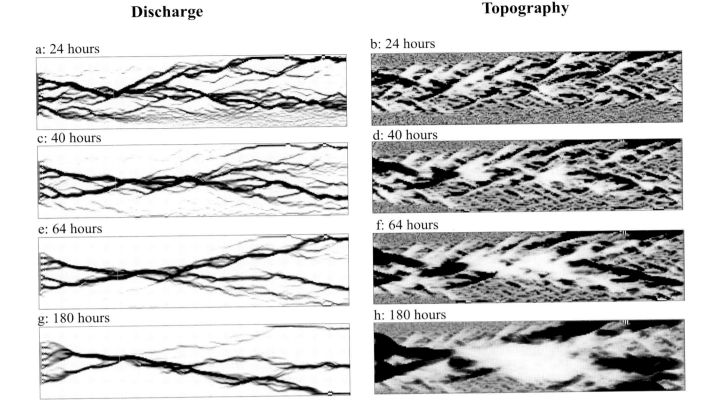

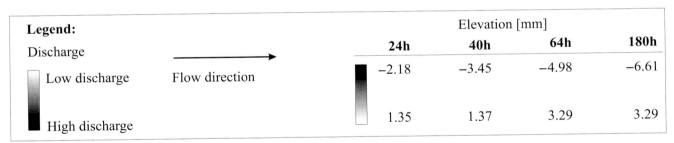

Fig. 4 Digital elevation models of the discharge and topography of the MP modelled river. The overall slope was removed before plotting.

gradually decreased until the MP modelled river was no longer braided.

Comparison of the characteristic scales of the laboratory and the modelled river

For most of the measures, the agreement between the measured values and the values predicted by empirical studies for the laboratory river is strong (Table 2). The measured total width exceeds the predicted width by a factor greater than two, but lies still within the 95% confidence limits. In contrast, the discrepancy between the characteristic scales of the MP modelled river and those of the laboratory river is very high. Compared with the laboratory river, the MP modelled river has a larger number of narrower channels, which also split and join more frequently. Consequently, the modelled river has a higher frequency of pools and bars than the flume. Even after 180 h simulated flume time, the continuously increasing magnitudes and areas of bars and pools are on average an order of magnitude smaller than those in the flume experiment. Similarly, the total sinuosity in the laboratory and

Table 2 Characteristic scales estimated from empirical studies and observed in the laboratory and modelled river, respectively

Measure	Empirical estimate*	Laboratory river	Modelled river†
Total channel width	280 mm	620 mm	710 mm
Average number of channels per cross-section	2.8	5.3	
Average number of confluences per metre	0.43	1.65	
Total sinuosity	2.9	3.4	19.3
Average maximum bar height	8 mm	3.18 mm	0.99^+ mm
Average maximum scour depth	−14 mm	−21.22 mm	-1.75^- mm
Topographic variation	4.9	4.8	4.4^-
Total link length/number of channel confluences	28 mm	32 mm	130 mm
Time to develop to fully braided state		3 h	30 h

*The following formula were used: total channel width, Van den Berg (1995); total sinuosity, Robertson-Rintoul & Richards (1993); average maximal scour depth and bar height as well as topographic variation, Zarn (1997); total link length/number of confluences, Ashmore (2000).
†Superscripts + and − indicate increasing and decreasing tendency with time, respectively.

the MP modelled river differs by an order of magnitude. The time-scale associated with the evolution of braiding is considerably longer in the MP modelled river. Development towards the statistical equilibrium in the planform structure differs by a factor of ~10.

Influence of the raster structure of the computational grid on the spatial scales

In traditional computational fluid dynamics models, the choice of the appropriate grid structure is an essential part of model testing, since it affects the accuracy of the model predictions (Patankar, 1980). To determine the impact of the dimension of the computational grid on the scales of the MP modelled river, the model outputs associated with the same parameter combinations, but various grid dimensions, were compared. Surprisingly, the spatial scales of the model predictions, when measured in cell units, show little dependence on the grid structure. Figure 5 shows that channel width and the spacings between successive confluences and diffluences vary little between different grid dimensions. For cell lengths ranging from half to four times the chosen cell lengths of the raster structure for laboratory flume, the average channel width varies between 4.18 and 4.32 cells; a drop in the channel width occurs only for extremely coarse grid structures. The average spacing between adjacent nodes varies between 22.5 and 28 cells.

The invariance of the average channel width and nodal spacing in terms of cell units has several implications. First, similar planform scales to those observed in the laboratory river could be obtained by the MP model through a change of the cell dimensions in the computational lattice. In this case, model assessment using exclusively planform patterns would give misleading results, since only the scales of the topographic features indicate substantial differences between the MP modelled and the laboratory river. Second, the observed invariance suggests that the scales of the planform characteristics in the model runs are strongly dependent on the spatial resolution at which the river is represented, instead of hydraulic forces as observed in natural braided rivers. This infers further that the MP model laws operate only on a certain scale, and thus that the model would perform poorly if the modelled river is not represented by a grid structure that matches this preferred scale. This is further supported by experiments that test whether the MP model predicts a realistic evolution of the laboratory river in an already braided state, which show a strong dependence of the model performance on the resolution of the computational grid (Doeschl, 2002; Doeschl & Ashmore, 2004).

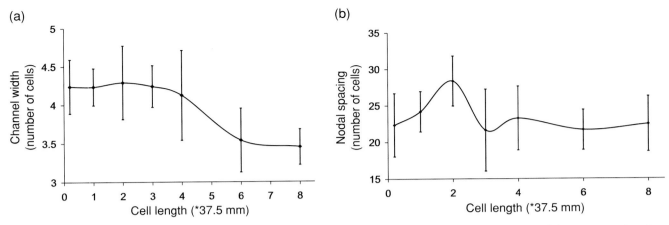

Fig. 5 Dependence of (a) the average channel width and (b) the nodal spacing on the dimension of the computational grid in the MP modelled river.

Implications

The above results show that a proper representation of the setup of the physical experiments in the MP model does not lead to similar spatial and temporal scales for characteristic features as those observed in the physical experiment. Whereas the spatial scales associated with the laboratory river generally agree with the scales predicted by empirical studies, the MP modelled river produces different spatial scales in both planform and topographic patterns. The temporal scales associated with the evolution towards a statistical equilibrium state and structural morphological changes are different for the laboratory and MP modelled river.

The model's failure to generate a statistically stable topography and to produce characteristic patterns of the expected scales leads to the inference that the model does not represent sufficiently all physical components that are necessary to produce a realistic braided behaviour. Two model variables, i.e. water and sediment flux, and local bed gradients as the driving forces for planform and topographic evolution may not suffice for this purpose, causing the observed intrinsic scales of the generated patterns.

The suspicions arising from the present studies are supported by results obtained in complementary studies, in which the MP model was applied to topographies of natural and laboratory rivers and catchment systems (Coulthard *et al.*, 1999, 2000; Thomas & Nicholas, 2002; Doeschl & Ashmore, 2004). It was found that the MP model, without modifications, could not predict a realistic evolution of the corresponding braided rivers. However, in both computational experiments, model results improved considerably through the incorporation of additional model variables and more detailed representation of the physical laws.

Assessment of scale-invariant properties of the modelled rivers

Planform characteristics

Figure 6a shows that the cumulative frequency distribution of standardized link-lengths of the modelled river, calculated over various time intervals during the fully braided stage, shares the main characteristics such as skewness and link-length-range with those of various natural and laboratory rivers. The Kolmogorov-Smirnov test does not identify the distributions of the modelled river and any of the physical rivers as significantly different at a 95% confidence level. The overall similarity in the link-length distributions thus supports the results of previous studies, i.e. that the modelled planform patterns resemble those of real braided rivers when represented by link-length distributions or other scale-invariant quantities.

Topographic characteristics

Transect graphs and statistical measures

The transect graphs of the topographies of the Sunwapta River and the laboratory river show that

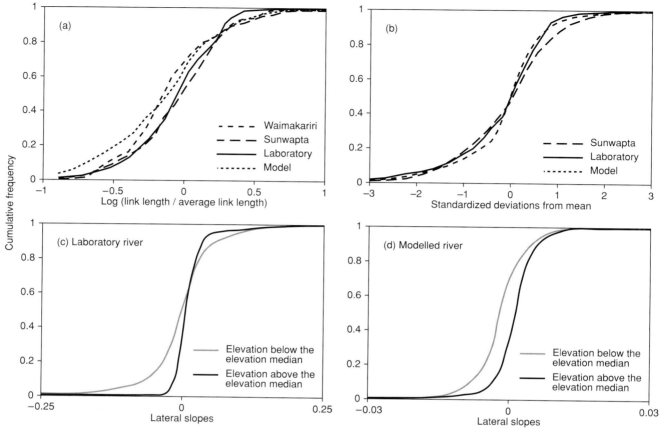

Fig. 6 Cumulative frequency distributions for natural, laboratory and MP modelled braided rivers. (a) Distribution of standardized link length. (b) Distribution of standardized deviations of elevation from the median elevation. (c & d) Cumulative distributions of lateral slopes for areas above the elevation median and for areas below the elevation median for (c) the laboratory river and (d) the MP modelled river.

areas above the elevation median are predominantly wide and flat, whereas areas below the median are more narrow and sharply peaked. Areas above the elevation median with relatively steep lateral slopes occur seldom in the flume cross-sections and are located primarily at the boundaries of the braid plain in the Sunwapta River. Transects of the MP modelled topographies do not share these properties, as many cross-sections are characterized by the coexistence of sharp peaks above and below the elevation median (Fig. 2e & f). Consequently, the mean elevation is more likely to exceed the median elevation in modelled versus physical rivers. Under the given constant flow conditions, the appearance of sharp peaks above the elevation median in the braid plain centre cannot be attributed to rising and falling flow stages, but is more likely a result of the cellular nature of the model. The distribution of flow and sediment according to local slopes between adjacent cells, and the lack of representation of flow momentum, allow rapid direction changes in the flow and the occurrence of abrupt deposition of large quantities of sediment as an immediate consequence of this shift in direction.

Degree of structure in the modelled topography

The temporal trend in the topographic structure of the modelled river, which became evident from the visual inspection of the modelled topographies at sequential times, is captured by the relative frequencies of sign changes in the deviations of cell elevations from the cross-sectional median. During fully braided states, the cross-sectional averages of the MP modelled river decreased from 0.26 relative sign changes at 30 h flume time, to 0.18 relative sign

Robertson-Rintoul, M.S.E. and Richards, K.S. (1993) Braided channel patterns and palaehydrology using an index of total sinuosity. In: *Braided Rivers* (Eds J.L. Best and C.S. Bristow), pp. 113–118. Special Publication 75, Geological Society Publishing House, Bath.

Rosatti, G. (2002) Validation of the physical modelling approach for braided rivers. *Water Resour. Res.*, **38**(12), 31/1–31/8.

Sapozhnikov, V. and Foufoula-Georgiou, E. (1995) Study of self-similar and self-affine objects using logarithmic correlation integral. *J. Phys. A Math. Gen.*, **28**, 559–571.

Sapozhnikov, V. and Foufoula-Georgiou, E. (1996) Self-affinity in braided rivers. *Water Resour. Res.*, **32**, 1429–1439.

Sapozhnikov, V. and Foufoula-Georgiou, E. (1997) Experimental evidence of dynamic scaling and indications of self-organized criticality in braided rivers. *Water Resour. Res.*, **33**(8), 1983–1991.

Sapozhnikov, V., Murray, A.B., Paola, C. and Foufoula-Georgiou, E. (1998) Validation of braided-stream models: spatial state-space plots, self-affine scaling and island shape comparison. *Water Resour. Res.*, **34**(9), 2353–2364.

Stojic, M., Chandler, J., Ashmore, P. and Luce, J. (1998) The assessment of sediment transport rates by automated digital photogrammetry. *Photogram. Eng. Remote Sens.*, **64**(5), 387–395.

Thomas, R. and Nicholas, A.P. (2002) Simulation of braided river flow using a new cellular routing scheme. *Geomorphology*, **43**, 179–195.

Tubino, M. (1991) Growth of alternate bars in unsteady flow. *Water Resour. Res.*, **27**(1), 37–52.

Van den Berg, J.H. (1995) Prediction of alluvial channel pattern of perennial rivers. *Geomorphology*, **12**, 259–279.

Westaway, R.M., Lane, S.M. and Hicks, D.M. (2003). Remote survey of large-scale, gravel-bed rivers using digital photogrammetry and image analysis. *Int. J. Remote Sens.*, **24**, 795–815.

Zarn, B. (1997) Einfluss der Flussbettbreite auf Wechselwirkung zwischen Abfluss, Morphologie und Geschiebetransportkapazitaet. *Mitt. Versuchsanst. Wasserbau, Hydro. Glaziol. Eidg. Techn. Hochschule Zuerich*, **154**, 71–75.

Bed load transport in braided gravel-bed rivers

CHRISTIAN MARTI and GIAN RETO BEZZOLA

Laboratory of Hydraulics, Hydrology and Glaciology (VAW), Swiss Federal Institute of Technology (ETH) Zurich, Gloriastr. 37–39, ETH Zentrum, CH-8092 Zurich, Switzerland (Email: marti@vaw.baug.ethz.ch; bezzola@vaw.baug.ethz.ch)

ABSTRACT

Historically, river training measures in the European Alps usually resulted in rivers being confined within relatively narrow, lined channels. Today, river management seeks, whenever the required space is available, to widen river reaches and so restore the original, braided character of Alpine rivers. Assessment of the effect of channel widening on bed-level changes, river morphology and the flood hazard requires an understanding of the bed load capacity of braided reaches. The experimental study described in this paper focuses on bed load transport and channel morphology in relatively steep braided rivers with a wide bed material grain-size distribution. Results of the study and a comparison with previous investigations show a need to improve existing approaches for describing bed load transport in braided gravel-bed rivers. An improved bed load equation is presented that allows the effects of wide grain-size distributions to be taken into account. In addition, the influence of varying discharge conditions on bed load transport and morphology in braided rivers is discussed.

Keywords Gravel-bed braided rivers, bed load transport, flume study, rewidening.

INTRODUCTION

As a result of river training works over the past two centuries, most rivers in the European Alps region have been constricted between embankments. The aim of most training works was to increase the discharge and sediment transport capacity of rivers. However, an emerging consciousness of river ecology strongly influences current river management philosophy. Thus, where possible, space is being given back to rivers by widening their bed and thus allowing the development of a morphology that more closely resembles the natural character of the river (Fig. 1).

However, exploitation of hydropower, channel stabilization, bed load retention in tributaries and gravel extraction from rivers for construction purposes has changed the discharge and bed load regimes of most rivers. These changes, to a large degree, explain why the originally desired incision, triggered by the training works, still progresses in most Alpine rivers. In many cases, the erosion has now reached an extent that necessitates interventions against further incision.

Widening of river reaches improves the ecological conditions, as generally there will be a transition from mostly flat bed conditions towards a braided morphology, and it may also prevent further erosion as bed load transport capacity decreases with increasing river bed width (Hunzinger & Zarn, 1997; Hunzinger, 1999). To avoid a change from erosion to uncontrolled aggradation, the width and thus the transport capacity of a widened reach has to be carefully adjusted to the bed load supply from upstream. Therefore, a reliable approach to quantify the bed load transport capacity for wide and braided rivers is a key element when assessing the effects of river widening on bed-level changes and thus on flood risks. However, most common bed load transport formulae are derived from model or prototype data gathered in straight, single channels and are not directly applicable to wide or braided river reaches.

PREVIOUS INVESTIGATIONS

Pickup & Higgins (1979) tried to estimate bed load yield over a 2 yr period in the Kawerong River,

Fig. 1 Rewidening for ecological reasons at the River Moesa near Lostallo, Grisons, Switzerland. Photographs taken in (a) 1998 before and (b) 2000 after the rewidening measure. Flow direction from top to bottom. (Photographs courtesy: Tiefbauamt Canton GR, Switzerland.)

Papua New Guinea. They used an elaborate algorithm with a random-number-based simulation and empirically determined probability distributions for the hydraulic variables, together with the bed load transport formula of Meyer-Peter & Müller (1948), and obtained satisfactory results. Ashmore (1988), using bed load data from his flume study on braided rivers, tried to reproduce the experimental results with the Meyer-Peter and Müller formula. However, to obtain an agreement between observed and calculated values, he required a modification of the original Meyer-Peter and Müller (1948) equation. This demonstrated that a direct application of bed load transport equations, without any possibility of calibration, will usually fail. Young (1989) found good agreement between his bed load transport data and the transport formula of Bagnold (1980). However, in comparison with other flume studies of braided rivers (Ashmore, 1988; Warburton & Davies, 1994; Zarn, 1997) the bed load transport rates found by Young (1989) are astonishingly high.

Carson & Griffiths (1989) used the Meyer-Peter and Müller formula to predict bed load yield in the Waimakariri River, New Zealand. For their analysis the authors assumed that bed load transport is confined to the main anabranches, which they approximated by a channel of 100 m width. A more sophisticated approach to estimate bed load yield in the same river section was presented by Nicholas (2000). He described the variation of flow depth at braided river cross-sections as a function of discharge using a gamma probability density function. Nevertheless, short-term prediction (over one or a few flood peaks) of bed load yield often failed. Additionally, the works cited above dealt with wide generally aggrading braided rivers. However, in densely populated Alpine areas braided river sections will always be limited in width and usually show a trend to erosion (Surian & Rinaldi, 2003).

Studies carried out at our laboratory have focused on the situation of braiding rivers with a limited width. Zarn (1997) derived two methods to calculate the bed load transport in braided rivers based on the transport formula by Meyer-Peter & Müller (1948). One of these methods, the optimum width approach, will be discussed in detail later in this paper. Hunzinger (1998) developed general design criteria for local widenings and river bank protection in widened reaches. The bed load transport relations presented by Zarn (1997) are based on flume experiments, mostly performed under steady-state conditions, with slopes ranging from 0.2 to 1.5%. Only one bed material mixture with a relatively narrow range of grain sizes was used. Hence, the derived relations are valid for river reaches with modest longitudinal slopes. Experience in practice has shown that difficulties may occur when calculating bed load transport for flood hydrographs (Zarn, 2003).

AIM OF THE PRESENT STUDY

The focus of this study was to gather bed load transport and detailed topographic data under equilibrium conditions for rivers with steeper slopes and a wider range of bed material grain sizes in order to enlarge existing data sets. This enabled us to test, and if necessary improve, existing bed load transport approaches for braided rivers. For this purpose, experiments with constant discharge and bed load supply were conducted until an equilibrium state was attained. The equilibrium state is herein defined as when the spatially averaged bed slope is approximately constant, and the bed load output mass as well as the grain-size distribution equals the input during the time span observed. Further, the effects of flood peaks and long periods with lower bed load input on the morphological development and bed load transport were investigated. Periods with little bed load input are common in Alpine rivers, mostly due to anthropogenic factors.

EXPERIMENTAL CONCEPT AND DATA AQUISITION

The experiments presented here were conducted in a concrete flume 28.5 m long and 3.2 m wide (Fig. 2). The slope of the flume was 2.1%, but vertically moveable sills at the inflow and outflow sections allowed for bed slopes between 1.2 and 2.8%. The river width could be varied by laterally moveable side elements. The maximum bed width reached 3.0 m. A computer-controlled valve allowed the simulation of any desired hydrograph and the sediment feeder at the upstream end of the flume controlled the bed load input.

With $\sigma = (D_{84}/D_{16})^{0.5} = 3.5$ the model sediment mixture used had a rather wide range of grain sizes and represented the typical grain-size distribution of braided Alpine rivers. The median grain size D_{50} was 2.9 mm and D_{90} equalled 9.1 mm.

Each experiment was started with an initially planar bed and lasted for between 150 and 300 h, during which the discharge and sediment input were held constant. Over the whole duration of an experiment the discharge, the bed load input and the bed load output, caught in a filtering basket at the downstream end of the flume, were continuously measured and weighed. It was thus possible to control whether an equilibrium state was reached or not. Equilibrium was attained when the cumulative bed load output curve was similar to the cumulative input curve (Fig. 3).

Each experiment was divided into sequences of about 15 h to allow for the filtering basket to be emptied and to refill the bed load feeder. Between

Fig. 2 Overview of the laboratory flume: 1, laterally moveable side elements; 2, water inlet; 3, sediment feeder; 4, xyz-positioning system; 5, laser and ultrasonic sensors; 6, filtering basket with automatic scales.

each sequence the bed topography was surveyed in a dry state. This was done by laser distance sensors installed on an xyz-positioning system. The grid spacing for this topography recording was 2 cm (lateral) by 10 cm (longitudinal). An example, referred to as snapshot no. 17, of such a measurement in the dry state is given in Fig. 4a for run S3-1, which shows differences in elevation compared with the initially planar bed.

After the start of the next sequence, the experiment measurements were repeated with water

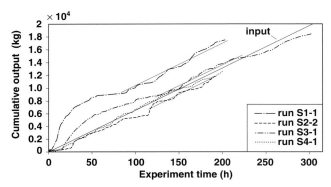

Fig. 3 Cumulative bed load output for four selected runs compared with the cumulative bed load input (solid straight line). Equilibrium, in the sense of equal bed load input and output, is reached when the cumulative bed load output curve (of the corresponding run) runs more or less parallel to the input line.

flowing along the flume. Ultrasonic sensors detected the water level and, at the same time, laser sensors detected the bed level. The laser measurement was corrected for refraction effects. Turbidity and waviness of the water surface additionally affected the bed level measurements. However, calibration enabled determination of the flow depth as the difference between the simultaneously measured water and bed levels (Fig. 4b). It even appeared possible to detect zones of intense bed load transport by analysing the standard deviation of the laser signal. Zones with a high standard deviation, representing areas of intense bed load transport, are shown in Fig 4c. In contrast to Fig. 4a, Figs 4b & 4c represent measurements over a time span, beginning shortly after snapshot 17 at the left side and ending 2.5 h later at the downstream end. In addition to these measurements by ultrasonic and laser sensors, the morphological changes along an 11 m section of the lower flume were documented by a digital camera over the full duration of the experiment (one shot per minute).

When an equilibrium state with constant discharge and constant bed load supply was attained, the then braided bed was subjected to a flood peak. To maintain the equilibrium state during the flood peak, bed load was supplied at a rate according to the current discharge as computed by a bed load approach by Zarn (1997). The flood peak was then followed by a sequence with the same constant discharge as used to obtain the equilibrium state, but with the bed load supply rate reduced to 20%, to simulate the period of low bed load input (e.g. graph in Fig. 5). This sequence of a flood peak followed by a phase with low sediment input was repeated two or three times with flood peaks of different duration.

Four series of such experiments were conducted each with a different initial slope. Each series consisted of one to three runs, in which different conditions of discharge and bed load supply were used until equilibrium state was reached. The first phase of each run was followed in most cases by up to three hydrograph runs. Apart from one reference experiment with a restricted bank-full width of 0.3 m, all runs were carried out with the maximum possible flume width of 3 m. The key parameters of all runs are listed in Appendix Tables A1–A3 and the main results concerning bed load transport are summarized in Tables A4 & A5.

GENERAL OBSERVATIONS

Experiments with constant discharge and constant bed load supply

For the four runs shown in Fig. 3 the equilibrium state was attained after 70 to 150 h. In one run (S1-2) channel development lasted 290 h and it was uncertain whether an equilibrium state had been achieved. However, even after attaining an equilibrium state, the bed load output from the flume showed significant fluctuations in all runs. Similar fluctuations have also been observed in other studies (Warburton & Davies, 1994; Hoey et al., 2001).

The significant fluctuations in bed load output can be correlated with morphological changes in the flume. As described by Marti (2002), low output is observed as long as several branches exist, which refers to a high braiding index, and output peaks occur when a dominant channel migrates laterally. As an example, Figs 4a & 4b show the topography and flow depths during a period with low bed load output. Towards the lower part of the flume at this time the braiding index was rather high. These observations contrast with those of Warburton & Davies (1994) and Nicholas (2000), but are in agreement with Ashmore (1988) and Warburton (1996). Whether a higher braiding index corresponds to an increase or decrease of bed

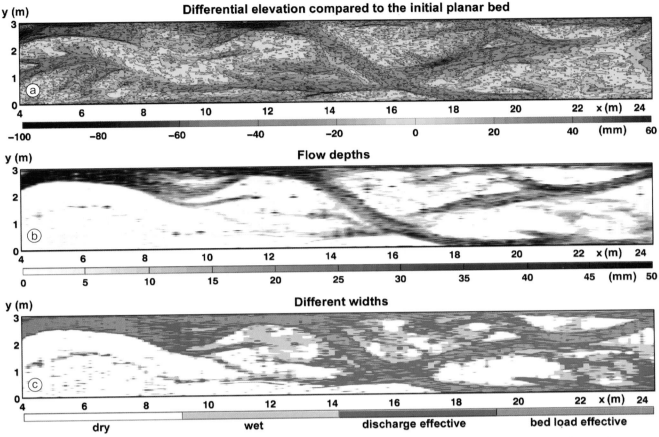

Fig. 4 Examples of results from the laser and ultrasonic measurements for run S3-1 snapshot 17. (a) Differential elevations compared with the initially planar bed, measured in dry state by laser. The blue areas mark erosion zones, while red areas mark aggradation zones. (b) Calculated flow depth (difference between laser and ultrasonic measurement in wet state). (c) Characteristic areas derived from standard deviation of the laser signal. White is the dry bed, clear blue the wetted areas, dark blue the discharge effective width of the channels and brown the bed load transport effective area. Flow direction is from left to right.

load transport may depend on the current state of the system (aggradation, equilibrium or erosion). However, in most cases, bed load transport was observed only in a few dominant anabranches of a braided river. This situation is shown in the right half of Fig. 4c, where bed load transport seems to occur in three anabranches. Counting only the transport-active anabranches, Ashmore (2001) introduced an 'active' braiding index, which is usually between 1 and 2, with maximum values up to 2.5. In this study, where the situation with a limited river width is examined, the active braiding index was usually lower than 1.5. In comparison with the entire river width w_{bf}, the active bed load width was very small. In Fig. 4c, for example, the active bed load width is only about 25% of the entire river width.

Influence of hydrographs and reduced bed load input

As mentioned above, a series of flood peaks divided by periods with low bed load input was simulated after having attained the equilibrium state. During such hydrograph tests, significant changes of the river bed morphology, especially during the period with low bed load input, could be observed. This development is exemplified by run HS3-1A (Fig. 5).

In the upper graph of Fig. 5, water discharge, bed load input rate and bed load output flux are shown during one flood peak and the subsequent period with low bed load input. The morphological changes during this test are highlighted with topographic measurements taken at the indicated

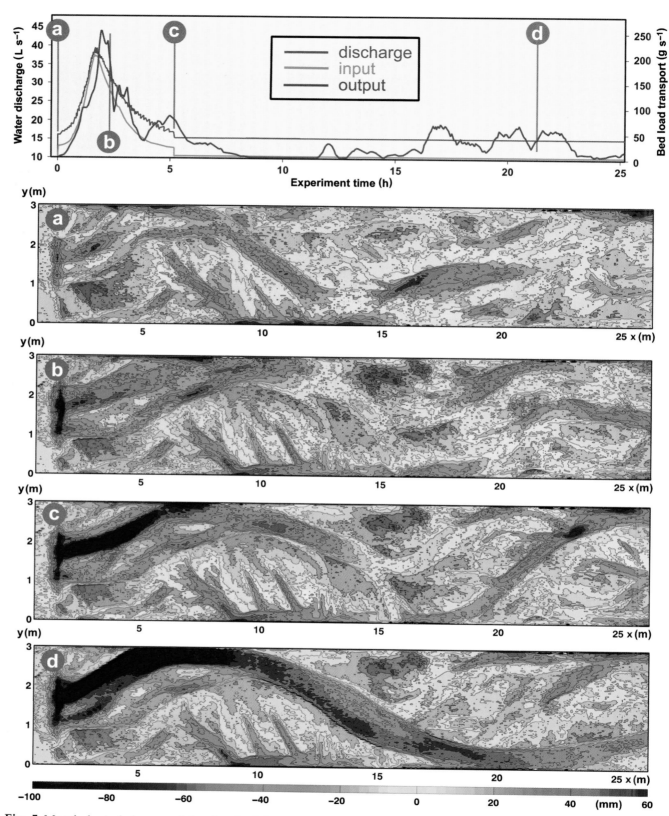

Fig. 5 Morphological changes of the river bed during hydrograph run S3-1A. In the uppermost graph the water discharge (blue), the bed load input rate (green) and the bed load output flux (brown) are shown. Flow direction is from left to right and a–d indicate times where topographic measurements were carried out. (a) Topography at the beginning of the hydrograph test, (b) shortly after the flood peak and (c) at the end of the flood peak. (d) After the phase with bed load input reduced to 20%, which caused a pronounced single channel.

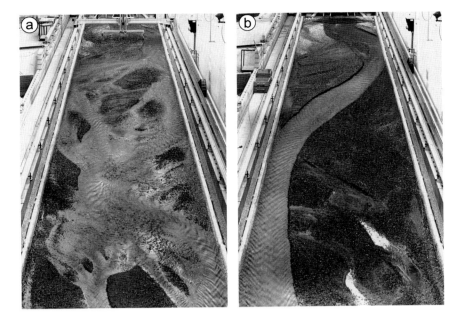

Fig. 6 Morphological changes in the river bed due to a longer period with reduced bed load input during hydrograph run S3-1A. (a) Braided pattern before the start of the hydrograph run. (b) The resulting single channel at the end of the run. Both photographs were taken at a discharge of 16 L s^{-1}, which refers to the channel forming discharge during the experiment. Flow direction is from top to bottom.

times a–d. Bed elevation changes were compared with the initial planar bed: blue areas mark erosion zones, whereas red areas mark aggradation zones. During the flood peak, the first 5 h of the hydrograph run, the bed load output was approximately 30% higher than the input, therefore moderate erosion took place. Comparing the topography at the beginning of the test (Fig. 5a) with that surveyed shortly after the peak discharge (Fig. 5b), erosion is recognizable near the inlet. Not far downstream, significant aggradation occurred as a substantial part of the eroded material was deposited here. Until the recession of the flood wave (Fig. 5c), the erosion progressed downstream. At 15 m, a bifurcation is recognizable and near the downstream end of the flume three or four anabranches were still present. In this area, the river bed still showed a braided pattern at this time. After a flood peak, slight erosion beginning at the inlet, a general flattening of the bed topography in the middle part of the flume and a tendency to braiding near the outlet could be observed in all experiments.

Much more pronounced erosion occurred during the subsequent phase, with constant discharge and reduced bed load input. During this phase massive erosion led to the formation of a single incised channel (Fig. 5d). This change from a braided morphology to a single incised channel is also documented in Fig. 6. Figure 6a shows the braided pattern before the start of the hydrograph run HS3-1A and Fig. 6b shows the resulting single channel after 21.5 h. Both pictures were taken at a discharge of 16 L s^{-1}, which corresponds to the channel-forming discharge of the initial phase of the experiment during which an equilibrium state was attained.

It is interesting to note that despite reduced bed load input the time-averaged bed load output during the development of the incised channel was always higher than the transport capacity at equilibrium. Only towards the very end of the experiment did the output flux decrease to a value similar to the bed load input rate. This is due to the armouring effects in the single channel. The spatially averaged slope of the initially braided river bed of 2.12% was reduced to a slope of 1.92% measured along the thalweg of the single channel.

Another interesting aspect results from a comparison of the width of the developed channel with different predictions for the equilibrium width of a single channel. The developed channel width is 1.7 times larger than the optimum width w_{opt}, but on the other hand it is 4.7 times smaller than the calculated equilibrium width according to Ikeda *et al.* (1988).

BED LOAD TRANSPORT FORMULA FOR BRAIDED RIVERS WITH LIMITED WIDTH

For a given slope and discharge, the transport capacity of a single channel varies with bed width. This

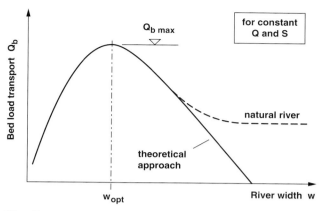

Fig. 7 Bed load transport capacity as a function of the river width for given discharge and slope.

relationship is depicted qualitatively in Fig. 7: for a certain width w_{opt}, the transport capacity Q_b reaches a maximum value, herein referred to as Q_{bmax}. For widths smaller than w_{opt}, the influence of the lateral banks causes a reduction of bed shear stress and thus of the transport capacity. Similarly, for widths larger than w_{opt}, the water depth decreases, which causes a reduction of bed shear stress and transport capacity. For very wide channels with plane beds, the transport capacity theoretically reduces to zero as the bed shear stress becomes smaller than the critical stress for the initiation of motion.

However, for large widths in natural rivers the river bed does not remain plane and a transition to either a meandering or braided morphology occurs. Especially in braided rivers the transport capacity, except for low flows, remains larger than zero because in individual channels there will still be sufficient discharge to allow for bed load transport.

Optimum width approach by Zarn (1997)

Zarn (1997) derived an empirical approach to determine the bed load transport capacity Q_b of a wide braided river as a function of the maximum bed load transport capacity Q_{bmax} of a corresponding single channel having the optimum width w_{opt}. To determine Q_{bmax} and w_{opt} for given values of discharge, slope and grain size, the theoretical curve shown in Fig. 7 is calculated using the flow law of Keulegan (1938) and a slightly modified version of the Meyer-Peter & Müller (1948) transport formula:

$$Q_b = w \frac{5\sqrt{g}\rho_s}{s-1}[R_b S - \theta_{cr}(s-1)D_m]^{3/2} \quad (1)$$

For a given discharge and width, the hydraulic radius R_b of the bed section in Eq. 1 is computed iteratively using the sidewall correction procedure proposed by Einstein (1934, 1950). A detailed description of this procedure can be found in Smart & Jaeggi (1983).

With the calculated values of w_{opt} and Q_{bmax} for the corresponding channel, the bed load transport capacity Q_b of a braided river can then be computed with the empirical relationship

$$\frac{Q_b}{Q_{bmax}} = 3.65 e^{-0.65U} - 4 e^{-1.5U} + 0.35 \quad (2a)$$

obtained by Zarn (1997) from a fit through his experimental data points. The parameter

$$U = \frac{w_{bf}}{w_{opt}} \quad (2b)$$

corresponds to the ratio between bank full width w_{bf} of the braided river and the optimum width w_{opt} of the corresponding single channel.

In Fig. 8, Eq. 2 is compared with the data from the flume experiments of Zarn (1997) and Ashmore (1988), the hydraulic model experiments for the Lombach River in Switzerland (see Tables A6 & A7) and data from the presented study. The last of these were obtained from both the experiments with constant discharge and sediment supply and from the experiments with hydrographs. It can be seen that only the data of Zarn and Ashmore are predicted with satisfactory accuracy. The data from the Lombach case study and the present investigation show rather poor agreement with Eq. 2. Additional effects, not taken into account by the Zarn (1997) approach, thus seem to influence the bed load transport capacity in wide braided rivers.

Improved approach

Examination of the experimental data presented in Fig. 8 shows two fundamental differences between the data sets:

1 the data from the Lombach project and the present study were obtained from experiments with much

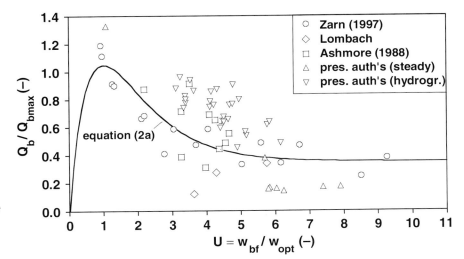

Fig. 8 Comparison of the optimum width approach of Zarn (1997) with experimental data. Only the data of Zarn (1997) and Ashmore (1988) are predicted with satisfying accuracy. The data for constant discharge and sediment supply of the present study are overestimated, whereas the data of the hydrograph experiments are underestimated.

wider grain-size distributions than the data of Zarn (1997) and Ashmore (1988);
2 the relative bed shear stresses τ/τ_{cr} (where τ_{cr} represents the critical shear stress for the initiation of motion) vary remarkably within the experimental data considered here.

In the approach proposed by Zarn (1997), measured transport rates Q_b obtained from braided river experiments are normalized with computed transport rates Q_{bmax} of a corresponding single channel. The rate Q_{bmax} usually reflects a situation with higher relative bed shear stresses, as the width of the corresponding single channel is smaller than the width of the braided river, whereas discharge and slope are equal for both. The value of Q_{bmax}, therefore, usually falls within the domain of fully developed transport and is relatively unaffected by armouring processes. If, however, braided river experiments are conducted at low relative bed shear stresses, the measured transport rates Q_b fall into the domain commonly referred to as weak transport and hence are influenced by armouring processes. Only rates obtained from experiments conducted at higher relative bed shear stresses are thus directly comparable to the transport rates Q_{bmax} of a corresponding single channel.

The differences in grain-size distributions and relative shear stresses may possibly explain the data scatter in Fig. 8. This can be considered by use of a factor $f_1(\sigma, Q/Q_0)$, where σ is a measure of the applicable grain-size distributions and Q/Q_0 is an alternative measure of the shear stress ratio τ/τ_{cr}.

Furthermore, at high discharges, usually occurring during flood peaks, the flow pattern in a braided river with limited width becomes similar to the one in a single channel. Therefore a U-dependent correction according to Zarn (1997) will underestimate the bed load transport rates and a factor $f_2(Q_0/Q)$ has to be used to compensate for this effect. Consequently only Q/Q_0 has to be determined for both factors.

The threshold discharge Q_0 was determined for only a small number of experiments during the present study. The experimental runs were divided into single sequences, between which the test had to be interrupted and the discharge and the sediment supply turned off. To restart the test, discharge and sediment supply were slowly increased up to values defined for the experiment. During this phase of discharge increase, the moment at which the sediment output from the flume began was determined by visual observation and the corresponding discharge was interpreted as the threshold discharge Q_0. To obtain Q_0 for all other tests considered here, the threshold discharge can in principle be computed by means of the relatively complicated procedure proposed by Zarn (1997). Here, a simplified approach is proposed, where the threshold discharge Q_0 in a braided river is related to the threshold discharge $Q_0(w_{opt})$ of the corresponding single channel with the optimum width w_{opt}. Figure 9 shows that Q_0 computed according to Zarn (1997) can be well expressed using $Q_0(w_{opt})$ by the polynomial expression:

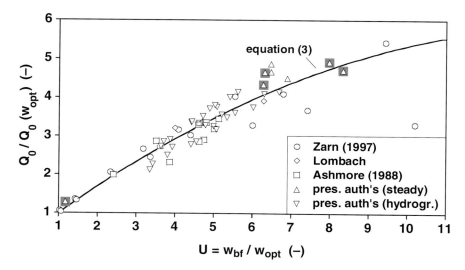

Fig. 9 Ratio between the bed load threshold discharge in the braided river and the threshold discharge in the corresponding single channel $Q_0/Q_0(w_{opt})$ as a function of the standardized river width U. Highlighted in grey are the data points where Q_0 was experimentally evaluated.

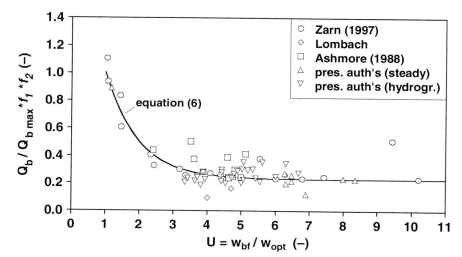

Fig. 10 Comparison of the new bed load transport formula with all experimental data. By incorporating the standard deviation of the grain-size distribution and the ratio Q/Q_0, the new approach allows for a good estimation of all experimental data.

$$\frac{Q_0}{Q_0(w_{opt})} = -0.027 \cdot U^2 + 0.78 \cdot U + 0.24 \quad (3)$$

The data for Q_0 determined experimentally within the present study are also shown in Fig. 9 and confirm the applicability of Eq. 3.

To develop an improved approach for the transport capacity of wide braided rivers, the factors f_1 and f_2 can now be introduced in a general form as in Eq. 4, where the threshold discharge Q_0 at initiation of motion is calculated with Eq. 3:

$$\frac{Q_b}{Q_{bmax}} = f\left(\sigma, \frac{Q}{Q_0}, U\right) \quad (4)$$

Several functions were examined with the aim of allowing the experimental data to coincide with a more distinct curve as in Fig. 8. The best results were obtained when, instead of the transport rate Q_b/Q_{bmax}, the dimensionless transport rate was used:

$$\frac{Q_b}{Q_{bmax}} \cdot f_1\left(\sigma, \frac{Q}{Q_0}\right) \cdot f_2\left(\frac{Q_0}{Q}\right)$$

$$= \left(\frac{Q_b}{Q_{bmax}}\right) \cdot 8^{(\sigma-1)(8e^{-3(Q/Q_0)})} \cdot \left(\frac{Q_0}{Q}\right)^{(1.95 - 2.48e^{-0.24U})} \quad (5)$$

In Fig. 10 the experimental data normalized according to Eq. 5 are plotted against the dimensionless

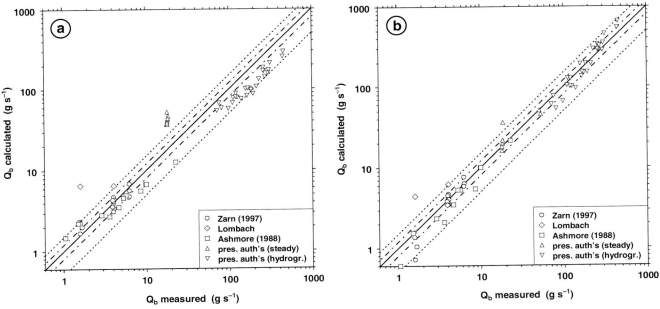

Fig. 11 Comparison between measured and calculated values for (a) the approach of Zarn (1997) and (b) the new proposed Eq. 6. The solid line shows perfect agreement; the inner dotted lines represent confidence intervals of ±20% and the outer dotted lines of ±50%.

width $U = w_{bf}/w_{opt}$. It can be seen that, compared with Fig. 8, the scatter within the data is clearly reduced. The data normalized in this way may be described by the fitted empirical relationship:

$$\frac{Q_b}{Q_{bmax}} \cdot f_1\left(\sigma, \frac{Q}{Q_0}\right) \cdot f_2\left(\frac{Q_0}{Q}\right) = 2.21 \cdot e^{-1.05 \cdot U} + 0.23 \quad (6)$$

As Fig. 10 shows, Eq. 6 allows a much better description of the experimental data for $2 \le U \le 12$. To use Eq. 6, Q_{bmax} and w_{opt} have to be evaluated first with the same procedure as described for the approach of Zarn (1997). In this context, the analysis has shown that the use of the original Meyer-Peter & Müller (1948) formula given as

$$Q_b = w\frac{8\sqrt{g}\rho_s}{(s-1)}\left[\left(\frac{k_{Stb}}{k_{Stg}}\right)^{3/2} R_b S - \theta_{cr}(s-1)D_m\right]^{3/2}, \text{ with}$$

$$\left(\frac{k_{Stb}}{k_{Stg}}\right)^{3/2} = 0.85 \quad (7)$$

and of a value for the equivalent sand roughness of $k_s = 1.5D_{90}$ in the flow law of Keulegan (1938), led to better results than the modified form of the Meyer-Peter & Müller formula and the value of $k_s = 2D_m$ proposed by Zarn. Compared with the approach of Zarn (1997), Eq. 6 yields a better agreement between measured and calculated values of bed load transport capacity (Fig. 11). More than 67% of the measured data can be predicted with an error of less than 20%, which is good for bed load transport data and for only four experiments is the deviation larger than 50%.

As can be seen from Fig. 12, Eq. 6 also allows a much better prediction of the bed load transport during flood hydrographs. The measured total bed load output for run HS3-1C, for example, can be computed with an error of only −3.6%. For the nine hydrograph runs performed in this study the deviations between calculated and measured output mass lie in a range of −16% and +23%, and the average deviation over all runs is +2.1%.

CONCLUSIONS AND OUTLOOK

Based on the optimum width approach of Zarn (1997) an improved empirical bed load transport formula for braided gravel-bed rivers is proposed.

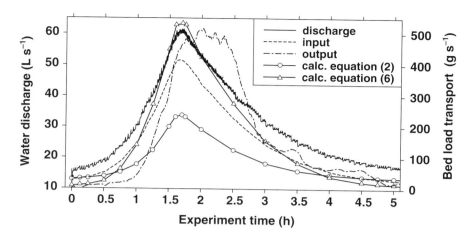

Fig. 12 Water discharge, bed load input and output during hydrograph run S3-1C. The calculated bed load output is indicated, calculated using the optimum width approach of Zarn (1997; Eq. 2) and the new formula presented here (Eq. 6).

The new formula considers the influence of wide grain-size distributions and of the bed load transport level, expressed by the ratio of Q/Q_0. Taking these two factors into account, good agreement between the calculated and measured data for experiments with constant discharge and bed load supply is achieved. The new formula also allows for a good estimation of the total bed load during hydrograph events.

The new formula is applicable for slopes between 0.2% and 2.5%, ratios of w_{bf}/w_{opt} between 1 and 12, and a standard deviation of grain-size distribution up to 4. A comparison of the new approach with prototype data should be based on a data set recorded over a longer period of time, because the bed load flux shows significant variability.

The most significant fluctuations in the bed load output flux measured in the experiments can be explained by morphological changes. Low output is observed as long as several anabranches exist, whereas extremely high output takes place during lateral shifting of a dominant single channel. The observations during periods of low bed load input also showed a strong influence on the morphological pattern. Reduction of bed load input leads to incised channels, and no braiding occurs. An ability to determine both the final slope and the final channel width, which are developed during such an erosion phase, is of considerable practical interest, and an analysis of all of the data from the present study may enable an approach to be developed. However, the high temporal and spatial resolution of the gathered data and the possibility of distinguishing the effective zones of bed load transport might provide a better understanding of the link between morphological changes and bed load transport. Furthermore, this study also provides consistent data as input and for calibration of numerical models.

ACKNOWLEDGEMENTS

This study was funded by the ETH-Zurich and the experimental work is supported by the workshop staff of VAW. Considerable translation assistance was provided by Urs Keller.

NOMENCLATURE

D_m	mean grain size according to Meyer-Peter & Müller (1948) ($\sim D_{60-65}$)
D_{50}	median grain size
D_x	characteristic grain size, $x\%$ by weight of the bed material are finer
g	gravitational acceleration
k_s	equivalent sand roughness according to Nikuradse (1933)
k_{Stb}	coefficient of total bed roughness according to Strickler (1923)
k_{Stg}	coefficient of grain roughness according to Strickler (1923)
k_{Stw}	coefficient of wall roughness according to Strickler (1923)
h	flow depth
Q	water discharge

Q_0	threshold discharge for initiation of bed load transport	t_{peak}	duration of flood peak	
		t_{tot}	total experiment time	
Q_{peak}	maximum discharge at flood peak	U	$= w_{bf}/w_{opt}$, standardized river width	
Q_b	bed load discharge	w	channel width	
$Q_{b\ in}$	bed load input	w_{bf}	bankfull width	
$Q_{b\ peak}$	maximum bed load input at flood peak	w_{opt}	optimum width	
$Q_{b\ red}$	reduced bed load input	w_w	wetted width	
Q_{bmax}	maximum bed load discharge	ρ	fluid density	
R_b	hydraulic radius related to bed section	ρ_s	bed material density	
$(s-1)$	$= (\rho_s/\rho - 1)$, relative submerged specific weight of bed material	θ_{cr}	dimensionless critical bed shear stress	
		τ	bed shear stress	
S_{in}	initial spatially averaged bed slope	τ_{cr}	critical bed shear stress	
S_m	mean spatially averaged bed slope	σ	$= (D_{84}/D_{16})^{0.5}$, geometrical standard deviation of grain-size distribution	
t_{eq}	experiment time during equilibrium state			

APPENDIX

Table A1 Key parameters of runs with constant discharge and constant bed load supply

Run	w_{bf} (mm)	S_{in} (%)	Q (L s^{-1})	$Q_{b\ in}$ (g s^{-1})	t_{tot} (h)	t_{eq} (h)
S1-1	3000	1.855	26.48	18.0	202.8	129.3
S1-2	3000	1.855	21.37	18.0	316.6	23.9
S1-3	3000	1.855	26.16	53.0	98.0	38.5
S2-1	3000	1.504	29.81	18.0	105.1	28.8
S2-2	3000	1.504	29.21	18.0	181.2	64.9
S3-1	3000	2.205	15.69	18.0	301.8	82.3
S4-1	3000	2.550	12.71	18.0	191.5	71.8
Ref	300	1.855	9.60	51.0	37.9	25.1

Table A2 Key parameters of hydrograph runs

Run	w_{bf} (mm)	S_{in} (%)	Q_{basis} (L s^{-1})	Q_{peak} (L s^{-1})	$Q_{b\ basis}$ (g s^{-1})	$Q_{b\ peak}$ (g s^{-1})	$Q_{b\ red}$ (g s^{-1})	t_{tot} (h)	t_{peak} (h)
HS1-1	3000	1.497	26.45	66.3	18.0	236.1	3.6	80.1	5.1
HS1-2A	3000	1.958	21.43	53.8	18.0	288.7	3.6	70.2	5.2
HS1-2B	3000	1.806	21.48	52.3	18.0	286.0	3.6	73.1	13.0
HS2-2A	3000	1.463	29.23	66.3	18.0	240.8	3.6	65.2	5.2
HS2-2B	3000	1.429	29.26	66.3	18.0	240.8	3.6	44.9	13.0
HS2-2C	3000	1.437	29.23	66.3	18.0	342.7	3.6	69.9	12.8
HS3-1A	3000	2.116	15.75	38.3	18.0	196.6	3.6	41.5	5.2
HS3-1B	3000	2.117	15.80	60.3	18.0	309.0	3.6	53.1	13.0
HS3-1C	3000	1.955	15.79	60.8	18.0	408.5	3.6	39.6	5.1

Table A3 Bed material parameters

Variable	Unit	Value
D_m	mm	3.90
D_{16}	mm	0.62
D_{50}	mm	2.98
D_{84}	mm	7.54
D_{90}	mm	9.10
ρ_s	kg m^{-3}	2650
θ_{cr}	–	0.047
k_{Stw}	m$^{1/3}$ s^{-1}	58

Table A4 Results of bed load transport in runs with constant discharge and constant bed load supply. Values are the mean values over the whole observation time during equilibrium conditions (t_{eq})

Run	Q (L s^{-1})	S_m (%)	w_w (mm)	h (mm)	Q_b (g s^{-1})	Q_{bmax} (g s^{-1})	w_{opt} (mm)	U (–)	$Q_0 (w_{opt})$ (L s^{-1})	Q_0^* (L s^{-1})
S1-1	26.48	1.554	2182	23.8	18.23	153.32	464	6.47	5.04	23.52
S1-2	21.37	1.918	2457	18.5	18.45	174.57	435	6.89	3.52	15.85
S1-3	26.16	1.749	2147	24.0	53.20	189.21	478	6.28	4.38	18.99
S2-1	29.81	1.399	1470	30.0	17.99	145.27	463	6.47	5.84	28.42
S2-2	29.21	1.473	1986	25.6	18.29	156.72	476	6.30	5.57	25.92
S3-1	15.69	2.120	1960	18.9	17.68	143.24	376	7.98	2.65	13.05
S4-1	12.71	2.432	1953	16.4	17.91	142.53	360	8.33	2.09	9.81
Ref	9.60	1.787	300	42.3	51.30	52.28	257	1.17	2.40	3.08

*Observed mean values (five out of eight, the others are estimated by the surrogate channel approach of Zarn (1997)).

Table A5 Results of bed load transport in hydrograph runs. The values are time-averaged values for three selected time windows of 5 min for each run

Run	Q (L s^{-1})	S_m (%)	w_w (mm)	h (mm)	Q_b (g s^{-1})	$Q_{b\,max}$ (g s^{-1})	w_{opt} (mm)	U (–)	$Q_0\,(w_{opt})$ (L s^{-1})	Q_0^* (L s^{-1})
HS1-1	40.60	1.508	ND	ND	140.46	248.88	592	5.07	6.60	24.83
HS1-1	63.00	1.508	ND	ND	295.05	425.07	774	3.88	8.52	24.83
HS1-1	34.40	1.508	ND	ND	80.85	202.27	534	5.62	5.99	24.83
HS1-2A	34.50	1.982	ND	ND	159.36	330.65	601	4.99	4.56	14.90
HS1-2A	53.60	1.982	ND	ND	314.24	552.63	789	3.80	5.94	14.90
HS1-2A	29.20	1.982	ND	ND	112.44	271.18	542	5.54	4.13	14.90
HS1-2B	34.50	1.806	ND	ND	165.95	281.68	578	5.19	5.01	17.83
HS1-2B	53.60	1.806	ND	ND	300.83	475.06	764	3.93	6.55	17.83
HS1-2B	40.00	1.806	ND	ND	215.76	336.40	628	4.78	5.42	17.83
HS2-2A	44.50	1.509	ND	ND	122.43	279.19	635	4.72	7.05	24.88
HS2-2A	69.00	1.509	ND	ND	259.47	474.11	827	3.63	9.07	24.88
HS2-2A	37.60	1.509	ND	ND	75.86	226.48	554	5.41	6.20	24.88
HS2-2B	44.50	1.429	ND	ND	126.62	252.83	597	5.03	7.17	27.34
HS2-2B	69.00	1.429	ND	ND	265.05	432.41	810	3.70	9.58	27.34
HS2-2B	51.60	1.429	ND	ND	177.28	303.81	679	4.42	8.10	27.34
HS2-2C	44.50	1.437	ND	ND	120.93	255.43	614	4.89	7.31	27.13
HS2-2C	69.00	1.437	ND	ND	258.87	436.55	810	3.70	9.51	27.13
HS2-2C	51.60	1.437	ND	ND	169.68	306.87	678	4.43	8.03	27.13
HS3-1A	24.50	2.116	ND	ND	110.70	246.32	499	6.01	3.48	13.07
HS3-1A	38.10	2.116	ND	ND	230.34	414.56	661	4.54	4.56	13.07
HS3-1A	20.80	2.116	ND	ND	71.70	202.10	449	6.68	3.15	13.07
HS3-1B	31.50	2.117	ND	ND	185.77	332.20	593	5.06	4.11	13.05
HS3-1B	60.80	2.117	ND	ND	455.59	707.83	897	3.34	6.14	13.05
HS3-1B	39.80	2.117	ND	ND	272.65	436.40	680	4.41	4.69	13.05
HS3-1C	31.50	1.969	ND	ND	194.94	293.56	562	5.34	4.32	15.10
HS3-1C	60.80	1.969	ND	ND	444.48	631.60	877	3.42	6.64	15.10
HS3-1C	23.90	1.969	ND	ND	98.32	210.29	476	6.30	3.69	15.10

ND = not determined.
*Estimated by the surrogate channel approach of Zarn (1997).

Table A6 Results of bed load transport for the Lombach project conducted at the VAW-Laboratories

Run	Q (L s^{-1})	S_{in} (%)	w_w (mm)	h (mm)	Q_b (g s^{-1})	$Q_{b\,max}$ (g s^{-1})	w_{opt} (mm)	U (–)	$Q_0\,(w_{opt})$ (L s^{-1})	Q_0^* (L s^{-1})
V_2	2.37	2.070	629	11.6	4.05	18.41	170	4.71	0.46	1.56
V_3	4.02	1.370	763	15.5	1.60	15.76	200	4.00	0.96	3.10
V_4	1.20	2.850	560	8.4	3.94	14.42	127	6.28	0.22	0.86

*Estimated by the surrogate channel approach of Zarn (1997).

Table A7 Bed material parameters of the Lombach project

D_m (mm)	D_{16} (mm)	D_{50} (mm)	D_{84} (mm)	D_{90} (mm)	ρ_s (kg m^{-3})	θ_{cr} (–)	k_{Stw} (m$^{1/3}$ s^{-2})
2.00	0.65	1.43	3.20	3.77	2650	0.045	64

REFERENCES

Ashmore, P.E. (1988) Bed load transport in braided gravel-bed stream models. *Earth Surf. Process. Landf.*, **13**, 677–695.

Ashmore, P.E. (2001) Braiding phenomena: statics and kinetics. In: *Gravel Bed Rivers V* (Ed. M.P. Mosley), pp. 95–121. New Zealand Hydrological Society, Wellington.

Bagnold, R.A. (1980) An empirical correlation of bed load rates in natural rivers. *Proc. R. Soc. London, Ser. A*, **372**, 453–473.

Carson, M.A. and Griffiths, G.A. (1989) Gravel transport in the braided Waimakariri River: mechanisms, measurements and predictions. *J. Hydrol.*, **109**, 201–220.

Einstein, H.A. (1934) Der hydraulische oder Profil-Radius. *Schweiz. Bauz.*, **103**(8), 89–91.

Einstein, H.A. (1950) *The Bed-load Function for Sediment Transportation in Open Channel Flows*. Technical Bulletin 1026, U.S. Department of Agriculture, Washington, DC.

Hoey, T., Cudden, J. and Shvidchenko, A. (2001) The consequences of unsteady sediment transport in braided rivers. In: *Gravel Bed Rivers V* (Ed. by P.M. Mosley), pp. 121–140. New Zealand Hydrological Society, Wellington.

Hunzinger, L.M. (1998) *Flussaufweitungen—Morphologie, Geschiebehaushalt und Grundsätze zur Bemessung*. Mitteilung Nr. 159, Versuchsanstalt für Wasserbau, Hydrologie und Glaziologie, ETH Zürich, Zürich, 206 pp.

Hunzinger, L.M. (1999) Morphology in river widenings of limited length. In: *28th International Association of Hydraulic Research Congress*, Graz, Austria, 22–27 August. (Paper on CD-Rom)

Hunzinger, L.M. and Zarn, B. (1997) Morphological changes at enlargements and constrictions of gravel bed rivers. In: *3rd International Conference on River Flood Hydraulics*, Stellenbosch, South Africa, pp. 227–236.

Ikeda, S., Parker, G. and Kimura, Y. (1988) Stable width and depth of straight gravel rivers with heterogeneous bed materials. *Water Resour. Res.*, **24**(5), 713–722.

Keulegan, G.H. (1938) Laws of turbulent flow in open channels. *J. Res. Nat. Bur. Stand.*, **21**, 707–741. (Research Paper RP1151)

Marti, C. (2002) Morphodynamics of widenings in steep rivers. In: *River Flow 2002: Proceedings of the International Conference on Fluvial Hydraulics*, Vol. 2, Louvain la Neuve, Belgium, 4–6 September (Eds D. Bousmar and Y. Zech), pp. 865–873. Balkema, Lisse.

Meyer-Peter, E. and Müller, R. (1948) Formulas for bed load transport. In: *2nd Conference of the International Association for Hydraulic Research*, Stockholm, Sweden, pp. 39–64.

Nicholas, A.P. (2000) Modelling bedload yield in braided gravel bed rivers. *Geomorphology*, **36**, 89–106.

Nikuradse, J. (1933) *Strömungsgesetze in rauhen Rohren*. Forschungsheft 361, VDI Verlag GmbH, berlin.

Pickup, G. and Higgins, R.J. (1979) Estimating sediment transport in a braided gravel channel—the Kawerong river, Bougainville, Papua New Guinea. *J. Hydrol.*, **40**, 283–297.

Smart, G.M. and Jaeggi, M.N.R. (1983) *Sediment Transport on Steep Slopes*. Mitteilung Nr. 64, Versuchsanstalt für Wasserbau, Hydrologie und Glaziologie, ETH Zürich, Zürich, 191 pp.

Strickler, A. (1923) *Beiträge zur Frage der Geschwindigskeitformel und der Rauhigskeitszahlen für Ströme, Kanäle und geschlossene Leitungen*. Mitteilung Nr 16, Amt für Wasserwirtschaft, Eigenössisches Departement des lunern, Bern.

Surian, N. and Rinaldi, M. (2003) Morphological response to river engineering and management in alluvial channels in Italy. *Geomorphology*, **50**, 307–326.

Warburton, J. and Davies, T.R.H. (1994) Variability of bed load transport and channel morphology in a braided river hydraulic model. *Earth Surf. Process. Landf.*, **19**, 403–421.

Warburton, J. (1996) Active braidplain width, bed load transport and channel morphology in a model braided river. *N. Z. J. Hydrol.*, **35**(2), 259–285.

Young, W.J. (1989) *Bedload Transport in Braided Gravel-bed Rivers: a Hydraulic Model Study*. Lincoln College, Canterbury, New Zealand, 187 pp.

Zarn, B. (1997) *Einfluss der Flussbettbreite auf die Wechselwirkung zwischen Abfluss, Morphologie und Geschiebetransportkapazität*. Mitteilung Nr. 154, Versuchsanstalt für Wasserbau, Hydrologie und Glaziologie, ETH Zürich, Zürich, pp. 240

Zarn, B. (2003) Alpeurhein (Kapitel 5.7.2). In: Fetstafftransport modelle für Fliessgewässer, ATV-DVWK-Arbeitsbericht, Henef, Germany.

Sediment transport in a microscale braided stream: from grain size to reach scale

P. MEUNIER and F. MÉTIVIER

Laboratoire de Dynamique des systèmes géologiques, IPGP, 4 place Jussieu, 75252 Paris cedex 05, France
(Email: meunier@ipgp.jussieu.fr)

ABSTRACT

This paper explores the influence of grain size and floodplain length on transport equations in a microscale braided stream. One-hundred and fifty-nine experiments were conducted with varying slope, water discharge, input sediment discharge, grain size and floodplain length. Bed load transport at the outlet is correlated with both stream power index and input flux for all the grain sizes. A general dimensionless transport equation is derived that relates transport efficiency of a braided river to its effective stream power. Experiments have been run to evaluate the critical stream power index for the inception of sediment motion and include it in the transport relationship. The influence of different characteristic length scales (flow depth, floodplain length) on the transport dynamics is also examined, and it is shown that bed load transport rate is probably related to the ratio of water-depth:grain-size. In the present experiments, the input flux tends to reduce the bed load rate by forcing the river to braid, and this reduction increases with grain size and decreases with floodplain length. This suggests that the braiding process is significantly enhanced with coarse material but that dispersion of the sediment wave occurs to counteract this influence.

Keywords Braided river, microscale river, bed load, grain size, mass balance.

INTRODUCTION

The pattern of alluvial rivers is linked to the movement of bed material, and amongst these different river patterns, braided rivers are those for which bedload is the greatest component of the total load. Expressed in another way, in braided streams the fraction of time during which bed load material moves is long enough to ensure bed destabilization. Field measurements on proglacial braided streams have shown that bed load transport, although intermittent, can account for a significant portion of the total load (e.g. Wohl, 2000). In the case of the Ürümqi River in the Chinese Tian Shan, measurements have shown that the bedload fraction can vary between 30 and 60% of the total load (Métivier *et al.*, 2004). Previous work has suggested that braiding can be viewed as a succession of sediment waves, of different wavelength, moving downstream (Ashmore, 1988; Hoey & Sutherland, 1991; Warburton & Davies, 1994). These results suggest that the sediment flux measured at a given cross section should correlate with its value upstream. Following this approach, Meunier & Métivier (2000) and Métivier & Meunier (2003) showed that in an experimental braided stream the bed load transport is a linear function of both stream power index and the input sediment flux injected into the flume. However, the relationship of Métivier & Meunier (2003) was established with beds made of a uniform particle distribution in a fixed length flume. A new set of experiments, performed with different grain sizes and different flume lengths, should allow extension of the transport law of Métivier & Meunier (2003) to take account of these new parameters. However, the use of different grain sizes raises the problem of the inception of bed movement for a given grain size. This paper thus begins with discussion of work on the incipient motion of streambed materials, and experimental results are presented and a new transport relationship is derived. To explore the dependence of braiding on grain size, experiments were performed in a straight channel to establish, for each grain size, a macroscopic critical stream power index which is incorporated into the transport relations. The influence of different characteristic length scales, flow depth and flume length, on the transport dynamics in braided streams is examined in order to propose a new relationship. Finally, the implications of these results are discussed.

Grain size and bed movement inception

The effect of grain size on bed load transport dynamics was introduced in a quantitative way by Shields (1936) who focused on the conditions required for the inception of movement in streambeds. The transport system can be entirely described by several physical parameters, which are the density of the fluid and sediment grains, respectively denoted by ρ_f and ρ_s, the fluid kinematic viscosity ν, the stress exerted on the particle by the fluid τ_0, and the particle diameter D. A range of experimental runs with different fluids (varying ρ_f and ν) and different particles (varying ρ_s and D) allowed Shields (1936) to establish a relation between the critical shear stress τ_{*cr} defined as:

$$\tau_{*cr} = \frac{\tau_0}{(\rho_s - \rho_f)D} = \frac{\rho_f u_*^2}{(\rho_s - \rho_f)D} \quad (1)$$

where $u_* = \sqrt{\tau_0/\rho_f}$ is the shear velocity of the fluid, and the grain Reynolds number Re_*, defined as:

$$Re_* = \frac{u_* D}{\nu} \quad (2)$$

This empirical curve (Fig. 1) is still used and remains the most stable relation to describe the conditions required for the inception of transport. However, a fundamental problem remains in the definition of what is exactly considered to be incipient motion. Evaluation of the inception of movement by visual observation depends on the quantity of moving grains, the size of the observed area and the duration of the observation (Buffington & Montgomery, 1997). Transport theoretically takes place when one particle begins to move but, because this condition is very difficult to observe, Shields and most later studies concerning the initiation of movement (Buffington & Montgomery, 1997) generally proceeded by extrapolation of transport data to the point at which movement vanishes (Shields, 1936; Day, 1980; Parker & Klingeman, 1982; Ashworth et al., 1992)]. However, this method is sensitive to the extrapolation used. In order to dismiss this ambiguity, Parker & Klingeman (1982)

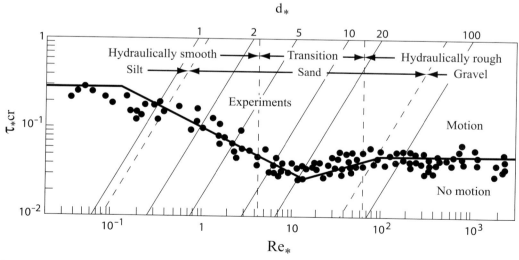

Fig. 1 Shields curve that expresses the critical Shields stress, τ_{*cr}, required for bed load motion as a function of the grain Reynolds number, Re_*, or the dimensionless grain size $d_* = d[g(\rho_s/\rho_f - 1)\nu^2]^{1/3}$ (from Julien, 1995).

suggested the use of a dimensionless transport criterion of the form:

$$W_* = \frac{Q_s}{\sqrt{gD}D^2} \quad (3)$$

where Q_s is the volumetric sediment flux. More importantly, Parker & Klingeman (1982) assumed that the initiation of transport takes place for a small, but given, value of W_*. The choice of this transport condition is totally arbitrary but offers the advantage of clarity and simplicity. Finally, note that the use of a single parameter, D, to describe a bed relies on the assumption that the bed is, to the first order, composed of single sized spherical grains. Because of the complexity of natural bed material, several studies have modified the study of Shields by taking into account grain-size distributions and the effects of sorting or packing (Buffington & Montgomery, 1997).

Transport relations

The problem of the movement threshold in a global sediment transport relation has been expressed by Bagnold (1973), making use of the work of Shields (1936). Indeed, Bagnold (1973) defined the sediment transport rate per unit width as a function of the product of the bottom velocity, u_b, and the excess shear rate:

$$q_s = \frac{\beta u_b}{\Re}(\rho_0 - \rho_{cr}) \quad (4)$$

where $\Re = (\rho_s - \rho_f)g/\rho_f$ and β is a non-dimensional factor depending on grain size. This relation is equivalent to:

$$Q_s \propto \xi(\Omega - \Omega_c) \quad (5)$$

where Q_s is sediment transport rate over the total width, Ω and Ω_c are stream power and the critical stream power, defined as $\Omega = \rho_f gqS$ and $\Omega_c = \rho_f gqS_c$, and ξ is another factor (see Métivier & Meunier (2003) for a detailed derivation). In the case of constant discharge, the threshold is given by a critical slope S_c required for the initiation of transport. Most studies of bedload transport start from this type of relation to describe transport at the river scale (e.g. the Meyer-Peter and Müller equation; Yalin, 1992). The fact that the values of bed load transport rate, measured from a given stream power, can vary by one to two orders of magnitude (Pickup et al., 1983; Ashmore, 1988; Hoey & Sutherland, 1991) leads the present investigation to consider transport as an intermittent wave-like process. In this case, a missing parameter that appears to be significant for transport is the upstream sediment flux. In order to express this dependence, several experiments have been run in a microscale flume with varying slope, discharge and input sediment flux (Métivier & Meunier, 2003). This study showed that, for steady state conditions (constant discharge and input flux), the output sediment flux tends toward a steady-state while a stable braiding pattern develops (Fig. 2). Relating the sediment discharge to both the stream power and input sediment flux, Q_e, produces a linear correlation. The transport efficiency is thus defined as $Q_* = Q_s/Q_e$ as a function of the effective stream power index $\Omega_* = \Omega/(\rho_f gQ_e)$. The dimensionless relation is expressed as:

$$Q_* = E\Omega_* - S_f \quad (6)$$

where E is the stream effectiveness and S_f is the source factor. In the case of the experiments conducted by Métivier & Meunier (2003), $E = 0.47$ and $S_f = 0.58$. Equation 6 is useful for describing the dynamic equilibrium of the system, which is the aggradation/degradation state of the stream over

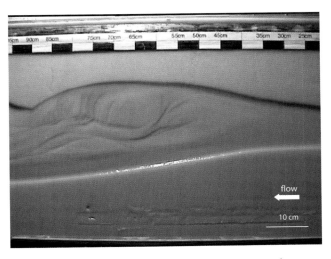

Fig. 2 View of braiding developed during a run for $D_{50} = 259$ μm, $Q = 1.5$ L min^{-1} and $S = 3.4.10^{-2}$.

a characteristic length-scale defined by the injection point (Q_e) and the measurement point (Q_s) which, in the present experiments, is the length of the flume. This state is given by the value of Ω_* in comparison to an equilibrium value (here 3.36; Métivier & Meunier, 2003). These results have been obtained using an almost uniform grain size with a D_{50} value of 477 μm. In order to study the grain-size dependence, this work has been continued with different size distributions. The main purpose of the present study is to determine if Eq. 6 remains valid for varying size distributions and can be generalized in order to include a grain-size dependence. The introduction of different grain sizes leads to consideration of the different conditions required for the initiation of sediment motion. These conditions are expressed in terms of critical stream power for a given grain size, and several experiments were used to establish this critical stream power. Two methods for the detection of incipient motion have been used: (i) visual observation and (ii) analysis of the output bed load transport rate. Equation 6 has thus been changed to a threshold relation which takes into account Ω_{*c}. In the discussion that follows, several scalings are proposed to explain the evolution of transport rate with grain size. Finally, experimental results for varying flume length are reported to show the limit of the relation by pointing out a damping effect on the correlation between the input and output fluxes.

EXPERIMENTAL METHODOLOGY AND RESULTS

Four different grain sizes have been used successively that range from 260 μm to 1110 μm. The size distribution for each mixture follows a Gaussian form with a rather small standard deviation (5–13%; Fig. 3). The density of each mixture was measured with a pycnometer, and the values of the median size, D_{50}, the standard deviation, σ, and density values for each mixture are given in Table 1. Approximately 100 runs were performed in the same flume following a protocol identical to that used in the first set of experiments (see Métivier & Meunier, 2003), and values of the main physical parameters are reported in Table 2. For each value of the effective stream power, the output sediment flux was measured. Values of Q_s were deduced from the linear fit of the output mass curve assuming that, after a given duration, the transport tends toward a steady state so that the sediment flux carried by the stream is constant (see Métivier & Meunier (2003) for a discussion). Correlations of Q_* with Ω_* for the four grain-size distributions (Fig. 4) show that the shape of Eq. 6 is clearly verified in each case. Plots resulting from experiments made with the largest sand size (1110 μm, Fig. 4B, triangles) seem to divide into two distinct linear regressions. One of these (Fig. 4; set 1, open triangles) almost merges with the set of plots resulting from the experiments performed with the 878 μm diameter

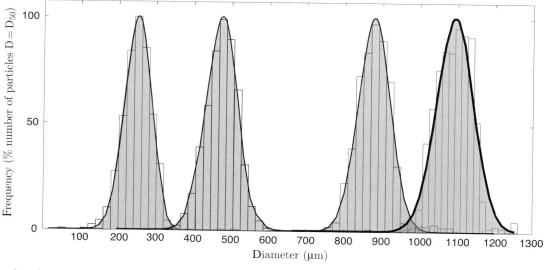

Fig. 3 Size-distribution curves of the sediment used. The distributions follow a Gaussian law; note the small overlap between the distributions.

Table 1 Median diameter and density of the four grain sizes used

Number of particles analysed	Median size (D_{50}) (μm)	Standard deviation	Density (kg m^{-3})
1235	259	23	2485
1624	477	65	2485
1246	878	42	2587
1475	1110	67	2583

run with stream power index values significantly lower ($\Omega = 1$–1.5 g s^{-1}) than in the second set of experiments ($\Omega = 2.7$ g s^{-1}). For low values of Ω, only the smaller particles are transported, thus shifting the size distribution of the bed load from 1110 μm to a lower value (around 1000 μm). It is interesting to note that, even with a tightened distribution of diameters, the system is able to select a subclass with the only condition that the amount of grains of this subclass is great enough to provide a continuous flux.

This condition is ensured by the overlap of the size distribution curves of the 878 μm and 1110 μm classes (Fig. 3). Starting from the linear regressions, the transport efficiency can be expressed as:

sand. This can be explained by analysing the average value of the stream power index for these two sets. The first set of data corresponds to experiments

$$Q_* = E(D)\Omega_* - S_f(D) \qquad (7)$$

Table 2 Variables and physical parameters used in this study. Factors and empirical constants are not included

Symbol	Definition	Dimension	Range in experiment
ρ_f	Density of water	[M][L]$^{-3}$	1000 kg m^{-3}
ρ_s	Density of sediment	[M][L]$^{-3}$	2485–2587 kg m^{-3}
g	Acceleration of gravity	[L][T]$^{-2}$	9.81 m s^{-2}
γ_s	Reduced gravity	[L][T]$^{-2}$	14.71 m s^{-2}
ν	Kinematic viscosity of water	[L]^2[T]$^{-1}$	1–1.13 × 10^{-6} m^2 s^{-1}
q	Fluid discharge	[L]^3[T]$^{-1}$	0.0086–0.065 L s^{-1}
$\rho_f q$	Fluid discharge	[M][T]$^{-1}$	8.6–65 g s^{-1}
Q_e	Input flux of sediment	[M][T]$^{-1}$	0.01–0.7 g s^{-1}
Q_s	Output flux of sediment	[M][T]$^{-1}$	0.004–0.88 g s^{-1}
S	Slope	NA	0.026–0.093
D	Grain size	[L]	259–1110 × 10^{-6} m
h	Flow depth	[L]	0.1–1 × 10^{-2} m
W_c	Channel width	[L]	1–10 × 10^{-2} m
W	Total flow width	[L]	3.5–30 × 10^{-2} m
L	Braidplain length	[L]	1–2 m
u	Flow velocity	[L][T]$^{-1}$	0.2–0.5 m s^{-1}
u_*	Shear velocity	[L][T]$^{-1}$	NM
u_b–u_*	Bottom velocity	[L][T]$^{-1}$	NM
τ_0	Bottom shear stress	[M][L]$^{-1}$[T]$^{-2}$	NM
S_c	Critical slope	NA	NM
v_s	Fall velocity of sediment	[L][T]$^{-1}$	0.025–0.16 m s^{-1}
Ω	Stream power	[M][L][T]$^{-3}$ or Watt m^{-3}	2.5–60 mW m^{-3}
Ω_c	Critical stream power	[M][L][T]$^{-3}$ or Watt m^{-3}	1.76–4.5 mW m^{-3}
Ω_*	Dimensionless stream power index	NA	2–200
Q_*	Dimensionless transport efficiency	NA	0.02–35

NA = not applicable; NM = not measured.

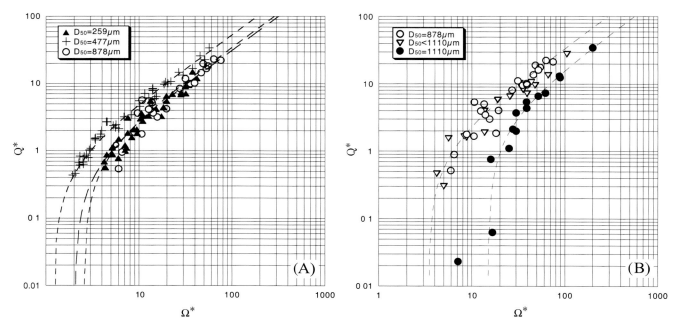

Fig. 4 The transport efficiency, $Q_* = Q_s/Q_e$, related to the effective stream power index, $\Omega_* = Q_l S/Q_e$, for different grain sizes. The 1110 μm experiments are divided into two sets of plots: the higher set (noted ∇, $D_{50} < 1110$ μm in B) is shifted to the 878 μm curve, and is provided by experiments made with a low stream power index. The dashed lines show fit according to Eq. 6.

where $E(D)$ and $S_f(D)$ are grain-size-dependent coefficients; $E(D)$ is the 'stream effectiveness coefficient' expressing the influence of the stream power index, while $S_f(D)$ can be defined as a 'source factor' expressing the influence of the input sediment flux. Variation of both $E(D)$ and $S_f(D)$ with grain size (Fig. 5) shows the stream effectiveness coefficient to exhibit a non-monotonic evolution whereas the source factor strictly increases with grain size.

DISCUSSION ON THE DEPENDENCE OF BED LOAD TRANSPORT ON GRAIN SIZE, FLOW DEPTH AND STREAM LENGTH

The two coefficients computed above are shown to be grain-size dependent. The following discussion presents different approaches that study the evolution of $E(D)$ and $S_f(D)$ with grain size in order to explain this dependence.

Comparison of incipient motion conditions

Considering the range of values of Re_* in the present experiments (from 3 to 50), the corresponding critical shear stress given by the Shields curve follows a non-monotonic function in the area considered (Fig. 1). Thus, a possible relation between $E(D)$ and τ_{*cr} has to be re-enforced by several incipient motion studies. Following Bagnold's (1973) description of bed load (Eq. 5), the transport rate does not simply depend on the stream power, but on the difference between the stream power index and a critical value corresponding to the inception of motion. This critical stream power index, Ω_c, is grain-size dependent and its determination enables description of the influence of grain size on the incipient motion conditions.

Several experiments were performed to establish the critical stream power index for each grain size. In these experiments, flow was concentrated in an initial channel 2.5 cm wide and 1 cm deep. The bed slope was kept constant during the run (Fig. 6). After a pre-wetting phase during which the granular medium was saturated with water, the discharge was

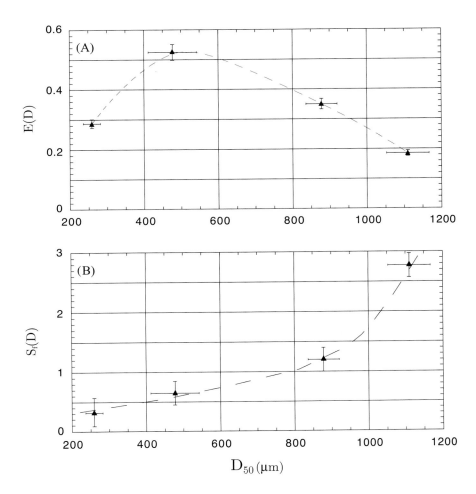

Fig. 5 Evolution of (A) the stream effectiveness coefficient $E(D)$, and (B) the source factor $S_f(D)$ with grain size.

increased in successive steps until the grains moved within the channel. For each grain size, three runs were performed using different slopes and the observations were made at the outlet of the flume. Two independent methods have been used to determine the inception of motion: visual observation and continuous bed load measurements using a bed load trap (see Métivier & Meunier (2003) for details). The critical value is reached when the measured bed load at the outlet becomes different from zero for an unlimited time duration. This threshold thus corresponds to the inception of motion along the entire length of the reach. Both methods show that the value of Ω_c increases with the grain size (Fig. 7). Values of Ω_c determined visually are significantly lower than when measured by bed load detection (Fig. 7), with the difference being ~7.10^{-2} g s^{-1}. The bed load transport rates were used as the value of Ω_c because it is an objective and simple criterion, although it does correspond to the length integrated transport threshold in a straight channel rather than a braided one. Using a single channel geometry ensures that river width is controlled and is almost stable at the beginning of motion. Thus, the changes in channel geometry and hydraulic conditions should be minimized with the main varying parameter being grain size.

The transport efficiency, Q_*, is then related to the differential stream power index $\Omega - \Omega_c$ divided by the input sediment flux Q_e. The form of the transport relation is changed by taking Ω_c into account. The transport efficiency is now expressed as:

$$\frac{Q_s}{Q_e} = E'(D)\frac{\Omega - \Omega_c}{Q_e} - S'_f(D) \quad \text{or}$$
$$Q_* = E'(D)(\Omega - \Omega_c) - S'_f(D) \tag{8}$$

Fig. 6 The bed configuration for incipient motion detection runs. An initial channel 2.5 cm wide by 1 cm deep was cut and the discharge increased in successive steps until grains were transported out of the flume.

The values of $E'(D)$ and $S'_f(D)$ are thus computed from the new correlations between Q_* and $\Omega_* - \Omega_{*c}$ and are presented in Fig. 8.

The relations of these new coefficients to grain size are not significantly changed by the presence of the critical stream power index in the relations. More particularly, the stream efficiency coefficient still evolves in a non-monotonic manner with grain size. Consequently, this behaviour is not due to the effect of incipient motion, and several experiments have therefore been run with systematic measurements of the average fluid velocity and average channel width, in order to investigate the changes in stream morphology with grain size. Velocity was measured several times during the runs using a blue dye injected into the stream and images, taken at 25 Hz, allowed measurements of the velocity and width of the dye.

Flow velocity comparison

The preceding results suggested another explanation for the non-monotonic evolution of the coefficient $E(D)$ (or $E'(D)$). First, it could be useful to study how water velocity is modified by the grain size of the bed for a given stream power. Figure 9 shows the average velocity and total stream width measured during four runs with different grain sizes. The stream power index was fixed at 0.87 g s^{-1}. All the runs were performed without input sediment fluxes and the bed was degrading. The data dispersion is important, but on average the fluid velocity

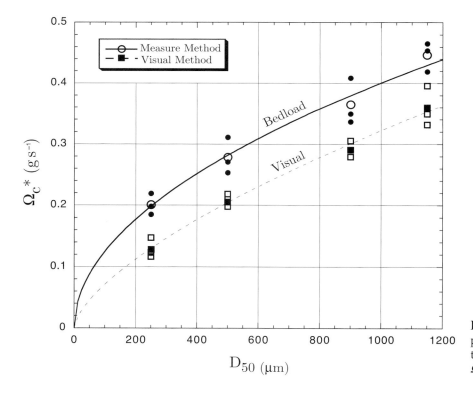

Fig. 7 Values of the critical stream power index measured with the two methods. Both estimates of Ω_c increase with D_{50}.

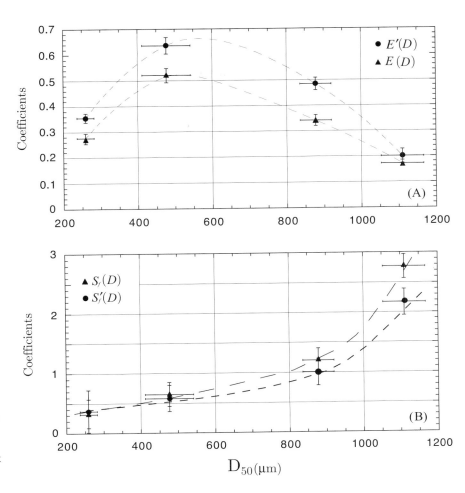

Fig. 8 Changes in (A) the stream effectiveness coefficient $E'(D)$, and (B) the source factor $S'_f(D)$, with grain size. Comparison is presented with the coefficients computed without use of the critical stream power index ($E(D)$ and $S_f(D)$).

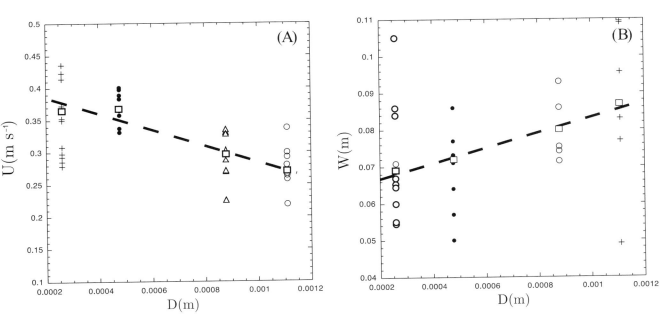

Fig. 9 (A) Average fluid velocity and (B) the average total width for four runs plotted against grain size. For all runs, Ω was held at 0.87 g s^{-1}.

decreases with grain size while the total stream width increases. Transport of coarser grains implies a greater volume of bed motion and thus increased widening. The pattern of velocity, and thus shear stress on the bed, does not explain the decrease of the stream effectiveness $E'(D)$ for the finest grains.

Relative submergence

The possible influence of water depth is now discussed. Consider a one-dimensional particle of diameter D immersed in a flow depth, h. If the water velocity profile is given by the law-of-the-wall, the average velocity applied on a particle with a diameter equal to D is

$$\bar{U}_D = \frac{u_*}{(D - z_0)} \int_{z_0}^{D} \frac{1}{\kappa} \ln\left(\frac{z}{z_0}\right) dz \quad (9)$$

if $D \leq h$, and the average velocity over the total depth is

$$\bar{u} = \frac{u_*}{(h - z_0)} \int_{z_0}^{h} \frac{1}{\kappa} \ln\left(\frac{z}{z_0}\right) dz \quad (10)$$

If z_0 is defined as $z_0 = aD$, the velocity ratio can be expressed by the relation:

$$\frac{\bar{U}_D}{\bar{u}} = \frac{\frac{h}{D} - a}{1 - a} \frac{\ln\left(\frac{1}{ea}\right) + a}{\frac{h}{D}\left(\ln\left(\frac{h}{D}\right) + \ln\left(\frac{1}{ea}\right)\right) + a} \quad (11)$$

while $0 < D \leq h$. When $D > h$ the velocity ratio is directly expressed as $\bar{U}_D/\bar{u} \propto h/D$.

The ratio h/D is computed from velocity and width measurements using the relationship for the average depth $h = Q/W\bar{u}$. Figure 10 plots the theoretical evolution of the velocity ratio and $E'(D)$ values with h/D. Note that h/D is very close to unity for the two largest grain classes, is ~2 for the 477 μm class and can reach 5 for the smallest class. The finest grain-size experiments have the largest range of h/D and thus the finer grains are submitted to a lower velocity flow and move slower than the larger ones. In this case, the stream effectiveness coefficient $E'(D)$ has a similar role to the ratio $\bar{U}_D/\bar{u}$. Thus, the transport relation can be expressed as:

$$Q_* = \psi\left(\frac{\bar{U}_D}{\bar{u}}\right)(\Omega_* - \Omega_{*c}) - S'_f(D) \quad (12)$$

when expressed in terms of velocity dependence, or

$$Q_* = \Psi\left(\frac{h}{D}\right)(\Omega_* - \Omega_{*c}) - S'_f(D) \quad (13)$$

when expressed in terms of size dependence. Both the theoretical velocity ratio and $E'(D)$ two curves show the same tendency (Fig. 10), but the $E'(D)$ maximum value that is different from unity is reached for $h/D = 2$. The analytical form for the transport equation computed from the data gives:

$$\begin{cases} Q_* = 0.66\left(\frac{h}{D} - 1\right)(\Omega_* - \Omega_{*c}) - S'_f(D) & \text{for } h/D \leq 2 \\ Q_* = \dfrac{1.646\dfrac{h}{D} - 1.8}{\left(\dfrac{h}{D} - 1\right)\ln\left(\dfrac{h}{D} - 1\right) + 1.408\dfrac{h}{D} - 1.317}(\Omega_* - \Omega_{*c}) \\ \quad - S'_f(D) & \text{for } h/D > 2 \end{cases} \quad (14)$$

This calculation of $\bar{U}_D/\bar{u}$ is only approximate and several comments should be borne in mind:

1 Velocity profiles are assumed to be logarithmic in the above analysis. However, this is not exactly the case in the present experiments, especially when $h/D \approx 1$. Although parabolic profiles should not change significantly the shape of the theoretical transport curve, for flat velocity profiles this analysis is no longer strictly correct.
2 Velocity measurement is averaged over several channels. Inactive channels (flowing without transporting any grains) are accounted for in the averaging process. In reality, moving grains that protruded above the flow were not observed, and the h/D ratio may thus not reach a value of unity in channels where transport is active.
3 Another problem appears when $h/D \approx 1$, when the three-dimensional fluid velocity field may be perturbed by the presence of large grains and the evaluation of $\bar{u}$ is difficult and uncertain.
4 The calculation of velocity ratio was made for grains in contact with, and totally protruding from, the bed. When in movement, such particles will experience varying drag force depending on their vertical posi-

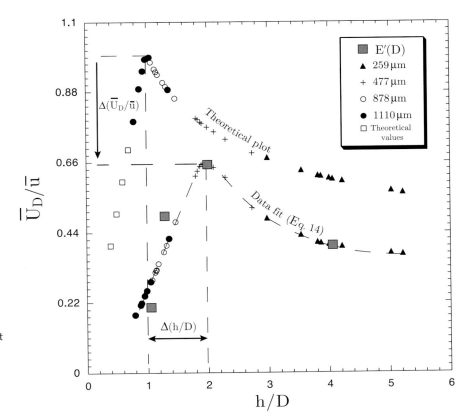

Fig. 10 Coefficient $E'(D)$ and theoretical ratios $\bar{U}_D/\bar{u}$ plotted against h/D. The lower curve corresponds to the fit of $E'(D)$ according to Eq. 14. The values of h/D computed from water depth measurements for each grain size are given on each curve.

tion in the flow. As has been clearly demonstrated by Francis (1973), both the particle trajectory and the fluid drag depend on h/D.

Points 1 to 3 above may in part explain the vertical shift $\Delta(\bar{U}_D/\bar{u})$ of the velocity ratio (Fig. 10) towards measured values of $E(D)$, whilst points 2 and 4 may explain the lateral $\Delta(h/D)_{max}$ shift of the velocity ratio. Point 4, especially, will most probably result in a shift of the $\bar{U}_D/\bar{u}$ curve towards higher h/D ratios (Francis, 1973). The restrictions mentioned above thus highlight the need for further experiments and modelling of particle–fluid interactions in order to constrain the mechanisms involved. Additionally, most of the analysis developed above assumes that the particle concentration is dilute, and this is addressed below.

Sediment flux considerations

Another way to understand the variation in the stream effectiveness coefficient is to consider transport as a flux of particles. Both Q_s and Q_e can be expressed as fluxes of round particles of diameter D. The transport efficiency can thus be expressed as:

$$Q_* = \frac{N_s}{N_e} E_n(D) \frac{\Omega - \Omega_{cr}}{N_e}, \qquad (15)$$

where N_e and N_s are the fluxes of particles entering and leaving the flume respectively. In this form, the coefficient E_n becomes dimensional and is similar to a volumetric concentration (number of particles per unit volume). The relation between N_s/N_e and Ω_* (Eq. 14) and the resulting values of $E_n(D)$ (by fitting the data) are given in Fig. 11A. It is important to note that the volumetric concentration, and thus the number of moving grains, is a strictly decreasing function of grain size. Coefficient $E_n(D)$ reaches 13,000 particles cm^{-3} of fluid for finer grains. It should be remembered that $E_n(D)$ is strictly equivalent to $E'(D)$ but expressed in different dimensions. However, when expressed as a fraction of the fluid volume (dimensionless coefficient $E'(D)$), the evolution is no longer monotonic (Fig. 11). The volumetric concentration corresponding to a linear

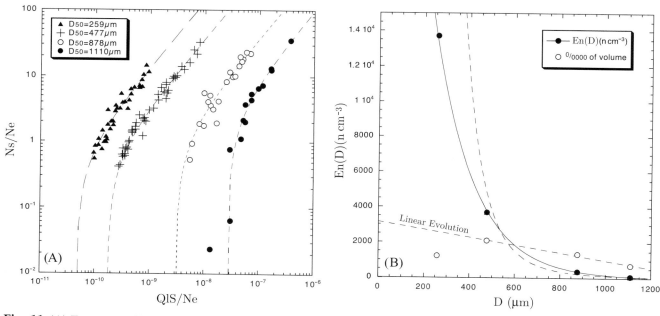

Fig. 11 (A) Transport efficiency expressed as a ratio of moving particles, $Q_* = N_s/N_e$, to the effective stream power index, Q_lS/N_e. (B) Evolution of the volumetric concentration $E_n(D)$ (denoted ●) showing that finer moving particles are numerous. Nevertheless, when expressed as a volumetric fraction (similar to $E(D)$, denoted ○), the function declines for finer grains. The dashed curve presents the volumetric concentration required to provide a monotonic and linear evolution of the volumetric fraction with D.

evolution of the volume fraction is shown by a dashed line to allow comparison. The large number of finer particles in a given water volume leads to the conclusion that momentum coupling effects could influence the grain velocity. Further experiments are needed to precisely quantify the influence of particle concentration on the average grain velocity.

Source factor evolution and floodplain length scaling

Examination of the source factor evolution $S_f'(D)$, and its constant increase with grain size (Fig. 8) leads to consideration of whether information is transmitted from upstream to downstream when grains are larger. This fact, which can appear paradoxical, is considered in terms of river morphology. The larger the grains are, the more difficult it is for the stream to transport them for a given sediment flux, and thus with increasing D the input sediments stay longer in the braidplain. The stream morphology is related to the residence time of the sediment, since grains settling in the braidplain induce an increase in the braiding intensity, which results in a reduction of the stream power index and total transport ability. Several past studies have also highlighted the decrease in bed load transport with increasing braiding intensity (Hoey & Sutherland, 1991; Warburton & Davies, 1994; Marti, 2002). When considering the 259 μm class, this effect is counterbalanced by the increase of h/D, which makes grains move slower and thus stay longer in the braidplain. This may explain why $S_f'(D)$ shows a damping in the 259–577 μm range (Figs 5 & 8) and suggests that braiding processes are better developed in coarse material.

The influence of the source input leads to consideration of how changes in the distance separating the inlet and the bed load measurement point may affect the transport law. To answer this question, several experiments were run with a varying length flume and with one grain-size class of 577 μm. Three different flume lengths were tested: 100, 160 and 180 cm. These experiments reveal that the relation between transport rate and stream power index weakens with an increase in stream length (Fig. 12A). Following the same protocol, the values of $E(L)$ and $S_f(L)$ (Fig. 12B) show that, while

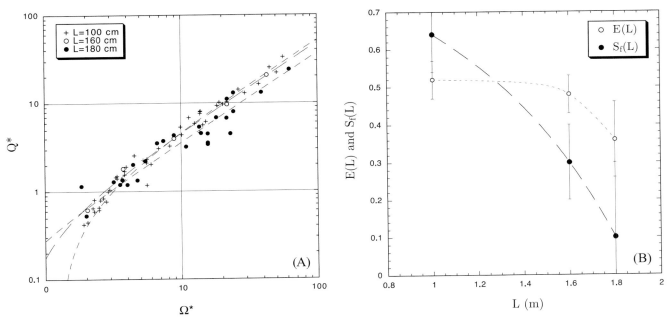

Fig. 12 (A) Transport efficiency, $Q_* = Q_s/Q_e$, related to the effective stream power index $\Omega_* = Q_l S/Q_e$ for different flume lengths. For greater flume lengths, the correlation becomes poorer. (B) Values of the coefficients $E(L)$ and $S_f(L)$ drawn from the regressions. Note that the influence of the input flux disappears for the greater flume length.

$E(L)$ does not change significantly with increasing flume length, $S_f(L)$ abruptly falls and becomes very small for the 180 cm length runs. These experiments show the limit of the correlation between the sediment source and bed load. Once a characteristic distance is reached, the river has completely absorbed the input sediment flux by changing its topography and the bed load is thus no longer modified by the flux. Variations in the source factor with both D and L are plotted in Fig. 13 which presents the following relation:

$$S_f = S_{f0}\left(\frac{L}{D}\right)^{-\frac{3}{2}} \quad (16)$$

where S_{f0} is arbitrary and equal to 64,800. This relationship may be dependent on the width of the flume, and this damping effect could be strongly modified by a narrow braidplain where braiding would be confined. This point is important because the lateral divergence of flow is the principal characteristic of braiding processes. Nevertheless, the width:length ratio of the present flumes (0.4 and 0.5) assures a fully braided pattern and small sidewall effects.

CONCLUSIONS

Experimental results confirm that the macroscopic transport equation of Métivier & Meunier (2003) can be extended to varying bed material and floodplain size. The grain-size dependence has been partially explored using the relation between grain velocity and the ratio of water-depth:grain-diameter. This work still ignores the effects of particle interactions, which are shown to be possibly significant. Moreover, uncertainties concerning the average water velocity should be raised in future work in order to quantify the ratio between particle and water velocity. The correlation of the input/output solid fluxes is shown to be enforced with coarse material, with high input sediment loads tending to decrease the bed load transport rate by forcing the river to braid. The idea that increasing braiding intensity is a morphological response to a high sediment load and coarse grain-size confirms previous work by Hoey & Sutherland (1991), Warburton & Davies (1994) and Marti (2002). However, the present paper shows this process has a spatial limitation, and that damping of the source factor influence illustrates the dispersion of the sediment wave

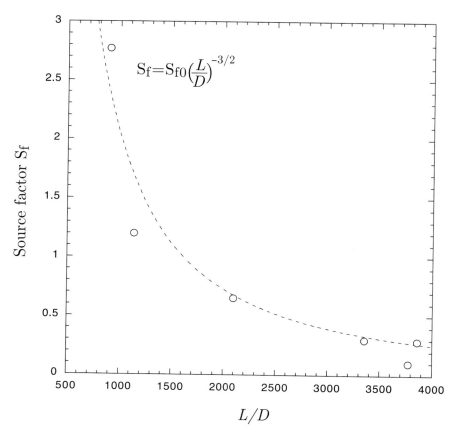

Fig. 13 Evolution of S_f with the ratio of the flume-length:grain-size, L/D. The trend follows a power law relationship with an exponent of $-3/2$.

coming from the source. In spite of the necessity to scale this length by the floodplain width, these results suggest the existence of a characteristic length scale in which the river absorbs a certain amount of sediment by changing its morphology, this being in agreement with previous work concerning sediment wave dispersion (Pickup et al., 1983; Ashmore, 1988; Hoey & Sutherland, 1991; Warburton & Davies, 1994). Future development of measurement methods for bed topography should prove useful to investigate this characteristic length further.

ACKNOWLEDGEMENTS

This study was supported by French research programmes PNSE and PNRH, and by IPGP. We benefited from the technical assistance of G. Bienfait and Y. Gamblin. T.B. Hoey and C. Marti made constructive and useful comments.

REFERENCES

Ashmore, P. (1988) Bed load transport in gravel-bed stream models. *Earth Surf. Process. Landf.*, **13**, 677–695.

Ashworth, P.J., Ferguson, R.I., Ashmore, P.E., Paola, C., Powell, D.M. and Prestegaard, K.L. (1992) Measurements in a braided river chute and lobe 2. Sorting of bed load during entrainment, transport and deposition. *Water Resour. Res.*, **28**, 1887–1896.

Bagnold, R. (1973) The nature of saltation and of 'bed load' transport in water. *Proc. R. Soc. Lond.*, **332**, 473–504.

Buffington, J.M. and Montgomery, D.R. (1997) A systematic analysis of eight decades of incipient motion studies, with special reference to gravel-bedded rivers. *Water Resour. Res.*, **33**, 1993–2029.

Day, T.J. (1980) *A Study of the Transport of Graded Sediments*. Report IT190, Hydraulic Research Station, wallingford, 10 pp.

Francis, J.R.D. (1973) Experiments on the motion of solitary grains along the bed of a water-stream. *Proc. R. Soc. London*, **332**, 443–471. Communicated by R.A. Bagnold.

Hoey, T. and Sutherland, A. (1991) Channel morphology and bedload pulses in braided rivers: a laboratory study. *Earth Surf. Process. Landf.*, **16**, 447–462.

Julien, P.W. (1995) *Erosion and Sedimentation*. Cambridge University Press, Cambridge.

Marti, C. (2002) Morphodynamics of widenings in steep rivers. In: *River Flow 2002: Proceedings of the International Conference on Fluvial Hydraulics*, Vol. 2, Louvain la Neuve, Belgium, 4–6 September (Eds D. Bousmar and Y. Zech), pp. 865–873. Balkema, Lisse.

Métivier, F. and Meunier, P. (2003) Input and output flux correlations in an experimental braided stream. Implications on the dynamics of the bed load transport. *J. Hydrol*, **271**, 22–38.

Métivier, F., Meunier, P., Moreira, M., Crave, P., Chaduteau, C., Ye, B., Liu, G. (2004) Transport dynamics and morphology of a high mountain stream during the peak flow season: the Ürümqi river (Chinese Tien Shan). In: *River Flow 2004: Proceedings of the Second International Conference on Fluvial Hydraulics*, Vol. 1, pp. 769–777. Balkema, Lisse.

Meunier, P. and Métivier, F. (2000) Permanence des flux de masse d'une rivière en tresses expérimentale. *C. R. Acad. Sci. Paris, Sci. Terre Planèt.*, **331**, 105–110.

Parker, G. and Klingeman, P. (1982) On why gravel bed streams are paved. *Water Resour. Res.*, **18**, 1409–1423.

Pickup, G., Higgins, R. and Grant, I. (1983) Modeling sediment transport as a moving wave—the transfer and deposition of mining waste. *J. Hydrol.*, **60**, 281–301.

Shields, A. (1936) Anwendung der Aehnlichkeitsmechanik und der Turbulenzforschung auf die Geschiebebewegung. Wasserbau und Schiffbau. *Mitt. Preuss. Versuchsanst.*

Warburton, J. and Davies, T.R.H. (1994) Variability of bed load transport and channel morphology in a braided river hydraulic model. *Earth Surf. Process. Landf.*, **19**, 403–421.

Wohl, E. (2000) *Mountain Rivers*. American Geophysical Union, Washington, DC, 320 pp.

Yalin, M. (1992) *River Mechanics*. Pergamon Press, Oxford, 219 pp.

Morphological analysis and prediction of river bifurcations

GUIDO ZOLEZZI, WALTER BERTOLDI *and* MARCO TUBINO

Department of Civil and Environmental Engineering, University of Trento, via Mesiano 77, 38050 Trento, Italy
(Email guido.zolezzi@ing.unitn.it)

ABSTRACT

Braiding is a complex and highly dynamic process, the evolution of which is at present hard to predict, even on short time-scales. In order to improve the understanding and modelling of braided rivers, there is a need to gain a better insight into the processes of channel bifurcation, which is widely recognized as being important for the onset of braiding. Bifurcations are crucial components of braided rivers that control the adjustment of braiding intensity, since they largely determine the distribution of flow and sediment discharge along the downstream channels. Natural bifurcations show repetitive features in different types of gravel braided rivers. This paper presents a quantitative description of such features, based on observations performed on several bifurcations at two different field sites, and these features are related to the hydraulic conditions of the upstream incoming flow. The observed imbalance of flow partitioning at the bifurcation, which is also reflected by asymmetry in the width of the downstream anabranches, is found to be related to local bed aggradation in the main channel. This is immediately followed by a sudden step in the main channel profile, while the slope of the downstream anabranches remains almost equal and approximately constant downstream. The flow invariably concentrates towards the external bank of the largest downstream channel, enhancing bank erosion over a distance of several channel widths. The resulting data set allows a first assessment of theoretical predictions based on a recent model of bifurcation equilibrium and stability, which is essentially based on a simple one-dimensional formulation. In spite of its approximate character, the model seems to capture, at least qualitatively, the gross recurring features of these bifurcations.

Keywords Bifurcation, braiding, inlet step, field measurements, discharge partition, asymmetry.

INTRODUCTION

The intermittent processes of channel change and the related changing modes of sediment transport in braided rivers are responsible for the continuous rearrangement of the morphological patterns that create the appearance of braiding as a system of interwoven single-channel streams. This suggests that an important influence on the dynamics of braiding is exerted by single channels, which may even turn out to be dominant since the number of channels that are morphologically active at a time, even in intensely braided networks, is usually quite small (Stoijc *et al.*, 1998; Ashmore, 2001). In fact, it is often observed that, for braiding indices of ~10, most of the total discharge (from 65 to 85%) is carried by the largest one or two channels (Mosley, 1983). The dynamics of braiding are, however, much more complex than those of single-thread meandering streams. A major ingredient responsible for this complexity is the fact that the partitioning of water and sediment discharge in the active channels is mainly controlled by the dynamic processes of bifurcation. In spite of its relevance for the overall dynamics of braiding (Bristow & Best, 1993; Bolla Pittaluga *et al.*, 2001), the process of bifurcation is still poorly understood, and this gap in knowledge greatly limits the present ability to model and predict the behaviour of braided streams on medium and long time-scales (Jagers, 2003).

Bifurcations have been identified by the field observations of Leopold & Wolman (1957) as the primary cause of braiding. Subsequently, laboratory models have for many years been the preferred tool to investigate their dynamics. The experimental observations of Ashmore (1991), among others, have clearly identified four main mechanisms through which bifurcations set the onset of braiding:

1 chute cutoff;
2 central bar initiation and emergence;
3 transverse bar conversion;
4 multiple bar braiding.

Among these mechanisms, chute cutoff and transverse bar conversion (more commonly referred to as 'chute and lobe') have been detected as the most frequent, and have also been documented in the field. Chute cutoff has been observed as a relevant process responsible for the reduction of the overall sinuosity in both meandering (Johnson & Painter, 1967; Carson, 1986; Erskine et al., 1992; Gay et al., 1998) and braided streams (e.g. Krigström, 1962; Klaassen & Masselink, 1992). The presence of chute-and-lobe units in natural braided streams was originally documented by Southard et al. (1984). Ferguson et al. (1992) measured the evolution of one chute and lobe unit and showed how a sediment wave could affect the bed topography, inducing a modification of the water distribution in the two downstream branches. Further insight into the process of chute cutoff has been gained through the recent flume observations of Bertoldi & Tubino (2004). These authors documented and quantified the interaction between bed features and planform evolution of a laterally unconstrained channel before bifurcation, and determined an objective criterion to define the conditions for the occurrence of bifurcation. Finally, Richardson & Thorne (2001) have described the flow field in a newly formed bifurcation in the sandy Jamuna River, Bangladesh, showing that flow instability can generate a multi-thread current and therefore trigger the bifurcation process.

The above papers mainly focus on how bifurcations are generated, and do not provide substantial information on how an existing bifurcation evolves. Hence, despite its crucial role, understanding of such processes is still poor. The situation is almost reversed for the complementary processes at channel confluences, for which various distinctive features, such as the planform evolution (Ashmore, 1993), the amplitude of scour (Ashmore & Parker, 1983; Best, 1988; Klaassen & Vermeer, 1988), the distribution of secondary flows (Best & Roy, 1991) and their combination within a bifurcation in a 'confluence–diffluence unit' (Ashworth, 1996), have been characterized in detail.

A 'Y-shaped' bifurcation configuration, consisting of two branches connected with the upstream channel through a central node, can respond to formative conditions by either keeping both branches open or modifying the distribution of water and sediment in the downstream channels until one of them eventually closes. Federici & Paola (2003) have examined the role of the initial flow divergence on the bifurcation dynamics, finding the conditions that may lead to abandonment of one of the downstream branches; a similar process has been investigated in the field by Mosselmann et al. (1995). A theoretical model of channel bifurcations has been proposed by Wang et al. (1995) and improved by Bolla Pittaluga et al. (2003), through the introduction of a suitable nodal point condition able to account for the transverse exchange of sediments at the bifurcation. Furthermore, several attempts have been made to describe the evolution of chute cutoff in meandering streams on the basis of probabilistic (Howard, 1996), analytical (Biglari, 1989; Klaassen & Van Zanten, 1989) and numerical (Jagers, 2003) models. A major limitation in the assessment of model results is the lack of comprehensive field data sets on channel bifurcations that would allow testing of the applicability of such models in a natural context. Available field studies generally describe the evolution of a single bifurcation unit due to some formative event. A more systematic description of the morphological characteristics of bifurcations within a braided river reach, their state after formative conditions and an assessment of the relationship between morphodynamics and hydraulic parameters has not been presented to date.

The present paper reports two series of field observations from two different gravel braided rivers which are characterized by different size and degree of braiding intensity. Through detailed measurement of bed topography, grain-size distribution, transverse velocity profiles and partition of water discharge, the study characterizes quantitatively the flow and morphology of seven bifurcations and attempts to relate these properties to the hydraulic conditions of the incoming flow. The resulting dataset enables a first assessment of the ability of state-of-art mathematical models, like that of Bolla Pittaluga *et al.* (2003), to replicate these observed features.

FIELD MEASUREMENTS

The field sites

Field data reported in the present work have been collected within the framework of two distinct field campaigns that have been devoted to build a comprehensive dataset on two braided river reaches: (i) the braided reach of Ridanna Creek at Aglsboden, northeast Italy, which was monitored from summer 2002 to autumn 2004, and (ii) the upper braided reach of the Sunwapta River, Jasper National Park, Alberta, Canada, which was surveyed in summer 2003. In both studies specific attention also has been paid to the analysis of active bifurcations within the study area.

The two field sites are located in mountainous areas, and lie approximately at the same distance from the present position of an upstream glacier snout. Ridanna Creek is a proglacial gravel-bed river in South Tyrol, northeast Italy (Fig. 1), which displayed many braided reaches before being subject to a series of channelization works in its lower part during the middle of the last century. At present, braiding develops mainly in the study reach, which is located after a series of waterfalls, within an area that is approximately 600 m long and has a maximum width of 350 m; the river lies at an average elevation of 1750 m above sea level (Aglsboden, see Fig. 2). The study site in the Sunwapta River (Fig. 3) is located approximately 3.5 km downstream from the Athabasca glacier snout and has a longitudinal extent of ~150 m (Fig. 4). The river shows a braided pattern, with a braidplain width of ~100–150 m and unvegetated bars in between the channels. The braided reach occurs downstream of an incised channel where the river flows from Sunwapta Lake. In its upper part, which includes both the straight, incised reach and the study site, the Sunwapta River receives a lateral tributary, the meltwater stream of the Dome Glacier, a few kilometres upstream of the surveyed site.

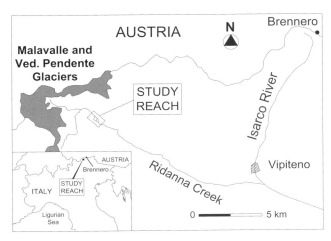

Fig. 1 Study location map showing the Ridanna Creek field site.

Fig. 2 View of the Ridanna Creek study reach from upstream. The main channel is approximately 13 m across. Flow towards top of photograph.

The braided reach of Ridanna Creek was monitored in detail for the first time during this field campaign (Bertoldi *et al.*, 2003); the site was chosen because it represents a good combination of easy accessibility and good predictability of morphological movement. Moreover, it offers the additional advantage of being surrounded by high cliffs, a condition that

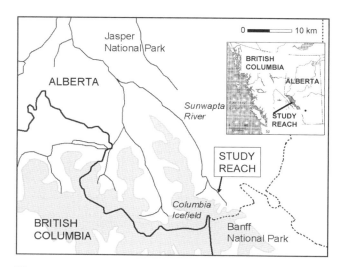

Fig. 3 Location of the Sunwapta River field site.

Fig. 4 View of the study reach in the Sunwapta River. Flow towards observer. Valley floor is approximately 1.5 km across.

enables easy acquisition of remote sensing digital images with a camera inclination that is high enough to guarantee an accuracy of the order of the average grain size. The Sunwapta River has also been repeatedly surveyed in recent years, due to a favourable combination of accessibility and good predictability of morphological changes during the summer snowmelt period (Ferguson et al., 1992; Goff & Ashmore, 1994; Chandler et al., 2002).

Methods of field data collection

Datasets from the two sites include water discharge, flow partitioning at bifurcations, longitudinal velocity and bed profiles in individual channels, geometry of cross sections, grain-size distributions at different locations, path lengths of individual sediment grains (only for Sunwapta River) and orthorectified images of the rivers from oblique digital photogrammetry and remote sensing procedures. Here, the methods used to collect the data that were used for the quantitative characterization of channel bifurcations at both sites are described; the description of other procedures is beyond the scope of the present study but can be found in Bertoldi et al. (2003).

Discharge and velocity measurements

Discharge measurements were obtained to determine both the values of water discharge feeding the braided reaches and the partition ratio at the surveyed bifurcations.

Measurement of total discharge

At both sites, a monitoring system was established that allowed reconstruction of the time sequence of water discharge flowing into the networks. This was obtained by converting automatic free-surface level measurements into discharge data by means of flow rating curves, estimated from direct discharge measurements at relatively stable cross-sections at the two sites. Figure 5 shows the location of the gauging stations in the two study reaches. In Ridanna Creek, a pressure sensor was placed to obtain a continuous survey of the free-surface level, with acquisition time-steps of 5 min. In the same section, a flow rating curve was obtained by means of direct discharge measurements using the salt dilution method. The salt concentration was measured using hand-held conductivity meters from both bank sides, a procedure that ensured the achievement of complete transverse mixing of diluted salt (Fig. 6). In the Sunwapta River, a gauging station was established upstream of the study reach, immediately downstream of the Dome confluence, where the river displays a single thread pattern. The free-surface elevation was measured through an ultrasonic distance gauging (UDG) device, collecting data every 10 min with an automatic correction to account for the effect of air temperature. To determine a suitable rating curve, a series of velocity profiles was measured in all channels across a transverse section within the braided reach, using four current-meters

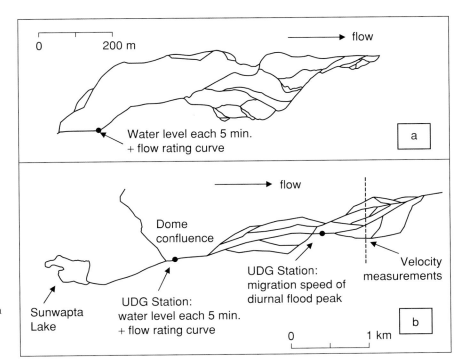

Fig. 5 Sketch illustrating the position of discharge monitoring systems in (a) Ridanna Creek and (b) Sunwapta River.

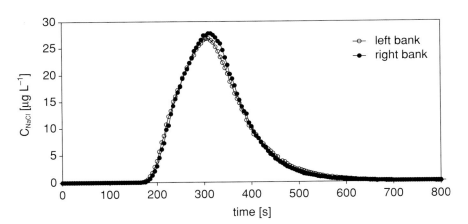

Fig. 6 Salt dilution measurements at Ridanna Creek: salt concentration was measured at both banks of the gauging section.

with 5 cm diameter propellers. Since the UDG station was located about 2.4 km upstream with respect to the study reach, it was necessary to estimate the time delay between the two gauging points in order to relate the estimated values of flow discharge to free-surface measurements. The migration speed of the diurnal flood wave was determined by setting another ultrasonic device in the braided reach, just upstream the surveyed area, within a major branch carrying most of the discharge (Fig. 5). The comparison between the results of the two instruments yielded an average time delay of ~25 min.

Flow partitioning at bifurcations

The distribution of water discharge in the downstream branches of the surveyed bifurcations was measured under different flow conditions, with the aim of establishing a relationship with the hydraulic conditions of the upstream channel. Three different techniques were employed in Ridanna Creek:

1 salt dilution, where the channel was long enough to ensure full transverse mixing;
2 velocity measurements with a 5 cm propeller current-meter;

3 velocity measurements with a Seba electromagnetic current-meter of 12 cm diameter.

In the braided reach of Sunwapta River, the length of individual channels was not long enough for complete lateral mixing to occur, and hence velocity measurements were always performed with propeller current-meters.

The velocity measurements were conducted in each of the three branches of the selected 'Y-shaped' bifurcations, except when flow conditions did not allow personnel to stand safely within the channel. Data were collected on a regular grid in each cross section, with a transverse spacing of 1 m. Flow velocity was gauged according to the standard procedure recommended for the instruments, namely performing a single measurement at a height equal to 40% of the local depth, when this was lower than 0.4 m, or, for deeper flows, three measurements within the vertical direction, at heights corresponding to 20%, 40% and 80% of the water depth.

Besides allowing an estimate of flow partitioning in the downstream anabranches, the overall dataset also provided detailed information on the local flow structure at the bifurcations. Estimated values of flow discharge obtained through different procedures and instruments generally showed a fairly good agreement. Comparison between the measured discharge in the upstream channel and the sum of the downstream branch discharges for Ridanna Creek showed a relative error ranging between 1% and 10% for the different flow conditions. Velocity measurements in very shallow channels were less accurate due to the presence of large pebbles, which induced local flow recirculations and strongly affected the local velocity measurements.

Grain-size measurements and topographic survey

Grain-size measurements were carried out using the 'Wolman Count' procedure (Wolman, 1954), a size-by-number method with random sampling of surface particles, measured and classified through a standard gravelometer. Grain-size sampling was performed at different locations along the channels: in order to quantify sorting effects, samples of at least 100 grains were collected across transverse sections, with a longitudinal spacing of 25 m, following a regular grid that was defined before the sampling.

The topographic survey included several cross sections of the braided reaches, longitudinal free water surface and bed profiles of the monitored channels. Measurements of the free water surface level for each wet channel at given cross sections were used to estimate the free water surface slope. A Leica TPS 700 total station was used in Ridanna Creek, while a similar automated total station was employed together with two levels in the Sunwapta field study. After each relevant morphological change, 30 cross sections in the Ridanna network were surveyed, mainly located in the central region; in the Sunwapta River, 13 cross sections were monitored daily, in the morning at low flow from 23 July to 4 August 2003. In both cases, the longitudinal spacing of the sections was ~10 m while the acquisition step was 1 m in the transverse direction, with additional points corresponding to significant changes in bed topography. More accurate measurements were performed close to the selected bifurcations, where cross section spacing was determined according to the channel width of individual channels (whose range was between 5 and 15 m): therefore the surveyed cross sections were spaced longitudinally from 3 m to 5 m along the three branches of each bifurcation.

Site characterization

The two braided networks are primarily fed by glacier meltwater in the summer, when the water discharge follows a daily cycle driven by solar radiation. Daily oscillations up to 10 $m^3 s^{-1}$ are common in the warmer periods. Through interpolation of the direct discharge measurements, the following flow rating curves for the two sites were obtained:

$$h = 0.3129 Q^{0.39} \quad (Ridanna\ Creek) \qquad (1)$$

$$h = 1.62 - 0.028Q \quad (Sunwapta\ River) \qquad (2)$$

where Q is water discharge and h is the free-surface elevation, measured with respect to the lowest bottom level of the gauging section in Ridanna Creek and to the elevation of the UDG probe in the Sunwapta River. Equation 1 was derived in a channel of approximately trapezoidal cross section which was ~12 m wide under bankfull conditions. In the Sunwapta River, the gauging section where

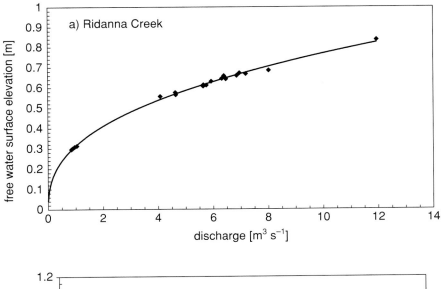
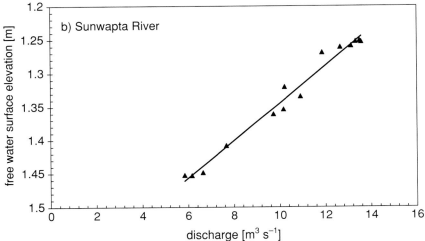

Fig. 7 Flow rating curves for Ridanna Creek and Sunwapta River at the gauging stations shown on Fig. 5.

Eq. (2) was derived had a more rectangular shape with an average width of 20 m. Equation (2) is valid between 6 and 14 m³ s⁻¹, including the range of interest for the present study; lower values of water discharge fall below the threshold for sediment transport. The flow rating curves given in Fig. 7 were constructed from measurements conducted under different flow conditions, and cover approximately the observed discharge range.

The frequent channel changes observed during the field campaigns provide an indication of the formative values of the total discharge feeding the networks. The contribution of glacier meltwater is only able to determine formative conditions in the Sunwapta River, where the threshold total discharge for sediment mobility ranges around 12–14 m³ s⁻¹. In Ridanna Creek, this range increases up to 15–20 m³ s⁻¹; in this case formative conditions are met only at relatively high values of the incoming discharge, which can only be reached through the occurrence of intense rainfall that adds to the contribution of glacier meltwater.

Both reaches are gravel bed, with sediment transport occurring mainly as bedload. Both sediments are heterogeneous in grain size, depending on location along the channel or bank, and on the presence of sorting patterns driven by the morphology at different scales. Figure 8 shows two reach-averaged distributions of sediment size for the two reaches: the bed material of the Sunwapta River is finer on average and is characterized by a weaker heterogeneity. In the Sunwapta River, the bed material

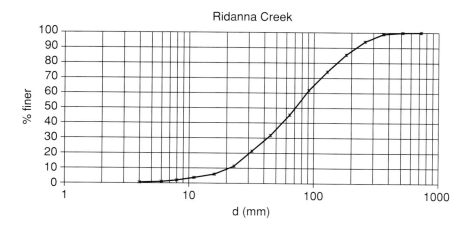

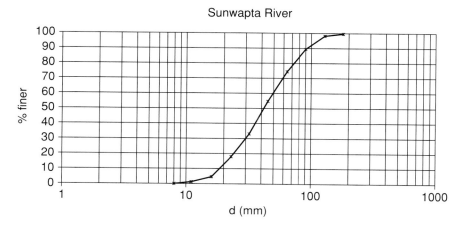

Fig. 8 Typical grain-size distributions for the study reaches.

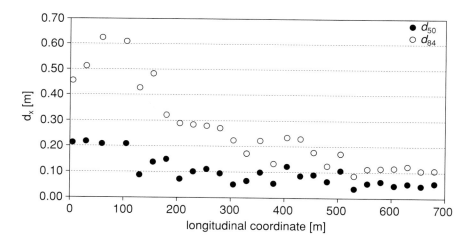

Fig. 9 Longitudinal distribution of d_{50} and of d_{84} in the Ridanna study reach.

mainly consists of pebbles with mean diameter, d_{50}, of 0.04 m and a d_{90} of 0.09 m: these values do not vary significantly in the longitudinal direction. In Ridanna Creek, a strong downstream fining is observed, with d_{84} decreasing from 0.6 m to 0.1 m and d_{50} changing from 0.25 m to 0.05 m in the study reach (Fig. 9).

The grain-size difference between the two rivers is also reflected by the free-surface slope, which remains almost constant (around 1.5%) in the

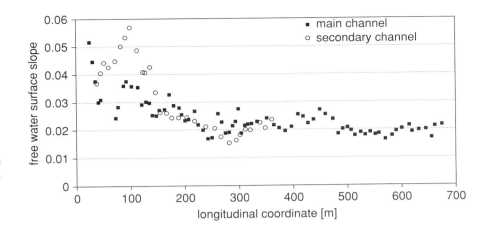

Fig. 10 Free water surface slope in the Ridanna study reach, measured in the main channel and in the left downstream channel of the first bifurcation upstream.

Sunwapta River, but in Ridanna Creek decreases from 4% in the upstream part of the study reach to a constant value of approximately 2.2% in the central region (Fig. 10). These longitudinal variations in grain size and valley gradient strongly affect the development of braiding in Ridanna Creek, which mainly occurs in the central region of the reach, with the upstream parts being characterized by a main channel that displays a single bifurcation. By contrast, braiding intensity is almost constant in the surveyed reach of the Sunwapta River (see also Figs 2 & 4).

MORPHOLOGICAL DESCRIPTION OF BIFURCATIONS IN BRAIDED NETWORKS

The section below describes the morphology and hydraulic characteristics of seven bifurcations that have been monitored on the two river reaches, namely four bifurcations in Ridanna Creek and three in the Sunwapta River. Each bifurcation is represented by a Y-shaped configuration (Fig. 11), where the upstream channel is denoted by the letter 'a', the main downstream branch with 'b' while 'c' is used for the second branch: in the following, a subscript (a, b or c) will relate any quantity to the corresponding channel in a bifurcation. Moreover, b denotes channel width, S the longitudinal slope, Q the water discharge, D the average flow depth, d_{50} the mean bed material size and $r_Q = Q_c/Q_b$ the discharge ratio between the smaller channel and main downstream channel.

The three bifurcations surveyed in the Sunwapta River are named I, II and III (Fig. 12). The main

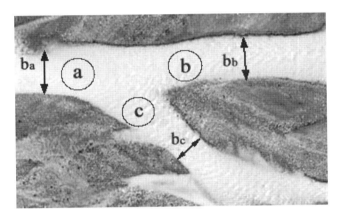

Fig. 11 Notation used to describe bifurcation morphology: $b_a = 10$ m.

bifurcation in the study reach is I: its upstream channel carries approximately 65% of the overall discharge and many morphological changes were observed during the field study. Bifurcation II is located immediately downstream of I (channel II_a coincides with channel I_b) and is markedly unbalanced: the left channel II_c carries nearly one quarter of the discharge in channel II_b. Channel II_c is very shallow and is located on the external bank of channel II_a; it appears as the result of an overtopping of the external bank of the bend. The angle between the two downstream branches of bifurcation II is ~72°. The third bifurcation analysed (III) is the smallest: the incoming discharge is nearly 30% of the total discharge; bifurcation III was not included in the area covered by the daily surveyed cross sections, but displays most of the common features observed in the other bifurcations. Measurements of bifurcations I and II were repeated six times, from 22 July

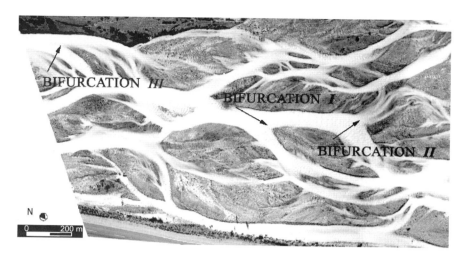

Fig. 12 Location of the three monitored bifurcations (*I–III*) in the Sunwapta River.

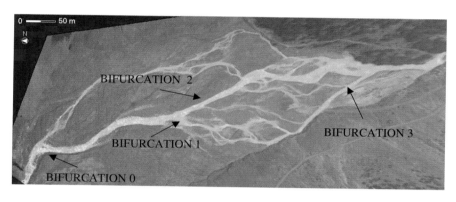

Fig. 13 Location of the four monitored bifurcations (0–3) in Ridanna Creek.

to 1 August, with different values of the total discharge flowing in the network.

The four bifurcations monitored on Ridanna Creek are denoted as 0, 1, 2 and 3 and their position in the study reach is shown in Fig. 13. Bifurcation 0 is located at the beginning of the braided reach, where the valley slope and sediment size are much greater than the other bifurcations. The other three bifurcations belong to the central region where the network is more active; morphological changes took place in summer 2003 and 2004 and mainly involved bifurcations 1 and 2. Bifurcation 3 is located in a region with finer sediments and is characterized by lower values of the incoming water discharge.

An example of the morphological and hydraulic data (channel widths, longitudinal slopes, water discharges and average depths) of the three channels at each surveyed bifurcation is reported in Tables 1 & 2; measured data for each bifurcation refer to the same flow conditions. Tables 1 & 2 also show the resulting values of the discharge ratio, r_Q, and the measured values of the transverse bed anomaly, Δz, whose definition and evaluation procedure are described below.

The main recurring feature of the bifurcations studied herein is the strong asymmetry of their morphological configuration, which appears from analysis of most of the collected data. A first indicator of such asymmetry is the flow distribution in the downstream branches, which has been found to be invariably unbalanced, with values of the discharge ratio ranging from 0.65 in the Sunwapta River to 0.1 in Ridanna Creek. Measurements of the discharge partition refer to different upstream conditions, i.e. different water stages in the main channel. Analysis of the data reveals a quasi-linear relationship between the discharge ratio, r_Q, and the upstream discharge, Q_a, for each dataset, with a larger imbalance for lower values of Q_a (Fig. 14). Lower values of the angular coefficient of the linear regression in Fig. 14 reflect a more pronounced

Table 1 Summary of the bifurcation data measured in the Sunwapta River

Variable	Bifurcation		
	I	II	III
b_a (m)	10	8.5	8
b_b (m)	8.5	10	6.5
b_c (m)	7.5	8	6
S_a	0.012	0.014	0.009
S_b	0.015	0.019	0.020
S_c	0.021	0.012	0.012
Q_a (m³ s⁻¹)	9.0	5.5	4.5
Q_b (m³ s⁻¹)	5.5	4.4	2.7
Q_c (m³ s⁻¹)	3.5	1.1	1.8
D_a (m)	0.45	0.36	0.38
D_b (m)	0.37	0.23	0.27
D_c (m)	0.28	0.08	0.19
d_{50} (m)	0.04	0.04	0.04
Δz (m)	0.24	0.32	0.15
r_Q	0.64	0.25	0.66

Table 2 Summary of the bifurcation data measured in Ridanna Creek

Variable	Bifurcation			
	0	1	2	3
b_a (m)	9.8	16.7	9.6	6.5
b_b (m)	6.7	10	8.7	4.9
b_c (m)	5.4	5.5	5	2.4
S	0.029	0.022	0.017	0.011
Q_a (m³ s⁻¹)	1.00	6.0	5.4	0.86
Q_b (m³ s⁻¹)	0.95	5.4	5.1	0.80
Q_c (m³ s⁻¹)	0.05	0.6	0.3	0.06
D_a (m)	0.70	0.65	0.70	0.31
D_b (m)	0.72	0.76	0.66	0.29
D_c (m)	0.29	0.29	0.22	0.21
d_{50} (m)	0.21	0.1	0.08	0.05
Δz (m)	2.00	0.50	0.55	0.20
r_Q	0.05	0.11	0.06	0.08

tendency towards an unbalanced configuration, as for bifurcations 0 and 2.

Other recurring features reflecting the channel asymmetry at the bifurcation can be identified through analysis of the longitudinal bed and free-surface profiles close to the bifurcations. The longitudinal profiles of channels I_a, I_b, III_a and III_b (Fig. 15) show, in both cases, that the bed displays aggradation close to the bifurcation, probably resulting from the enhanced tendency towards deposition induced by backwater effects due to local flow divergence. As a consequence, the water surface slope decreases just upstream from the bifurcation and increases downstream. Such an effect was observed at all of the monitored bifurcations. At the downstream end of this local aggradation, a sudden bed degradation invariably occurred in the 'b' branches (Fig. 15). Indeed, the longitudinal slope of the downstream branch carrying most of the discharge was locally much higher at the channel inlet close to the bifurcation; the above effect characterizes the bed profile only locally, since the longitudinal slope of the downstream branches almost coincides if measured from a cross-section located a few widths downstream of the bifurcation region. This is clearly seen in Fig. 16 which shows the monitored bifurcations in Ridanna Creek.

Such local steepening in the main downstream branch leads to the establishment of a transverse gap between the local bed elevations of the downstream channels in the neighbourhood of their inlet. This 'inlet step', which is denoted in the following as Δz, is clearly detectable from analysis of the cross-sectional geometry close to the bifurcation. In Fig. 17, a sequence of cross sections surveyed near bifurcations I (Sunwapta) and 1 (Ridanna) is plotted: both the bed elevation and the free-surface level are reported. Notice that in bifurcation I, the water level is different in the two downstream channels, with a higher level in the smaller branch, though the maximum water depths are almost coincident. Establishment of such an inlet step between the two downstream channels has been observed in all the monitored bifurcations: the mean and the lowest bed elevation of the two downstream branches have been invariably found to differ, with the smaller channel always exhibiting a higher elevation.

The size of the inlet step mainly depends on the asymmetry of the water distribution and scales with the mean upstream flow depth. It is worthy of note that the degree of transverse asymmetry between the two sides of the inlet step clearly changes in the longitudinal direction; hence, a

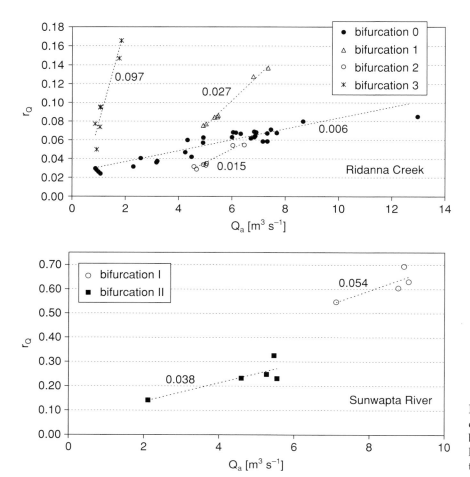

Fig. 14 Measured values of the discharge ratio r_Q in the monitored bifurcations. The coefficients of the linear regressions are reported close to each trendline.

suitable procedure must be established in order to define a representative value for a bifurcation. The observed values reported in Tables 1 & 2 have been derived according to two different methods that produced similar results. The first procedure calculates Δz as the difference of bed elevation between the lowest point of corresponding cross sections in the 'b' and 'c' branches, immediately downstream of the local degradation, for instance at the cross section where the average downstream slope is established. The second procedure is based on the observation that the linear interpolation of the longitudinal profiles of channels 'b' and 'c' provides a very small difference in slope; Δz is then obtained as the relative distance between these two lines.

Besides the occurrence of an inlet step between the downstream channels, the cross-sectional geometry (Fig. 17) reveals two further related phenomena, which are also clearly detectable in the other surveyed bifurcations:

1 the inlet step corresponds to the spatial development of a transverse bed inclination in the bifurcation region which extends over a length that is typically of the order of a few channel widths—this is mainly responsible for the transverse displacement of flow which diverts the main fluid flux towards channel 'b';
2 the outer bank of the larger downstream channel (channel 'b') is invariably subject to erosion over a distance of few channel widths, as witnessed by its relatively high steepness.

The mutual relationship between the above recurring features can be better understood through analysis of the transverse profiles of the longitudinal component of flow velocity for bifurcation 3 (Fig. 18), together with the cross-sectional geometry and the free-surface water level. At an upstream location, with respect to the bifurcation node, the locus of high velocity shifts to the side corresponding to the outer bank of the larger downstream channel (Fig. 18).

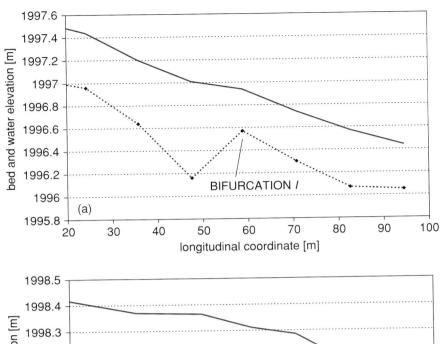
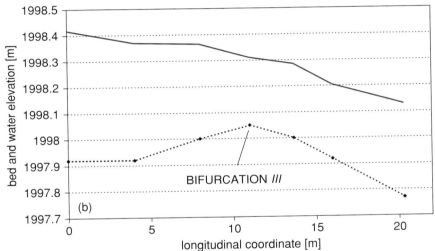

Fig. 15 Longitudinal profiles of bifurcations *I* (a) and *III* (b). The dashed lines with filled symbols correspond to bed elevation; solid lines denote water surface level.

A THEORETICAL MODEL

As highlighted in the introduction, the dataset collected in the field campaigns can be used both to assess the ability of theoretical models to replicate the observed phenomena, and to identify those ingredients that must be included to ensure a physical basis for predictive, albeit simplified, models. In this section, the main features of the model of Bolla Pittaluga *et al.* (2003; hereinafter referred to as BRT), which has been developed to investigate the equilibrium configuration and stability of channel bifurcations in gravel bed rivers, will be reviewed. The BRT model provides a complete formulation, and assumes that sediment is transported as bedload under formative conditions,

which implies that sediment transport does not cease, at least in the upstream channel. The analysis considers a single loop (Fig. 19) generated by the bifurcation of an upstream 'a' channel into two branches ('b' and 'c'), whose lengths are denoted by L_b and L_c, that rejoin downstream. A simple rectangular geometry for the channel cross section is used, and the banks are assumed to be fixed such that the channel width b does not change in time (an improved version of the model, which includes the effect of channel width adjustment to the flowing discharge, has been recently proposed by Miori *et al.*, 2004). The mathematical problem for the flow and bed topography in each channel is cast within the framework of a classic one-dimensional formulation.

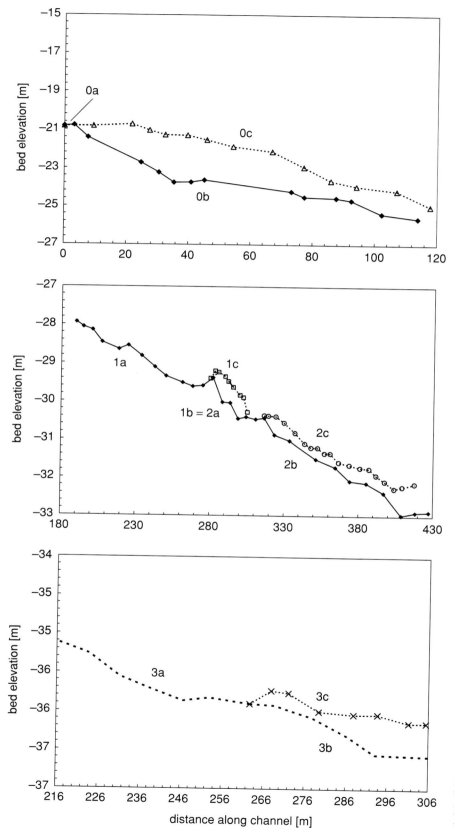

Fig. 16 Longitudinal bed profiles of bifurcations 0, 1, 2 and 3 and their location in the Ridanna Creek study reach.

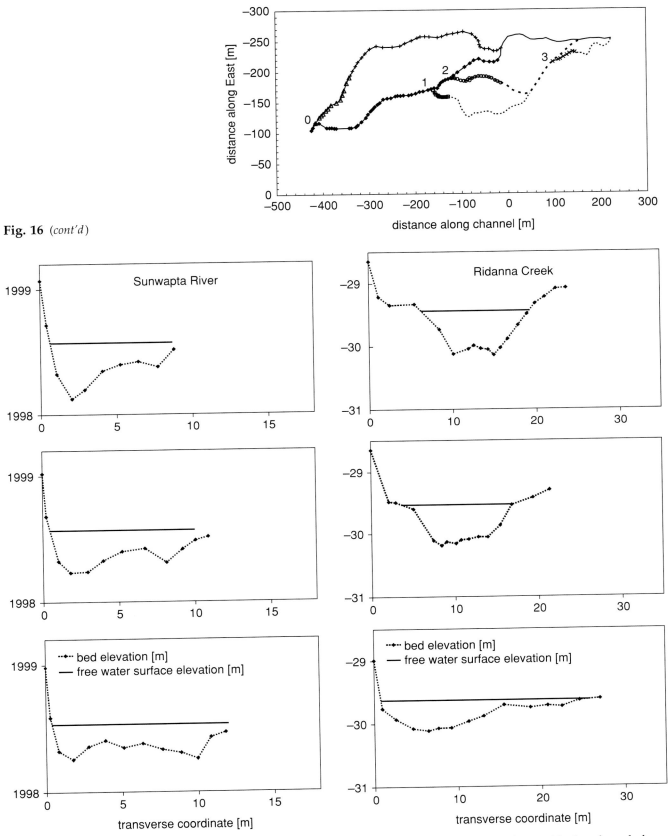

Fig. 16 (cont'd)

Fig. 17 Cross-sectional geometry in the neighbourhood of bifurcations *I* and 1. The dashed lines with closed symbols correspond to bed elevation; solid lines denote the water surface level.

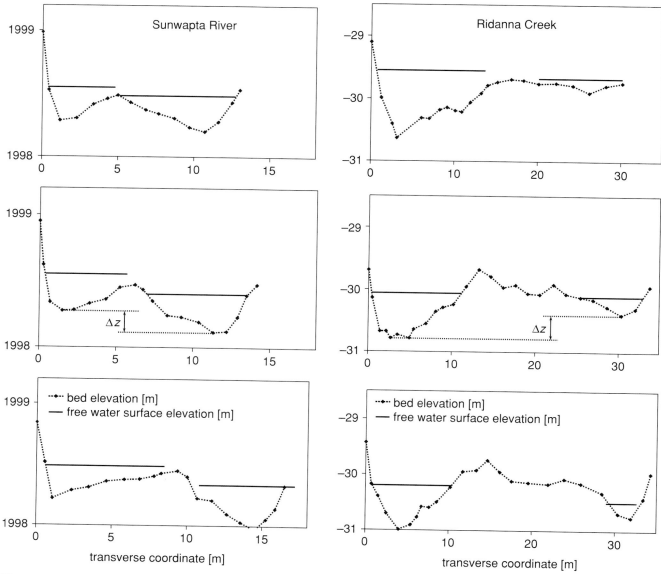

Fig. 17 (cont'd)

The differential problem requires nine boundary conditions, three for each channel. The 'external' conditions impose the water (Q) and sediment (Q_s) discharges that feed the upstream channel 'a' and the water levels at the outlet of the downstream branches (which coincide in the configuration adopted by BRT). Furthermore, five 'internal' conditions must be assigned at the node: BRT imposed the constancy of the free-surface water level at the node, the continuity of water discharge and two further conditions which control the partitioning of sediment discharge in the downstream branches. The latter conditions crucially affect the equilibrium bed configuration and stability of the bifurcation, when treated through a one-dimensional formulation, as pointed out by Wang et al. (1995). This implies that such conditions require a physical basis in order to limit their dependence on empirical parameters that cannot be related to any physical condition, which would thus strongly reduce the applicabil-

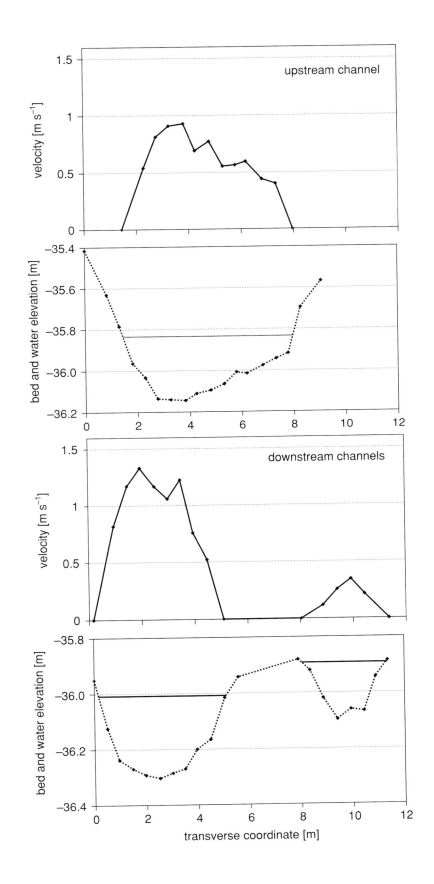

Fig. 18 Transverse profiles of longitudinal velocity at bifurcation 3.

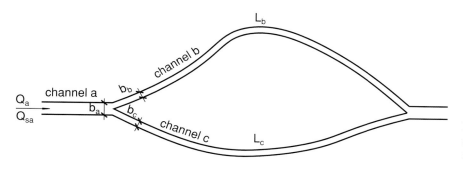

Fig. 19 Sketch of the bifurcation loop analysed in the Bolla Pittaluga *et al.* (2003) model.

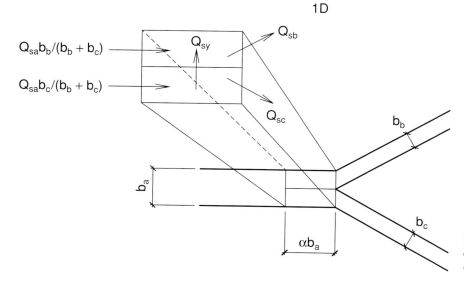

Fig. 20 Sketch of the nodal point condition in the Bolla Pittaluga *et al.* (2003) model.

ity of one-dimensional models (Jagers, 2003). To overcome the above difficulty, BRT introduced a quasi-two-dimensional formulation at the node that is able to include, in the simplified form imposed by the overall one-dimensional formulation, the topographic effect responsible for the transverse displacement of flow and sediment discharge at the bifurcation (Fig. 20). In particular, the nodal relationships introduced by BRT assume that a transverse bed-step may be established at the downstream end of the incoming channel 'a', within a short reach whose length is given by $L = \alpha b_a$, where α is an O(1) coefficient. As reported above, such inlet steps have been invariably observed in the field.

According to the model of BRT, the equilibrium configurations of the loop and their stability can be related, for given values of the channel widths, to the hydraulic parameters of the incoming flow in the upstream channel 'a', namely the aspect ratio β_a and the Shields stress θ_a:

$$\beta_a = \frac{b_a}{2D_a}$$
$$\theta_a = \frac{\tau_a}{(\rho_s - \rho)g d_s}$$
(3)

where D is the flow depth, τ is the bed shear stress, ρ_s and ρ are the sediment and water density respectively, g is acceleration due to gravity and d_s is sediment diameter.

For relatively high values of θ_a (>0.1), the only stable equilibrium solution for a symmetrical loop (same widths and slopes in branches 'b' and 'c') corresponds to an equal flow distribution within the downstream branches ($r_Q = 1$). At lower values of θ_a, the balanced configuration becomes unstable, even for symmetrical loops, while two further stable equilibrium configurations appear that are characterized by an unbalanced distribution of flow discharge in the downstream branches ($r_Q < 1$) and by establishment of an inlet step Δz

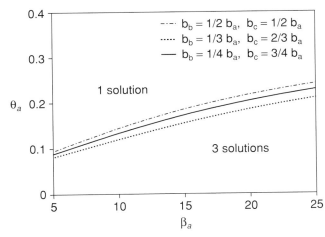

Fig. 21 The threshold value of θ_a for the occurrence of the symmetrical solution as a function of β_a for different widths in channels 'b' and 'c'. (From Bolla Pittaluga et al., 2003.)

at the bifurcation. The threshold value for the occurrence of the symmetrical solution increases as β_a increases, such that increasing values of the aspect ratio of the upstream channel promote establishment of unbalanced configurations (see Fig. 21).

These theoretical results, cast in dimensional form, suggest that for a given width of the upstream channel, and provided that the incoming sediment transport is not vanishing, the degree of asymmetry of the bifurcation, namely the size of the inlet step and the uneven partitioning of flow discharge,

is likely to be enhanced at relatively low stages. It is worthy of note that within the range of values of Shields stress and aspect ratio typical of gravel-bed braided rivers, the BRT model generally predicts an uneven distribution of flow in the downstream branches. Additionally, the role of the inlet step in the nodal relationships is crucial; neglecting its effect would invariably imply establishment of the symmetrical solution, regardless of the value attained by β_a and θ_a.

The above results are slightly sensitive to the choice of the empirical parameter α, which provides a measure of the longitudinal distance where the presence of the inlet step is felt. The value α has been estimated by BRT as an $O(1)$ parameter on the basis of laboratory investigations, as the upstream influence of the bifurcation on bed topography has been observed to extend over a distance of one to two channel widths. Figure 22 shows the dimensionless amplitude, A_1, of the transverse bed deformation, scaled by its value at the bifurcation, plotted versus the upstream distance, s, measured from the nodal point. The field observations reported in the preceding section provide further support to the above estimate.

COMPARISON AND DISCUSSION

The dataset concerning the surveyed bifurcations in the two field campaigns allows a first assessment

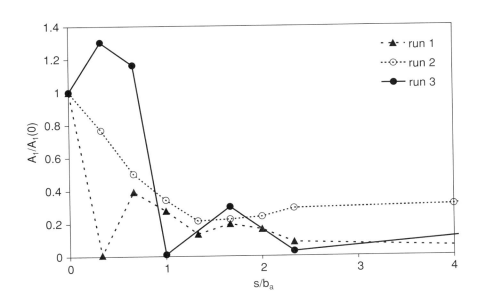

Fig. 22 Experiments of Bolla Pittaluga et al. (2003): the transverse dimensionless bed deformation is plotted versus the dimensionless upstream distance from the nodal point. (From Bolla Pittaluga et al., 2003.)

of the theoretical model of Bolla Pittaluga et al. (2003). As highlighted above, the equilibrium configurations of the bifurcation are computed in the model once the relevant parameters of the upstream flow are given, namely the discharge Q_a, the average slope S_a, the channel width b_a and the mean grain size d_{50}; moreover, the widths b_b and b_c of the downstream branches are also required. According to the model formulation, the upstream flow parameters must correspond to channel-forming conditions. During field work on the Sunwapta River, the measured peak values of the discharge were able to determine the rearrangement of the network configuration, with the formation of new bifurcations and readjustment of several channels. Hence, it can be assumed that the discharge data collected in the field campaign fall within the range of values representing channel-forming conditions.

By contrast, however, measurements on Ridanna Creek were conducted at relatively low flow, under non-formative conditions. During the first part of the survey period (until the end of July 2003), the four surveyed bifurcations did not display any morphological change, with values of the Shields stress falling below the threshold for sediment transport. Hence, the range of values of flow discharge measured during the field campaign at Ridanna Creek cannot be used to test the BRT model: in this case, an extrapolation procedure has been introduced on the basis of the observed rating curve in order to obtain a gross estimate of the formative conditions. Furthermore, suitable values of the total discharge have been determined by extrapolating a *bankfull* value from the cross-sectional geometry. The corresponding values of channel width and depth have been calculated as those corresponding to an equivalent rectangular cross section that would carry the same total sediment discharge as the original section.

Input data for each bifurcation have been chosen through a suitable reach-averaging procedure, involving several (four to six) surveyed cross-sections with a mean longitudinal spacing of 3–5 m. Table 3 shows the input values used in the model for some bifurcations; the subscript bf denotes bankfull conditions. It is worth noting that the value of Q_{bf} computed for bifurcation *I* in the Sunwapta River approximates well to the value that has been observed to correspond to the occurrence of formative events, which suggests the suitability of the procedure adopted to compute the input data. From Table 3 it also appears that the Shields parameter, θ_a, in the upstream channel is always <0.1: under such conditions, the BRT model invariably predicts that the equilibrium configuration of the bifurcation is characterized by an unbalanced flow distribution in the downstream branches, as has been observed for the monitored bifurcations described above.

A direct comparison between the measured and predicted quantities can only be attempted in terms of the bed morphology at the bifurcation. It is not possible to test the theoretical results for the flow parameters, such as the discharge ratio r_Q, since field measurements in most cases do not refer to formative conditions. Figure 23 shows a comparison between the observed and predicted values

Table 3 Input data for the Bolla Pittaluga et al. (2003) model corresponding to bankfull conditions in channel 'a'

Variable	Ridanna Creek bifurcations			Sunwapta River bifurcations		
	1	2	3	I	II	III
Q_{bf} (m^3 s^{-1})	24.7	19.5	6.3	10.3	7.5	4.5
D_{bf} (m)	0.60	0.68	0.45	0.53	0.44	0.38
b_{bf} (m)	17.4	12.9	9.9	11.3	9.9	8.0
θ_{bf}	0.075	0.079	0.053	0.088	0.079	0.058
S_a	0.022	0.017	0.011	0.012	0.014	0.010
d_{50} (m)	0.10	0.08	0.05	0.04	0.04	0.04

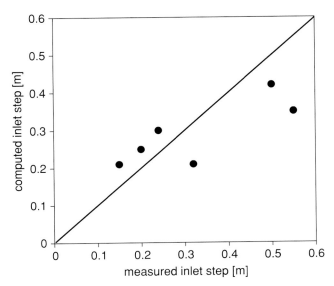

Fig. 23 Comparison between the measured and computed values of the inlet step elevation of natural bifurcations.

of the transverse inlet step, Δz, for different bifurcations; calculations have been performed setting $\alpha = 1$. In spite of the approximate character of the model, which is based on a simple one-dimensional approach, the comparison appears satisfactory. However, exact agreement cannot be expected because:

1 the model predictions refer to equilibrium conditions, whose occurrence cannot be assessed in the field—'equilibrium' is a useful concept here that stands for the benchmark towards which the system may display a tendency in its evolution;

2 the one-dimensional equilibrium formulation accounts for the effect of transverse bed deformation induced by the bifurcation, whose intensity changes in the longitudinal direction, in terms of a localized step Δz—however, the formulation ignores the effect of flow non-uniformity that may become important in the bifurcation region, where a relatively rapid local bed aggradation is followed by a sudden step.

Though a comparison in terms of the discharge ratio, r_Q, cannot be performed, it is nevertheless of interest to examine the dependence of r_Q on the hydraulic conditions of the upstream channel 'a'. Figure 24 plots the values of the discharge ratio reported in Fig. 14 using the estimated values of bankfull discharge, Q_{bf}, for the bifurcations examined in order to scale the upstream discharge, Q_a. The discharge ratio is seen to increase as formative conditions are approached. Although different bifurcations show slightly different behaviours, Fig. 24 suggests the existence of a possible relationship between flow partitioning into the downstream channels and the hydraulic conditions in the upstream channel, which is also embodied in the results of the BRT model. It is worth noting that the establishment of such a relationship between upstream conditions and flow partitioning at the

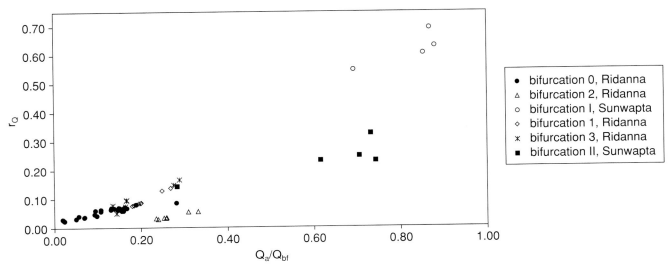

Fig. 24 Measured discharge ratios, r_Q, as a function of the magnitude of discharge in relation to formative conditions for several bifurcations, Q_a/Q_{bf}.

bifurcation constitutes a fundamental requirement to improve knowledge on the dynamics of braiding (Ashmore, 2001; Jagers, 2003).

FINAL REMARKS

Field observations reported in the present paper highlight various common features displayed by seven bifurcations belonging to two gravel-bed braided rivers. For values of the Shields stress typical of gravel bed rivers, bifurcations show a recurring asymmetrical configuration, which is reflected by (i) an unbalanced water distribution in the downstream branches; (ii) the presence of a transverse inlet step, which characterizes the bed topography at the bifurcation, and (iii) the lateral shift of the main flow towards the external bank of the main downstream channel, where erosion is mainly concentrated. The degree of this asymmetry can be related to the hydraulic conditions of the upstream channel. The above features are also mutually related, since the presence of a lateral bed slope gives rise to topographic steering of the main flow that is diverted towards the largest downstream channel, thus enhancing bank erosion and creating an unbalanced water distribution.

The theoretical model of Bolla Pittaluga et al. (2003), by virtue of the nodal point condition adopted that also accounts for the lateral exchange of sediment transport due to the topographic effect at the bifurcation, is able to reproduce, at least qualitatively, the phenomena described from the field. Application of the Bolla Pittaluga et al. (2003) model to the bifurcations surveyed also suggests that various limitations should be removed in order to improve the ability of the model to reproduce the dynamics of river bifurcations. In particular, relaxing the condition of fixed banks would allow the channels to adjust their width to the discharge; in this case, the bifurcation model would provide a theoretical estimate of downstream channel widths in terms of the upstream flow conditions, instead of requiring the widths to be fixed a priori. Furthermore, the behaviour of the model on shorter timescales should be investigated with respect to those required to achieve the equilibrium conditions, which may turn out to be quite large when compared with the scale of formative events (a recent attempt in this direction has been pursued by Hirose et al., 2003). This aim would also benefit from extension of the present dataset from both field and laboratory studies. Additional measurements of bifurcation morphology and flow fields in natural bifurcations would also provide the opportunity to estimate relevant flow parameters especially close to formative conditions, thus allowing a closer comparison with theoretical models. Finally, the properties of the relationships linking bifurcation parameters with the upstream flow conditions can be better quantified through investigation of bifurcation dynamics in controlled laboratory conditions, and a first attempt in this direction has been conducted by Bertoldi (2004).

ACKNOWLEDGEMENTS

The field activity reported herein has been carried out under the framework of several projects: RIMOF, COFIN 2001, COFIN 2003, and under the umbrella of CUDAM, University of Trento. The field campaign on the Sunwapta River was carried out by an international group of researchers belonging to various institutions and coordinated by Professor Peter Ashmore, University of Western Ontario, London, Canada. Special thanks to Rossella Luchi, who brilliantly managed the field activity in the Ridanna Creek and decisively contributed to the processing of data. The manuscript was greatly improved by reviews from M.G. Kleinhans, S. Lane and J. Best. The authors are deeply thankful to all the people who joined and participated in the field activities and data processing, and in particular to Gianluca Vignoli, Emanuele Belloni, Tommaso Pinter, Gianluca Antonacci, Rodolfo Repetto, Luigino Bonuti, Stefano Miori and Thomas Buffin-Bélanger.

NOMENCLATURE

A_1 Dimensionless amplitude of bed deformation
b Channel width (m)
C_{NaCl} Salt concentration (mg L^{-1})
d Sediment diameter (m)
d_x x percentile of grain-size distribution
d_s Sediment diameter in the BRT model (m)
D Flow depth (m)
g Acceleration due to gravity (m s^{-2})

h	Free water surface elevation (m)
L	Length of channel in the BRT model (m)
Q	Water discharge (m³ s⁻¹)
r_Q	Discharge ratio between the smaller channel and the main downstream channel at a bifurcation
s	Upstream longitudinal distance (m)
S	Longitudinal slope
α	Coefficient in BRT model
β	Channel half width to depth ratio
Δz	Transverse bed elevation of inlet step (m)
ρ_s, ρ	Density of sediment and water respectively (kg m⁻³)
θ	Shields parameter
τ	Bed shear stress (N m⁻²)
subscripts a,b,c	Location of channel in a 'Y-shaped' bifurcation (see Fig. 11)
subscript bf	Bankfull conditions

REFERENCES

Ashmore, P.E. (1991) How do gravel-bed rivers braid? *Can. J. Earth Sci.*, **28**, 326–341.

Ashmore, P.E. (1993) Anabranch confluence kinetics and sedimentation processes in gravel-braided streams. In: *Braided Rivers* (Eds J.L. Best and C.S. Bristow), pp. 129–146. Special Publication 75, Geological Society Publishing House, Bath.

Ashmore, P.E. (2001) Braiding phenomena: statics and kinetics. In: *Gravel Bed Rivers V* (Ed. M.P. Mosley), pp. 95–121. New Zealand Hydrological Society, Wellington.

Ashmore, P.E. and Parker, G. (1983) Confluence scour in coarse braided stream. *Water Resour. Res.*, **19**, 392–402.

Ashworth, P.J. (1996) Mid-channel bar growth and its relationship to local flow strength and direction. *Earth Surf. Proc. Landf.*, **21**, 103–123.

Bertoldi, W. (2004) *River bifurcations*. PhD thesis, Department of Civil and Environmental Engineering, University of Trento, Trento, Italy, 123 pp.

Bertoldi, W., Pinter, T., Vignoli, G., Zolezzi, G. and Tubino, M. (2003) Morphological characterization of a braided reach of the Ridanna Creek. *Proceedings of the 3rd International Conference on River, Coastal and Estuarine Morphodynamics*, Barcelona, Spain, 1–5 September, pp. 807–818.

Bertoldi, W. and Tubino, M. (2004) Bed and bank interaction of bifurcating channels. *Water Resour. Res.*, **42**, w07001, doi 10.1029/2004WR003333.

Best, J.L. (1988) Sediment transport and bed morphology at river channel confluences: *Sedimentology*, **35**, 481–498.

Best, J.L. and Roy, A.G. (1991) Mixing layer distortion at the confluence of channels of different depths. *Nature*, **350**, 411–413.

Biglari, B. (1989) *Cutoffs in curved alluvial rivers*. Unpublished Master's thesis, IHE Delft, Delft, The Netherlands, 201 pp.

Bolla Pittaluga, M., Federici, B., Repetto, R., Paola, C., Seminara, G. and Tubino, M. (2001) The morphodynamics of braided rivers: experimental and theoretical results on unit processes. In: *Gravel Bed Rivers V* (Ed. M.P. Mosley), pp. 143–181. New Zealand Hydrological Society, Wellington.

Bolla Pittaluga, M., Repetto, R. and Tubino, M. (2003) Channel bifurcation in braided rivers: equilibrium configurations and stability, *Water Resour. Res.*, **39**, 1046–1059.

Bristow, C.S. and Best, J.L. (1993) Braided rivers: perspectives and problems. In: *Braided Rivers* (Eds J.L. Best and C.S. Bristow), pp. 1–9. Special Publication 75, Geological Society Publishing House, Bath.

Carson, M.A. (1986) Characteristics of high-energy 'meandering' rivers: the Canterbury plains, NZ. *Geol. Soc. Am. Bull.*, **97**, 886–895.

Chandler, J.H., Ashmore, P., Paola, C., Gooch, M. and Varkaris, F. (2002) Monitoring river channel change using terrestrial oblique digital imagery and automated digital photogrammetry. *Ann. Assoc. Am. Geogr.*, **92**, 631–644.

Erskine, W., McFadden, C. and Bishop, P. (1992) Alluvial cutoffs as indicators of former channel conditions. *Earth Surf. Process. Landf.*, **17**, 23–37.

Federici, B. and Paola, C. (2003) Dynamics of bifurcations in noncohesive sediments. *Water Resour. Res.*, **39**, 1162/3–1 3–15.

Ferguson, R.I., Ashmore, P.E., Ashworth, P.J., Paola, C. and Prestegaard, K.L. (1992) Measurements in a braided river chute and lobe. 1. Flow pattern, sediment transport and channel change. *Water Resour. Res.*, **28**, 1877–1886.

Gay, G.R., Gay, H.H., Gay, W.H., Martinson, H.A., Meadem, R.H. and Moody, J.A. (1998) Evolution of cutoffs across meander necks in Powder River, Montana, USA. *Earth Surf. Proc. Landf.*, **23**, 651–662.

Goff, J.R. and Ashmore, P.E. (1994) Gravel transport and morphological change in braided Sunwapta river, Alberta, Canada. *Earth Surf. Proc. Landf.*, **19**, 195–212.

Hirose, K., Hasegawa, K. and Meguro, H. (2003) Experiments and analysis on mainstream alternation in a bifurcated channel in mountain rivers. *Proceedings of the 3rd International Conference on River, Coastal* and *Estuarine Morphodynamics*, Barcelona, Spain, 1–5 September, pp. 571–583.

Howard, A.D. (1996) Modelling channel evolution and floodplain morphology. In: *Floodplain Processes* (Eds M.G. Anderson, D.E. Walling and P.D. Bates), pp. 15–62. Wiley, Chichester.

Jagers, H.R.A. (2003) *Modelling planform changes of braided rivers*. Unpublished PhD thesis, University of Twente, The Netherlands, 318 pp.

Johnson, R.H. and Painter, J. (1967) The development of a cutoff on the River Irk at Chadderton, Lancashire. *Geography*, **52**, 41–49.

Klaassen, G.J. and van Zanten, B.H.J. (1989) On cutoff ratios of curved channels. *Twenty-third Congress of the International Association on Hydrological Research*, August 21–25, Ottawa, Vol. B, pp. 121–130.

Klaassen, G.J. and Masselink, G. (1992) Planform changes of a braided river with fine sand as bed and bank material. *Fifth International Symposium on River Sedimentation*, April, Karlsruhe, pp. 459–471.

Klaassen, G.J. and Vermeer, K. (1988) Confluence scour in large braided rivers with fine bed material. *Proceedings of the International Conference on Fluvial Hydraulics*, Budapest, Hungary, pp. 81–394.

Krigström, A. (1962) Geomorphological studies of sandur plains and their braided rivers in Iceland. *Geogr. Ann.*, **44**, 328–346.

Leopold, L.B. and Wolman, G. (1957) River channel patterns: braiding, meandering and straight. *Physical and Hydraulic Studies of* Rivers. *U.S. Geol. Survey Spec. Publ.*, **14**, 283–300.

Miori, S., Repetto, R. and Tubino, M. (2004) Configurazioni di equilibrio di biforcazioni in canali a fondo mobile e sponde erodibili. *Proceedings of 'IDRATN', XXIX Convegno di Idraulica e Costruzioni Idrauliche*, Trento, September 7–10. (In Italian.)

Mosley, M.P. (1983) Response of braided rivers to changing discharge. *N.Z.J. Hydrol.*, **22**, 18–67.

Mosselman, E., Huisink, M., Koomen, E. and Seijmonsbergen, A.C. (1995) Morphological changes in a large braided sand-bed river. In: *River Geomorphology* (Ed. E.J. Hickin), pp. 235–249. Wiley, Chichester.

Richardson, W.R. and Thorne, C.R. (2001) Multiple thread flow and channel bifurcation in a braided river: Brahmaputra-Jamuna River, Bangladesh. *Geomorphology*, **38**, 185–196.

Southard, J.B., Smith, N.D. and Kuhnle, R.A. (1984) Chutes and lobes: newly identified elements of braiding in shallow gravelly streams. In: *Sedimentology of Gravels and Conglomerates* (Eds E.H. Koster and R.J. Steel), pp. 51–59. Memoir 10, Canadian Society of Petroleum Geology, Calgary.

Stojic, M., Chandler, J., Ashmore, P. and Luce, J. (1998) The assessment of sediment transport rates by automated digital photogrammetry. *Photogr. Eng. Remote Sens.*, **64**, 387–395.

Wang, Z.B., Fokkink, R.J., De Vries, M. and Langerak, A. (1995) Stability of river bifurcations in 1D morphodynamics models. *J. Hydraul. Res.*, **33**, 739–750.

Wolman, M.G. (1954) A method of sampling coarse riverbed material. *Trans. Am. Geophys. Union*, **35**, 951–956.

Braided river management: from assessment of river behaviour to improved sustainable development

HERVÉ PIÉGAY*, GORDON GRANT†, FUTOSHI NAKAMURA‡ and NOEL TRUSTRUM§

*UMR 5600—CNRS, 18 rue Chevreul, 69362 Lyon, cedex 07, France (Email: piegay@univ-lyon3.fr)
†USDA Forest Service, Corvallis, USA
‡University of Hokkaido, Japan
§Institute of Geological and Natural Sciences, Lower Hatt, New Zealand

ABSTRACT

Braided rivers change their geometry so rapidly, thereby modifying their boundaries and floodplains, that key management questions are difficult to resolve. This paper discusses aspects of braided channel evolution, considers management issues and problems posed by this evolution, and develops these ideas using several contrasting case studies drawn from around the world. In some cases, management is designed to reduce braiding activity because of economic considerations, a desire to reduce hazards, and an absence of ecological constraints. In other parts of the world, the ecological benefits of braided rivers are prompting scientists and managers to develop strategies to preserve and, in some cases, to restore them.

Management strategies that have been proposed for controlling braided rivers include protecting the developed floodplain by engineered structures, mining gravel from braided channels, regulating sediment from contributing tributaries, and afforesting the catchment. Conversely, braiding and its attendant benefits can be promoted by removing channel vegetation, increasing coarse sediment supply, promoting bank erosion, mitigating ecological disruption, and improving planning and development. These different examples show that there is no unique solution to managing braided rivers, but that management depends on the stage of geomorphological evolution of the river, ecological dynamics and concerns, and human needs and safety. For scientists wishing to propose 'sustainable' solutions, they must consider the cost-benefit aspects of their options, and the needs and desires of society. This requires an interdisciplinary approach linking earth scientists and social scientists concerned with environmental economics, planning, and societal and political strategies, in order to fully evaluate the economic and social validity of different options for different time-scales.

Keywords Human impacts, river restoration, flooding risk management, channel adjustment, ecological conservation, geomorphological sensitivity, braided rivers.

INTRODUCTION

Braided rivers are strongly influenced by high sediment delivery from nearby sources (e.g. glacial outwash, torrential tributaries) coupled with lower sediment throughput due to hydraulic conditions (primarily gentle slopes). They are sensitive to changes in their flood regime or sediment influx, and can completely modify their geometry over a few decades (Ferguson, 1993). Common braided river adjustments to changing environmental conditions typically include both narrowing (Kondolf, 1997; Nakamura & Shin, 2001; Liébault & Piégay, 2002; Surian & Rinaldi, 2003), and widening (Lyons & Beschta, 1983; Page et al., 2000). Moreover, rivers can shift from other planforms to a braided pattern when human activities accelerate sediment delivery processes. At the same time, climatic conditions and human influences that reduce sediment production can have the opposite effect, with braiding slowly diminishing through time. Thus the dynamic nature of braided rivers makes it difficult for societies to both predict the direction of their evolution and maintain nearby and associated infrastructure.

Braided channels are rarely in a steady state and are indicative of a valley bottom still actively undergoing construction. In undeveloped floodplain areas, braided rivers are considered part of the natural environment and are typically preserved because of their associated ecological richness. However, when permanent infrastructure is built in such active floodplains many problems can occur. This problem is particularly acute in central Japan, where rivers drain steep mountains with unstable geology, large-scale landsliding and intense rainfall, which is coupled with densely populated and sensitive alluvial plains (Nishiyama & Miyazaka, 2001; Maita and Shizuoka River Work Office, 2001).

In response to these problems, scientists and engineers have sought to reduce braiding activity, particularly where braiding results from human impacts or degrades resources, ecological diversity and human safety. Examples of this approach can be found in the northern island of New Zealand or Japan (Marutani et al., 2001). Conversely, in other parts of the world, the high ecological value of braided rivers has led to technical approaches and management strategies to preserve and, in some cases, restore braided conditions (Bornette & Amoros, 1991; Gilvear, 1993; Ward et al., 1999). Some braided rivers are extremely productive biologically, such as the Fraser River in British Columbia or the Waitaki River on the southern island of New Zealand, where Chinook salmon fisheries are of national significance (James, 1992). This paradox of opposing management strategies for braided rivers can be seen with reference to Pine Creek in Idaho, USA and the Drôme River in France (Kondolf et al., 2002) (Fig. 1).

The aim of this paper is to propose a conceptual model for the temporal evolution of braided rivers, consider the implications of this behaviour from social and ecological perspectives, and provide several contrasting case studies from around the world that illustrate the way managers and scientists approach the issues of braided river management in different landscape, societal and ecological contexts.

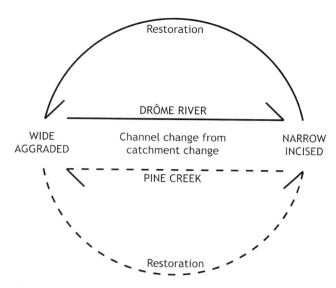

Fig. 1 Contrasting directions of channel planform evolution due to changes in catchment sediment yield, and consequent directions of intended restoration in Pine Creek, Wyoming, USA and the Drôme River, southern France (From Kondolf et al., 2002; with permission from Elsevier).

The conceptual model of braided river temporal trajectory

Braided rivers can be distinguished according to their stage of evolution and the human and ecological benefits they provide. These attributes, in turn, result directly from the relation between sediment supply and braiding intensity (Fig. 2). A braided river typically aggrades and widens when sediment supply is high, thereby progressively occupying more and more of the valley bottom; this is referred to as the *expansion phase*. The active channel shifts laterally, either by bank erosion or avulsion. Considerable economic costs can be incurred if the valley bottom is developed during this phase. On the other hand, under conditions when sediment supply or peak flows are reduced, referred to here as the *contraction phase*, braided rivers typically narrow and incise their beds. This degradation, in turn, can destabilize bridges, dams, erosion control works and other infrastructure. It is proposed, therefore, that the social and economic costs and benefits associated with braiding therefore follow a bell-shaped curve, with the greatest human benefits occurring in the middle

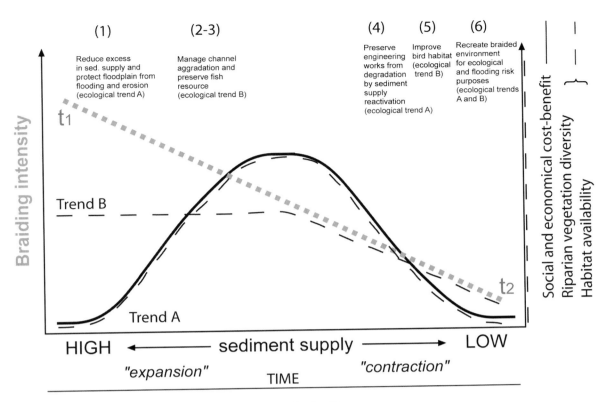

Fig. 2 Conceptual framework for understanding the relationship between braiding intensity as a function of sediment supply and the ecological and human benefits derived from braided rivers. Braiding intensity (dotted grey line) decreases with decreasing sediment supply: t1 and t2 represent time-points along a trajectory of evolution from high sediment loads, high braiding intensity ('*expansion phase*') to low sediment supply, low braiding intensity ('*contraction phase*'). Social and economic benefits of braided rivers (solid black line) describe a bell-shaped curve, with intermediate levels of braiding providing the most benefits. Ecological benefits (dashed black lines) of braided rivers differ from one system to another according to other parameters (low flow conditions, turbidity, α versus β diversity). Following trend A, ecological diversity has a peak in intermediate levels of braiding, mostly in riparian environments. Some rivers can also follow trend B (e.g. Fraser River, Platte River) providing ecological value not only at the intermediate levels of braiding, but all along the braiding stage; their ecological interest diminishes once the gravel surface area decreases, and the associated optimal habitat conditions are no longer present.

section of the trajectory (Fig. 2). Ecological benefits associated with the trend from high to low braiding intensity are more complex (Fig. 2). In some of the examples discussed below, extremely low or high braiding intensity is associated with poor riparian and aquatic habitat conditions, while in other examples the full spectrum of braiding conditions promotes ecological diversity.

The above conceptual model suggests that it is important to understand both the stage and trajectory of river evolution before managing a braided river to solve immediate problems such as loss of property due to bank erosion, flooding, or to improve ecological conditions. Such an approach allows managers to evaluate the potential efficacy of proposed actions and likely future problems. This conceptual model therefore helps managers to anticipate changes and consider management options across longer time-scales.

To amplify these points, examples drawn from around the world, of braided rivers in both their expansion and contraction phases, are presented. Drawing on the conceptual model, it is emphasized how the geomorphological dynamics of these systems in different phases affect both ecological functions and social costs and benefits.

Examples of expanding braided systems: the Waiapu and Waiapoa Rivers in New Zealand

The Waiapu and Waiapoa Rivers (ca. 1800 and 2000 km^2 respectively) are located on the East Coast of the North Island of New Zealand and represent good examples of braided rivers in their expansion phases with associated negative consequences. In this setting, the intrinsically high rates of geomorphological activity and associated erosion and sedimentation have been accelerated by the extensive deforestation associated with European settlers as they began pastoral farming between 1880 and 1920 (Hicks *et al.*, 2000). In particular, the clearing of native forest initiated gullying in the headwaters of the Mangatu and Waipaoa Rivers, resulting in a cumulative eroded area of 290 ha in 1939 and 1250 ha in 1960 (DeRose *et al.*, 1998).

The rivers have undergone substantial changes over this period. As reported in the Tairawhiti—Conservation Quorum (1998, issue 14, pp. 5–6), Stanley Tait, who was born in Whatatutu in 1897, described the Waipaoa early in the last century as *'hard and full of huge boulders. The water was clear and sweet and it ran fast. Children swam in the clear pools and there were eels, native trout and freshwater mussels'*. Brush had been burned between 1890 and 1910 for sheep grazing but the roots still held the soils for several years. In the early 1930s, earthflow and gully development increased suspended and bed-load sediment input to channels, inducing channel aggradation and widening. A reach 60 m wide in 1896 attained 400 m in width by 1939. Aggradation buried fertile river terraces, tracks and buildings, and greatly reduced the navigability of rivers. Photographs show how rapid sediment delivery buried a large debris trap built in the Te Weraroa stream in the Waipaoa river basin (Peacock, 1998) (Fig. 3). Active aggradation has also been reported at the Rip bridge crossing the Raparapariri stream (Hudson, 1998). In March 1988, after Cyclone Bola, the bed of the river buried the lower third of the piers. By 1994, the deck of the bridge was barely above water. In 1995, the deck of the bridge was removed and by 1998 no trace remained of the bridge. A new bridge was constructed downstream.

The lower reaches of the Waiapu River and its main tributaries are still aggrading and widening today, affecting infrastructure, private property and ecological diversity. With the Waipaoa River aggrading

A

B

Fig. 3 (A) Large debris trap built during the 1950s in the Te Weraroa Stream, New Zealand. (B) The same structure overwhelmed by high sediment loads derived from an active gully immediately upstream.

2 to 5 cm yr^{-1}, flood control schemes and reservoirs are steadily losing capacity, increasing the risk of flooding in townships such as Te Karaka. High suspended sediment concentrations, low discharges, and shallow flows disrupt and limit fish habitat. Few species and low abundance of fish are observed in these rivers. Due to aggradation and widening, the active channel occupies most of the valley bottom, eroding the foot of the valley sides. Where the valley bottom still exists, channel aggradation is so rapid that riparian vegetation cannot easily become established on channel bars and floodplains.

These two rivers clearly display negative effects associated with the rapid increase in braiding intensity due to high introduced sediment loads. Although the floodplains in this region are relatively undeveloped by global standards, impacts on human infrastructure and ecosystems are still severe.

Examples of contracting braided systems: the French Alpine rivers

Both expansion and contraction phases have been observed in French braided rivers that aggraded during the modern period (17th to 19th centuries), and then degraded after 1880 (Bravard, 1989; Kondolf et al., 2002; Liébault, 2003). The initial expansion phase was driven by human land-use changes due to increased population density in rural areas of the French Alps, which reached a maximum during the mid-19th century (Taillefumier & Piégay, 2002). By the end of that century, hillslope farming and overgrazing had increased both sediment delivery to the rivers and the frequency of peak flows. Photographs taken at the end of the 19th century reveal extensive braided rivers that generally occupied all of the valley bottoms without riparian forests, and with channels directly in contact with active alluvial fans (Fig. 4). This trend occurred during a period when economic development and increasing technical and financial capacity promoted the construction of important infrastructure, such as bridges and dykes, to improve transportation and prevent floods.

Since the end of the 19th century, the cumulative effect of human activities has been to reverse this trend. Many braided reaches have been destroyed by embankment protection measures, as discussed by Bravard et al. (1986) for the upper Rhône.

Fig. 4 Landscape of the valley bottom of the Ubaye River immediately downstream from Barcelonnette, France, in (A) 1894 and (B) 1996. The Rioux Bourdoux is the prominent tributary to the Ubaye; this tributary was called 'the Monster' 100 yr ago. (The photograph of 1894 has been provided by Restouration des terrains de montagne (RTM), Digne.)

Moreover, changing land-use activities, in-channel mining, upland afforestation and dam construction in headwater areas have resulted in a large sediment deficit in braided rivers, inducing a progressive metamorphosis and incision (Bravard et al., 1999). For example, the Fier and the Arve Rivers in the northern French Alps incised up to a maximum of 12 to 14 m (Peiry et al., 1994). Of the few braided rivers still active in the French Mediterranean piedmont area, many have undergone significant narrowing. Some of those located in the northern piedmont area, such as the Ain River, changed from a braided to a sinuous single-thread and locally meandering channel (Marston et al., 1995).

The consequences of these changes are that specific channel environments, such as braid bars and islands, are being lost (Bravard & Peiry, 1993) and that preserving the last braided corridors will require deliberate action and management. Underscoring the need for such action is the recognition that braided rivers support valuable ecosystems (Michelot, 1994) and represent our natural heritage. Another problem concerns the stability of infrastructure, for example, dyke complexes built a century ago are being undermined by channel degradation of 2 m or more. This is a fairly common situation along many reaches, and can lead to dyke failure during floods. On the Drôme River, 7.5 km of dykes are now destabilized (12% of the total length). Bridges are also destroyed during floods by pile undermining, requiring protection by weirs. In the case of the Drôme catchment, 64 bridges (one of which was destroyed in 1995 and another destabilized in November 2003) are protected by weirs that require regular maintenance. Where channel degradation is ongoing, some weirs have been restored or protected by a second weir located immediately downstream. This is the case in the Arve River where a weir was built in 1984 to stop incision, then restored in 1996 and protected by a new one downstream in 2001.

Ecology and temporal evolution of braided rivers: theory and examples

Although the bell-shaped cost-benefit model appears reasonable for issues surrounding human impacts associated with braided rivers (e.g. bank erosion, flooding, and risk to property in valley bottoms), the form of the trend is more complex when considering ecological impacts and benefits.

It is recognized that intermediate braiding activity can provide high ecological values for riparian ecosystems. Braided channels shift course within the floodplain, destroying old riparian units, and creating new ones with a variety of physical conditions, thereby favouring higher plant diversity (Hirakawa & Ono, 1974; Pautou & Wuillot, 1989). Theoretically, diversity of riparian tree species will be maximized where braiding intensity is somewhere between the extreme expansion phase, where seedling establishment is inhibited and plant mortality high, and the contraction phase, where flood disturbance is insufficiently frequent to reset and renew riparian surfaces (Fig. 2—trend A). This intermediate disturbance condition may provide both exposed mineral soil conditions for pioneer species to become established, and relatively stable 'safe sites' where mid- and late-successional species can survive. This model corresponds to the Intermediate Disturbance Hypothesis developed by Connell (1978).

Braided rivers can, however, provide ecological value throughout their temporal trajectory. In particular, during their expansion phase, braided rivers provide unique animal habitats, such as gravel bars for avian resting, or in-channel rearing and spawning areas for fish species (Fig. 2—trend B). Ecological potential, therefore, does not follow a simple bell-shaped curve but may be high for rivers with significant braiding and then diminish once the gravel surface area decreases, and optimal habitat conditions are no longer present.

An example from the Platte River in central Nebraska, USA, illustrates such an ecological trend. The Platte River has historically been a major resting and breeding site for many species of migratory birds. It currently provides critical habitat for the migration of endangered whooping cranes (*Grus americana*) and 90% of the world's population of lesser sandhill cranes (*Grus canadensis*), along with millions of ducks and geese, over 200 other species of migratory birds, and a range of other endangered or threatened plant, clam and fish species (Johnsgard, 1991; Farrar, 1992). Over the past 100 yr, dams and diversions have reduced peak flows and sediment supply to the braided reaches of the Platte, while irrigation returns have increased base flows (Williams, 1978; Nadler & Schumm, 1981). Reduced peak flows do not scour sandbars, thereby preventing woody vegetation establishment. Instead, the channel has narrowed and incised, and woody vegetation, primarily cottonwood (*Populus deltoides*) and willow (*Salix exigua*), has colonized the once active braid bars and meadows, further restricting lateral migration of the river. The river went from being 450 m wide in the early 1900s to a single thread channel just 100 m wide by 1970 (Fig. 5). Reduced instream flows have also affected adjacent wet meadows and mesic prairie ecosystems by lowering groundwater tables and surface flow (Sidle *et al.*, 1989). These planform and groundwater changes have so altered the structure of riparian habitats that the

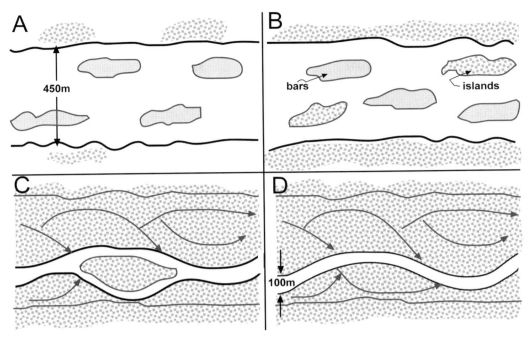

Fig. 5 Transformation of the South Platte River from a multi-thread braided river to a single thread meandering river. (A) Early 1800s: flows are intermittent and transient bars dominate channel. (B) Late 1800s, discharge is perennial due to irrigation returns, vegetation is stabilizing bars and encroaching on floodplain. (C) Early 1900s, low flow during droughts allows vegetation to establish below mean annual high water level, bars become stable, vegetated islands and a single-thread channel forms. (D) The modern single-thread channel and densely wooded floodplain. Braided patterns on floodplain are vestiges of historical channels. (Modified from Nadler & Schumm, 1981.)

sustainability of migratory and resident birds and other biota is in question.

Along with birds, other studies have demonstrated the ecological importance of braided river patterns for fish and invertebrate communities because of the diversity of in-channel habitats (Church *et al.*, 2000; Rempel *et al.*, 2000). Church (2001) described how the braided Fraser River in British Columbia, Canada is one of the most productive rivers in the world, providing complex channel habitat for spawning and rearing fish. Around 30 fish species such as the chum (*Oncorhynchus keta*), pink salmon (*O. gorbuscha*), coho (*O. kisutch*), chinook (*O. tshawytscha*) or sockeye salmon (*O. nerka*) were identified as utilizing braided reaches of the Fraser. Similarly, valuable habitat for fish with associated high sport fishing value is found in the braided rivers of southern New Zealand (Mosley, 1982). For example, sockeye salmon, brown trout and rainbow trout utilize the braided Ohau river. Graynoth *et al.* (2003) stated that because of its width and large number of braids the lower Waitaki River might support more trout per kilometre than most New Zealand rivers (Table 1). Key habitat features include large riffles, runs and braids suitable for spawning by Chinook salmon, while brown trout spawn in the smaller main channel braids as well as the tributaries. The Waitaki River also provides suitable habitat for several species of native fish, such as upland bullies in most braids, and long finned eels in braids where vegetation and debris provide cover. Jellyman *et al.* (2003) indicated that the value of this river is mainly in its middle reach, and depends on a wide range of habitats and water types accessible for a variety of fish species to use. As braiding intensity decreases due to changing discharge and bedload delivery, the braided river structure may retain enough internal heterogeneity to maintain a given level of biodiversity until the braided structure disappears entirely (Mosley, 1982).

The ecological value of braided rivers requires consideration in any plan for sustainable management. As discussed in the next section, the challenge

Table 1 Subjective assessment of the relative importance (%) of the tributaries and different sections of the lower Waitaki River for fish spawning, rearing and fishing (From Graynoth et al., 2003, p. 34)

Species	Function	Tributaries	Section 1	Section 2	Section 3
Brown trout	Spawning	50	5	45	0
	Juvenile rearing	40	5	50	5
	Adult rearing	30	10	50	10
	Fishing	20	10	50	20
Salmon	Spawning	12	6	82	0
	Juvenile rearing	5	3	82	10
	Fishing	0	0	40	60
Rainbow trout	Spawning	90	10*	0	0
	Juvenile rearing	35	20	40	5
	Adult rearing	9	40	50	1
	Fishing	9	40	50	1

*Assumes some juveniles originate from Lake Waitaki.

in such schemes is to find a reasonable equilibrium between satisfying immediate human needs and environmental preservation, which in the long run will guarantee sustained human needs. Such an equilibrium is not easily reached when braiding is both a social problem and has high ecological values. The two objectives are easier to reconcile during the contraction phase of evolution.

SUSTAINABLE MANAGEMENT OPTIONS FOR BRAIDED RIVERS

Balancing sustainable development of braided river valleys without compromising natural river ecosystem functions is a complex task. In this section strategies for achieving this, drawing on global examples, are established.

Expansion phase

Protecting development against flooding at the reach scale

One popular approach to reducing the flooding and bank erosion risks of an actively braided river is to remove excess gravel; this may also prevent rapid avulsions. Kondolf (1994) reported the case of the Waimakariri River near Christchurch in New Zealand, which transports 150,000 m^3 yr^{-1} of sediment from the Alps and aggraded an average of almost 3 m over the downstream-most 15 km between 1929 and 1973. During this period, a reach-averaged depth of almost 6 m of gravel and sand was extracted, saving the town of Christchurch from channel avulsion (Basher et al., 1988).

Where there is high regional demand for gravel, most braided rivers have been heavily mined because of the high gravel quality, ease of mining, and proximity between sites of production and consumption. Gravel extraction is an effective solution where both flood risk and demand for gravel is high, but excess gravel extraction may cause other problems, as discussed below. In France, where gravel extraction has been prohibited since 1994, along with other Mediterranean countries, excess gravel mining has resulted in dramatic channel degradation (Kondolf, 1994; Rinaldi et al., 2005).

Using gravel extraction as a solution to heightened flood risk is more problematic in regions where there is no demand for gravel as a resource, and the cost of gravel mining and storage must therefore be directly borne by local inhabitants. In such situations, gravel extraction is expensive and difficult to maintain, and involves complex trade-offs among different stakeholders and users. Ultimately, the issue becomes how to best satisfy the legitimate needs of flood protection, without

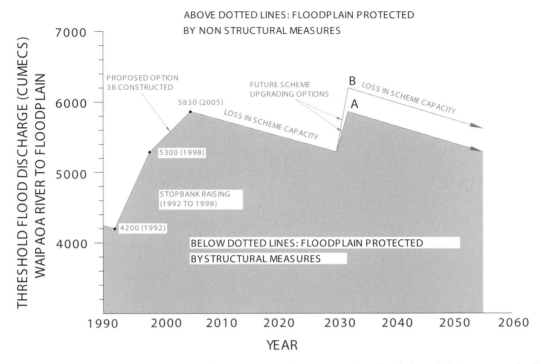

Fig. 6 Potential models of change in the level of protection to the Poverty Bay floodplain, Waipaoa mouth, East Coast region, North Island, New Zealand. (From Peacock, 1998.)

wasting a gravel resource of high quality and degrading the river's conditions to the extent that other users and resources are compromised.

The immediate risk of flooding can often be prevented by developing engineering structures. This approach may be necessary if existing conditions or time lags prevent the use of other countermeasures. For example, as described by Peacock (1998), by 1995 about 30% of the upper Waipaoa in New Zealand was afforested, resulting in a 5% reduction in gravel entering the river. However, afforestation is only an effective strategy on a long-term (multidecade) basis, and a large quantity of gravel had already been delivered to the channel. This material is temporarily stored in the upper reaches of the river and will progressively move downstream for years. The best solution in the short to medium term (years to decades) is therefore to upgrade the dykes in order to protect settlements, coupled with the construction of diversion cuts (Fig. 6). In the long term, because of the negative environmental consequences and recurring costs of such works, new approaches must be developed.

The case of New Zealand's Waiho River is a good example of an alternative approach (Rouse *et al.*, 2001). To combat continuing river aggradation, which increased bed elevation by 5 m in the past 20 yr, with the associated risks of avulsion and damage to local infrastructure, one option was to raise the levees or stop banks. Studies showed that the aggradation was not natural but associated with the reduction of the apex sector angle of the river fan by the river-control stop-banks (Davies & McSaveney, 2001). Some argued that raising the levees would increase the chance of seepage or piping-driven collapse of banks, with increased risk of flooding, and would also increase the costs of maintaining development in the protected area. An alternative option was the continued protection of the right bank but removal of the left one and relocation of its inhabitants.

Reducing sediment delivery to decrease braiding intensity

Excess sediment produced by unstable montane hillslopes in the European Alps sparked development of major mountain restoration programmes in France, Italy, Germany and Austria. These followed the pioneer works of Demontzey (1882) in the Ubaye catchment, where a torrent called the 'monster'—the Rioux Bourdoux—received most of the attention (Fig. 4). In the Drôme catchment in southern France, 16% of the basin (ca. 13,000 ha) was afforested by the Mountain Restoration Services mainly between 1887 and 1917. Between 1863 and 1887, foresters prepared the area by building engineered structures designed to reduce the river's slope and decrease gravel output (i.e. check-dams of various heights constructed with wood or piers). Valley walls were also stabilized by wooden structures. Between 1887 and 1917, the previously stabilized slopes were afforested (Liébault, 2003). These pioneering approaches led to the construction of 'sabo dams' to trap sediment from debris flows, and broad flats at the base of the slopes to dissipate flow energy and promote sediment retention. Such structures have been built all over the world, most notably in Japan, where the first structures were established on the Ishikari River in Hokkaido (Marutani *et al.*, 2001).

Afforestation policies to reduce sediment input from landsliding are actively pursued in the Gisborne–East Coast region of New Zealand. About 16% of the region is planted in exotic forest with an average annual planting rate of ca. 6500 ha. In 1955, the government approved a plan to buy the worst eroding land from the farmers in order to plant closed-canopy forest. Erosion control planting began in 1960. Research demonstrates that the landslide rate under forest is about ten times less than under pasture (Hicks, 1991; Marden & Rowan, 1994). Following replacement of grazed areas by forest, sediment production from 11 gullies in a 4 km^2 area in the Waipaoa headwater area decreased from 2480 t ha^{-1} yr^{-1} during the 1939–1958 period to 1550 t ha^{-1} yr^{-1} between 1958 and 1992. Among the 315 gullies observed at Mangatu forest in the Waipaoa catchment in 1960, only 102 gullies remain today. Reforestation has both contributed to stabilization of many small gullies and prevented formation of new ones.

In pursuing an afforestation approach to reducing sediment delivery to streams, knowledge of land type and storm magnitude–frequency zones enables better targeting of landslide susceptible areas and therefore more effective restoration strategies. For example, a reduction of 50% in landslide-

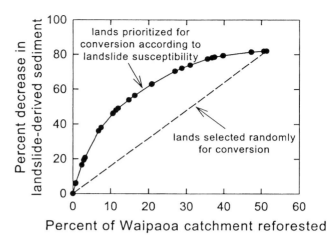

Fig. 7 Percentage of difference in landslide-derived sediment for targeted versus random reforestation strategies. (From Reid & Page, 2002, with permission from Elsevier.)

derived sediment can be achieved when only 12% of the catchment area is forested if priority is given to areas sensitive to landsliding (Fig. 7). On the other hand, the same reduction would require treating 30% of the catchment area if reforestation was distributed randomly. Sediment reduction through reforestation of landslide areas is a long-term solution, however, and may require additional measures for short- to medium-term protection. For example, sediment production from the two major landslides of the Waipaoa catchment, the Mangatu (27 ha) and the Tarndale (20 ha), generated 73% of the total basin output between 1939 and 1958. From 1958 onward, reforestation decreased their surface area by only 5% and 30%, respectively, and although sediment production was decreased the two landslides were still producing 95% of the sediment between 1958 and 1992. Reducing the negative consequences of high sediment transport and braiding in downstream rivers therefore requires additional countermeasures.

Balancing human needs in active braided systems

Geomorphological analysis coupled with careful management and engineered interventions can achieve multiple objectives in braided rivers. The Fraser River in central British Columbia, Canada, for example, poses significant flood hazards to those communities and urban areas that border it,

Fig. 2 Test structures: (a) permeable groyne at Kamarjani; (b) falling apron (left), launching apron (centre) and stone pitching (right) of bank revetment at Ghutail.

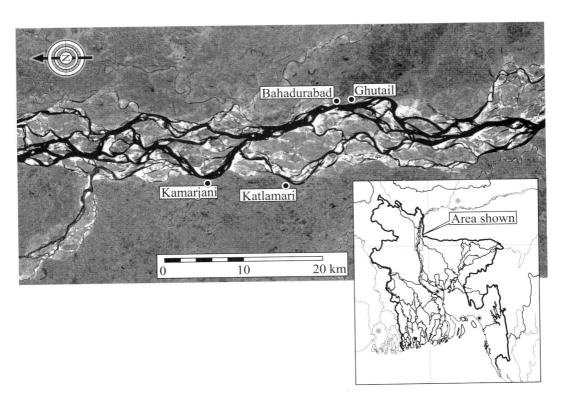

Fig. 3 Location of project test sites.

apron and a falling apron. Aprons are flexible protection layers that unfold along the underwater slope as the river erodes the land in front of the structure (Fig. 5b). The elements of launching aprons are interconnected, whereas those of falling aprons are loose. The upper slope protections were made of brick mattresses, wiremesh mattresses filled with stones, concrete cement (CC) blocks, interlocking CC slabs and riprap as cover layer over different grades of geotextile filters and granular filters.

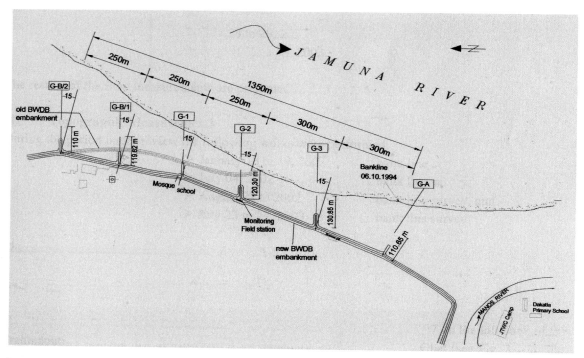

Fig. 4 General layout of groynes at Kamarjani.

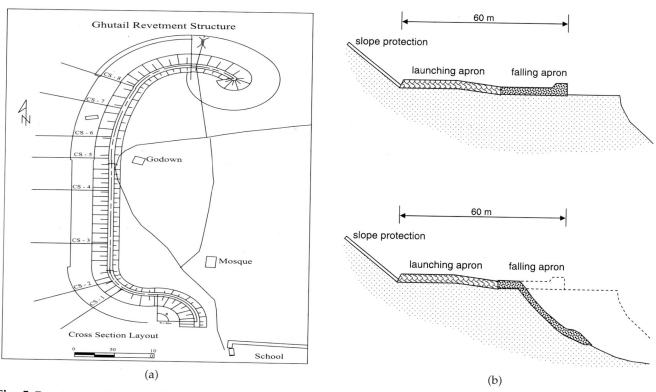

Fig. 5 Revetments: (a) design layout at Ghutail; (b) unfolding of launching and falling apron.

Fig. 6 Component FAP22 recurrent measures: (a) high-water bandals at Katlamari; (b) floating screens mounted between country boats at Katlamari.

The launching aprons were made of dumped CC blocks, riprap and different types of mattresses over geotextile filters. The falling aprons comprised CC blocks, riprap, geo-sand containers, stone-filled gabion sacks and selected boulders.

The structures stopped bank retreat effectively, even though the groyne test structure was damaged during the first year. In the subsequent low-water season, the groynes were repaired and improved by increasing their transverse lengths as well as the lengths of their piles. No other damage occurred.

Indirect interventions to control bank retreat were developed and tested in FAP22. The underlying idea was to influence the morphological development of the river in such a way that aggressively eroding channels located adjacent to the mainland floodplain would be closed by silting their entrances using recurrent measures. The tests took place in the Katlamari channel along the west bank of the river, inducing siltation by a combination of fixed screens or 'bandals' (Fig. 6a), floating screens (Fig. 6b) and an erodible cross-dam or 'sand plug'. The principle of the screens is that they separate the sediment-laden flow near the bed from the clearer water near the surface. The sediment-laden water near the bed passes under the screens in a cross-direction, whereas the clearer water near the surface is guided in a direction parallel to the screens. The helical water motion thus induced persists over a certain distance downstream. The separation produces a sediment overload in the cross-flow moving away from the screens. The resulting siltation in the area of the measures was substantial, but a new entrance to the Katlamari channel was formed further downstream. A more complete closure was thus found to require deployment of the recurrent measures over a larger area, including the seasonally flooded bars and islands, zones locally known as 'chars'.

LESSONS LEARNED

The 10 yr of research and 'learning by doing' has produced a wealth of knowledge regarding structural design, construction materials, construction techniques, planning, procurement, monitoring, morphological predictions and river response. This paper cannot do justice to all the lessons learned, but highlights selected topics that have a strong relation with the science of fluvial morphodynamics.

Techniques of bank protection and river training

The FAP21 structures proved to be capable of stopping bank erosion along the Brahmaputra–Jamuna River. Table 1 shows how their costs compare with those of similar structures along the same river.

Table 1 Costs of structures built on Brahmaputra–Jamuna River between 1994 and 2000 (Courtesy Dr Knut Oberhagemann, Northwest Hydraulic Consultants)

Location	Structure	Total cost (million US$)	Protected length (km)	Cost per protected length (US$/m)
Jamuna Bridge	Two guide bunds with revetment	288.2	6.4	45,000
Sirajganj	Revetment	73.5	2.5	29,300
Sariakandi	Groynes	52.1	4.5	11,500
Bahadurabad	Revetment	8.1	0.8	10,100
Bhuapur	Revetment	6.8	1.55	4300
Ghutail	Revetment	3.2	0.55	5800
Kamarjani	Permeable groynes	11.8	1.7	6900
Kazipur	Groynes	5.1	3.0	1700
Shovgacha	Groynes	1.9	1.4	1300

Considering that the FAP21 costs included substantial investments for additional research that should be subtracted in a proper comparison, the FAP21 structures can be concluded to be cost-effective. The design process revealed, however, serious gaps in existing textbook design methods. Traditionally, design conditions are related to extreme discharges with a certain prescribed return period, assuming that the severest loads on hydraulic structures occur at the highest discharges. However, the severest conditions for bank erosion may occur at lower discharges around or below bankfull stage, for example: (i) at certain unfavourable near-bank channel configurations, such as deep channels, channel confluences and sharp bends; and (ii) following rapid scour hole development or a rapid fall of the water level. Return periods for the occurrence of unfavourable near-bank channel configurations would be a better basis for design than return periods for extreme discharges. The problem is that the bank protection structures themselves make these configurations more unfavourable.

Observations in the largely untrained Brahmaputra–Jamuna River are insufficient to generate reliable statistics on structure-affected channel configurations, and the present state-of-the-art in mathematical modelling is still not sufficiently advanced to compute structure-induced scour in a process-based manner. Nonetheless, FAP21 produced some progress in process-based modelling of two mechanisms by which the mere presence of bank protection structures increases the loads on these structures: (i) the deeper bend scour due to the stopping of bank migration (Mosselman et al., 2000); and (ii) the attraction of channels and associated flow attack towards scour holes (Mosselman & Sloff, 2002). The stopping of bank erosion is assumed to produce deeper bend scour through: (i) prevention of bank sediment supply; (ii) channel narrowing due to retarded point-bar growth; (iii) bend deformation due to local prevention of channel migration; and (iv) vortices generated by flow impingement. As the mechanisms by which stabilized banks increase the loads on the structures could not be reproduced well by the extensive mobile-bed physical modelling of the project either, engineering judgement remained an important ingredient for the final design of the FAP21 structures. The explicit identification and approximate quantification of the two mechanisms nonetheless represent a step forwards with respect to existing textbook design methods.

Technically, the FAP22 recurrent measures were a success as well. Practically, however, the recurrent measures were unsuccessful. The international donors would rather allocate funds to more permanent and prominent structures. The local authorities were less committed to recurrent measures and even somewhat resented what was perceived as 'second-class' technology, as if Bangladesh was not good enough for real bank protection structures. Nonetheless, recurrent measures were concluded to hold promise for the future, in particular as auxiliary measures in a river training strategy primarily based on more permanent structures. In the form of vanes and screens, recurrent measures are currently a topic of experimental research at Bangladesh University of Engineering and Technology (BUET/DC, 2001).

Measurement techniques

Satellite remote sensing provided the most prominent advances in the measurement of the rapid morphological changes of the Brahmaputra–Jamuna River during the 1990s. It allowed rapid data acquisition and easily covered the 200 km long and 10–18 km wide braid belt of the river in Bangladesh. Multi-temporal overlays of satellite images revealed annual changes (Fig. 1) as well as long-term trends (Fig. 7). Furthermore, a new technique of float

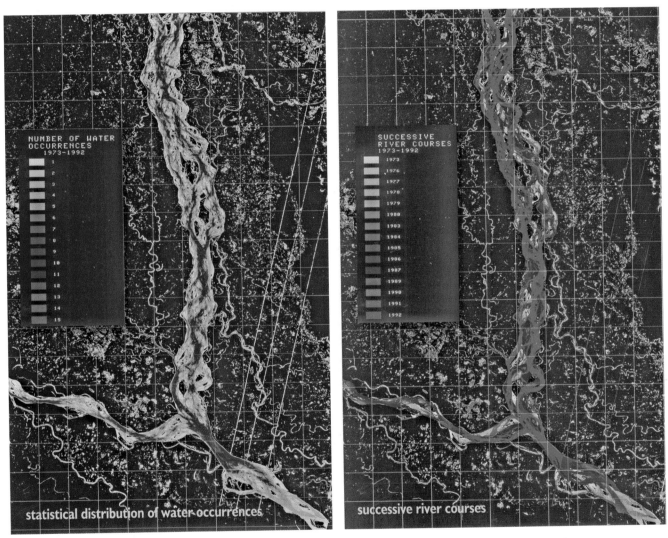

Fig. 7 Multi-temporal low-water planforms of the Brahmaputra–Jamuna River, 1973–1992. Number of water occurrences as a measure for low-water channel persistence (left) and successive river courses (right). Grid squares have 10 km sides. (Courtesy National Aerospace Laboratory, The Netherlands.)

tracking was developed under FAP21/22, using drifter floats with drogues, GPS receivers and data loggers (through involvement of Hochschule Bremen, Labor für Wasserbau, Germany). The FAP21/22 project also benefited from availability of field data supplied by the River Survey Project (FAP24), in which flows, sediment transport and bed topographies of the Brahmaputra–Jamuna River were measured all year round, including during the summer monsoon floods.

Morphology prediction methods

A prerequisite for the testing of structures constructed under FAP21/22 was that those structures had to be attacked by the river within the time frame of the project, which originally was assumed to span 5 yr. At the same time, the test sites should not be eroded during the first years of land acquisition and construction. This led to the need to predict bank erosion and planform changes 2–3 yr ahead. For the rapidly changing Brahmaputra–Jamuna with its sensitive dependence on small perturbations, such a prediction span is too long to be covered by either simple extrapolation or detailed two- or three-dimensional mathematical modelling. Therefore, a new prediction method was developed by Klaassen *et al.* (1993), based on empirical laws derived from a large set of satellite images (Klaassen & Masselink, 1992; Klaassen *et al.*, 1993). Having been improved further by Mosselman *et al.* (1995) and Sarker & Khayer (2002), the method is now operational on a routine basis at the Center for Environmental and Geographic Information Services in Bangladesh (www.cegisbd.com).

Jagers (2001, 2003) implemented the prediction method in a computer model and tested it against observations. He also constructed and tested an artificial neural network for the prediction of low-water planform changes in the Brahmaputra–Jamuna. Despite some fair results, it appeared that the predictive powers of the computer model and the neural network are limited. The original prediction method has nonetheless proven to result in a good selection of sites for bank protection structures.

In the 1990s, two-dimensional depth-averaged mathematical modelling of river morphology became a standard tool in river engineering. In Bangladesh, it is operational at Institute of Water Modelling (IWM) in Bangladesh (www.iwmbd.org).

Jagers (1999, 2003) applied two-dimensional depth-averaged mathematical modelling to chute cutoffs in the Brahmaputra–Jamuna River. Having explored so many different model concepts in depth, the work of Jagers (2003) now stands as probably the most thorough state-of-the-art in the mathematical modelling of braided rivers.

Understanding the causes of braiding

In 1992, detailed knowledge was already available about how rivers braid, with such mechanisms as: local sedimentation in a lateral flow expansion (mid-channel bar growth); chute dissection of bars; and avulsive channel shifting (e.g. Ashmore, 1991). However, knowing *how* rivers braid does not necessarily imply knowing *why* rivers braid. Why do some in-channel bars grow while other bars are eroded away? Why do some channels develop in-channel bars while other channels do not? These questions have been answered by the mathematical technique of linear stability analyses (Engelund & Skovgaard, 1973; Parker, 1976; Fredsøe, 1978; Colombini & Tubino, 1991). In these analyses, flow strength, width-to-depth ratio and sediment calibre appear to be the key parameters. Assuming furthermore that the river width results from a balance between flow strength and bank strength (Ikeda *et al.*, 1981), braiding can be understood to depend basically on a balance between flow strength, sediment calibre and bank strength (cf. Ferguson, 1987). It was also known that tectonic subsidence (Gregory & Schumm, 1987) and sediment overloading (Germanoski & Schumm, 1993) enhance the braiding intensity, but are not necessary conditions for braiding (Parker, 1976).

Despite all of this knowledge, one important question remained in 1992. If linear stability analyses provide such a good explanation for braiding, then why do these analyses predict the number of channels per cross-section to be much larger than observed in the Brahmaputra–Jamuna River? It was speculated at the beginning of FAP21/22 that non-linear interactions and non-uniform bank-line forcing could provide answers, but insight into the effect of these factors was only obtained in the subsequent decade. Numerical computations by Enggrob & Tjerry (1999) show that, initially, when starting from a flat bed, the number of channels is high, after which non-linear interactions reduce the

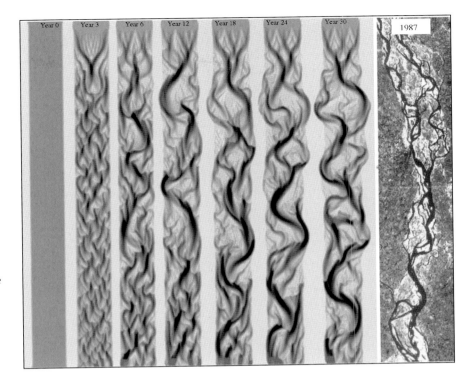

Fig. 8 Non-linear interactions reduce the number of channels as braiding evolves in a numerical computation (Enggrob & Tjerry, 1999). The input parameters were based on the Brahmaputra–Jamuna River, shown in the last panel for reference.

number even if the river widens (Fig. 8). In theory, non-uniform bank-line forcing can both increase and decrease the number of channels. It may play a role in the decrease in the number of channels computed by Enggrob & Tjerry (1999), whereas Repetto *et al.* (2002) demonstrated that it can increase the number of channels (Fig. 9).

It is worth noting that the method of Bettess & White (1983) was found to predict the number of channels in the Brahmaputra–Jamuna River better than linear stability theories. In this method, the total discharge and sediment load are distributed over 'n' equal channels, each channel satisfying equations for flow and sediment transport. However, successful application requires a priori knowledge of the empirical regime width predictor for individual channels. Furthermore, the method involves the assumption that rivers tend to minimize their channel slopes. This assumption is an extremal hypothesis and hence questionable; because extremal hypotheses do not represent fundamental laws of physics, they do not lead to better predictions than alternative hypotheses such as the Principle of Twenty Percent Inefficiency (Mosselman, 2004), and they may even lead to conclusions that are incompatible with observations (Griffiths, 1984).

In the method of Bettess & White (1983), however, the assumption could be rephrased less controversially by stating that valley slopes impose tight limits to channel slopes in low-sinuosity rivers that easily adjust their planforms.

CONCLUSIONS

Bank protection structures and indirect interventions to control bank retreat have been developed and tested along the braided Brahmaputra–Jamuna River in Bangladesh in the period between 1992 and 2003. This has produced a wealth of knowledge regarding structural design, construction materials, construction techniques, planning, procurement, monitoring, morphological predictions and river response. This paper highlights selected topics that have a strong relation with the science of fluvial morphodynamics.

Gaps have been identified in existing textbook design methods. Textbooks usually relate design conditions to extreme discharges with a long return period. However, the severest conditions for bank erosion occur at discharges around or below bank-full stage, and are related to certain unfavourable

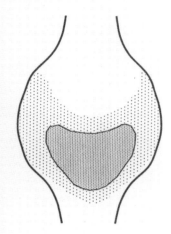

Fig. 9 Mid-channel bars forced by non-uniform width as an example of increased number of channels due to planform forcing (Repetto *et al.*, 2002).

near-bank channel configurations, such as deep channels, channel confluences and sharp bends. Moreover, these unfavourable channel configurations appear to be created by the mere presence of bank protection structures, which implies that their frequencies of occurrence near stabilized banks are underestimated if they are inferred from observations in a largely untrained river such as the Brahmaputra–Jamuna. The mechanisms by which the mere presence of bank protection structures creates unfavourable channel configurations are still poorly known, but the project did produce some progress in the process-based modelling of two of these mechanisms:

1 the deeper bend scour due to the stopping of bank migration;
2 the attraction of channels and associated flow attack towards scour holes.

The project has initiated the development of a new, appropriate morphology prediction method that is now operational on a routine basis in Bangladesh. Other morphology prediction methods for braided rivers have been explored and improved as well. Contemporary studies have deepened the insight into the ability of linear stability analyses to predict aspects of braiding; Linear stability analyses explain why rivers braid; they also discriminate between braiding and non-braiding bed topographies in straight uniform channels. However, they fail to predict the number of channels correctly in real braided rivers because the number of channels is affected by non-linearities and non-uniform bank-line forcing.

ACKNOWLEDGEMENTS

The Bank Protection and River Training Pilot Project (FAP21/22) was funded by KfW (Kreditanstalt für Wiederaufbau, Germany) and CFD (Caisse Française de Développement, France). It was carried out by Jamuna Test Works Consultants: a consortium of RRI Beller (Germany), Professor D. Lackner & Partners (Germany), CNR (Compagnie Nationale du Rhône, France), WL/Delft Hydraulics (The Netherlands) and BETS (Bangladesh Engineering and Technical Services, Bangladesh). I thank the University of Birmingham for co-funding the oral presentation of this paper at the second Braided Rivers Conference. The overview of costs in Table 1 is an update by Dr Knut Oberhagemann of information compiled originally in the CADP River Erosion Prevention and Morphology Study, executed by DHV with participation of Northwest Hydraulic Consultants and RRI Beller staff. The satellite imagery in Figs 1 & 7 was processed by the National Aerospace Laboratory, The Netherlands.

REFERENCES

Ashmore, P.E. (1991) How do gravel-bed rivers braid? *Can. J. Earth Sci.*, **28**, 326–341.

Bettess, R. and White W.R. (1983) Meandering and braiding of alluvial channels. *Proc. Inst. Civ. Eng.*, Part 2, **75**, 525–538.

BUET/DC (2001) *Capacity Building in the Field of Water Resources Engineering and Management in Bangladesh*. Inception Report, Bangladesh University of Engineering and Technology and Delft Cluster, November 2001.

Colombini, M. and Tubino, M. (1991) Finite amplitude multiple row bars: a fully non-linear spectral solution. In: *Proceedings of Euromech 262, Wallingford, 26–29 June, 1990* (Eds R. Soulsby and R. Bettess), pp. 163–169. Balkema, Rotterdam.

Elahi, K.M., Ahmed, K.S. and Mafizuddin, M. (Eds) (1991) *Riverbank Erosion, Flood and Population Displacement in Bangladesh*. Riverbank Erosion Impact Study, Jahangirnagar University, Dhaka, Bangladesh.

Engelund, F. and Skovgaard, O. (1973) On the origin of meandering and braiding in alluvial streams. *J. Fluid Mech.*, **57**(2), 289–302.

Enggrob, H.G. and Tjerry, S. (1999) Simulation of morphological characteristics of a braided river. *Proceedings of the IAHR Symposium on River, Coastal and Estuarine Morphodynamics*, Genova, 6–10 September, Vol. I, pp. 585–594.

Ferguson, R.I. (1987) Hydraulic and sedimentary controls of channel pattern. In: *River Channels: Environment and Process* (Ed. K. Richards), pp. 129–158. Basil Blackwell, Oxford.

Fredsøe, J. (1978) Meandering and braiding of rivers. *J. Fluid Mech.*, **84**(4), 609–624.

Germanoski, G. and Schumm, S.A. (1993) Changes in braided river morphology resulting from aggradation and degradation. *J. Geol.*, **101**, 451–466.

Gregory, D.I. and Schumm, S.A. (1987) The effect of active tectonics on alluvial river morphology. In: *River Channels: Environment and Process* (Ed. K. Richards), pp. 41–68. Basil Blackwell, Oxford.

Griffiths, G.A. (1984) Extremal hypotheses for river regime: an illusion of progress. *Water Resour. Res.*, **20**(1), 113–118.

Ikeda, S., Parker, G. and Sawai, K. (1981) Bend theory of river meanders, Part 1, linear development. *J. Fluid Mech.*, **112**, 363–377.

Jagers, H.R.A. (1999) Numerical analysis of cutoff development. *Proceedings IAHR Symposium on River, Coastal and Estuarine Morphodynamics*, Genova, 6–10 September, Vol. I, pp. 553–562.

Jagers, H.R.A. (2001) A comparison of prediction methods for medium-term planform changes in braided rivers. *Proceedings of the 2nd IAHR Symposium on River, Coastal and Estuarine Morphodynamics*, 10–14 September, Obihiro, Japan, pp. 713–722.

Jagers, H.R.A. (2003) *Modelling planform changes of braided rivers*. PhD thesis, University of Twente, ISBN 90-9016879-6.

Klaassen, G.J. and Masselink, G. (1992) Planform changes of a braided river with fine sand as bed and bank material. *Proceedings of the Fifth International Symposium on River Sedimentation*, 6–10 April Karlsruhe, pp. 459–471.

Klaassen, G.J., Mosselman, E. and Brühl, H. (1993) On the prediction of planform changes in braided sand-bed rivers. In: *Advances in Hydro-Science and—Engineering* (Ed. S.S.Y. Wang), pp. 134–146. Publ. University of Mississippi, University, Mississippi.

Mosselman, E. (2004) Hydraulic geometry of straight alluvial channels and the principle of least action (discussion). *J. Hydraul. Res.*, **42**(2), 219–220,222.

Mosselman, E. and Sloff, C.J. (2002) Effect of local scour holes on macroscale river morphology. In: *River Flow 2002: Proceedings of the International Conference on Fluvial Hydraulics*, Vol. 2, Louvain la Neuve, Belgium, 4–6 September (Eds D. Bousmar and Y. Zech), pp. 767–772. Balkema, Lisse.

Mosselman, E., Huisink, M., Koomen, E. and Seymonsbergen, A.C. (1995) Morphological changes in a large braided sand-bed river. In: *River Geomorphology* (Ed. E.J. Hickin), pp. 235–249. Wiley, Chichester.

Mosselman, E., Shishikura, T. and Klaassen, G.J. (2000) Effect of bank stabilization on bend scour in ana-branches of braided rivers. *Phys. Chem. Earth*, Part B, **25**(7–8), 699–704.

Parker, G. (1976) On the cause and characteristic scales of meandering and braiding in rivers. *J. Fluid Mech.*, **76**(3), 457–480.

Repetto, R., Tubino, M. and Paola, C. (2002) Planimetric instability of channels with variable width. *J. Fluid Mech.*, **457**, 79–109.

Sarker, M.H. and Khayer, Y. (2002) *Developing and Updating Empirical Methods for Predicting Morphological Changes in the Jamuna River*. EGIS Technical Note Series, Environment and GIS Support Project for Water Sector Planning (now CEGIS, cf. Table 2), Ministry of Water Resources, Government of Bangladesh.

Morphological response of the Brahmaputra–Padma–Lower Meghna river system to the Assam earthquake of 1950

MAMINUL HAQUE SARKER* and COLIN R. THORNE[†]

*CEGIS, House 6, Road, 23/c, Gulshan 1, Dhaka 1212, Bangladesh (Email: msarker@cegisbd.com)
[†]School of Geography, University of Nottingham, Nottingham NG7 2RD, UK (Email: Colin.Thorne@nottingham.ac.uk)

ABSTRACT

The braided river-systems of Bangladesh annually consume large areas of floodplain and make thousands of people landless through bank erosion. Severe bank retreat associated with widening of major rivers such as the Jamuna, Padma and Lower Meghna during the past 50 yr has greatly increased this loss of land. Channel widening indicates that these rivers are not operating in dynamic equilibrium, at least at the decadal scale, although the causes of this unstable behaviour remain contested. Identifying the causes is of great interest to river scientists, engineers and planners at the national level attempting better to manage the natural and human resources of Bangladesh.

This study puts forward the hypothesis that morphological changes have occurred in response to disturbance of the fluvial system by the 1950 Assam earthquake. Landslides triggered by the earthquake generated about 45×10^9 m^3 of sediment, much of which entered the Brahmaputra River and its tributaries. It is suggested that the fine fraction of this sediment (silt and clay) travelled quickly through the system, without disturbing the morphology of the channels, before settling in the Meghna Estuary. In contrast, the coarser fraction (sand) has taken half a century to travel through the system, moving as a wave of bed material load, with a celerity between 16 and 32 km yr^{-1}. Analyses of historical maps and satellite images, together with records of discharge, water level, sediment transport and cross-sectional form, reveal a sequence of morphological change in the Jamuna–Padma–Lower Meghna system, with a downstream phase lag that is commensurate with the celerity of the coarse sediment wave.

A conceptual process–response model has been developed to elucidate the relationship between downstream propagation of the sediment wave and morphological response in the river channels. The model is mainly based on records and observations of the Brahmaputra River in Assam and the Jamuna, Padma and Lower Meghna rivers in Bangladesh. When fully validated, the model could be helpful in predicting the future morphological evolution of the major rivers of Bangladesh as these systems respond to new drivers of change, such as climate, neo-tectonics, river engineering and the decline of distributary rivers.

Keywords Brahmaputra River, Padma River, Lower Meghna River, braiding, process–response model, earthquake.

INTRODUCTION

The Brahmaputra is one of the largest rivers in the world, ranking fifth in terms of discharge (mean flow 20,200 m^3 s^{-1}) (Thorne et al., 1993). It originates in China and flows for several hundred kilometres through Assam in India before entering Bangladesh. The river flows for 240 km through Bangladesh, where it is called the Jamuna, to its confluence with the River Ganges (Fig. 1). The combined flow forms the River Padma, which joins the Upper Meghna approximately 100 km downstream to form the

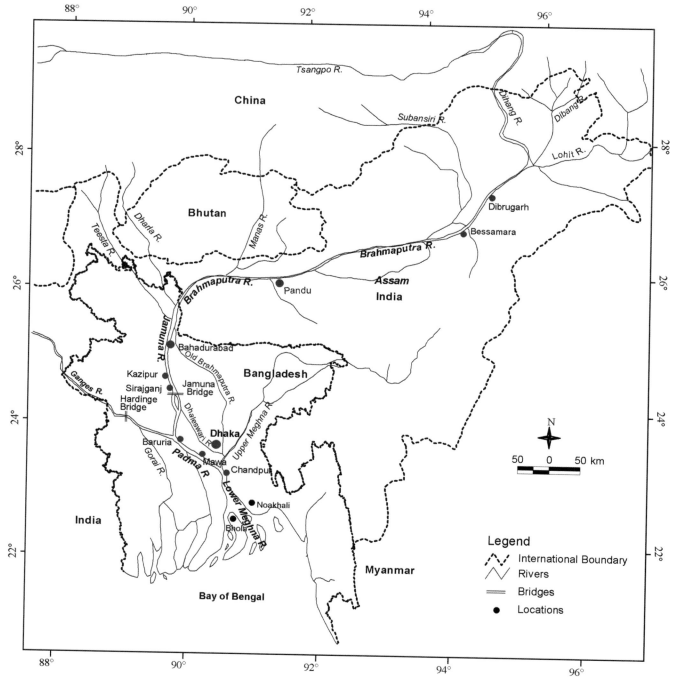

Fig. 1 Brahmaputra, Jamuna, Padma and Lower Meghna river system.

Lower Meghna River. The Lower Meghna enters the Bay of Bengal about 160 km to the south of the confluence.

The major physical characteristics of these rivers are presented in Table 1. The flow regimes of the rivers are largely dominated by runoff generated by monsoon precipitation, with high flows occurring between July and September and low flows from January to March. The 30-yr mean water level hydrograph for the Jamuna River at Bahadurabad

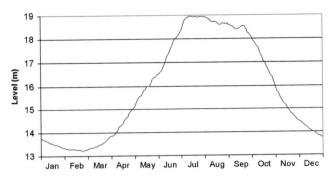

Fig. 2 A mean water level hydrograph of the Jamuna River at Bahadurabad.

(Fig. 2) represents the annual variation of water levels in the main rivers of Bangladesh. The difference between summer and winter water levels at Bahadurabad is more than 6 m, the magnitude of this variation decreasing with distance downstream. For example, in the River Padma at Mawa, mean annual variation in water level is 4 m.

The large discharges and heavy sediment loads carried by the rivers result in highly variable and dynamic channel morphologies characterized by rapid adjustments to the cross-sectional geometry, bankline positions and planform attributes (Coleman, 1969). A number of studies have been carried out, particularly concerning morphological processes in the Jamuna since the late 1960s. These include the academic work of Coleman (1969) and Bristow (1987), and consultancy studies carried out under the Jamuna Bridge Authority and Flood Action Plans (FAP 1, FAP 4, FAP 9B, FAP 19 and FAP 24). The general conclusion drawn in most of these studies is that, while the main rivers of Bangladesh exhibit short-term instability, they may be considered to be broadly in dynamic equilibrium. It follows from this that morphological changes may be explained on the basis of short-term departures from a form that is generally unchanging.

Recent observations by the Center for Environmental and Geographic Information Services (CEGIS), based on analyses of a 30-yr time-series of satellite images, reveal that the Jamuna River during the 1970s and 1980s widened at a rate of approximately 160 m yr^{-1}. Widening continued during the 1990s, although the rate decreased to about 50 m yr^{-1} (EGIS, 2000; Sarker et al., 2003). The result of widening is that during the past three decades the river has destroyed a net area of 70,000 ha of floodplain. During the late 1980s and 1990s, the Padma and Lower Meghna rivers widened at rates even higher than those exhibited by the Jamuna, resulting in net destruction of 30,000 ha of floodplain. The human impact of the net loss of around 100,000 ha of floodplain is extremely serious, with around 800,000 people becoming landless during the past 30 yr. These high and persistent rates of widening suggest that these rivers are not in dynamic equilibrium, at least at the decadal time-scale.

Recognizing the non-stationary trends of morphological change in the Jamuna–Padma–Meghna system, several past studies have developed short and long-term empirical methods to predict future morphological changes in the Jamuna (Klaassen & Masselink, 1992; Klaassen et al., 1993; Mosselman et al., 1995; Thorne et al., 1995). Similarly, bank-line migration in the Padma and Lower Meghna was predicted during the National Water Management Plan (Halcrow & Mott MacDonald, 2000) for the decadal scale. In most cases, future morphological changes were predicted by extrapolating past and prevailing migration trends, assuming that these will continue in the coming decades. These approaches implicitly assume either that the rivers are in dynamic equilibrium, or that they are in disequilibrium but that past trends and patterns of change will continue in the future. Their predictions may be inaccurate if, in fact, the Jamuna, Padma and Lower Meghna rivers are in the process of *adjustment* rather than dynamic equilibrium or disequilibrium.

Brammer (1995) first suggested that the observed widening in the upstream reach of the Jamuna River was probably due to the effects of the 1950 Assam earthquake. More recently, EGIS (2000) and Goodbred et al. (2003) attempted to relate observed morphological changes, such as the widening of the Jamuna River, to the propagation of coarse sediment generated by the 1950 Assam earthquake. They suggested a possible process for transporting the huge amount of coarse sediment through the fluvial system, but did not provide any evidence to support their arguments.

In view of the immense suffering to hundreds of thousands of people and the significant loss of land and infrastructure caused by river instability during the past three decades, research was initiated on morphological changes (at the macroscale) of the main rivers of Bangladesh with the aim of identifying the main causes of channel change. This paper reports the findings of that research, which provides evidence to support the hypothesis that changes in the Jamuna, Padma and Lower Meghna rivers may be attributed to disturbance of the system by the 1950 Assam earthquake. The paper also presents a conceptual process–response model to explain the ongoing morphological changes in the river system that, once validated, could provide the basis on which to predict the future morphological development of the system.

HYPOTHESIS

Working hypothesis

In 1950, an earthquake of magnitude 8.6 occurred in the Indian State of Assam. The earthquake delivered approximately 45×10^9 m^3 of sediment to the Brahmaputra and its tributaries in the Assam valley, raising the riverbed at Dibrugarh by 3 m. The fine fraction of this huge sediment input passed quickly to the Bay of Bengal, but the coarser fraction, moving as bed material load, has taken decades to travel downstream as a sediment wave. Passage of the wave through the system generates a particular sequence of morphological adjustments through changes in bed elevation, width, braiding intensity and planform pattern. The same sequence of morphological responses to the passage of the sediment wave may be observed in the Jamuna, Padma and Lower Meghna rivers, but the timing is lagged with increasing distance from Assam, reflecting downstream propagation of the sediment wave.

Celerity of the disturbance in a riverbed

The theory of river mechanics provides a basis from which to approximate the maximum speed at which the sediment wave generated by the Assam earthquake could propagate downstream. The celerity of a small disturbance in the riverbed can be expressed as:

$$C = \frac{ns}{h(1 - Fr^2)} \cong \frac{ns}{q} u \qquad (1)$$

where u is depth-averaged velocity (m s^{-1}), n is the exponent of 'u' in the simplified sediment transport equation (–), s is sediment transport per unit width of a channel (m^2 s^{-1}), h is water depth (m) and q is discharge per unit channel width (m^2 s^{-1}). Equation 1 provides an upper bound to the speed with which the disturbance of the bed associated with a sediment wave introduced in the Assam valley could travel downstream through the Jamuna–Padma–Lower Meghna system.

To apply Eq. 1 to the Jamuna, it is necessary to specify representative values for the discharge, mean velocity, bed material transport rate and velocity exponent in the sediment transport equation. The mean discharge of the Jamuna River at Bahadurabad is 20,200 m^3 s^{-1} (Table 1). The continuity equation, together with representative values for channel width and depth, is used to calculate

Table 1 Physical characteristics of the main rivers of Bangladesh

Parameters	Jamuna (Bahadurabad)	Ganges (Hard. Bridge)	Padma	Upper Meghna (B. Bazar)	Lower Meghna (Chandpur)
Catchment area (10^3 km^2)	573	1000	1573	77	1650
Mean annual discharge (m^3 s^{-1})	20,200	11,300	30,000	4600	~34,000
Mean annual flood (m^3 s^{-1})	70,000	52,000	95,000	13,700	–
Slope (cm km^{-1})	8.5–6.5	5	5	2	5
Mean annual sediment load (Mt yr^{-1})	590	550	900	13	–
Mean annual bed material load (Mt yr^{-1})	200	195	370	–	–
Median bed material size (D_{50}) (mm)	0.20	0.15	0.12	0.14	0.09

a typical mean velocity for the river. According to Delft Hydraulics & DHI (1996a), average annual bed material load in the Jamuna River during the 1960s was about 200 Mt, falling to approximately 100 Mt in the 1980s. Delft Hydraulics & DHI (1996b) also estimated the value of 'n' to be 3.66 for the Jamuna River.

Substituting the above data into Eq. 1 gives $C = 32$ km yr^{-1} during the 1960s, falling to 16 km yr^{-1} during the 1980s. The wave celerity in the Brahmaputra in Assam would be close to or higher than 32 km yr^{-1}, while typical values for the Padma and Lower Meghna rivers would be similar to or less than that estimated for the Jamuna River.

Although these results are only indicative at best, they serve to demonstrate that a wave of coarse sediment introduced into the fluvial system in Assam would, according to the theory of river mechanics, require not less than 30 yr, and probably about 50 yr, to travel the 1000 km from Assam to the Bay of Bengal. This establishes a basis from which to examine the record of channel change in the Jamuna–Padma–Lower Meghna system since the 1960s for evidence of morphological response to the propagation of a sediment wave downstream from Assam.

THE ASSAM EARTHQUAKE OF 1950

At just after 7:30 pm on 15 August 1950, what was at that time the strongest earthquake ever recorded occurred in the Indian State of Assam. The magnitude of the earthquake was 8.6 on the Richter scale. The epicentre was near the Chinese border at latitude 28.6°N, longitude 96.5°E, and the focus was 14 km below the Earth's surface. About 52,000 km^2 of territory in Assam were seriously affected by the tremors and subsequent floods (Tillotson, 1951).

Kingdon-Ward (1951) provided a vivid description of the earthquake based on his direct observations of its effects from a vantage point on the left bank of the Lohit River at approximately latitude 28.5°N, longitude 97°E, 'the first feeling of bewilderment—an incredulous astonishment that these solid-looking hills were in the grip of a force which shook them as a terrier shakes a rat—soon gave place to stark terror.... The destruction extended to the very top of the main ranges—15,000–16,000 ft. above sea-level.... No wonder the mountain torrents begun to flow intermittently as the gorges became blocked, followed later by the breaking of the dam; whereupon a wall of water 20 ft. high would roar down the gulley, carrying everything before it and leaving a trail of evil-smelling gray mud.'

The extent of the area affected by severe landslides was 15,550 km^2 and the estimated volume of landslide debris was about 45×10^9 m^3 (Verghese, 1999). Extensive landslides on the Himalayan slopes temporarily blocked the courses of the Dibang, Dihang and Subansiri rivers (Fig. 1). Bursting of these dams, 3 to 4 days later, released huge amounts of water and caused devastating floods downstream (Goswami, 1985). Immediately following the earthquake the riverbed was raised by 1.5 m at Dibrugarh (Fig. 1), and in the months and years following the earthquake a vast amount of debris and sediment continued to pour into the Brahmaputra River. As a result, the bed of the river at Dibrugarh had risen by more than 3 m by 1955, altering the regime of the river (Krug, 1957; Verghese, 1999). Based on the rapid decline in the rate of aggradation during the mid-1950s, Krug concluded that the riverbed probably reached its maximum elevation in 1956. Krug did not have any data for changes in the Jamuna River, but he reported his suspicion that within a few years the effects of the earthquake would be observed in the downstream reaches and that a few decades would be required for the huge volume of sediment to be transported through the river system. Similarly, Verghese (1999) noted that fine sediment generated by the earthquake had been washed downstream rapidly, but coarser material had been transported slowly.

Working in the 1960s, Brammer (personal communication, 2000) learned from interviews with local inhabitants that an unusually large amount of sedimentation had occurred on the Jamuna floodplain around Dhaleswari during the monsoon flood of 1950. He also identified rapid accretion in the Meghna Estuary in the southern part of Noakhali District from aerial photographs taken in 1952. These observations provide evidence to support Verghese's (1999) statement concerning the rapidity with which fine sediment moved through the system. On the basis of a rise in peak water levels in the Jamuna River at Bahadurabad, detected in hydrological surveys conducted during the late 1960s, Brammer also suspected that this might be attributable to raising of the channel bed by the coarse sediment generated by the Assam earthquake.

DATA SOURCES AND PROCESSING

A wide variety of data from different sources were used to establish the timing, spatial distribution and sequencing of geomorphological change in the Jamuna–Padma–Lower Meghna fluvial system over the past four decades. Information on the Brahmaputra River in Assam was extracted from published sources such as Goswami (1985). Data on water levels, discharges, cross-sectional changes and sediment loads for the river system within Bangladesh were assembled from gauging and river survey records provided by the Bangladesh Water Development Board (BWDB). Changes in the planform of the river system were obtained using a time-series of optical satellite images between 1973 and 2000.

Temporal resolution of the images used in the research varies from river to river. For the Jamuna, 17 images were used to study channel changes over the 27-yr record period, while six and seven images were used for the Padma and Lower Meghna, respectively. The spatial resolution of the images from the 1970s and 1980s is 80×80 m, improving to 30×30 m for the 1990s. All images used in the research were acquired between January and March when the flow is at its lowest (Fig. 2) and the banks and bars are clearly visible. This facilitates comparison of the planform features observed in different years. Image processing and geo-referencing were performed at CEGIS.

It is accepted by most researchers that the Jamuna, Padma and Lower Meghna Rivers may all be classified as braided (Coleman, 1969; Thorne et al., 1993; Thorne, 1997; Ashworth et al., 2000). It was, however, argued by Bristow (1987) that the rivers actually consist of multiple channels of different sizes (hundreds of metres to kilometres in width), each of which displays a different pattern (braided, meandering or anastomosing). To define the overall channel width in a consistent fashion, while recognizing the complexity of planforms exhibited by the rivers, a special technique developed in the mid-1990s was used to delineate the bank lines (Fig. 3A & B). The technique is fully reported in Hassan et al. (1997) and EGIS (1997). The approach of Howard et al. (1970) was then used to calculate the braiding intensity in the dry season satellite images (Fig. 3B).

MORPHOLOGICAL CHANGES

Changes in sediment load

There are no continuous, long-term measurement records of sediment transport in the Brahmaputra–Padma–Lower Meghna system. Hence, to determine the history of sediment transport, reference can only be made to a few short periods of measurement at just two locations. Goswami (1985) reported that measured total sediment loads in the Brahmaputra at Pandu, about 390 km downstream of Dibrugarh, were very high during the period 1955 to 1963, but relatively low from 1969 to 1976 (Fig. 4A).

A similar pattern of decreasing sediment trans-port in the Jamuna at Bahadurabad, about 270 km downstream of Pandu, was noted by Delft Hydraulics & DHI (1996a). They analysed sediment measurements performed by BWDB, to show that sediment transport and especially bed material load was very high in the late 1960s but gradually decreased in the 1970s and 1980s (Fig. 4B). Comparison of bed material load rating curves in Fig. 4B indicates that by the late 1980s the bed material load was only about 40% of its 1960s value. Delft Hydraulics & DHI (1996a) reported that sediment transport decreased further to a minimum in the early 1990s, and has since increased somewhat, although the data to support this statement were not presented.

Changes in bed elevation

Krug (1957) noted that the bed of the Brahmaputra at Dibrugarh rose by more than 3 m by the mid-1950s due to the accumulation of sediment supplied during and after the Assam earthquake. Raising of the riverbed at Pandu (390 km downstream of Dibrugarh) as a result of sediment influx from the Assam earthquake was inferred by Goswami (1985) through specific gauge analysis (Fig. 5A). Records suggest that the riverbed rose by 2 m, reaching its

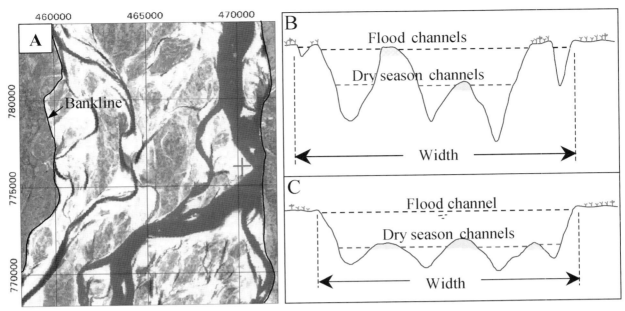

Fig. 3 (A) Bank-lines and dry season channels of the Jamuna River as shown in Landsat TM images 1997. (B) A typical cross-section of the Jamuna River; braiding intensity was determined based on the dry season channel. (C) Another type of cross-section of a braided river, as defined by Schumm (1977), i.e. a single channel bed-load river which at low water has islands of sediment or relatively permanent vegetated islands, and has a single flood channel and a number of dry season channels, in contrast to multiple channel rivers.

highest level in 1969 and degrading through the 1970s. A specific gauge analysis of the Jamuna River at Bahadurabad (660 km downstream of Dibrugarh) indicates aggradation during the 1960s and early 1970s followed by degradation (Fig. 5B). The bed was raised in total by about 1 m, reaching its highest elevation in 1974. The period between the mid-1970s and 1990 was characterized by degradation that returned the bed close to its elevation in the early 1960s. The 1990s feature a second aggradational trend at Bahadurabad. This pattern of aggradation, followed by degradation and then a shorter period of aggradation, in Fig. 5B is similar to the damped oscillation in bed elevations observed in disturbed fluvial systems in the USA by Simon (1989).

These records of bed elevation change at three locations supply two pieces of evidence that support the working hypothesis concerning morphological responses to the earthquake. First, they establish that the magnitude of bed response is at a maximum at Dibrugarh, close to the point where most of the earthquake-derived sediment entered the Brahmaputra River, and that the magnitude of riverbed response decreases downstream to 2 m at Pandu and 1 m at Bahadurabad. This finding is consistent both with Simon's (1989) empirical observation that maximum bed-level changes occur close to the point of greatest disturbance and with different analytical solutions. These suggest that the magnitude of bed elevation rise due to overloading declines downstream with a parabolic/hyperbolic pattern (Ribberink & Van Der Sande, 1985) for aggradation in the rivers due to overloading. Second, the records reveal a clear downstream phase lag in the timing of bed-level rises. The riverbed reached its maximum elevation in 1956 at Dibrugarh, 1969 at Pandu and 1974 at Bahadurabad.

Changes in width

The way that the width of an alluvial river is defined may vary, depending on the planform of the river, the type of data available (historical maps, cross-sections, aerial photographs, satellite images) and the criteria used to delineate the bank lines. While every effort has been made to use a consistent basis for comparing widths and width changes, this must be borne in mind when assessing the evidence for width change in the Brahmaputra–Padma–Lower Meghna system.

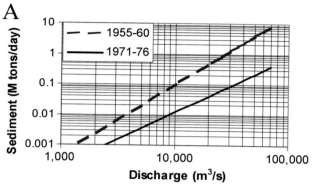

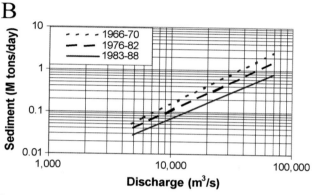

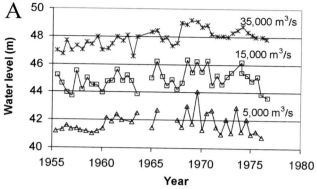

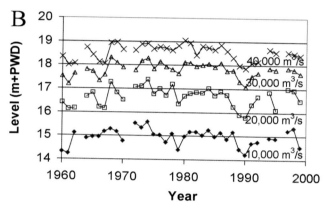

Fig. 4 (A) Changes in the relation between sediment (bed material + wash load) and water discharges of the Brahmaputra River at Pandu, and (B) changes in the relation between sediment (bed material) and water discharges in the Jamuna River at Bahadurabad.

Fig. 5 (A) Specific gauge analysis of the Brahmaputra River at Pandu. (B) Specific gauge analysis of the Jamuna River at Bahadurabad.

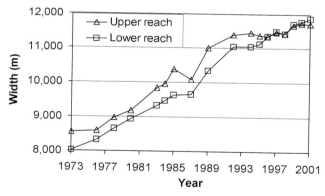

Fig. 6 Changes in the reach-averaged width of the Jamuna over time.

Goswami (1985) derived changes in the width of the Brahmaputra in Assam by analysing cross-sections surveyed between Dibrugarh and Bessamara for the years 1971, 1977 and 1981. He found an average widening rate of 220 m yr^{-1} during 1971–77, which reduced to 37 m yr^{-1} during 1977–81. Unfortunately, no information is available on widening prior to 1971.

In this study, the time series of satellite images was used to establish rates of widening for the river system in Bangladesh. For width analysis, the Jamuna was divided into two 120 km long reaches. The upper reach extends from the Indian border to Kazipur and the lower reach from Kazipur to the Ganges confluence. Both reaches widened throughout the period 1973 to 2000 (Fig. 6). Between 1973 and 1992, the upper reach widened at a rate of about 150 m yr^{-1}, but the rate decreased to only 30 m yr^{-1} after 1992. The lower reach has been widening at about 140 m yr^{-1} throughout the period 1973–2000. However, it

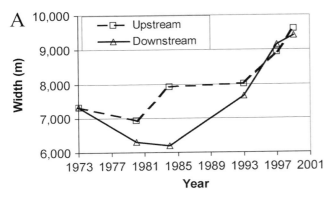

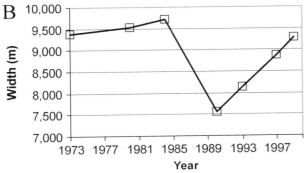

Fig. 7 Changes in the reach-averaged width of the (A) Padma and (B) Lower Meghna over time.

appears that the rate of widening in the lower reach has decreased markedly since 2000.

No information is currently available to the authors concerning the date when rapid widening of the Brahmaputra and Jamuna Rivers began. However, it is possible to infer a downstream phase lag from the dates at which rapid widening ceased in the three reaches. The time lag in the reduction in widening rate between Dibrugarh to Bessamara and the upper Jamuna reaches is about 15 yr. The river distance between these reaches is about 620 km. The time lag between the upper and lower reaches of the Jamuna is 8 yr. The distance between the central points of these reaches is 120 km.

In contrast to the Brahmaputra and Jamuna, information is available concerning the date when rapid widening began in the Padma and Lower Meghna. Satellite images show that the upstream 50 km of the Padma River started widening in 1980, while the downstream reach has been widening since 1984 (Fig. 7A) the Lower Meghna began widening in 1990 (Fig. 7B). Both rivers continue to widen rapidly up to the present day. Interestingly, satellite images for both the Padma and the Lower Meghna reveal that both rivers actually narrowed significantly immediately prior to the onset of widening. Narrowing occurred in both reaches of the Padma during the 1970s and during the 1980s in the Lower Meghna (Fig. 7A and B). The dates of the onset of both narrowing and widening episodes in these rivers display a downstream phase lag that is consistent with the passage of a disturbance in the fluvial system.

It is possible that the Jamuna River also narrowed during the 1960s prior to widening rapidly during the 1970s and 1980s. Although data to support or refute this proposition are sparse, one piece of supporting evidence comes from EGIS (1997). Based on cartographic analysis of maps published in 1955, they estimated the average width of the Jamuna River to be 9.5 km, which is noticeably greater than the average width of 8.3 km in the satellite image of 1973.

Comparing the records of width change in the river system to the changes in sediment transport and bed level reported earlier, it is interesting to note that widening tends to continue even after sediment transport has decreased to a post-high minimum and riverbed levels have stabilized. This suggests that major adjustments to sediment transport and bed level may be achieved more rapidly than those to width.

Changes in braiding intensity

The intensity of braiding is an important morphological parameter in rivers with multi-thread planforms. The braiding index defined by Howard *et al.* (1970) has been found to be a suitable indicator of the degree of division of the flow by bars and islands (Bridge, 1993; Thorne, 1997). As braiding intensity is strongly stage-dependent (see Fig. 3B & C) (Bridge, 1993; Nanson & Knighton, 1996), indices for different dates can be compared only if they represent similar stages. Satellite images of the rivers at similar stages were used in this study, which provide a reliable record of morphological changes over time.

In the Jamuna River, the braiding intensity of the upper reach increased throughout the early 1970s and 1980s, before declining markedly in the 1990s, from 1992 onwards (Fig. 8A). In contrast, braiding intensity declined in the lower reach during the

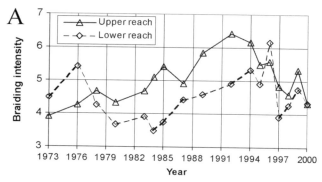
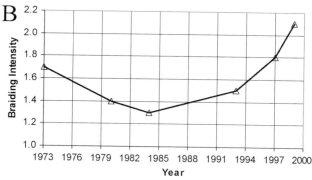
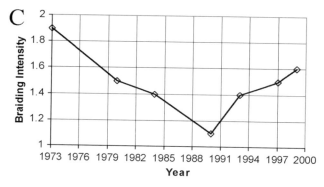

Fig. 8 Changing in the braiding intensity of the (A) Jamuna, (B) Padma and (C) Lower Meghna rivers.

1970s, reaching a minimum in the early 1980s, after which braiding intensity increased, peaking in 1996. In the late 1990s braiding became less intense in both reaches, resulting in their convergence to a common value in 2000.

The Padma and Lower Meghna rivers are much less intensively braided than the Jamuna (Fig. 8B & C). Typical braiding indices for the Jamuna lie in the range 4 to 6, while those for the Padma and Lower Meghna very rarely exceed 2, and both rivers have approached a straight, single-channel planform configuration on one occasion during the past three decades. The Padma River became concentrated in a single channel in 1984, while adoption of this pattern in the Lower Meghna occurred 6 yr later, in 1990. The record of change in the Padma is similar to that in the lower reach of the Jamuna, with braiding intensity first decreasing and then increasing (Fig. 8B). The pattern in the Lower Meghna is also similar, but features a downstream phase lag of 6 yr (Fig. 8C).

Comparison of the changes in braiding intensity with available records for sediment transport and bed-level changes suggests that when sediment transport started to decrease the braiding intensity of the river started to increase. In many braided rivers, changes in braiding intensity are positively correlated with changes in width. This is also the case in the Padma and Lower Meghna Rivers, which have relatively simple planforms. It is interesting to note, however, that the intense braiding and great complexity of the planform of the Jamuna River results in situations where the braiding index decreases even while the width is increasing (compare Figs 6 & 8A. for the lower reach) thus changes in braiding intensity need not always be associated with changes in width.

Celerity of the observed morphological changes

The record of observed morphological change in the river system recounted above allows calculation of the celerity with which turning points in the trends of bed-level change, width adjustment and braiding index alteration have migrated downstream. The results are summarized in Table 2 for different reaches of the Brahmaputra–Jamuna–Padma–Lower Meghna system. The celerity of changes in the sediment flux cannot be estimated as the records for Pandu and Bahadurabad are insufficiently long to include any turning points.

Observed values of celerity are much higher for the Brahmaputra River, where events occurred earlier, and lower for the Jamuna, Padma and Lower Meghna rivers, where events occurred later. These results are credible in that average observed celerity values for the Brahmaputra (45 km yr^{-1}) and the Jamuna–Padma–Lower Meghna (13 km yr^{-1}) are of the same order of magnitude as those calculated from the theory of river mechanics (32 and 16 km yr^{-1}).

Table 2 Celerity of observed morphological changes in the river system

Reach	Celerity of morphological change (km yr^{-1})		
	Bed level	Width	Braiding index
Dibrugarh to Pandu	43		
Bessamara to Bahadurabad		37	
Pandu to Bahadurabad	54		
Upper Jamuna to Lower Jamuna		17	15
Upper Padma to Lower Padma		12	
Padma to Lower Meghna		10	12

CONCEPTUAL PROCESS–RESPONSE MODEL

Background

Based on the available accounts of the volume and timing of sediment introduced to the Brahmaputra River by the Assam earthquake and the observed changes in sediment transport, bed level, width and braiding intensity in the river system downstream, a conceptual process–response model was constructed to explain how morphological adjustments may be linked to downstream propagation of a sediment wave. The conceptual model (Fig. 9) presented here builds on and extends the approaches of previous authors, particularly Lane (1955), Schumm (1969) and Simon (1989). The upper reach of the Jamuna River around Bahadurabad provides the basis for model development as it has the most complete record of bed material load and channel adjustments (Figs 4B, 5B, 6 & 8A).

Slope adjustments are often considered to be a key element of dynamic process-response and have been used to describe why complex response occurs in disturbed fluvial systems (Schumm, 1977). However, in the rivers studied in this project the huge scale of their channel and floodplain systems diminishes the significance of slope adjustment in driving morphological response. This point may be illustrated by considering the slope adjustment associated with bed-level changes at Pandu and Bahadurabad. In 1969 the riverbed had aggraded by 2 m at Pandu (Fig. 5A) and 0.5 m at Bahadurabad (Fig. 5B). This indicates a change in bed gradient of $1.5/270 \times 10^3$ or 0.5 cm km^{-1} in the reach from Pandu to Bahadurabad. As the slope of the upper reach of the Jamuna River is 8.5 cm km^{-1} (Table 1), the increase in slope was only about 5%. Given the great variability in local downstream and cross-stream slopes observed in braided rivers, it seems unlikely that such a small change would have a significant impact on the fluvial processes responsible for driving channel changes. Changes in slope in the system downstream of Bahadurabad would be even smaller, given that the maximum change in bed level was only 1 m. Consequently, slope was not considered as a variable in the conceptual process–response model.

Similarly, changes in the characteristic size of the bed material are often found to be significant in influencing process-response in disturbed rivers, especially those with gravel-beds (Simon & Thorne, 1996). The study rivers are characterized by sand beds, with only a slight downstream fining trend (Table 1). Recorded changes in bed material grain size of the Jamuna River during the 1960s to the 1990s are insignificant and certainly would not significantly alter hydraulic roughness or mobility of bed material. Consequently, change in bed material size was not considered in the model.

The contribution of the wash load to promoting morphological change is also neglected, and variations in the supply of bed material load to a reach are considered to drive morphological adjustments. As recognized by Knighton (1998), morphological response is not instantaneous and a reaction time

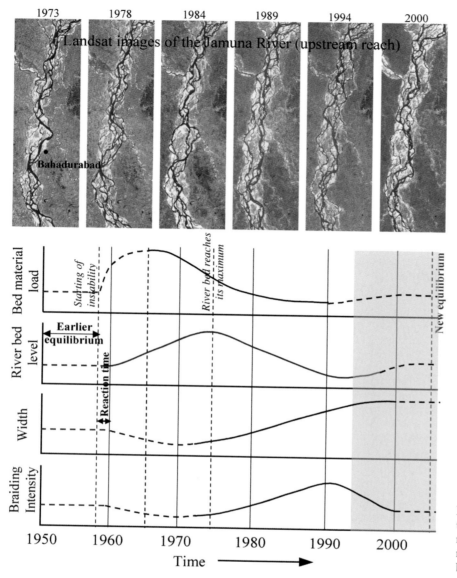

Fig. 9 Conceptual model showing the propagation of the different morphological changes in the upper reach of the Jamuna River around Bahadurabad over time.

is included between the time that the input bed material load changes and the time that channel adjustments become detectable.

Upper Jamuna around Bahadurabad

Unfortunately, the sediment transport record at Bahadurabad does not extend back beyond 1966, by which time the input of bed material load was already high (Fig. 4B). Hence, it is necessary to assume a date for the arrival of the leading edge of the bed material load pulse at Bahadurabad. The available data show that the bed material load peaked in the late-1960s and then decreased until the end of the 1980s (Fig. 4B).

The specific gauge analysis at Bahadurabad (Fig. 5B) indicates that the bed started to rise in the early 1960s because the increased input of bed material load was greater than the local transport capacity. Aggradation continued until the mid-1970s by which time the transport capacity matched the supply (which was by that time decreasing). From the mid-1970s to the late-1980s, the bed level decreased as the bed material input fell below the available transport capacity. During the 1990s the bed rose again, but to a lesser degree, probably in response to the damped oscillation in the bed material supply.

As mentioned earlier, cartographic analysis of the 1955 map suggests that the Jamuna narrowed during the late-1950s and early-1960s. However, the channel widened rapidly throughout the period 1973 to 1992, but since then the rate of increase in the upper reach has decreased dramatically (Fig. 6).

It is known that the braiding intensity of the upper reach increased throughout the early 1970s and 1980s, before declining markedly after 1992 (Fig. 8A). Based on observations in the lower Jamuna immediately downstream, it is hypothesized that the braiding index in the upper reach decreased slightly in the early 1960s when the width was narrowing. Increasing braiding intensity during the 1970s and 1980s was associated with rapid widening at a time when the bed material supply was decreasing and the bed was degrading. After 1992, when the bed began to aggrade the braiding intensity began to decline.

Padma River

The Jamuna River is not the only source of water and sediment entering the Padma as the Ganges also feeds the Padma (Fig. 1). Records of sediment transport in the Ganges River at Hardinge Bridge, about 100 km upstream of the Padma, show that changes in the load during the 1970s and 1980s were small compared with those in the Jamuna at that time (Delft Hydraulics & DHI, 1996a). Consequently, morphological changes in the Padma can be considered to be attributable to changes in sediment supply from the Jamuna River and the conceptual model for the Padma ignores any influence due to changing sediment inputs from the Ganges (Fig. 10).

Measured bed material loads at Baruria, just downstream of the Ganges–Jamuna confluence, show that input to the Padma was high between 1968–69 and 1976–77, but that during the 1980s and early 1990s it was much reduced (Fig. 11). Bearing in mind that Baruria is located at the upstream end of the 100 km long Padma reach, it would be expected that the reaction time for morphological response in the river would require a few years.

No information on bed levels was available, but good records of changes in width and braiding intensity are. The planform of the Padma River in 1973 indicates that the river responded to the high bed material input by narrowing and decreasing its braiding intensity (Figs 7A & 8B). By 1980, the width and braiding intensity were both approaching their minimum values, at the time that the bed material input to the Padma was probably around its maximum. The satellite image for 1980 shows that as a result of narrowing and a reduction in braiding intensity the channel had adopted a straight, single-thread planform. By 1984 the single-thread channel was responding to the decrease in bed material input by widening in the upper reach and showing a tendency towards meandering. According to the conceptual model, the riverbed reached its highest level at this time, although no bed-level data are available to verify this.

The image for 1993 shows that the upper reach of the Padma had widened through braiding while the lower reach had widened through meandering. The measured bed material load input at this time was low and probably approaching its minimum. According to the model the riverbed would have been lowering at this time. The image for 1999 shows that the width has increased through intensified braiding along the whole of the Padma. According to the model both bed material supply and bed level are at their minimum values in 1999, though they are expected to increase slightly due to the oscillating behaviour displayed by disturbed fluvial systems.

Lower Meghna River

The Lower Meghna River extends about 90 km from Chandpur to the northern tip of Hatia Island. Only the upper 50 km of the river is truly fluvial as the lower 40 km are strongly influenced by tidal processes. Therefore, application of the conceptual model is limited to the first 50 km of the river.

No sediment transport data are available for the Lower Meghna River and only limited information on bed-level changes, from the Meghna Estuary Study (MES II), has been obtained to date. The model is therefore applied primarily on the basis of the time series of satellite images (Fig. 12).

The 1973 image shows a relatively narrow, single-thread pattern immediately downstream of Chandpur, followed by a braided reach (Fig. 12). Both channels in the multi-threaded reach have well-

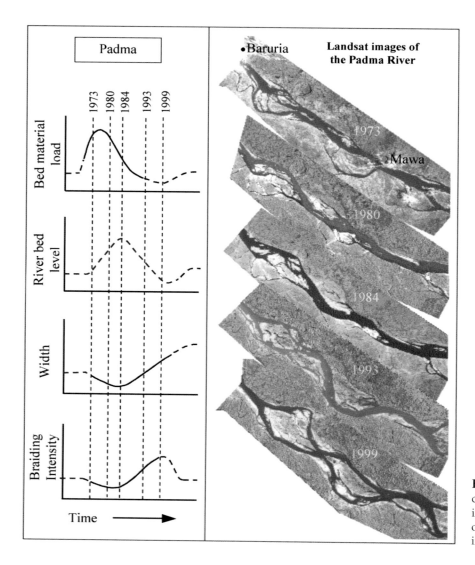

Fig. 10 Comparison between the conceptual model and the changes in the planform of the Padma as observed in the time-series satellite images.

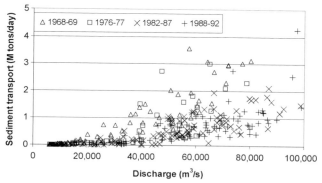

Fig. 11 Relation between sediment transport and discharge in the Padma River at Baruria.

defined channels at this time, suggesting a degree of stability. However, in the 1980 image, the left bank anabranch has enlarged at the expense of the right bank anabranch, resulting in decreases in both width and braiding intensity. In the conceptual model, these changes are interpreted as signalling that the wave of coarse sediment flux from the Assam earthquake reached the Lower Meghna River in the mid-1970s.

By 1990, the channel had become single-threaded throughout its length, reducing its width and braiding intensity to minimum values. According to the conceptual model the bed should have been aggrading to its highest level at this time. There are very few observations to confirm or refute this, although a small amount of data from the Lower

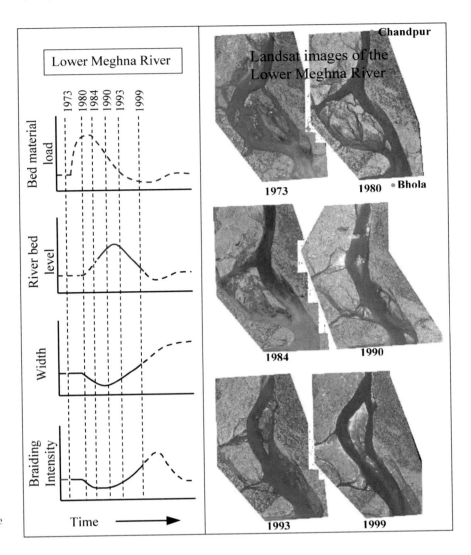

Fig. 12 Comparison between the conceptual model and the changes in the planform of the Lower Meghna as observed in the time-series satellite images.

Meghna River between Bhola and Gajaria (in the tidal reach immediately downstream of the reach shown in Fig. 12) does show net accumulation of 140×10^6 m^3 of sediment, indicating bed accretion at about 20 cm yr^{-1} around 1990 (Table 3).

By 1993 the river immediately downstream of Chandpur had begun to widen and increase its braiding intensity, processes that continued through to 1999. According to the conceptual model, the bed of the river should have been lowering through degradation during this period. The Meghna Estuary Studies (MES and MES II) bathymetric surveys of the Lower Meghna River between Chandpur to north of Hatia Island in 1997 and 2000 provide some support for this supposition (Table 3). The records show that net erosion of 400×10^6 m^3 occurred in the 90 km long reach during a 3-yr period, indicating degradation at 12 cm yr^{-1}.

DISCUSSION

Comparison of the conceptual model with previous process–response models

A major difference between the conceptual model derived here and previous process–response models lies in the initial morphological response to an increase in coarse sediment supply (bed material load input). It is proposed here that the increase of bed material load associated with the arrival of the wave of coarse sediment generated by the

Table 3 Estimated erosion/accretion in the Lower Meghna River (source: DHV Consultants BV, 2001)

Period	Reach	Reach length (km)	Data source	Accretion (+) or erosion (−) ($\times 10^6$ m^3)	Accretion/erosion rate (cm yr^{-1})
1986–1992	North Bhola to Gazaria	20	LRP	140	20
1997–2000	Chandpur to north of Hatia	90	MES and MES II	−400	−12

Assam earthquake generated bed aggradation, width *reduction* and a *decrease* in braiding intensity. This hypothesis contrasts with the model proposed by Schumm (1977; Fig. 3C), in which an increase of sediment input generates *increases* in width and braiding intensity. Also, Germanoski & Schumm (1993) observed that the number of braid bars, and hence the braiding intensity, was higher in an aggrading system than in equilibrium or degrading systems.

This apparent paradox may be resolved by considering differences between the planform configurations when the sediment wave arrives and the mechanism by which the morphology responds. In the Brahmaputra–Jamuna–Padma–Lower Meghna system, the sediment wave entered rivers that were already braided and, in the case of the Brahmaputra and upper reach of the Jamuna, intensely so. Aggradation, narrowing and a decrease in braiding intensity were achieved through siltation and abandonment of the smaller anabranches to raise the average elevation of the bed, reduce the width and lessen anabranch channels, simultaneously (Fig. 13). The situation conceptualized in Schumm's model is quite different, envisaging an initially single-thread or weakly braided channel that widens by eroding its banks, becoming more braided through aggradation that builds medial bars. The importance of the starting configuration in conditioning the initial morphological response to an increase in sediment supply can be illustrated by comparing the response of the Toutle and Cowlitz Rivers to the massive sediment input following the volcanic eruption of Mount St Helen (Bradley, 1983; Simon, 1992) to what happened in the upper Jamuna. Prior to the eruption, the Toutle and Cowlitz were meandering rivers and they responded by aggrading, widening and braiding.

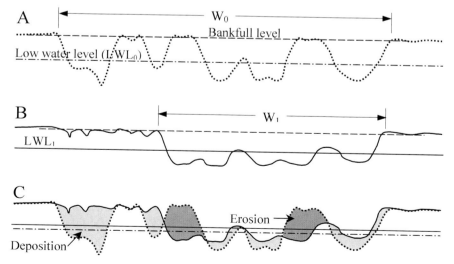

Fig. 13 Adjustment of a river cross-section (hypothetical) with an increase of bed material load. (A) Cross-section prior to the onset of the sediment wave. (B) Cross-section during the high bed material load ($W_0 > W_1$ and $LWL_1 > LWL_0$). (C) Sediment deposition and erosion during the adjustment (deposition > erosion). The process reverses when the bed material load decreases from its peak.

The Jamuna was already intensely braided and it responded by silting up its smaller anabranches, narrowing and becoming less braided.

In fact, there is no conflict between the conceptual model of Schumm and that proposed here. It is envisaged that, in the Jamuna, Schumm's (1977) morphological response model applies within individual anabranches, which did indeed start to aggrade and become more braided immediately after the sand wave arrived, due to sediment overloading. However, the existence of a higher order of braiding meant that instead of concentrating flow against the banks and driving anabranch widening, choking of the smaller anabranches with sediment simply diverted discharge from them into the larger anabranches, leading to their abandonment.

In the conceptual model, it is proposed that channel narrowing and reduced division of discharge as the flow concentrates into fewer, larger anabranches leads to an increased sediment transport capacity. This is consistent with the well-established rule that sediment transport capacity increases as a non-linear function of discharge. For example, bed material transport capacity is often found to vary as a power function of discharge in the form:

$$Q_s = aQ^b \qquad (2)$$

where Q is discharge, Q_s is bed material load, and a and b are empirically derived constants. Measurements of sand load reveal that the exponent b in Eq. (2) varies between 1.3 and 2.5 in the main rivers of Bangladesh (Deft Hydraulics & DHI, 1996a). Taking a conservative assumption that $b = 1.5$, conversion of a braided channel with two anabranches into a single-thread channel with the same discharge would increase the capacity to transport sand by 40%. If $b = 2.0$, which is not unreasonable, the transport capacity of the single-thread channel is double that of two anabranches. Although these calculations are purely indicative, Eq. (2) provides a rational basis for proposing that sediment transport increases with decreasing braid intensity in the conceptual model.

In the model, the bed continues to aggrade until the increasing bed material load transport capacity of the reach matches the bed material load input from upstream. This occurs after the coarse sediment input has peaked, as the supply declines with the approach of the trailing edge of the earthquake-related sediment wave. At this stage, the bed elevation momentarily stabilizes, but subsequently the declining coarse sediment input no longer matches local transport capacity and the channel begins to degrade.

During this degradational phase, width and braiding intensity increase because the enlarging channels erode and undercut their banks as well as their beds, attacking the floodplain at the margins of the braid belt and converting braid bars to semi-permanent islands, which are termed chars in Bangladesh. This phase of increasing braiding intensity continues for several years until the bed material transport capacity decreases sufficiently to match the supply of coarse sediment from upstream. The increase in width and braiding intensity observed in the upper Jamuna in 1970s and 1980s was indeed achieved through the development of the multiple channels separated by large chars. The process involved the development of anabranches with different planforms (sinuous, meandering and braided) leading some investigators to define the Jamuna as anastomosing at this time (Bristow, 1987). Certainly, the mechanism by which an increase in braiding intensity is achieved is manifestly different to that described by Schumm (1977) as occurring in response to an increase in sediment supply.

Complex response in the latter stage of the conceptual model

In the conceptual model, a further phase of morphological adjustment is proposed, due to a second increase in coarse sediment supply. This proposal stems from theories and observations put forward by (amongst others) Hey (1979), Simon (1989, 1992) and Richards & Lane (1997) that, following a single major disturbance, instability in the fluvial system does not decay monotonically, but follows a damped oscillation.

Evidence to support complex response in the rivers studied here comes from the marked reduction in the braiding intensity observed in the upper Jamuna during the 1990s (Fig. 8A), which shows that adjustments are not unidirectional and is consistent with morphological response to a secondary increase in sand load supply from upstream. In the model it is hypothesized that this second pulse of sand is much smaller than the first, earthquake-derived wave, and is likely to have reached its maximum at about the Millennium.

This part of the conceptual model is actually shaded in Fig. 9, because it is in fact a 'grey area'. Its veracity may be confirmed or refuted as time passes, further data become available and the patterns of ongoing morphological change are revealed. However, in predicting future change and evolution of the upper Jamuna, it must be borne in mind that the intensity and extent of morphological adjustments associated with this second, damped, oscillation in sand load will be much smaller than those driven by the first wave. In practice, morphological responses to the waning impact of the Assam earthquake are likely to be subsumed within responses to more recent and current drivers of channel change, including: neotectonics (warping, subsidence, lineament development), which may be altering the valley slope; engineering interventions such as the Jamuna Multi-purpose Bridge; and changes in the flood regime during the past few decades due to climate change and the decline of the Old Brahmaputra and Dhaleswari distributaries. The effects of these current drivers are not considered in the conceptual model presented here. However, once validated, the model could be used to predict the morphological response of the river to such disturbances.

Channel patterns and morphological responses

The apparent paradox between the conceptual model and earlier process–response models draws attention to the pre-disturbance planform pattern of the river as being crucial in determining the style and sequence of morphological responses triggered by an increase in the supply of bed material load. This highlights the need to be both accurate and comprehensive when defining channel planform. Early planform definitions, such as those of Leopold & Wolman (1957), classified all multi-threaded rivers as 'braided'. Subsequently, Brice (1975) and Schumm (1977) differentiated between multi-thread streams that were braided (unstable, bed-load dominated) and anastomosing or anabranching (more stable, suspended sediment dominated). Bridge (1993) defined anastomosing channels as those where the length of channel segments exceeds the width of the first-order channel and he pointed out that a braiding index cannot be used to define the intensity of channel division in an anastomosing channel. More recently, observations by Nanson & Knighton (1996) have confirmed that anastomosing rivers do differ from braided ones in that they have less energy, more erosion-resistant silt/clay boundary materials and, as a result, greater stability.

Following the definitions of Bridge (1993) the Lower Jamuna River is divided into braided, meandering and anastomosing. The observed responses in changing the planform with the changes in bed material load are presented in Fig. 14 and Table 4. Aerial photographs from 1967 show a predominantly single-thread, meandering channel, with some braiding. Many recently abandoned anabranches can be distinguished. This pattern is consistent with aggradation, width reduction and reduced braiding due to siltation of second- and third-order channels due to sediment overloading when the bed material wave entered this reach in the early to mid-1960s. This situation is analogous to that in the Padma in 1973. As the bed material load peaked in the early 1970s, a braided/anastomosed pattern replaced the meandering one (Table 4). Reduction in the bed material supply from its peak value resulted in a return to meandering from 1980. By the 1990s, the whole 120 km of the lower reach of the Jamuna River had become anastomosing, except for the location of the Jamuna Bridge, where the width of the river is constrained by guide bunds to 4.5 km.

According to Bridge (1993) the channels of the Padma and Lower Meghna were anastomosed prior to 1973 (Figs 11 & 12). It should be pointed out, however, that these rivers do not meet Nanson & Knighton's (1996) criteria for anastomosing as they are high energy, do not have silt/clay bed and bank materials and are very unstable. The rivers responded to an increase in the supply of bed material load (late-1960s for the Padma, mid-1970s for the Lower Meghna) by switching from a multi- to a single-thread planform through siltation and abandonment of one anabranch and enlargement of the other. When sand load subsequently decreased (early 1980s for the Padma, mid-1990s for the Lower Meghna), the single-thread channel first meandered and then resumed its anastomosed pattern.

The observations reveal a propensity for single-thread, meandering reaches to become anastom-

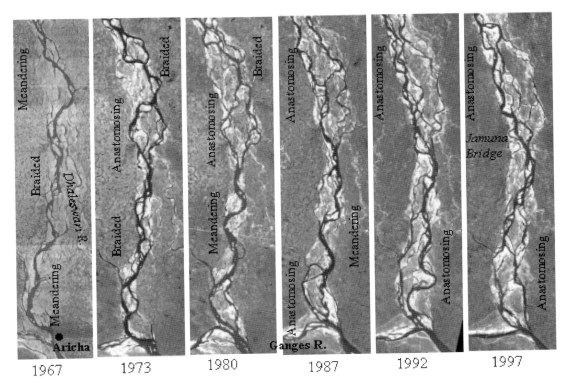

Fig. 14 Planform changes of the downstream reach of the Jamuna over time.

Table 4 Changes in planform of the downstream of the Jamuna River over time

Year	1967	1973	1980	1987	1992	1997
Sediment input	High	Very high	High	Low	Very low	Very low
Length of braiding (km)	40	95	25	–	–	–
Length of meandering (km)	80	–	55	40	10	20*
Length of anastomosing (km)	–	25	40	80	110	100

*Guide bunds of the Jamuna Bridge fixed the banks in 1995.

osed in response to a reduction in sediment load, instead of increasing their sinuosity. This phenomenon is displayed by the lower reach of the Jamuna, the Padma and Lower Meghna. It is most probably related to the relative ease with which sediment recently deposited in abandoned anabranches can be remobilized and the channel re-occupied, compared with the more difficult task of increasing sinuosity through eroding older, consolidated floodplain sediments. Or the high stream power factor (discharge × slope) of these rivers (Fig. 15) restricts the meandering beyond a certain range in the highly erodible floodplain.

SUMMARY AND RECOMMENDATIONS FOR FURTHER RESEARCH

The historical data and analyses developed here suggest that the Brahmaputra–Jamuna–Meghna system is not in a dynamic equilibrium, but has displayed a range of morphological changes during the

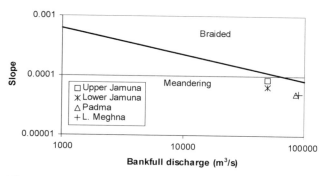

Fig. 15 Position of the Jamuna, Padma and Lower Meghna rivers in the planform classification diagram defined by Leopold & Wolman (1957).

past three decades. It is hypothesized that these changes result from morphological responses to the downstream propagation of a huge amount of relatively coarse sediment generated by the 1950 Assam earthquake. It is envisaged that this sediment has progressed downstream as a wave of bed material load, moving with a celerity in the upper Jamuna of about 30 km yr^{-1} for the peak of the wave to perhaps 16 km yr^{-1} for its trailing edge.

Historical records reveal sequential changes in the bed elevation, width, braiding intensity and planform of the study rivers. The timing of these changes is phase lagged downstream, in reasonable agreement with the estimated celerity of the sediment wave. Based on the working hypothesis and these observations a conceptual model has been developed to explain the process–response mechanisms responsible for the observed changes. The model highlights the importance of the pre-disturbance configuration of the channel in conditioning its response to disturbance. The model further raises issues concerning how the morphological response of multi-thread channels varies depending on whether they are braided or anastomosed.

If validated and verified, the conceptual model could be helpful in predicting future changes in the main rivers of Bangladesh. However, it may be the case that the response to disturbance due to the passage of coarse sediment derived from the Assam earthquake is now largely complete and that future evolution in the rivers will be driven by other more recent drivers such as neo-tectonics, climate change and the decline of distributaries such as the Old Brahmaputra and the Dhaleswari. Further progress in developing the conceptual model depends on acquisition of data on the morphological evolution of the Brahmaputra River in Assam, compilation of aerial photographs from the pre-satellite era to clarify pre-disturbance conditions in the study rivers, further detailed analysis of sediment transport and channel morphology records, and careful inspection of available data from the Meghna Estuary.

ACKNOWLEDGEMENTS

This research is being carried out as part of the PhD study of the first author. The University of Nottingham (UoN) and the Center for Environmental Geographic Information Services (CEGIS) are providing the financial support for the study, without which continuation of the research would not be possible. The support of Dr Erik Mosselman, Delf Hydraulics, The Netherlands, at the initial stage of the research is highly appreciated. The advice of Professor Nick Clifford of UoN and Dr Ainun Nishat of IUCN, Bangladesh during the research is kindly acknowledged. Thanks are due to the anonymous referees whose comments helped to improve the manuscript. Finally the co-operation of Md. Altaf Hossain, ATM Kamal, Nazneen Akhtar, Chitta Ranjan Gupta and Iffat Huque of CEGIS in the different phases of the research is kindly acknowledged.

REFERENCES

Ashworth, P.J., Best, J.L., Roden, J.E., Bristow, C.S. and Klassesn, G.J. (2000) Morphological evolution and dynamics of a large, sand braid-bar, Jamuna River, Bangladesh. *Sedimentology*, **47**, 533–555.

Bradley, J.B (1983) Transition of a meandering river to a braided system due to high sediment concentration flows. In: *River Meandering* (Ed. C.M. Elliott), pp. 89–100. American Society of Civil Engineers, New York.

Brammer, H. (1995) *Geography of Soils of Bangladesh*. The University Press, Dhaka, Bangladesh.

Brice, J.C. (1975) *Air Photo Interpretation of the Form and Behavior of the Alluvial Rivers*. Final Report to the U.S. Army Research Office.

Bridge, J.S. (1993) The interaction between channel geometry, water flow, sediment transport and deposition in braided rivers. In: *Braided Rivers* (Eds J.L. Best and C.S. Bristow), pp. 13–71. Special Publication 75, Geological Society Publishing House, Bath.

Bristow, C.S. (1987) Brahmaputra River: channel migration and deposition. In: *Recent Developments in Fluvial Sedimentology* (Eds F.G. Ethridge, R.M. Flores and M.D. Harvey), pp. 63–74. Special Publication 39, Society of Economic Paleontologists and Mineralogists, Tulsa, OK.

Coleman, J.M. (1969) Brahmaputra River: channel processes and sedimentation. *Sediment. Geol.*, **3**, 129–239.

Delft Hydraulics and DHI (FAP 24) (1996a) *Sediment Rating Curves and Balances*. Special Report No. 18, prepared for Water Resources Planning Organization, Dhaka, Bangladesh.

Delft Hydraulics and DHI (FAP 24) (1996b) *Sediment Transport Predictors*. Special Report No. 13, prepared for Water Resources Planning Organization (WARPO), Dhaka, Bangladesh.

DHV Consultants BV (MES II) (2001) *Analysis of Bathymetric Changes in the Meghna Estuary*. Technical Note MES-032, prepared for Bangladesh Water Development Board, Dhaka, Bangladesh.

EGIS (Environmental and GIS Support Project for Water Sector Planning) (1997) *Morphological Dynamics of the Brahmaputra–Jamuna River*. Prepared for Water Resources Planning Organization, Dhaka, Bangladesh.

EGIS (Environmental and GIS Support Project for Water Sector Planning) (2000) *Riverine Chars in Bangladesh: Environmental Dynamics and Management Issues*. The University Press, Dhaka, Bangladesh, 88 pp.

Germanoski, G. and Schumm, S.A. (1993) Changes in river morphology resulting from aggradation and degradation. *J. Geol.*, **134**, 367–382.

Goodbred, A.L., Kuehl, S.A., Steckler, M.S. and Sarker, M.H. (2003) Controls on facies distribution and stratigraphic preservation in the Ganges–Brahamaputra delta sequence. *Sediment. Geol.*, **155**, 317–342.

Goswami, D.C. (1985) Brahmaputra River, Assam, India: basin denudation and channel aggradation. *Water Resour. Res.*, **21**, 959–978.

Halcrow and Mott MacDonald (2000) *Draft Development Strategy*, Vol. 4, Annex C: *Land and Water Resources*. National Water Management Plan Project. Prepared for Water Resources Organization, Dhaka, Bangladesh.

Hassan, A., Huque, I., Huq, P.A.K., Martin, T.C., Nishat, A. and Thorne, C.R. (1997) Defining the banklines of the Brahmaputra–Jamuna River using satellite imagery. In: *Proceedings of the 3rd Conference on Flood Hydraulics, Stellenbosch, South Africa* (Ed. J. Watts), pp. 349–358. International Association of Hydrological Sciences, Wallingford.

Hey, R.D. (1979) Dynamic process–response model of river channel development. *Earth Surf. Process. Landf.*, **1**, 59–72.

Howard, A.D., Keetch, M.E. and Vincent, C.L. (1970) Topological and geometrical properties of braided streams. *Water Resour. Res.*, **6**, 1974–1988.

Kingdon-Ward, F. (1951) Notes on the Assam earthquake. *Nature*, **167**, 130–131.

Klaassen, G.J. and Masselink, G. (1992) Planform changes of a braided river with fine sand as bed and bank material. *Proceedings of the Fifth International Symposium on River Sedimentation*, Karlsruhe, pp. 459–471.

Klaassen, G.J., Mosselman, E. and Bruhl, H. (1993) On the prediction of planform changes of braided sand-bed rivers. *Proceedings of the International Conference on Hydroscience and Engineering*, Washington, DC, pp. 134–146.

Knighton, A.D. (1998) *Fluvial Forms and Processes: a New Perspective*. Arnold, London.

Krug, J.A. (1957) *United Nations Technical Assistance Programme, Water and Power Development in East Pakistan*, Vol. 1. Report of a United Nations Technical Assistance Mission for the Water and Power Development Board in East Pakistan.

Lane, E.W. (1955) *A Study of the Shape of Channels Formed by Natural Streams Flowing in Erodible Material*. Missouri River Division, Sediment Series No. 9, U.S. Army Division, Missouri River, Corps of Engineers, Omaha, Nebraska.

Leopold, L.B. and Wolman, M.G. (1957) River channel patterns: braided, meandering and straight. *U.S. Geol. Surv. Prof. Pap.*, **262B**, 39–85.

Mosselman, E., Huisink, M., Koomen, E. and Seymonsbergen, A.C. (1995) Morphological changes in a large braided sand-bed river. In: *River Geomorphology* (Ed. E.J. Hickin), pp. 235–249. Wiley, Chichester.

Nanson, G.C. and Knighton, A.D. (1996) Anabranching rivers: their cause, character and classification. *Earth Surf. Process. Landf.*, **21**, 217–239.

Ribberink, J.S. and Van Der Sande, J.T.M. (1985) Aggradation in rivers due to overloading—analytical approaches. *J. Hydraul. Res.*, **23**(3), 273–283.

Richards, K.S. and Lane, S.N. (1997) Prediction of morphological changes in unstable channels. In: *Applied Fluvial Geomorphology for River Engineering and Management* (Eds C.R. Thorne, R.D. Hey and M.D. Newson), pp. 269–292. Wiley, Chichester.

Sarker, M.H., Huque, I., Alam, M. and Koudstaal, R. (2003) Rivers, chars, and char dwellers of Bangladesh. *Int. J. River Basin Manag.*, **1**, 61–80.

Schumm, S.A. (1969) River metamorphosis. *Proc. Am. Soc. Civ. Eng., J. Hydraul. Div.*, **95**, 255–273.

Schumm, S.A. (1977) *The Fluvial System*. Wiley, New York.

Simon, A. (1989) A model of channel response in disturbed alluvial channels. *Earth Surf. Process. Landf.*, **14**, 11–26.

Simon, A. (1992) Energy, time and channel evolution in catastrophically disturbed fluvial system. *Geomorphology*, **5**, 345–372.

Simon, A. and Thorne, C.R. (1996) Channel adjustment of an unstable coarse-grained stream: Opposing trends of boundary and critical shear stress, and the applicability of extremal hypothesis. *Earth Surf. Process. Landf.*, **21**, 155–180.

Thorne, C.R. (1997) Channel types and morphological classification. In: *Applied Fluvial Geomorphology for River Engineering and Management* (Eds C.R. Thorne, R.D. Hey and M.D. Newson), pp. 175–222. Wiley, Chichester.

Thorne, C.R., Russel, A.P.G. and Alam, M.K. (1993) Planform pattern and channel evolution of the Brahmaputra River, Bangladesh. In: *Braided Rivers* (Eds J.L. Best and C.S. Bristow), pp. 257–275. Special Publication 75, Geological Society Publishing House, Bath.

Thorne, C.R., Hossain, M.M. and Russell, A.P.G. (1995) Geomorphic study of bank line movement of the Brahmaputra River in Bangladesh. *J. NOAMI*, **12**, 1–10.

Tillotson, E. (1951) The great Assam earthquake of August 15, 1950. *Nature*, **167**, 128–130.

Verghese, B.G. (1999) *Waters of Hope: from Vision to Reality in Himalaya–Ganga Development Cooperation*. The University Press, Dhaka, Bangladesh.

Use of remote-sensing with two-dimensional hydrodynamic models to assess impacts of hydro-operations on a large, braided, gravel-bed river: Waitaki River, New Zealand

D. MURRAY HICKS*, U. SHANKAR*, M.J. DUNCAN*, M. REBUFFÉ[†] and J. ABERLE[‡]

*National Institute of Water and Atmospheric Research, PO Box 8602, Christchurch, New Zealand
(m.hicks@niwa.co.nz)
[†]ENSAR, 65 Rue de Saint Brieuc, Rennes, France
[‡]Leichtweiss-Institut für Wasserbau—Abteilung Wasserbau und Gewässerschutz—Technische Universität, Braunschweig Beethovenstr., 51a D-38106, Braunschweig, Germany

ABSTRACT

The Waitaki River is a large gravel-bed river draining the eastern slopes of the Southern Alps, South Island, New Zealand. Hydroelectric power development since 1935, including dams, diversion canals and natural lake control, has damped flood runoff and reduced bed-material supplies to the lower, 60-km long, braided segment of river. A recent proposal aims to divert a substantial component of the lower river flow through a canal for further hydroelectric power generation. Investigations for this proposal have used remote-sensing techniques to help assess environmental impacts. Airborne laser scanning (ALS), in combination with vertical aerial photography and multispectral (MS) scanning, was used to survey river topography and ground cover to provide boundary data for a two-dimensional hydrodynamic model. Ground topography was mapped using last-return ALS. Submerged channel-bed topography was obtained by subtracting water-depth mapped from MS imagery from the water surface elevations surveyed by ALS. By comparison with the MS approach, the ALS proved a powerful means of classifying ground cover. The hydrodynamic modelling was used to assess issues such as habitat conditions in the river, the effectiveness of fine sediment flushing flows, the sensitivity of flood levels to riparian vegetation cover, and visualizing the future river environment.

Keywords Braided rivers, remote-sensing, ALS, LIDAR, vegetation classification, two-dimensional modelling.

INTRODUCTION

The biodiversity typically shown by gravel-bed braided rivers is underpinned by their diversity of physical habitats. This includes the runs, riffles and turbulent pools of the main channels, the quiescent pools and shallow riffles of backwaters and side braids, and the shady runs and undercut banks of channels flanked by riparian vegetation. However, this very diversity imposes challenges when the effects of human-influenced changes in the river regime need to be evaluated and managed. The first challenge stems from the need to adequately sample the riverbed to extract representative measurements. Typically, over recent decades braided river physical habitats have been sampled in the same way as have those in single-thread channels: by focusing field-based measurements at a few cross-sections deemed representative on the basis of their being randomly or expertly located (e.g. Bovee, 1982; MfE, 1998). Unfortunately, a rigorous specification of cross-section density on the basis of spatial variability has been lacking. Moreover, a comprehensive evaluation of cross-section based datasets from a selection of New Zealand braided rivers highlighted to Mosley (1983) how substantial the variability could be from section to section even within short reaches. The second challenge relates directly to scale, for example, undertaking cross-section based measurements over a partially vegetated riverbed several kilometres wide can be a daunting exercise.

High resolution two-dimensional or even three-dimensional hydrodynamic models, by permitting the calculation of flow depth and velocity in grid cells of order 1 m resolution, offer the potential to overcome these problems of sampling and scale, provided the models' key boundary condition requirements of high resolution topography and hydraulic roughness (or at least ground cover able to be converted into hydraulic roughness coefficients) are met. Remote sensing offers a means of providing such information in a cost-effective manner. Indeed, there is a growing list of hydrodynamic model studies of braided river channels (Beffa, 1996; Duncan & Carter, 1997; Nicholas & Sambrook Smith, 1999; Nicholas et al., this volume, pp. 137–150) that have utilized digital elevation models (DEMs) based on digital aerial photogrammetry.

In this paper, a combination of Airborne Laser Scanning (ALS, or LIDAR), aerial photography and aerial multispectral (MS) scanning is used to underpin a two-dimensional model-based environmental investigation of the lower Waitaki River—a large, gravel-bed braided river on the East Coast of New Zealand's South Island.

STUDY RIVER

The Waitaki River (Fig. 1) is New Zealand's largest braided river by discharge (mean discharge ~358 $m^3 s^{-1}$) and a major source of hydroelectric power. Hydropower works began in 1935, and include three dams along the middle, gorged section of the Waitaki Valley, and a network of canals, control structures and power stations that utilize the storage from three large natural lakes in the upper basin. The Lower Waitaki River (Fig. 2) flows ~70 km between Waitaki Dam and the coast. It is braided for all but the first few kilometres downstream of the dam, and has a sandy gravel bed-material with a median size of approximately 30 mm.

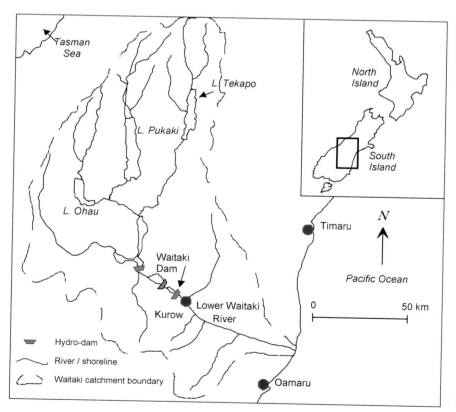

Fig. 1 The Waitaki basin, South Island, New Zealand.

Fig. 2 The Lower Waitaki River, looking upstream towards Kurow, July 2001.

The upstream hydroworks have resulted in a damped flood regime and a generally steadier flow in the lower river, while the hydro dams have reduced the supply of bed-material by approximately 50% (Hicks et al., 2002). The remaining bed-material is sourced from tributaries and reworking of the Pleistocene valley-fill. While some of the supply deficit has been recovered by degradation within a few kilometres downstream of Waitaki Dam, degradation along the braided reach is not obvious; indeed, it is likely that the effect of reduced sediment supply has been moderated by the reduced transport capacity of the flow regime.

The pre-dam riverbed of the Lower Waitaki was almost 2 km wide and was characterized by sparse willow trees, temporary islands vegetated mainly by native tussock and scrub, and shifting gravel bars and channels (Thompson et al., 1997). The onset of flow regulation was followed by an invasion of the riverbed by exotic vegetation, notably willow, broom and gorse. In consequence, the less resilient native vegetation was displaced and islands and bars became choked with the exotic vegetation and tended to stabilize, while flood breakouts along the riverbed margins became a hazard (Hall, 1984). Although a policy of de-vegetating a central 'fairway' or braidplain with spraying and machinery has been implemented since the 1960s, a net increase in vegetation cover remains traceable from aerial photographs. By 2001, the river's braidplain had been reduced to an average width of about 0.5 km (Hicks et al., 2002). Studies of aerial photographs (e.g. Hall, 1984; Hicks et al., 2002; Tal et al., 2004) have noted that this narrowing of the braidplain has been accompanied by a reduction in braiding activity and a tendency for flows to congregate in one or two principal braids. The Lower Waitaki is a popular fishing river for trout, salmon and indigenous species, and is equally popular for jet boating.

Substantial environmental investigations commenced in 2001 to assess the impacts of a proposal to divert flow from the Lower Waitaki at Kurow (Fig. 1) into a canal running along the right bank. The canal would contain a series of low-head hydropower stations and would eventually return the water back to the river ~6 km upstream of the coast. One key issue related to the minimum residual flow to be retained in the river to sustain the general habitat, fisheries and recreational amenities. Another was what special maintenance flow releases and ground-based operations would be required to maintain the riverbed in its current morphology, size grade and vegetation cover, and also to flush fine-grade sediment from the residual channel substrate. Unless vigorously controlled, the historical trend of encroaching riparian vegetation was expected to increase, thus there was also concern over how sensitive the channel's flood carrying capacity was to the state of vegetation cover.

Several of these issues were suited to analysis by a two-dimensional modelling approach, and a remote sensing package comprising ALS, vertical aerial photography, and MS scanning was duly designed to provide the necessary boundary data. Although not featured in this paper, there were numerous other uses made of the remotely sensed data, and all of the above issues were not exclusively investigated by the two-dimensional approach. For example, the ALS topography was also used to help engineering design work and to assess geomorphological issues at tributary confluences, canal intake and outlet sites, and at the coast. The riverbed topography was also used to provide three-dimensional visualizations of the future riparian landscape under different scenarios of minimum river flow and vegetation cover.

REMOTE SENSING

Strategy

The key data required of the remote sensing was a reliable DEM of the riverbed in an unvegetated dry state (i.e. with the vegetation and water burnt off), plus a map of ground cover that could be converted into a map of hydraulic roughness coefficients. For the DEM, a strategy similar to that developed by Westaway *et al.* (2003) was used to acquire topographic data for exposed and submerged areas of the riverbed. With this, ALS was used to secure the topography of exposed areas of riverbed and floodplain and also the topography of the water surface of the wetted channels. Geosynchronized airborne MS scanning, calibrated to ground-truth bathymetric measurements, was then used to map water depth, with the submerged topography calculated by differencing the bathymetry from the water surface topography. A bathymetry LIDAR system was not considered viable, either on the basis of cost or capability to resolve the relatively shallow water depths (largely <1 m). The primary strategy for mapping the riverbed and the floodplain ground-cover was by trained (i.e. supervised) classification of the MS imagery, although, as it turned out, it was found that trained classification of various data 'channels' generated from the ALS proved equally as useful for that purpose.

Data collection

The remotely sensed data were collected on one day (with three aircraft) during a managed low flow event in the river. On the day, the discharge was gauged at 90 $m^3 s^{-1}$. This low flow was required to maximize the area of riverbed exposed to the ALS. Also on this day, ground-truth data on water depths and riverbed levels were collected.

The MS scanning used a Daedalus 1268 Scanner (de Vries & Norman, 1994). This captures passive radiation within six bands across the visible light spectrum (0.42–0.75 µm wavelength), three near infrared bands (0.76–1.75 µm), one short-wavelength infrared band (2.08–2.35 µm), and one thermal infrared band (8.5–13.0 µm). The length of the lower river was covered by two overlapping swaths, each approximately 1 km wide, which allowed an on-the-ground pixel size of approximately 1.5 m. Half metre resolution orthophotographic imagery was used by an external contractor to georeference and rectify the MS imagery.

The ALS used an Optech ALTM1225 system. The area scanned for this study covered a swath approximately 1.8 km wide, centred on the river fairway, and extending from Waitaki Dam to the coast. This area was scanned twice, because turbulent flying conditions during the first run led to inadequate point spacing in some areas. The data from the combined runs yielded an average ground-point spacing of 1.2 m. The remote sensing was georeferenced into the NZMG coordinate system, with elevations referenced to the Lyttleton datum. For the ALS, this required developing a polynomial model of the local geoid across the project area from 25 control points of known orthometric height (Hicks *et al.*, 2001).

The vertical accuracy of ground elevation measurements was checked against 207 ground-surveyed test points over two separate areas, and showed a standard error of 5 cm for points that co-located with a laser 'shot' and 8 cm for points interpolated from nearby laser shots. At some locations where the laser scanning was affected by severe turbulence, the error degraded to several decimetres along discrete scan-tracks. Such tracks were clearly identifiable on shaded relief plots and were removed by manual DEM editing in critical areas, such as the two-dimensional model reach.

Coincident with the aerial remote sensing, in-river ground-truth data were collected using a jet-boat equipped with a Tritech echo-sounder, a Trimble Real Time Kinematic (RTK) GPS for navigation, and Trimble's HydroPro software system for logging bathymetric surveys. This delivered on-the-fly fixes of river bed-level, water surface elevation, and water depth 10 times per second and accurate to within a few centimetres. Some 30 km of river was surveyed along a zigzag track in order to sample a wide range of water depths.

The ALS gave good returns off the river water. Penetration of the near infrared light laser into the water was expected to be of the order of a few centimetres (D. Jonas, AAM Geoscan, personal communication), so the laser-indicated water 'surface' was a reasonably true rendition of the actual water surface. To check, the ALS data were interpolated along the jet-boat track at the sounding positions. This indicated that, after smoothing-out local-scale noise in the ALS data, the ALS and ground-based water surface elevations agreed to 13 cm or better over 30 km of river. The difference appeared to relate to drift between the geoid models adopted for the ALS and ground survey.

DEM AND ROUGHNESS EXTRACTION

DEM generation

The laser elevation data used for the DEM were based on 'last pulse' returns. With each incident cone-shaped laser pulse, where the ground is vegetated the earliest returns are received from vegetation strikes while the last returns have the best chance of being recovered from ground strikes (Kraus & Pfeifer, 1998). These last pulse returns were further processed numerically to filter-out non-ground returns (i.e. from understory strikes). The filtering algorithm rejects adjacent points if the slope from the current point exceeds a given angle, and is applied in several passes (D. Jonas, AAM Geoscan, personal communication). Beyond this, for our two-dimensional model study area we undertook further, manual editing of the laser data to remove any remnant tree-affected topography. This involved overlaying a detrended ALS-based DEM on the orthophotographs, 'masking' the tree-affected areas, and then applying to each masked area a minimum-value filter to set the detrended elevation of the ground under the trees to that of the adjacent exposed ground.

The DEM generation of the wetted channels required several steps using the GIS package ArcInfo. First, the MS imagery was classified to make a wet/dry 'mask'. A trained classification approach was used, and the resulting wet/dry boundaries were checked against orthophotographs taken on the same day. This checking led to some minor editing of the mask boundary. The wet mask was then overlaid on the ALS data so that the water surface elevations could be smoothed using a low-pass filter. Second, and following approaches similar to Winterbottom & Gilvear (1997) and Westaway (2003), a multiple-regression model was generated that related water depth to the intensity values of particular bands of the multispectral imagery. This involved interpolating the multispectral data at the locations of depth measurements. The multiple regression analysis found significant depth correlations with the blue (0.45–0.52 µm), green (0.52–0.60 µm), orange (0.60–0.63 µm), red (0.63–0.69 µm) and near-infrared channels (NIR1, 0.69–0.75 µm; NIR2, 0.75–0.90 µm). The blue and green channels were highly correlated, as were the two near-infrared channels, thus the green and second near-infrared channels were dropped from the regression model. The final model developed, relating depth h (m) to the natural logarithms of the channel reflectance intensities, was

$$h = 7.24 + 6.62 \ln(\text{blue}) - 8.02 \ln(\text{orange}) - 16.2 \ln(\text{red}) + 18.3 \ln(\text{NIR1}) \quad (1)$$

The standard error of the depth predicted from Eq. 1 was 29 cm while the adjusted model r^2 was 0.69. Predicted and measured depths are compared in Fig. 3.

The technique worked best at depths between about 0.5 and 2.4 m. At shallower depths, less than 0.5 m, the method was sometimes confused by the spectral signature of periphyton, which 'stained' the riverbed gravel brown on a patchy basis. This contributed to greater scatter in the depth predictions in this depth range (Fig. 3). For depths greater than

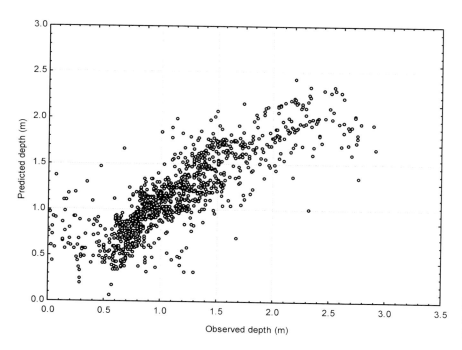

Fig. 3 Comparison of sounded depths and depths predicted from multispectral imagery.

2.4 m, the signature of the channel bed appears to have been lost altogether, due to the scattering of light in the water column. Because of this saturation effect, the method could not detect depths greater than this limit. This was of no great consequence, however, since the echo-sounding runs down the main channel showed that, on the day of the scanning, only small areas of the main channel exceeded this depth. Shadows cast from south-facing banks and trees also contributed to scatter in the regression relationship at all depths. Measured depths over periphyton and in shaded areas of channel were included in the regression analysis. Possibly, an improved depth-predictive model may have resulted from further classifying the shallower channels into areas with and without periphyton, and also by classifying shaded areas, then developing separate regression models for each of these classes.

The regression model was then applied to generate a depth map. This was further smoothed numerically and edited to correct for the effects of shadows (which generated falsely-large depths), overhanging trees (which hid the water, producing apparent zero-value depths), and the periphyton-induced confusions in shallow areas. The editing was done on an interpretative basis using ArcInfo tools, with the aid of the aerial orthophotographs.

Comparison of submerged bed levels generated by the remote sensing with those surveyed with the GPS and sounder indicated that the standard error of the remotely-mapped bed-levels was ±23 cm while the mean error was +8 cm. That is, the remotely sensed bed levels tended to be 8 cm higher than the directly measured levels on average. By comparison, the bed material D_{85} size was approximately 8 cm. The bias most likely followed from the 'saturation' effect discussed above. The improved standard error over that of the raw depth-mapping relation (Eq. 1) reflects the effects of numerical smoothing of the riverbed topography. By this process, a 2-m pixel DEM was generated for an approximately 4.5 km length of riverbed (Fig. 4).

Roughness mapping

The two-dimensional hydrodynamic model required spatially distributed information on hydraulic roughness. This was accomplished by mapping ground cover type, then assigning appropriate roughness coefficients. To meet project deadlines, and using a previously tested approach, the riverbed ground cover over the reach to be modelled was mapped from the georectified MS imagery using conventional trained classification techniques. However, a subsequent substudy was able to compare the efficacy of the MS-based ground cover classification with an independent

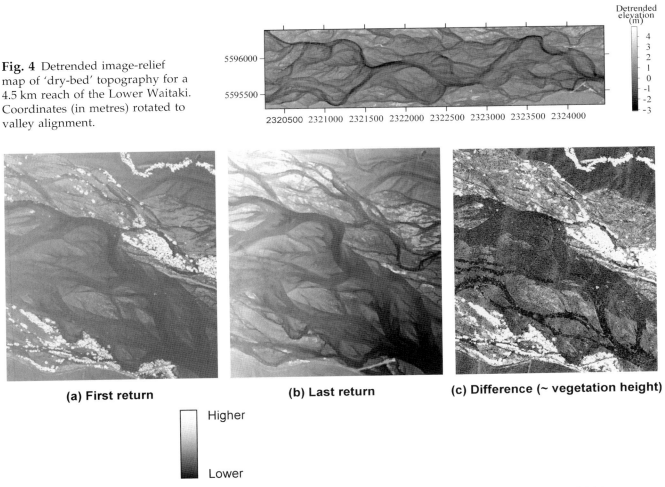

Fig. 4 Detrended image-relief map of 'dry-bed' topography for a 4.5 km reach of the Lower Waitaki. Coordinates (in metres) rotated to valley alignment.

Fig. 5 Image-relief maps generated from: (a) ALS first-return altimetry; (b) ALS last-return altimetry; and (c) by differencing last and first returns. Note willow trees showing as white (high) features in (a) and (c), but densely foliated scrub featuring in (b).

classification obtained from the ALS over a 2 km square test area, which is what is focused on here.

The ALS-based classification system involved defining a number of 'channels' or 'bands' generated from combinations of elevation and intensity values from the first and last laser returns. As discussed above, where bushy vegetation is present, the first return (Fig. 5a) is more likely to reflect off a canopy or higher branches, while the last return (Fig. 5b) is more likely to be a true ground-reflection, provided the ground is exposed through a discontinuous canopy. Thus, the difference in elevation values between the first and last returns (Fig. 5c) is a pseudomeasure of vegetation height. Indeed, this approach has been used directly for mapping vegetation height in forest surveys (e.g. Hardling et al., 2001; Naesset & Bjerknes, 2001). The last returns were also classified as either ground or non-ground with a numerical filter, thus the elevation difference between the first return and a ground-classified last return should provide an even better measure of vegetation height. The intensity of the laser returns depends on the proportion of the incident laser pulse light that is reflected from the target. This is diminished both by back scattering and by absorption, and varies with ground cover and canopy penetration (Hardling et al., 2001). For example, water tends to absorb much of the energy of the infrared laser, and so generates a low-intensity return, while gravel and concrete return a relatively high intensity (Fig. 6). The local variance, or 'texture', of both the ALS

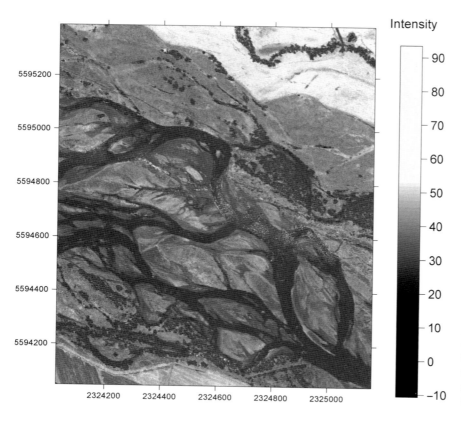

Fig. 6 Airborne laser scanning (ALS) first-return intensity map. Compare with photograph in Fig. 7. Coordinates are easting (X) and northing (Y) in New Zealand Map Grid (in metres).

elevations and intensities, defined as the variance of a nine-pixel window, provides further discrimination. For example, scrubby vegetation showed greater texture than relatively uniform cover such as grass and sediment.

In all, 11 potential channels were defined and gridded using a kriging routine. Various combinations were then evaluated in terms of ability to classify the ground cover in the test area into one of seven classes: (i) water, (ii) gravel-bar, (iii) scrub, (iv) willows, (v) green pasture, (vi) other grasses including rushes and (vii) fields in stubble or fallow. A 'ground-truth' classification was defined manually from the aerial photography by digitizing polygons of uniform ground cover onto the orthoimagery; this was aided by stereoviewing and a field inspection.

The most effective classification, as determined by the lowest root-mean-square error (RMSE) against the ground-truth classification (RMSE = 1.32%), was found using the following five channels: (i) first return elevation–ground-classed last-return elevation; (ii) non-ground last-return elevation–ground last-return elevation; (iii) first-return intensity; (iv) ground last-return intensity; and (v) non-ground last-return intensity. Respectively, these may be interpreted as: (i) vegetation height, (ii) understory height, (iii) vegetation canopy roughness, (iv) ground roughness and (v) understory roughness. The elevation-based channels efficiently discriminated between low and scrubby vegetation (broom and gorse) and willows, but the intensity channels proved essential for discriminating between the relatively low roughness classes (water, gravel, grass and green pasture). The main confusions arose between the green pasture, other grass and fallow classes, although for the purposes of determining flow resistance there was little practical difference between these classes anyway. More significantly from the flow resistance perspective, scrub (which develops a relatively high drag coefficient) and grass were sometimes confused when the scrub was too dense for much laser penetration through the canopy, and there was little elevation difference between first and last laser returns.

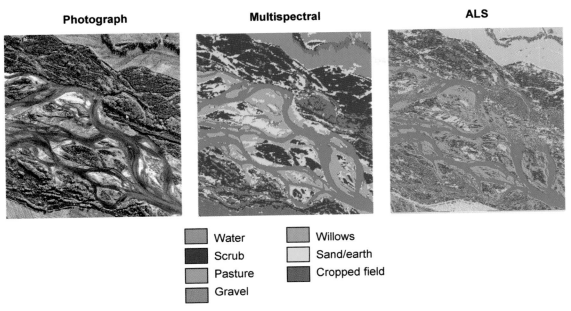

Fig. 7 Comparison of photograph, MS-based ground cover classification and ALS-based ground cover classification.

The ALS-based classification was very similar to the MS-based classification for the test area (Fig. 7); indeed the ALS classification had a better RMSE (1.3%) overall than did the MS classification (4%) when compared with the ground-truth. Other advantages of the ALS-based approach are:

1 the ALS ground-cover map is fully geosynchronized with the ALS DEM while the MS has to be painstakingly georectified in order to overlay the ALS DEM;
2 the ALS approach is based on an active radiation source while the MS approach relies on passive reflectance and so is subject to temporal and spatial variations relating to incident solar radiation and shade along the flight path;
3 by providing a direct measurement of physical roughness, the ALS product is more directly related to the end-objective map of hydraulic roughness.

For these reasons, it was concluded that ALS datasets hold great promise for mapping groundcover and roughness in riverbed environments. Asselman et al. (2002) reported similar success with using ALS for mapping floodplain roughness, although in that study the intensity channel was not used because it was found that intensity differences between different flight lines tended to exceed intensity variations caused by different vegetation types. This was not an issue with the Waitaki data, since it was collected in a single swath.

The roughness formulation used in the two-dimensional modelling varied with the ground-cover type (Aberle et al., 2002). For bare gravel, which represented the exposed and submerged riverbed, a value of k_s (equivalent sand roughness) equal to 130 mm was used. This was calculated from field measurements of depth and velocity and using preliminary model runs to derive water surface slope, and was approximately equal to 1.5 times the average D_{84} of 10 bed material surface samples collected from the modelled reach. Verification runs of the model indicated no need to adjust this k_s value further. A constant roughness coefficient was also assumed for short submerged vegetation (e.g. grass). The drag induced by willows and light and dense scrub (gorse, broom and lupin) on islands and on the floodplain was modelled using the approach of Lindner (1982). This relates a drag coefficient to the diameter and spacing of tree/bush stems. For this study, stem diameters and cross- and streamwise stem spacing were based primarily on data published by Dittrich (1998) from the River Murr in Germany, but were modified slightly to suit the vegetation characteristics of the Waitaki.

APPLICATIONS

Several issues were investigated with a two-dimensional hydrodynamic model set up along a 4.5 km long representative reach. The two-dimensional model, Hydro2de (Beffa, 1996; Beffa & Connell, 2001), solves the depth-averaged shallow-water equations for a grid using a finite volume scheme. Given riverbed elevation and hydraulic resistance across the model domain, and specifying a discharge across the upstream boundary, the model computes water depth, depth-averaged velocity, and bed shear at each grid cell. The model was set up in two forms:

1 covering a 1 km domain width with 2 m grid cells for higher resolution modelling of the braidplain channels at relatively low flows;
2 a 2.5 km wide, 5 m grid cell version for modelling flood flows.

Both versions were verified using a detailed set of measurements of depth-averaged velocity and depth at a discharge of 350 m^3 s^{-1}. These were collected across eight cross-sections using a jetboat. The flood model was also verified from floodwater edges extracted from oblique aerial photographs of an approximately 100-yr return period flood in November 1995 (~2700 m^3 s^{-1} peak discharge). The higher resolution model was then run for a range of discharges between 40 and 1000 m^3 s^{-1}, while the flood model was run for flows between 350 and 2700 m^3 s^{-1}.

Flood-sensitivity to riparian vegetation

The two-dimensional model was used to assess the sensitivity of flood levels to various scenarios of riverbed vegetation cover consistent with different levels of vegetation management. The scenarios ranged from no significant vegetation (i.e. bare gravel, very similar to the natural/unregulated state of the river prior to the 1930s), through the existing 'design' cover with a cleared 400–700 m wide fairway between belts of willow, to an extreme scenario whereby the main islands were 'allowed' to colonize with dense scrub (gorse and broom). In the bare gravel scenario, a k_s value of 130 mm (appropriate for gravel) was applied uniformly across the whole model domain. The existing cover scenario was based on the ground-cover classification derived from the MS data. For this, the same global k_s value was applied for surface roughness, but different coefficients for vegetation-induced drag were assigned to each vegetated polygon based on the stem spacing and diameter estimated for each vegetation class. The central cleared fairway that is associated with this scenario results from a regular programme of aerial spraying by the local regional council. The extreme scenario was considered the most likely development under the proposed project given no artificial vegetation control, that is, with no maintenance spraying. It largely involved gorse and broom establishing on gravel bars exposed for prolonged periods as a result of flow diversions from the river, but also included the closure of some small side channels by willow growth and encroachment. It was developed interpretatively with expert input from botanists.

A range of flood discharges was run through the model with each groundcover scenario. The results showed how, as the riverbed vegetation intensified, the conveyance of the central fairway became progressively reduced, flood depths increased, and more floodwater was routed across the marginal pastureland or focused in the main channel (Fig. 8). For a flood discharge of 2700 m^3 s^{-1}, the reach-averaged water level would increase by 0.5 m from the present situation if the vegetation was allowed to grow unchecked, while if all the riparian tree and bushy vegetation were cleared the flood level would be 0.4 m lower (Fig. 9). The specific discharge plots (Fig. 8) highlight the potential for suspended sediment entrapment on vegetated islands and the development of one or two main channels through flow focusing.

Physical habitat mapping and quantification

The model output maps of depth and velocity at given discharges were combined with habitat suitability functions for several species of fish and aquatic birds (i.e. functions that optimize physical habitat within parts of depth–velocity space) and integrated to determine the weighted usable area of channel as a function of water discharge. Passage connection through the braid network for salmon and recreational jetboaters at different discharges was assessed simply from depth maps. In addition, the breakdown of physical habitat between riffles,

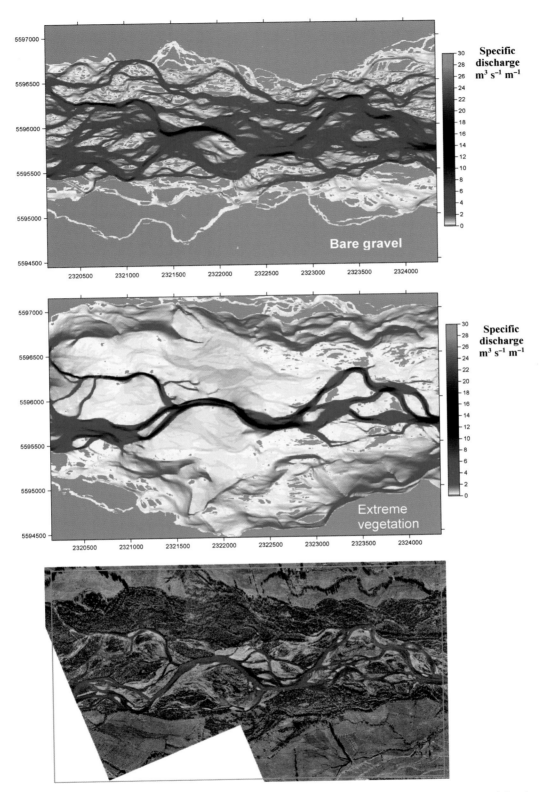

Fig. 8 Floodwater coverage and specific discharges at a flood discharge of 2700 m^3 s^{-1} for bare gravel (top) and extreme vegetation (middle) scenarios. Coordinates (in metres) rotated to valley alignment. Modelled reach is shown by red box on aerial photograph (bottom).

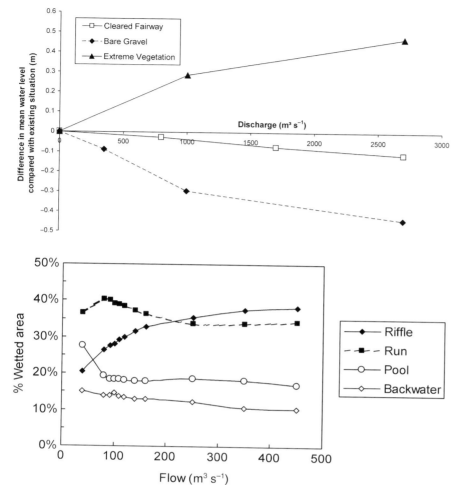

Fig. 9 Change in reach-averaged water-surface elevation from elevation under vegetation conditions existing in 2001.

Fig. 10 Relative areas of riffles, runs, pools and backwaters (as discriminated in depth–velocity space by Froude Number) in relation to total water discharge in the Waitaki River at the modelled reach.

runs, pools and backwaters at different discharges was mapped using Froude Number ($Fr = \bar{u}/\sqrt{gy}$, where $\bar{u}$ is depth-averaged velocity, g is gravitational acceleration and y is water depth) as a discriminator (Fig. 10). Pools, runs and riffles were associated with Fr values <0.18, between 0.18 and 0.41, and >0.41, respectively (after Jowett, 1993), while pools were further distinguished as backwaters where Fr was less than 0.05 (Dr B. Biggs, National Institute of Water and Atmospheric Research, personal communication). The aim of the habitat assessment was to search for any marked change in physical habitat within the proposed range of residual discharge in the river. For example, the relative percentages of pools and backwaters increased at the expense of riffles and runs as the discharge was reduced below about 100 m³ s⁻¹ (Fig. 10). This reflects the increasing control of the bed topography on the type of physical habitat as the discharge reduces.

Flushing-flow effectiveness

Riverbed stability in relation to total water discharge was evaluated by generating maps of bed shear stress. Bed shear, τ_o, was determined from the model results on depth and velocity by inverting the 'law of the wall', thus

$$\tau_o = \rho\left(\frac{\bar{u}\kappa}{\ln(11y/k_s)}\right)^2 \quad (2)$$

where ρ is water density and κ is von Karman's constant (assumed equal to 0.4). Critical bed shear stress at the threshold of motion for different grades of bed surface material was calculated using the relation of Komar (1989):

$$\theta_{ci} = \theta_{c50}(D_{50}/D_i)^{0.65} \quad (3)$$

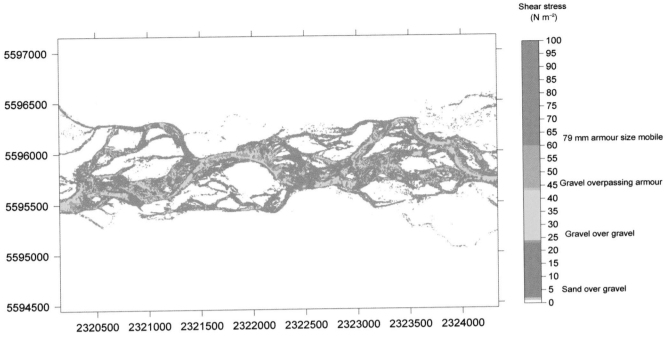

Fig. 11 Bed shear stress (N m^{-2}) at 450 m^3 s^{-1}, coded by size of sediment that should be mobile. Grid dimensions are in metres.

where θ_{c50} is the threshold dimensionless stress for the median diameter (D_{50}) and θ_{ci} is the threshold dimensionless stress for the ith grainsize percentile (D_i). A representative surface size distribution was obtained by averaging 10 'Wolman' samples from the model reach (Wolman, 1954).

With appropriate colour coding of threshold shear stress, the areas of riverbed over which representative grain sizes became mobile could be mapped and totalled. For example, at a discharge of 450 m^3 s^{-1} (Fig. 11), sand would be mobile through all channels but the gravel bed-material mode (31 mm) would be mobile only on a very patchy basis and would rarely overpass armoured sections. This confirms the suitability of 450 m^3 s^{-1} for an artificial 'flushing flow' release—it should be capable of flushing sand from the channel bed and off bars without significantly destabilizing the gravel-bed channels.

Landscape visualization

A further application of the remote sensing was for three-dimensional visualizations of the river landscape to aid public understanding and appreciation of predicted effects. This involved a merger of DEMs, projected vegetation coverage (on a GIS layer), orthophotography and two-dimensional model predictions of water coverage in order to render landscape stills and flybys.

DISCUSSION: EVALUATION AND PROJECTED USE OF REMOTE-SENSING OF BRAIDED CHANNEL ENVIRONMENTS

The high spatial variability shown by braided rivers such as the Waitaki (Fig. 2) highlights the need for a two-dimensional, spatially distributed basis for investigating flow distribution, sediment transport and associated habitat issues. Modern computational tools, such as two and three-dimensional numerical models, provide the basis of these investigations, but they themselves must be underpinned by data-intensive layers of boundary-condition information. On large rivers such as the Waitaki (order of 1–2 km wide by tens of kilometres long), aerial remote sensing currently provides the only practical means of delivering on demand data of adequate spatial density and local accuracy. For example, a systematic boat-based bathymetric survey of the Waitaki was regarded as slow, impractical and unsafe (due to the hazards from high velocities and numerous willow trees), while a satellite-based multi- or hyperspectral platform could not have delivered the required spatial detail and flexibility of service (which was vital when coordinating data collection within the constraints provided by daylight, weather and water scheduling within the Waitaki hydrosystem).

For our topography mapping, it was considered that the remote-sensing delivered results about as reliable as could be expected; moreover, this was regarded as adequate for the 2–5 m pixel size used with our hydrodynamic modelling. On dry areas of riverbed, the ALS was able to deliver interpolated elevation fixes to a standard error of ~8 cm under ideal conditions. This was degraded over vegetated areas, due to imperfect filtering of non-ground strikes, and also occasionally by excessive air turbulence (when the on-plane inertial guidance system could not keep up with the turbulent accelerations). However, there was generally sufficient redundancy in the dataset that elevation errors exceeding a few decimetres could be identified and removed by manual DEM editing techniques. The ~23 cm standard error on the elevations of wetted channel beds, down to a depth of ~2.4 m, was consistent with results from previous investigators (e.g. Westaway *et al.* (2003) who determined a standard error of 32 cm for wetted bed-levels fixed with photogrammetry and water colour mapped off air photographs). Lowering the flows to expose as much of the riverbed as possible during the remote sensing minimized the impact of the greater error in wetted channels. It is expected, however, that the shallow-water, dual-frequency bathymetry LIDAR systems currently under development will soon provide wet and dry topography of similar accuracy in the same pass, obviating the need for a mixed sensor approach.

In the future, the use of ALS/LIDAR may become the preferred option for mapping riverbed/floodplain ground cover and physical roughness as well as topography. As has been discussed, vegetation height measurement and laser return intensity appear to provide a powerful means of discriminating riparian ground cover, and ALS/LIDAR has the added advantages that the information is delivered fully synchronously and georeferenced with the ground altimetry information. While MS imagery can also be used to classify ground cover for the purpose of mapping hydraulic roughness, it requires careful rectification and georeferencing before being merged with elevation datasets. For this reason, the 'one stop' ALS/LIDAR package appears the more attractive.

The outstanding challenge remaining for remote sensing of large gravel-bed braided rivers relates to mapping the substrate, particularly within wetted channels. Accurate, quantitative information on mean size, variance and degree of fine-sediment embeddedness are the ideal products. Some progress is being made using air photography, radar and multispectral technologies. For example, Rainey *et al.* (2000) used a MS system to distinguish bed-material size on intertidal flats. Edge-detection techniques on photographic imagery and small-scale photogrammetry captured from portable platforms (Butler *et al.*, 2001; Smart *et al.*, 2002) have been proven at local scales but have yet to be up-scaled to reach dimensions. Verdu & Batalla (2003), by using a balloon platform to secure high-resolution photographic imagery at a pixel size similar to the bed-material median size, were able to correlate photographic image texture with gravel size at the reach scale. Nicholas *et al.* (2004) mapped physical roughness size in a gravel-bed braided river by inverting the hydraulic roughness determined by a fixed-lid three-dimensional hydrodynamic model.

CONCLUSIONS

This paper has shown that the spatially distributed information on riverbed topography and ground cover needed for two-dimensional hydrodynamic modelling of large braided rivers can be obtained from aerial remote-sensing platforms. A combined approach using both airborne laser scanning and multispectral scanning permits wetted channel topography to be mapped to ~23 cm accuracy in orthometric elevation. Use of first and last return information on elevation and intensity from airborne laser scanning permitted a classification of riverbed ground cover that was as good as, if not better than, a conventional trained classification off multispectral imagery. A two-dimensional hydrodynamic model proved a powerful tool for investigating environmental effects and management options on a large, braided gravel-bed river subject to hydroelectric power development.

ACKNOWLEDGEMENTS

The Lower Waitaki airborne laser scanning was undertaken by AAM Geoscan Pty Ltd, Kangaroo Point, Queensland; the multispectral scanning was done by Air Target Services Pty Ltd, Nowra, New South Wales; the multispectral image georeferencing and rectification was undertaken by Image Analysis & Mapping Pty Ltd, Canberra, Australia; air photography and photogrammetry were conducted by Precision Aerial Surveys Ltd, Auckland, New Zealand; and ground surveys were overseen by Scott Williams, Glasson Potts Fowler, Christchurch, New Zealand. We thank Meridian Energy Ltd for permission to publish results from the Lower Waitaki investigations. Preparation of this paper was supported in part by the Foundation for Research, Science and Technology under Contract No. C01X0308.

REFERENCES

Aberle, J., Hicks, D.M. and Smart, G.M. (2002) *Project Aqua: the Effect of Riverbed Vegetation on the Flooding Hazard in the Lower Waitaki River*. NIWA Client Report CHC01/111, National Institute of Water and Atmospheric Research, Christchurch, 52 pp.

Asselman, N.E.M., Middelkoop, H., Ritzen, M.R. and Straatsma, M.W. (2002) Assessment of the hydraulic roughness of river flood plains using laser altimetry. In: *The Structure, Function and Management Implications of Fluvial Sedimentary Systems* (Eds F.J. Dyer, M.C. Thoms and J.M. Olley), pp. 381–388. IAHS Publication 276, International Association of Hydrological Sciences, Wallingford.

Beffa, C. (1996) Application of a shallow water model to braided flows. In: *Proceedings of the Hydroinformatics 96 Conference* (Ed. A. Mueller), Zurich, 9–13 September, pp. 667–672.

Beffa, C. and Connell, R.J. (2001) Two-dimensional flood plain flow. I: model description. *J. Hydrol. Eng.*, **6**, 397–405.

Bovee, K.D. (1982) *A Guide to Stream Habitat Analysis using the Instream Flow Incremental Methodology*. Instream Flow Information Paper 12, Biological Sciences Program FWS/OBS-82/26, U.S. Fish and Wildlife Service, Washington, DC, 248 pp.

Butler, J.B., Lane, S.N. and Chandler, J.H. (2001) Automated extraction of grain-size data from gravel surfaces using digital image processing. *J. Hydraul. Res.*, **39**, 519–529.

De Vries, D.H. and Norman, R. (1994) Airborne multispectral scanner: Daedalus ATM (AADS 1268) commercial operations in Australia. In: *Proceedings of the 7th Australasian Remote Sensing Conference*, Melbourne, 1–4 April, pp. 128–135.

Dittrich, A. (1998) *Wechselwirkung Morphologie/Stroemung naturnaher Fliessgewaesser*. Habilitatonsschrift, Mitteilungen des Instituts fuer Wasserwirtschaft und Kulturtechnik, Universitaet Karlsruhe.

Duncan, M.J. and Carter, G.C. (1997) Two-dimensional hydraulic modeling of New Zealand rivers: the NIWA experience. In: *Proceedings of the 24th Hydrology and Water Resources Symposium*, Auckland, 24–28 November, pp. 493–497.

Hall, R.J. (1984) *Lower Waitaki River: Management Strategy*. A Waitaki Catchment Commission and Regional Water Board Report, Timaru, New Zealand.

Hardling, D.J., Lefsky, M.A., Parker, G.G. and Blair, J.B. (2001) Laser altimeter canopy height profiles: method and validation for closed-canopy, broadleaf forests. *Remote Sens. Environ.*, **76**, 283–297.

Hicks, D.M., Duncan, M.J., Shankar, U., Wild, M. and Walsh, J.R. (2002) Project Aqua: Lower Waitaki River geomorphology and sediment transport. NIWA Client Report CHC01/115, National Institute of Water and Atmospheric Research, Christchurch, 170 pp.

Hicks, D.M., Rice, J., Williams, S. and Turton, D. (2001) ALS stops the flow of water. *Scan. Horizons, AAMHatch Newsl.*, **11**, 1–3.

Jowett, I.G. (1993) A method for objectively identifying pool, run, and riffle habitats from physical measurements. *N. Z. J. Mar. Fresh.*, **27**, 241–248.

Komar, P.D. (1989) Flow-competence evaluations of the hydraulic parameters of floods: an assessment of the technique. In: *Floods: Hydrological, Sedimentological and Geological Implications* (Eds K. Bevan and P. Carling), pp. 107–132. Wiley, Chichester.

Kraus, K. and Pfeifer, N. (1998) Determination of terrain models in wooded areas with airborne laser scanner data. *J. Photogram.*, **53**, 193–203.

Lindner, K. (1982) *Der Stroemungswiderstand von Pflanzenbestaenden*. Mitteilungen Leichtweiss-Instituts fuer Wasserbau, Universitaet Braunschweig, Heft 75.

MfE (1998) *Flow Guidelines for Instream Values*, Vol. A. Ministry for the Environment, Wellington, New Zealand, 146 pp.

Mosley, M.P. (1983) Response of braided rivers to changing discharge. *N. Z. J. Hydrol.*, **22**, 18–67.

Naesset, E. and Bjerknes, K.O. (2001) Estimating tree heights and number of stems in young forest stands using airborne laser scanner data. *Remote Sens. Environ.*, **78**, 328–340.

Nicholas, A.P. and Sambrook Smith, G.H. (1999) Numerical simulation of three-dimensional flow hydraulics in a braided channel. *Hydrol. Process.*, **13**, 913–929.

Nicholas, A.P., Quine, T.A. and Thomas, R. (2006) Cellular modelling of braided river form and process. In: *Braided Rivers: Process, Deposits, Ecology and Management* (Eds G.H. Sambrook Smith, J.L. Best, C.S. Bristow and G.E. Petts), pp. 136–150. Special Publication 36, International Association of Sedimentologists. Blackwell, Oxford.

Rainey, M.P., Tyler, A.N., Bryant, R.G., Gilvear, D.J. and McDonald, P. (2000) The influence of surface and interstitial moisture on the spectral characteristics of inter-tidal sediments; implications for airborne image acquisition and processing. *Int. J. Remote Sens.*, **21**, 3025–3038.

Smart, G.M., Duncan, M.J. and Walsh, J.M. (2002) Relatively rough flow resistance equations. *J. Hydraul. Eng.*, **128**, 568–578.

Tal, M., Gran, K., Murray, A.B., Paola C. and Hicks, D.M. (2004) Riparian vegetation as a primary control on channel characteristics in noncohesive sediments. In: *Riparian Vegetation and Fluvial Geomorphology* (Eds S.H. Bennet, J.C. Collinson and A. Simon), pp. 43–58. Water Science and Application 8, American Geophysical Union, Washington, DC.

Thompson, S.M., Jowett, I.G. and Mosley, M.P. (1997) *Morphology of the Lower Waitaki River*. NIWA Client Report WLG97/55, National Institute of Water and Atmospheric Research, Wellington, 47 pp.

Verdu, J.M. and Batalla, R.J. (2003) Estimating grain size distribution of a gravel riverbed at reach scale from aerial orthophotos, geostatistitcs and digital image processing (Isabena River, Spain). *Conference Abstract Volume, Braided Rivers 2003*, University of Brimingham, 7–9 April, pp. 61.

Westaway, R.M., Lane, S.N. and Hicks, D.M. (2003) Remote survey of large-scale braided, gravel-bed rivers using digital photogrammetry and image analysis. *Int. J. Remote Sens.*, **24**, 795–815.

Winterbottom, S.J. and Gilvear, D.J. (1997) Quantification of channel bed morphology in gravel-bed rivers using airborne multispectral imagery and aerial photography. *Regul. River. Res. Manag.*, **13**, 489–499.

Wolman, M.G. (1954) A method of sampling coarse river bed material. *Trans. Am. Geophys. Union*, **35**, 951–956.

Effects of human impact on braided river morphology: examples from northern Italy

NICOLA SURIAN

Autorità di Bacino dei fiumi dell'Alto Adriatico, Dorsoduro 3593, 30123 Venice, Italy

ABSTRACT

Most Italian braided rivers have experienced considerable channel adjustments during the past 100 yr, particularly in recent decades. Channel adjustments have taken place mainly in response to sediment mining and the construction of dams—two interventions that have substantially altered sediment supply. This study examines recent and contemporary channel change along such braided rivers with a view to supporting and improving maintenance and management strategies. Three rivers in northeastern Italy, the Tagliamento, the Piave and the Brenta, represent different examples of channel adjustment to a decrease in sediment supply. Sediment mining has been very intense along all three rivers, but only the Piave and the Brenta have been strongly affected by the construction of dams. A historical analysis was performed on the three rivers using maps, aerial photographs, longitudinal profiles and cross sections. Morphological features that were examined included channel width, braiding index, bed elevation and channel configuration. Two types of channel adjustment have been dominant in these braided rivers: narrowing and incision. Channel-width reduction has been in the order of 50% (Tagliamento and Brenta rivers) or more (Piave River), whereas bed-level lowering has been up to 3–4 m in the Tagliamento and Piave rivers, and up to 5 m along the Brenta. Additionally, braiding intensity has decreased significantly in all of the rivers. Changes in channel pattern, from braided to wandering, have occurred mainly in the Brenta and the Piave rivers, but an upstream migration of the transition between braided and single-thread morphology has occurred along all three rivers. Temporal trends of channel adjustment show that the main phase of narrowing and incision has taken place since the 1950s or 1960s, and that now it may have finished, since other kinds of processes, such as widening and (local) aggradation, have occurred in the past 10–15 yr. For river management it is necessary to recognize the effects of channel adjustments on structures and the environment, but it is also necessary to understand the causes of these morphological changes, and in particular to recognize that sediment supply is a key issue in braided rivers.

Keywords Channel adjustment, braided rivers, gravel mining, dams, sediment supply, Italy.

INTRODUCTION

Braided rivers are very common in the north of Italy, draining both from the Alps and from the Apennines. There are also several examples of braided channels in the central and southern parts of the country. Considerable channel adjustments have occurred in Italian braided rivers (Pellegrini *et al.*, 1979; Castiglioni & Pellegrini, 1981; Maraga, 1989; Dutto & Maraga, 1994; Surian, 1999; Aucelli & Rosskopf, 2000), as in most Italian rivers (Surian & Rinaldi, 2003), during the past 100 yr. These adjustments are due to human interventions such as channelization, sediment mining and dam construction. Studies on braided rivers, as well as on single-thread rivers, have shown that channel

Present address: Dipartimento di Geografia, University of Padova, Via del Santo 26, 35123 Padova, Italy (Email: nicola.surian@unipd.it).

adjustments generally have significant effects on structures (e.g. bridges), the environment (e.g. groundwater resources, ecosystems, beaches) and river hydraulics (e.g. flood flows) (Wyzga, 1991, 1997; Bravard et al., 1999; Kondolf et al., 2002).

This paper examines the morphological response of three braided rivers (Tagliamento, Piave and Brenta) to human interventions, and in particular to gravel mining and impoundment. Knowledge of the magnitude and trends of channel adjustments and of factors controlling such adjustments along Italian Rivers needs to be improved (Surian & Rinaldi, 2003), and the aims of this study are: (i) to reconstruct channel adjustments and particularly their temporal trends; (ii) to explain adjustments in relation to changes in flow and sediment regimes; and (iii) to investigate the management implications of channel adjustments. A better understanding of channel dynamics is an essential requisite for present and future river maintenance and management.

GENERAL SETTING

The Tagliamento, Piave and Brenta rivers of northeastern Italy drain from the Alps to the Adriatic Sea (Fig. 1). Their drainage basins are mainly composed of sedimentary rocks. The main physiographic and hydrological characteristics of the three rivers are reported in Table 1. Floods occur mainly in the autumn, are relatively flashy and high in magnitude. The ratio between flood peak discharge and mean annual discharge is high, ranging between 34 (Brenta River) and 43 (Tagliamento River).

The Tagliamento and Piave have a braided morphology along more than half their courses, in the piedmont plain and also in their mountain reaches.

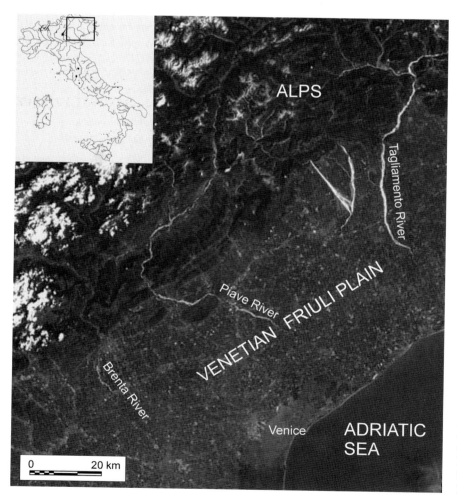

Fig. 1 Satellite image showing the courses of the Tagliamento, Piave and Brenta Rivers. The study reaches are in the upper part of the Venetian Friuli Plain.

Table 1 Physiographic and hydrological characteristics of the Tagliamento, the Piave and the Brenta Rivers

River	Drainage basin area (km^2)	Length (km)	Length of the study reach (km)	Basin relief (m)	Precipitation (mm yr^{-1})	Mean annual discharge (m^3 s^{-1})	Flood peak discharge* (m^3 s^{-1})
Tagliamento	2580	178	39	2696	2150	109	4650
Piave	3899	222	23	3162	1330	132	5300
Brenta	1567	174	18	3079	1386	71	2400

*Represents the largest discharge that has been measured or estimated and in all three rivers it occurred during the November 1966 flood, which has a recurrence interval of 100 yr at least.

Table 2 Human impacts in the Tagliamento, Piave and Brenta Rivers

River	Drainage area upstream from dams (%)	Date of dam closure	Date of intense sediment mining	Construction of levees and other bank protection structures	Reforestation and torrent control work in the drainage basin
Tagliamento	3	1950s	1970s–1980s	19th–20th century	20th century (?)
Piave	54	1930s–1950s	1960s–1980s	14th–20th century	Since the 1920s
Brenta	40	1954	1950s–1980s	19th–20th century	Since the 1920s

The Brenta has a braided configuration only in the piedmont plain. In this study, for a better comparison between the three rivers, only the braided reaches in the piedmont plain were analysed, since they have smaller lateral constriction. River beds are composed mainly of gravels. In the study reaches D_{50} and D_{84} are 20–30 mm and 35–50 mm respectively in the Piave River (Surian, 2002), and 20–30 mm and 40–60 mm in the Brenta River (Reniero, 1998); D_{50} and the largest particle are 20–60 mm and 80–150 mm respectively in the Tagliamento River (Petts et al., 2000). As for vertical variability of stream-bed material, no armouring was found at two sites along the selected reach of the Piave River (Surian, 2000), but these are the only available data.

HUMAN IMPACT

A range of human impacts (sediment mining, dams, construction of levees, reforestation) have taken place in the three rivers (Table 2). These interventions have both direct (e.g. levees) and indirect (e.g. reforestation) effects on channel dynamics. Most of the levees were constructed before the 20th century, and reforestation in the river basins took place during the 20th century, but no accurate analysis of the process is available. Two interventions that have a major impact on river dynamics are dams and gravel mining. Dams are very significant in the Piave and Brenta rivers but in the Tagliamento only 3% of the drainage area is impounded. Most of the dams were constructed in the 1950s and since then several million cubic metres of sediment has been trapped in the reservoirs. For instance, in the Piave river 15×10^6 m^3 of sediment was deposited in the Pieve di Cadore reservoir between 1949 and 1974 (drainage area upstream of this reservoir is 201 km^2), and 4.3×10^6 m^3 of sediment in the Mis reservoir between 1964 and 1989 (drainage area upstream of this reservoir is 107 km^2) (D'Alpaos & Dal Prà, 1996).

Gravel mining was particularly severe between the 1950s and the 1980s (Table 2). Though no complete records of the volumes of sediment mined are available, it may be estimated that some tens of millions of cubic metres were extracted along each of the three rivers during that period. For example, along the Tagliamento River and its main tributaries more than 24×10^6 m^3 of sediment was extracted from 1970 to 1991 (this value is an underestimate since it comes from official data, which commonly do not correspond to the actual volumes extracted, and also it does not include a reach of the river for which data were not available). Considering that the annual production of sediment in the Tagliamento basin has been estimated as 1.3×10^6 m^3 (Autorità di Bacino, 1998), it is evident that the volumes of sediment extracted will have had a significant effect on sediment fluxes.

It is likely that gravel mining and dams will have strongly altered the sediment regimes of these rivers. Dams and diversions have markedly reduced low flows, but no change has occurred in channel-forming discharges. For instance, in the Piave River the channel-forming discharge, which is approximately 400 m^3 s^{-1}, was equalled or exceeded on average 4 days a year in the period 1928–1959 (a period with low regulation of flows), and 5 days a year in the period 1991–2000 (a period with high regulation of flows) (Surian, 2003). Dams have had no major effects on such discharges (channel-forming discharges or higher floods) because the regulation schemes are not designed for flood control but for hydroelectric power generation and irrigation.

METHODS

A historical analysis of river morphology was performed using maps, aerial photographs, longitudinal profiles and cross sections. Morphological features that were examined included channel width, braiding intensity, bed elevation and channel configuration. Analysis of maps and aerial photographs was carried out using a geographical information system (GIS), which has the advantage of providing greater accuracy of measurements (channel width and braiding index) and repetition of measurements. The historical maps refer, approximately, to the period 1800–1960. The oldest map, which is at 1:26,000 scale, was surveyed by the Austrian army between 1801 and 1805, whereas more recent maps are those of the Istituto Geografico Militare Italiano (IGMI) at 1:25,000 scale. Aerial photographs were taken during the past 50 yr, and more recent ones in 1997 (Piave River), 2001 (Tagliamento River) and 2002 (Brenta River) by the Autorità di Bacino dei fiumi dell'Alto Adriatico. Topographical surveys (cross sections) have been carried out since the 1920s, the 1930s and 1960s, respectively in the Piave, Brenta and Tagliamento rivers. These sources allow a reliable historical analysis of morphological features, particularly over the past 50 yr. For instance, despite the excellent accuracy of the old maps, estimation of the braiding index is more reliable when it is performed on aerial photographs, which have been available only since the 1950s.

CHANNEL ADJUSTMENTS IN THE TAGLIAMENTO, PIAVE AND BRENTA RIVERS

Channel width

Channel width (width of the active braid belt) was measured on maps and aerial photographs. Width of the active braid belt was evaluated both including and excluding vegetated islands. Since there are some particular situations that could alter the average value of channel width (for instance the presence of a large island, 'Grave di Papadopoli', along the Piave River), the width measured without vegetated islands, that is the width of single channels plus the width of unvegetated or sparsely vegetated bars, will be considered here for a better comparison between the three rivers. Additionally, this measurement (width of channels plus width of bars) has a closer relationship with sediment transport along the rivers.

Average channel width has significantly decreased over the last two centuries (Fig. 2): in the Tagliamento from 1630 m (early 19th century) to 760 m (in 2001), in the Piave from 840 m (early 20th century) to 310 m (in 1997), in the Brenta from 480 m (early 19th century) to 220 m (in 2002). That is a width reduction of 53%, 63% and 54%, respectively for the Tagliamento, Piave and Brenta rivers. A large amount of channel width reduction has occurred in the past 50 yr: in the Tagliamento 51% of the whole width reduction (narrowing of 440 m in the period 1954–2001); in the Piave 73%

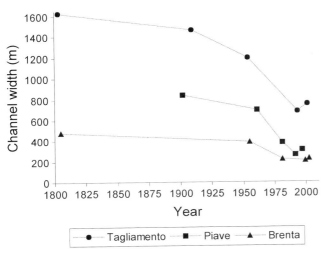

Fig. 2 Changes in channel width during the past 200 yr in the Tagliamento, Piave and Brenta Rivers.

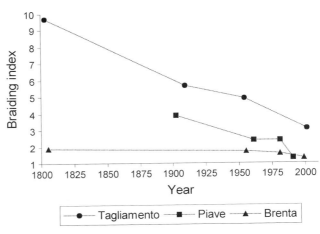

Fig. 3 Changes in braiding index during the past 200 yr in the Tagliamento, Piave and Brenta Rivers.

(narrowing of 390 m in the period 1960–1997); in the Brenta 62% (narrowing of 160 m in the period 1955–2002).

However, in the past few years there has been a reverse of this trend, with widening instead of narrowing (Fig. 2). In all three rivers, maximum channel narrowing was reached in the late 1980s or early 1990s (58%, 69% and 58% of width reduction, respectively in the Tagliamento, Piave and Brenta), and then channel widening has taken place during the past few years (widening of 70 m between 1993 and 2001 in the Tagliamento, of 50 m between 1991 and 1997 in the Piave, and of 20 m between 1999 and 2002 in the Brenta).

Braiding intensity

Braiding intensity, which is a fundamental characteristic of braided rivers, is typically described using simple measurements of the number or length of individual anabranches in a braided river (Thorne, 1997; Ashmore, 2001). For this study the braiding index proposed by Ashmore (1991) was adopted, which is defined by the average number of anabranches across the river. Historical trends in the braiding indices show that a remarkable decrease has taken place in all three rivers (Fig. 3): from 9.7 (early 19th century) to 3.1 (in 2001) in the Tagliamento; from 3.9 (early 20th century) to 1.3 (in 1991) in the Piave; from 1.9 (early 19th century) to 1.3 (in 1999) in the Brenta. In the Tagliamento and Piave Rivers therefore, where braiding intensity was higher, the average number of anabranches is actually about one-third the number at the beginning of the 19th or 20th century.

It is important to note that the measurement of braiding index is likely to be more affected by errors as compared with the measurement of channel width: first, because braiding index is dependent on water level and second, because historical maps could be less reliable than aerial photographs for braiding index determination. Thus the braiding indices shown in Fig. 3 could be slightly different from the 'true' values, especially in the 19th century and the first half of the 20th century, but it does not imply that the historical trends could be substantially different from those shown in the figure.

Bed elevation

Bed-level adjustments were evaluated from longitudinal profiles and cross-sections (Figs 4 & 5). In the Tagliamento and Piave Rivers bed-level lowering has been up to 3 m and 4 m, respectively (Fig. 4); in the Brenta incision has been up to 5 m (Castiglioni & Pellegrini, 1981). In all three rivers highest rates of incision are shown in the downstream reaches, at the transition between braided and single-thread planforms (see below for a description of these transitional reaches).

Even though there are some temporal gaps between surveys, available cross-sections allow

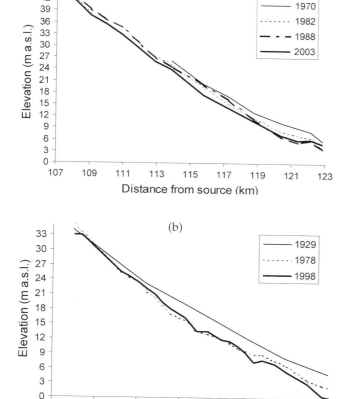

Fig. 4 Longitudinal profiles of the (a) Tagliamento River and (b) Piave River.

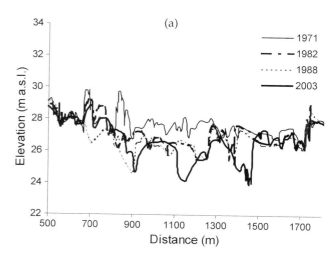

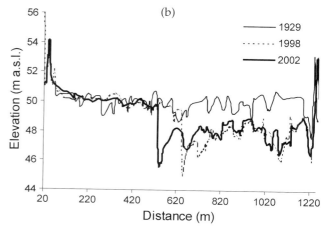

Fig. 5 Examples of cross-section changes in the (a) Tagliamento River and (b) Piave River.

reliable reconstructions of bed-level adjustment during the past decades (Fig. 5). The main phase of incision has likely occurred between the 1950s and the 1980s in the Piave and Brenta Rivers, and between the 1970s and the 1980s in the Tagliamento. Then, in the past 10–15 yr, there is evidence, in particular in the Piave and Brenta Rivers, that bed-level lowering has stopped. Cross sections, like that of Fig. 5b, show that locally there is some aggradation, but also a significant degree of bank erosion. Therefore erosional processes are still dominant in these rivers, even if presently erosion takes place mainly through bank erosion rather than bed-level lowering.

Channel pattern

Narrowing and incision have produced significant changes in the channel pattern of the three selected braided rivers. In the Brenta River major changes have taken place: at present in the upper reach (from Nove to S. Croce Bigolina) the river shows a wandering morphology, whereas in the lower reach (from S. Croce Bigolina to S. Giorgio in Brenta) a modification from braided to single-thread has commonly occurred (Fig. 6). On the other hand, notwithstanding significant channel adjustments, the Tagliamento River has preserved its braided morphology, as shown in Fig. 7. Due to the construction of a levee (in the 19th century) and to gravel mining (especially in the 1970s and 1980s) both channel width and braiding intensity have dramatically decreased during the last two centuries (channel width from about 2600 m to about 1200 m; braiding index from 12–15 to 3–5), but the river still retains a clear braided morphology. Changes

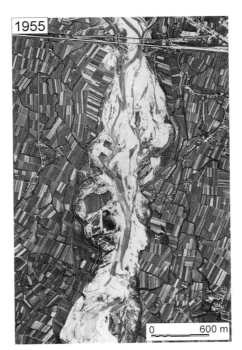

Fig. 6 Channel configuration in the Brenta River: change from braided to single-thread between 1955 and 1999.

of pattern in the Piave River are in between those of the Brenta and of the Tagliamento. In some reaches the Piave has retained a braiding morphology, even if with a much lower braiding intensity, whereas in other reaches a change from braiding to wandering has taken place.

Finally, it is worth describing the evolution in the transition reaches from a braided to single-thread morphology. The transition from braided to single-thread is relatively sharp in all the three rivers and, at least in the Tagliamento and Piave Rivers, is mainly due to channel slope (a remarkable decrease of the slope) rather than to sediment type or supply. The historical analysis shows an upstream migration of this transition (3–5 km) along all the three rivers (Fig. 8).

DISCUSSION AND CONCLUSIONS

Magnitude, temporal trends and styles of channel adjustments

As with most Italian rivers, in the last two centuries the three braided rivers analysed in this paper have undergone channel adjustments, particularly incision and narrowing. The magnitude of these adjustments is considerable: channel width reduction ranges between 53% (Tagliamento River) and 63% (Piave River); braiding index, which used to be much higher than 1 in all the three rivers, is at present not far from 1 in two cases out of three (Piave and Brenta); bed-level lowering has been up to 3 m, 4 m and 5 m, respectively in the Tagliamento, Piave and Brenta. These adjustments have also resulted in major changes to the channel pattern, from braided to wandering (Piave and Brenta) and from braided to single-thread (Brenta). In all three rivers, major change has also occurred in the transitional reaches with an upstream migration of the single-thread morphology. In contrast to the other two rivers, the Tagliamento River, which has always been characterized by the most pronounced braided morphology, has preserved its braided pattern, except for its lower reach where an upstream migration of the braided-single thread transition has taken place (not a downstream migration, as recently stated by Spaliviero (2003)).

Temporal trends of channel width suggest that channel adjustments have not been uniform during the last two centuries. Some adjustments have occurred during the 19th century and the first half of the 20th century, but major adjustments (narrowing and incision) have taken place since the 1950s or 1960s. Temporal trends (Fig. 2) and cross sections (Fig. 5b) also show that the major phase

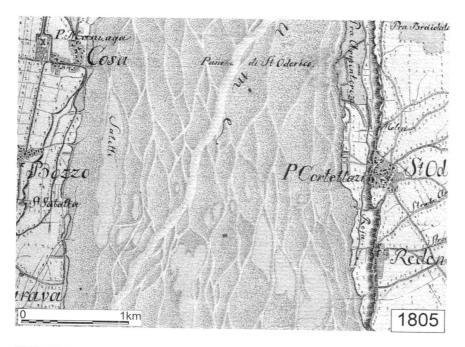

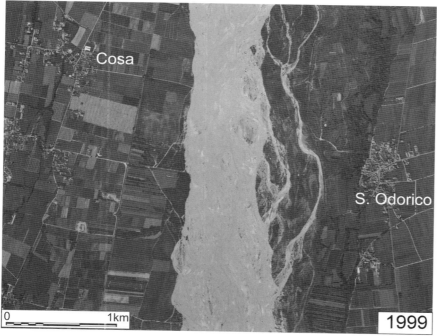

Fig. 7 Channel configuration in the Tagliamento River: conservation of the braided morphology notwithstanding, remarkable channel adjustments have taken place during the last two centuries.

of narrowing and incision should now be exhausted, since other kinds of processes, such as widening and (local) aggradation, have occurred in the last 10–15 yr.

The type of channel adjustments and the temporal trends described above allow some improvements of the classification scheme of channel evolution recently proposed by Surian & Rinaldi (2003) for Italian rivers (Fig. 9). The three rivers selected show that this scheme, which simply includes two stages (the initial stage and the final stage after major adjustments), could include four stages as regards braided rivers: (i) an initial stage, channel morphology at the beginning of the 19th century; (ii)

Fig. 8 Upstream migration of the transition from braided to single-thread morphology in the Piave River. The transition is clear in the 1954 aerial photograph (left), whereas it has migrated some kilometres upstream in 1999 (right).

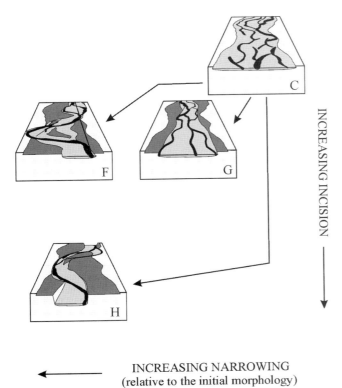

Fig. 9 Classification scheme of channel adjustments proposed by Surian & Rinaldi (2003) for Italian braided rivers: 'C' represents the initial morphology (in the early 19th or 20th century); 'F', 'G' and 'H' represent the present morphology due to variable degrees of incision and narrowing. (Modified from Surian & Rinaldi, 2003.)

a second stage, after a first phase of channel narrowing and, likely, of channel incision; (iii) a third stage, after the major narrowing and incision; (iv) a fourth stage (the present morphology), after channel widening. In the future this hypothesis (four stages of channel evolution) could be tested with more data regarding both the selected rivers and other braided rivers.

Causes of channel adjustments

The main causes of channel adjustments are: the construction of levees and other protection structures, causing the first phase of narrowing (the second of four stages mentioned above); dams and gravel mining, causing the major channel adjustments (the third and fourth stages). The construction of levees and other protection structures had a direct effect on river morphology, and in particular on channel width. Dams and gravel mining have had both direct and indirect effects on channel morphology: their major effect has been a remarkable decrease in sediment supply to river channels. Conversely, dams and diversions have not had a significant effect on channel-forming discharges. The relationship between those two interventions (dams and gravel mining) and channel adjustments seems straightforward: major phases of narrowing and incision correspond to the main periods of mining (between the 1950s and the 1980s) and began just after the closure of most of the dams (in the 1950s). However, since mining and dams have occurred concurrently and with similar magnitude (both interventions have caused a reduction in sediment supply of some tens of million cubic metres of sediment), evaluation of the relative importance of these two interventions is not straightforward. That said, in the study reaches of the three selected rivers gravel mining is likely to be more important than dams since (i) mining has been particularly intense in those reaches and (ii) dam effects should be less severe in the selected reaches (piedmont reaches) which are some tens of kilometres downstream of the dams.

The recent channel widening is connected to the formation of new depositional features (bars and islands), but this does not imply that rivers are now aggrading. In fact, cross sections show exactly the opposite, that is that erosional compared with depositional processes are still dominant (Fig. 5). Additionally, channel widening could be promoted by the change from a braided to a wandering pattern, especially in the Brenta and the Piave Rivers. The reduction of sediment mining could be one explanation for these widening processes; in fact mining has reduced considerably since the 1980s (in the Brenta and the Piave River) or the early 1990s (in the Tagliamento River) (unfortunately no complete records of the volumes of sediment mined are available, and therefore only a qualitative description of those interventions can be made). On the contrary, flows are unlikely to be a cause of channel widening. For example, in the Piave River no significant variation of flood magnitude or frequency took place before or during the widening period, which began, approximately, in the early 1990s (Fig. 10).

What about future conditions in these braided rivers? Considering that in Italy dams are still very important for the production of hydroelectric power, it is unlikely that dams will be removed in the near future. If all sediment mining is stopped (at present some mining is still carried out), then different morphological responses should be expected from rivers such as the Tagliamento, where dams have minor effects, and others, such as the Piave and the Brenta, where dams significantly alter sediment fluxes. Conversely, if sediment mining continues it is likely that these rivers are going to

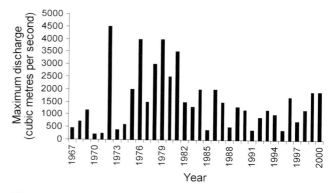

Fig. 10 Maximum annual discharge in the Piave River, at Nervesa barrage, between 1967 and 2000. Flow velocity measurements were not carried out at the barrage; discharges were estimated solely from water levels. (Modified from Canzian, 2001.)

show similar styles of channel adjustment in the near future.

In any case, it is worth remembering that there are other possible factors that could have had, or still have, some influence on channel evolution. These factors include climate changes (e.g. the Little Ice Age and recent global change) and land-use change (e.g. reforestation), and these should be taken into account when considering possible future scenarios of environmental conditions.

Management implications

The channel adjustments described in this study have important implications for river maintenance and management, since they affect structures, the environment and floods. The effect on structures is exemplified by the failure of two bridges along the Brenta River during the 1970s. With regard to the environment, two aspects are affected by channel adjustments: (i) groundwater resources and (ii) aquatic and riparian ecology. Since the recharge of the thick unconfined aquifers in the Venetian–Friuli Plain depends on river flows (seepage of a certain amount of these flows), there has been a significant loss of groundwater resources following channel narrowing and incision (for instance, incision causes a shortening of the river reaches where seepage takes place). As regards aquatic and riparian ecology there are several studies on the Tagliamento that indicate that the river still retains natural or semi-natural characteristics (Edwards *et al.*, 1999; Ward *et al.*, 1999; Arscott *et al.*, 2000; Gurnell *et al.*, 2001), whereas less information is available for the Piave and the Brenta. In any case, considering the magnitude of adjustments that occurred in these two rivers (e.g. changes in channel configuration), it is very likely that along the Piave and the Brenta, in contrast to the Tagliamento, there have been negative effects on aquatic and riparian ecology. Finally, channel adjustments have a significant effect on floods, since channel narrowing and incision cause a faster flood conveyance and, therefore, an increase of flood hazard in the downstream reaches where river channels are narrower and less steep than in the braided reaches.

For river management it is necessary to recognize the effects of channel adjustments in such braided rivers, and also the causes of these morphological changes. In the previous section it has been shown that the decrease of sediment supply to river channels has been the main cause for channel instability. This means that sediments are a key issue in these braided rivers, and that in the future river managers should deal much more with sediments, both in the drainage basins and river channels, if they want to avoid substantial alterations of natural processes, loss of environmental resources and increased risk to people.

Future research on braided rivers

This study suggests that there are still several issues about the braided rivers selected, and about braided rivers with similar conditions in Italy and in other countries, which require further research. Some of these issues are as follows.

1 Recent trends of channel adjustments: since there is evidence that in the past 10–15 yr channel processes have changed, it is important to reconstruct a detailed picture of these recent channel changes.
2 Prediction of future channel evolution: this prediction, which is fundamental for river managers, needs a detailed understanding of past and present channel adjustments and should be carried out according to different scenarios of human impact and, eventually, climate change.
3 Use of other approaches in the investigation of channel adjustments: for a better understanding of past channel evolution, and particularly for the prediction of future evolution, the historical analysis used in this study should be integrated with other types of approaches, for instance with laboratory experiments.

ACKNOWLEDGEMENTS

I wish to thank: Professor G.E. Petts and an anonymous referee for their helpful comments and suggestions; Marisa Spagnuolo for improving my English; and Renelda Stocco for helping with the analysis of mining data.

REFERENCES

Arscott, D.B., Tockner, K. and Ward, J.V. (2000) Aquatic habitat diversity along the corridor of an Alpine floodplain river (Fiume Tagliamento, Italy). *Arch. Hydrobiol.*, **149**(4), 679–704.

Ashmore, P. (1991) Channel morphology and bed load pulses in braided, gravel-bed streams. *Geogr. Ann.*, **73A**, 37–52.

Ashmore, P. (2001) Braiding phenomena: statics and kinetics. In: *Gravel Bed Rivers V* (Ed. M.P. Mosley), pp. 95–114. New Zealand Hydrological Society, Wellington.

Aucelli, P.P.C. and Rosskopf, C. (2000) Last century valley floor modifications of the Trigno River (Southern Italy): a preliminary report. *Geogr. Fis. Din. Quatern.*, **23**, 105–115.

Autorità di Bacino dei fiumi Isonzo, Tagliamento, Livenza, Piave, Brenta-Bacchiglione (1998) *Piano di Bacino del Fiume Tagliamento: Piano Stralcio per la Sicurezza Idraulica del Medio e Basso Corso*. Venice, 118 pp.

Bravard, J.P., Kondolf, G.M. and Piégay, H. (1999) Environmental and societal effects of channel incision and remedial strategies. In: *Incised River Channels: Processes, Forms, Engineering and Management* (Eds S.E. Darby and A. Simon), pp. 303–341. Wiley, Chichester.

Canzian, A. (2001) *Geomorfologia dell'alveo del Fiume Piave tra Nervesa e Ponte di Piave*. Unpublished thesis, University of Padova, Italy, 91 pp.

Castiglioni, G.B. and Pellegrini, G.B. (1981) Geomorfologia dell'alveo del Brenta nella pianura tra Bassano e Padova. In: *Il Territorio della Brenta* (Ed. M. Zunica), pp. 12–32. Amm. Prov. di Padova—Università di Padova.

D'Alpaos, L. and Dal Prà, A. (1996) *Studio per la Identificazione dei Vincoli e degli Aspetti Critici sia Idraulici che Naturalistici, Riguardanti le Escavazioni Potenziali dall'Alveo del Fiume Piave*. Autorità di

Bacino dei fiumi Isonzo, Tagliamento, Livenza, Piave, Brenta-Bacchiglione, 91 pp.

Dutto, F. and Maraga, F. (1994) Variazioni idrografiche e condizionamento antropico. Esempi in pianura padana. *Quaternario*, **7**, 381–390.

Edwards, P.J., Kollmann, J., Gurnell, A.M., Petts, G.E., Tockner, K. and Ward J.V. (1999) A conceptual model of vegetation dynamics on gravel bars of a large Alpine river. *Wetlands Ecol. Manag.*, **7**, 141–153.

Gurnell, A.M., Petts, G.E., Hannah, D.M., Smith, B.P.G., Edwards, P.J., Kollmann, J., Ward, J.V. and Tockner K. (2001) Riparian vegetation and island formation along the gravel-bed Fiume Tagliamento, Italy. *Earth Surf. Process. Landf.*, **26**, 31–62.

Kondolf, G.M., Piegay, H. and Landon, N. (2002) Channel response to increased and decreased bedload supply from land use change: contrasts between two catchments. *Geomorphology*, **45**, 35–51.

Maraga, F. (1989) Ambiente fluviale in trasformazione: l'alveo-tipo pluricursale verso un nuovo modellamento nell'alta pianura padana. *Proceedings of the International Congress on Geoengineering 'Suolosottosuolo'*, Torino, 27–30 September, pp. 119–128.

Pellegrini, M., Perego, S. and Tagliavini, S. (1979) La situazione morfologica degli alvei degli affluenti emiliani del Po. *Convegno di Idraulica Padana*, Parma, 19–20 October, 9 pp.

Petts, G.E., Gurnell, A.M., Gerrard, A.J., Hannah, D.M., Hansford, B., Morrissey, I., Edwards, P.J., Kollmann, J., Ward, J.V., Tockner, K. and Smith B.P.G. (2000) Longitudinal variation in exposed riverine sediments: a context for the ecology of the Fiume Tagliamento, Italy. *Aquat. Conserv. Mar. Freshwat. Ecosyst.*, **10**, 249–266.

Reniero, S. (1998) *Il Recupero Naturalistico delle Aree Fluviali. Prospettive Ecologiche e Limiti Idraulici per la Legge 183/89: Valutazione sul Medio Corso del Brenta*. Unpublished thesis, University of Padova, Italy, 234 pp.

Spaliviero, M. (2003) Historic fluvial development of the Alpine-foreland Tagliamento River, Italy, and consequences for floodplain management. *Geomorphology*, **52**, 317–333.

Surian, N. (1999) Channel changes due to river regulation: the case of the Piave River, Italy. *Earth Surf. Process. Landf.*, **24**, 1135–1151.

Surian, N. (2000) Sediment size in a gravel-bed river (Piave River, Italy): longitudinal, vertical and temporal variability. In: *Dynamics of Water and Sediments in Mountain Basins* (Ed. M. Lenzi). *Quad. Idronom. Mont.* (Special Issue), **20**, 131–143.

Surian, N. (2002) Downstream variation in grain size along an Alpine river: analysis of controls and processes. *Geomorphology*, **43**, 137–149.

Surian, N. (2003) Impatto antropico sulla dinamica recente del Fiume Piave (Alpi orientali). In: *Risposta dei Processi Geomorfologici alle Variazioni Ambientali* (Eds A. Biancotti and M. Motta), pp. 425–440. Glauco Brigati, Genova.

Surian, N. and Rinaldi, M. (2003) Morphological response to river engineering and management in alluvial channels in Italy. *Geomorphology*, **50**, 307–326.

Thorne, C.R. (1997) Channel types and morphological classification. In: *Applied Fluvial Geomorphology for River Engineering and Management* (Eds C.R. Thorne, R.D. Hey and M.D. Newson), pp. 175–222. Wiley, Chichester.

Ward, J.V., Tockner, K., Edwards, P.J., Kollmann, J., Bretschko, G., Gurnell, A.M., Petts, G.E. and Rossaro, B. (1999) A reference river system for the Alps: the Fiume Tagliamento. *Regul. River. Res. Manag.*, **15**, 63–75.

Wyzga, B. (1991) Present-day downcutting of the Raba River channel (Western Carpathians, Poland) and its environmental effects. *Catena*, **18**, 551–566.

Wyzga, B. (1997) Methods for studying the response of flood flows to channel change. *J. Hydrol.*, **198**, 271–288.

Ecology of braided rivers

KLEMENT TOCKNER*, ACHIM PAETZOLD*,[†], UTE KARAUS*, CÉCILE CLARET[‡]
and JÜRG ZETTEL[§]

*Department of Limnology, EAWAG, 8600 Dübendorf, Switzerland (Email: klement.tockner@eawag.ch)
[†]Catchment Science Centre, The University of Sheffield, Sheffield S37 1IQ, UK
[‡]IMEP, Ecologie des Eaux Continentales Méditerranéennes, Université Aix-Marseille III, 13397 Marseille cedex 20, France
[§]Institute of Zoology, University of Bern, 3012 Bern, Switzerland

ABSTRACT

Braided gravel-bed rivers are widespread in temperate piedmont and mountain-valley areas. In their pristine state, braided rivers are characterized by a shifting mosaic of channels, ponds, bars, and islands, since both flow and flood pulses create a diversity of habitats with fast turnover rates. Large wood has a major role in determining the geomorphology and ecological functioning of these rivers. Braided river habitats are colonized by a diverse fauna and flora adapted to their dynamic nature, including a significant proportion of highly endangered species. Animals exhibit high mobility, short and asynchronic life cycles, and ethological and phenological plasticity. Braided gravel-bed rivers also offer various categories of refugia such as shore areas, hypogeic and hyporheic habitats that are pivotal for maintaining diversity in the face of frequent disturbances. Today, however, most gravel-bed rivers bear little resemblance to their highly dynamic natural state due to anthropogenic modifications, and most braided rivers have been converted into incised single-thread channels. Gravel bars and vegetated islands are among the most endangered landscape elements worldwide. They are very sensitive to channelization, gravel extraction, and flow regulation. Therefore, more than for most other ecosystems, restoring braided rivers and their landscape elements means restoring their underlying hydrogeomorphological dynamics.

Keywords Biodiversity, island, pond, conservation, adaptation, floodplain, shifting habitat mosaic, large wood, restoration Tagliamento, Danube.

INTRODUCTION

Natural rivers are dynamic, and physically and biologically complex (Tockner & Stanford, 2002). They are characterized by a set of fluvial styles including straight, braided, wandering, and meandering channels (Richards *et al.*, 2002). Conditions that promote braided channel formation include: (i) an abundant supply of sediment; (ii) rapid and frequent variations in water discharge; and (iii) erodible banks of non-cohesive material (Church & Jones, 1992).

Braided gravel-bed rivers were once widespread in temperate piedmont and mountain-valley areas, primarily in regions containing young, eroding mountains (e.g. Alaska, Canada, New Zealand, the Himalayas, the European and Japanese Alps; Fig. 1a). Today, most gravel-bed rivers bear little resemblance to their highly dynamic natural state. A recent survey of all Austrian rivers with a catchment area of >500 km^2 showed that formerly braided sections have been particularly affected by channelization and flow regulation (Muhar *et al.*, 1998), with 25% of confined river sections, but only 1% of braided sections, remaining intact. Braided rivers are among the most endangered ecosystems (Sadler *et al.*, 2004). However, in Europe, Japan and in most parts of the USA, remaining braided rivers are among the very limited areas, in otherwise highly managed landscapes, where natural large-scale disturbances still are allowed to occur. Therefore, they serve as excellent model

Fig. 1 (A) The braided river mouth of the Kurobe River (Toyama Prefecture, Japan). Japanese rivers are steep, short and exhibit a flashy flow regime (Yoshimura et al., 2005). (B) Island-braided section along the Tagliamento River (northeast Italy) showing the complex mosaic of aquatic (channels, backwaters, parafluvial ponds) and terrestrial (gravel, large wood, vegetated islands) habitats. The active floodplain is ~900 m wide (Photograph: K. Tockner.)

systems to study the relationship between multiple disturbances (floods, droughts) and ecology, and to elaborate upon the complex relationship between habitat complexity and biodiversity. Braided rivers are also key areas for conservation and restoration since they provide habitat for a highly endangered fauna and flora (Tockner et al., 2003; Sadler et al., 2004). The understanding of their natural complexity and dynamics, however, forms the prerequisite for developing sustainable management schemes (Ward et al., 2001).

Though geomorphological knowledge of braided rivers is quite extensive (e.g. Billi et al., 1992; Best & Bristow, 1993; Richards et al., 2002, and references therein), only scattered information is available on their ecology (Plachter & Reich, 1998; Karrenberg et al., 2002; Robinson et al., 2002; Ward et al., 2002). This paper provides a general introduction to the ecology of braided rivers, primarily to those with gravel beds. Braided sand-bed rivers such as the famous Brahmaputra–Jamuna network are not included, although environmental conditions and ecological properties are supposed to be similar to gravel-bed rivers.

A brief description of the specific environmental conditions of braided rivers is followed by an overview of their fauna and flora, and how species are able to cope with the hostile conditions of highly dynamic systems. The importance of vegetated islands and parafluvial ponds in maintaining biodiversity is discussed, with a focus on shoreline communities and on the complex trophic linkages across the aquatic and terrestrial boundary. Finally, some perspectives for future research on braided rivers are provided. The Tagliamento River in north-eastern Italy, the largest remaining active gravel-bed river in central Europe (Fig. 1B), constitutes the main case study. On this river, an ongoing transdisciplinary research project integrates hydrology, geomorphology and ecology in order to understand the complexity and diversity of an Alpine braided river corridor (Ward et al., 1999b; Gurnell et al., 2001; Tockner et al., 2003; Francis et al., this volume, pp. 361–380).

THE ENVIRONMENTAL TEMPLATE OF BRAIDED RIVERS

Braided rivers consist of multiple channels with bars and islands, often with poorly defined banks of non-cohesive sedimentary materials. Cut-and-fill alluviation, channel avulsion, and production, entrainment and deposition of large wood (LW), coupled with ground- and surface-water interactions, create a complex and dynamic array of aquatic, amphibious and terrestrial landscape elements, which can be referred to as the shifting habitat mosaic (SHM; Stanford, 1998; Poole et al., 2002; Ward et al., 2002; Stanford et al., 2005). The SHM is composed of habitats, ecotones and gradients that possess biotic distributions and biogeochemical cycles that change in response to fluvial processes. This mosaic allows many species to coexist in the riverine landscape (Ward et al., 1999a; Robinson et al., 2002; Tockner & Stanford, 2002; Gurnell et al., 2005).

Habitat turnover

Frequently disturbed ecosystems with strong disturbance regimes such as braided rivers experience rapid turnover rates of the abiotic and the biotic components. This can create, for example, a high percentage of pioneer vegetation stages on freshly deposited sediments (e.g. Hughes, 1997). While the minimum age of the oldest floodplain sites on braided rivers, as a surrogate for turnover time, is often 100+ yr, habitats in the active braided tract are very young in comparison. Along the Tagliamento River, Italy, ~60% of the aquatic area and 30% of vegetated islands were observed to have been restructured in just 2.5 yr (Van der Nat et al., 2003b). The maximum age of vegetated islands was ~20 yr (Kollmann et al., 1999). Parafluvial ponds and backwaters were the youngest habitats, with half-life expectancies of <7 months. The degree of habitat change was determined by flood magnitude, time since the last flood, and the presence of large wood and vegetated islands. The presence of vegetated islands has led to an average increase in habitat age by providing more stable habitats (Gurnell & Petts, 2002). Overall, flood dynamics reconfigured the spatial environment, although the composition and diversity of aquatic habitats remained constant (Arscott et al., 2002). These high spatiotemporal dynamics conform very well to the shifting habitat steady-state concept proposed by Bormann & Likens (1979) and make braided rivers unique among terrestrial and aquatic ecosystems.

Expansion and contraction dynamics

Riverine floodplains are increasingly recognized as expanding, contracting and fragmenting ecosystems (Stanley et al., 1997; Malard et al., 1999; Tockner et al., 2000). The extent, composition and configuration of aquatic and terrestrial habitats vary in response to the pulsing of discharge. For example, in the braided Ohau River in New Zealand a gradual increase in flow from 26 to 507 $m^3 s^{-1}$ generated additional channels with the same characteristics as those existing at lower discharge, and the total number of channels at a cross-section remained constant (Mosley, 1982). This process had major implications for instream uses such as salmonid spawning because the area suitable for spawning remained constant over a wide range of flows. In this respect a braided river was considered morphologically more stable than a single-thread river.

In the Val Roseg (Switzerland) floodplain, a braided proglacial river, a distinct shift of individual water sources (groundwater, snowmelt water and glacial water) creates a complex mosaic of clear and turbid water patches during the seasonal expansion and contraction cycle (Malard et al., 1999, in press). The annual flow pulse is highly predictable and exhibits a unimodal pattern (Ward & Uehlinger, 2003). Only a small proportion of the total channel network (around 2.4 km of the 25 km channels) maintains benign environmental conditions (permanent flow, high substrate stability, low turbidity and relatively high temperature). Periods of benign environmental conditions are restricted to short periods in spring and autumn ('ecological windows of opportunity' *sensu* Ward & Uehlinger, 2003). These ecological windows are crucial for instream primary production, growth of benthic organisms, and maintaining species diversity (e.g. Burgherr et al., 2002). For high alpine areas future climate change scenarios predict an increase of streams with snowmelt- and rain-dominated flow regimes (Ward & Uehlinger, 2003). Concomitantly, there will be major shifts in the timing and duration of the occurrence of 'ecological windows'. In contrast to proglacial ecosystems, most braided rivers show polymodal inundation patterns. The Tagliamento River, for example, exhibits a flashy flow regime with frequent dry–wet cycles and a linear relationship between water level and inundated area (Van der Nat et al., 2002). On this river, inundation is primarily by lateral overspill of water from the main channel and by upwelling of alluvial groundwater.

During expansion and contraction cycles the boundary between water and land (shoreline) moves across the river–floodplain system (moving littoral sensu Junk et al., 1989). Shoreline length can be used as an index of habitat quality and availability (Tockner & Stanford, 2002). In dynamic systems (such as the Tagliamento River), shoreline length can be up to 25 km per river-kilometre and remains high throughout the annual cycle. In a regulated river (e.g. the Danube Alluvial Zone National Park, Austria) the availability of shoreline habitats strongly fluctuates during the annual cycles of expansion and contraction. In channelized rivers (e.g. the Rhône, Switzerland) shoreline length is only about 2 km per river-kilometre (Fig. 2). Decrease in shoreline length not only affects habitat availability for endangered communities but also impedes the exchange of matter and organisms between the river and its riparian area (see below).

Thermal heterogeneity

Temperature is a primary factor that regulates ecosystem processes, and therefore structures biotic communities (Ward, 1992). Lateral and vertical heterogeneity in temperature has been recognized as an important aspect in habitat conditions along rivers (Brunke & Gonser, 1997; Arscott et al., 2001; Ebersole et al., 2003; Uehlinger et al., 2003). Across a braided reach of the Tagliamento River average daily temperature difference between the coolest and the warmest water body ranged from 4°C to almost 17°C, with maximum diel amplitudes within

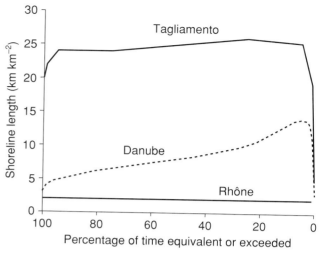

Fig. 2 Shoreline length (km km^{-2}) and duration relationship in natural (Tagliamento, northeast Italy), constrained (Danube, Alluvial National Park, Austria) and channelized (Rhône, Switzerland) river–floodplain systems. All flood plains are characterized by a dynamic hydrology (Van der Nat et al., 2002). The Rhône and the Tagliamento River are comparable in discharge and catchment area. In their pristine state, the Rhône and the Danube were morphologically similar to the present braided Tagliamento River.

an individual water body of up to 26°C (Karaus et al., 2005). A lateral thermal difference of up to 17°C on a given day corresponded to the observed difference along the entire 170-km long main river channel (Arscott et al., 2001). Water bodies in the braided river can be arranged along a gradient from low to high diel amplitudes and seasonal averages (Fig. 3). Factors that create and maintain thermal heterogeneity across the corridor include surface connectivity to the main channel, input of alluvial and hillslope groundwater, and the presence of vegetated islands and large wood. Water bodies fed by alluvial and hillslope groundwater exhibit relatively constant thermal conditions. Shallow ponds situated in the bare gravel matrix show very distinct diel and seasonal temperature amplitudes. In particular the presence of vegetated islands and large wood leads to a decrease in the diel temperature amplitudes in floodplain ponds (Fig. 3).

For a comprehensive thermal characterization of rivers both aquatic and terrestrial habitats must be simultaneously investigated. Exposed gravel sediments may exhibit large diel temperature differences, with maximum values at the surface of up to

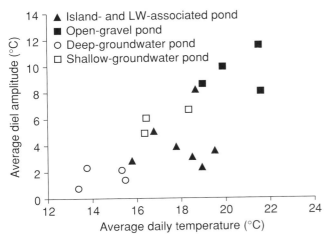

Fig. 3 Thermal heterogeneity (May to October 2001), expressed as average daily temperature and average diel amplitude, in 18 water bodies across a braided section along the Tagliamento River (northeast Italy). (After Karaus et al., 2005.)

50°C during a hot summer day, while groundwater upwelling provides relatively constant cold water areas. Therefore, a major challenge in the near future will be the application of non-invasive methods such as infrared (IR) cameras to map thermal heterogeneity at large scales. This would allow the identification of critical spots for aquatic (cold-water refugia, spawning areas for salmonids; Torgersen et al., 1999; Baxter & Hauer, 2002) and terrestrial (shading by vegetated islands) organisms, and to monitor thermal patch dynamics (a similar phenomenon to the shifting habitat concept). Methods used so far for thermal assessment in streams and rivers are characterized by either high spatial (IR-spectroscopy; Torgersen et al., 1999) or temporal resolution (temperature loggers; e.g. Arscott et al., 2001; Uehlinger et al., 2003). However, modern IR-thermography makes detailed (high spatiotemporal resolution) and precise temperature measure-ments possible. IR-thermography has already been successfully used to measure temperatures of landscape elements such as rock outcrops, block glaciers, and snow fields from distances of 100 m to several kilometres (Tanner, 1999), and to estimate mammal density in wetlands (Naugle et al., 1996).

Surface and subsurface exchange processes

Braided rivers are complex above and below ground (Fig. 4). Subsurface habitats across a river–floodplain transect include the hyporheic zone beneath the channel (alluvium areas saturated in water), parafluvial zones that extend lateral to the channel (saturated areas), and unsaturated hypogeic sediments. Geomorphological heterogeneity and highly permeable substrates, particularly those composed of open framework gravels, facilitate high hydrological connectivity. Considering different river styles, a braided pattern is predicted to allow maximal subsurface–surface exchanges, therefore braided rivers are likely to present the highest diversity of surface and subsurface exchange types (Malard

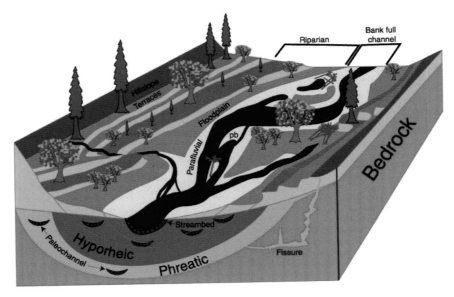

Fig. 4 An idealized three-dimensional view of a braided floodplain structure (pb, point bar), emphasizing dynamic longitudinal, lateral and vertical dimensions and the role of large wood in the natural–cultural setting of the catchment. (After Stanford, 1998.)

et al., 2002). Subsurface compartments may provide refugia for benthic organisms in unstable sites, but decrease in importance in more stable systems (Fowler & Death, 2001). As suggested by Pugsley & Hynes (1986), the hyporheic zone may offer many small-scale refugia for benthic organisms, thereby contributing to the stability (resilience) of the hydrosystem at a large scale. The boundaries of the subsurface zones fluctuate in response to changes in surface water discharge and groundwater pressure. Hence, the extent of the subsurface habitats varies greatly in space and time.

Interstitial sediments act as 'filters' for infiltrating surface water, and subsurface water that emerges to the surface may be of a different quality, with important consequences for surface processes. A more stable temperature regime (see Fig. 3) is one of the most obvious cues for subsurface influence on surface water providing thermal refugia for stenothermic species (Arscott *et al.*, 2001; Malard *et al.*, 2001; Poole, 2002). Qualitatively, upwelling DOM (dissolved organic matter) is mainly biodegradable, while refractory fractions seem to be physically adsorbed on sediments and the biofilm matrix (Claret *et al.*, 1998). In addition, subsurface water that upwells is generally enriched in inorganic nitrogen and phosphorus. Therefore, upwelling water influences algae productivity, benthic assemblages and locations of fish spawning (Brunke & Gonser, 1997; Stanford, 1998). Alluvial springbrooks, for example, provide thermal refugia for fish in addition to functioning as refugia during erosive floods in the main channel (Baxter & Hauer, 2000). In the Nyack floodplain (Montana, USA), increased water availability (hydration) and nutrient delivery (fertilization) in upwelling areas is associated with significantly higher diversity and productivity of riparian plants compared with losing areas (Harner & Stanford, 2003). Subsurface–surface exchange also increases water residence time and enhances hydrological nutrient retention within the stream corridor. For example, in some streams the volume of the hyporheic zone exceeds that of the surface channel and may extend laterally for several kilometres in rivers with large alluvial floodplains (e.g. Stanford *et al.*, 1994).

The significance of surface–subsurface exchanges on biodiversity is not limited to animals living beneath the surface, but may also concern surface organims. Subsurface inputs of water to the stream are often patchily distributed contributing to high spatial heterogeneity of habitat conditions in surface waters. Upwelling of subsurface water may maintain water levels or available moisture during dry periods, provides clear water as fine particles are trapped within the sediments, and adds cooler and more nutrient-rich water to the river.

While the hyporheic zone beneath the wetted channel is well known as an active component of the stream ecosystem, the unsaturated zone is a 'black box', at least from the ecological point of view. However, extensive layers of unsaturated gravel, often several metres thick, are a key feature of braided rivers (hypogeic zone). If only parts of the extensive hypogeic crevasse system are accessible for riparian invertebrates, it is likely to be the most extensive habitat within braided rivers (Plachter & Reich, 1998). Unsaturated sediments differ from saturated sediments in their physical characteristics. The top layer of the hyporheic zone is often clogged by fine particles and the hydraulic connectivity of unsaturated sediments is one to three orders of magnitude lower than that of saturated zones (Huggenberger *et al.*, 1998). During the short inundation periods in high-gradient river sediments may remain unsaturated beneath the water surface. This unsaturated zone may be crucial for the survival and the rapid recolonization of terrestrial arthropods after flood and drought events. Little is known about which organisms are using these habitats and how they respond to fluctuating water tables. Dieterich (1996) exposed sediment cages at different depths within a gravel bar, and found a diverse invertebrate community comprising aquatic (oligochaeta, larvae of midges and stoneflies) and terrestrial (mites, rove beetles, ground beetles) species. High densities of terrestrial invertebrates occurred in winter, which underpins the potential role of unsaturated sediments as refuge areas during the cold season (e.g. for hibernation). Future research has to focus on: (i) the physico-chemical characterization (microclimate, sediment structure, organic matter content) of the unsaturated zone; (ii) its functional role for the transformation of organic matter and nutrients (e.g. Claret *et al.*, 1997); and (iii) its importance as habitat and refuge areas for aquatic and, in particular, terrestrial invertebrates.

LIFE IN BRAIDED RIVERS

From an ecological perspective, braided rivers are considered as hostile environments as: (i) they are very dynamic systems shaped by frequent floods and periods of water stress; (ii) their channels flow through very unproductive areas of exposed gravel and sand (low organic content); and (iii) they are characterized by high fluxes of temperature and humidity (particularly on bare gravel surfaces). Although braided rivers are extreme environments located on the descending limb of a harshness–diversity curve, they show a very high overall biodiversity, and are particularly important regionally (Burgherr et al., 2002; Robinson et al., 2002; Tockner & Stanford, 2002). The high species richness and diversity in braided rivers can be explained by small-scale habitat mosaics encompassing aquatic habitats as well as terrestrial ones, by the presence of ecotones from the freshly created shoreline to mature riverine forests, and by multiple subsurface–surface exchange areas.

Rivers and floodplains are supposed to represent the ancestral conditions for many aquatic and terrestrial species, and braided rivers are expected to be important centres of biological diversification. The speciation of groundwater crustaceans is believed to be favoured by the shifting of river channels that lead to the isolation of formerly connected channels (e.g. cyclopids in the alluvial aquifer of the braided Danube River; P. Pospisil & D. Danielopol, personal communication). Fittkau & Reiss (1983) suggested that dynamic river–floodplain systems belong to those aquatic ecosystems where biota of lentic areas (standing water bodies) started their evolution. The temporal continuity of riverine systems and their associated disturbance regimes allowed the permanent presence of lentic water bodies throughout time, which would have facilitated such evolution. Ancient crustacean orders such as Conchostraca and Notostraca (e.g. *Chirocephalus* spp. *Lepidurs apus*) still live in river–floodplain systems. An example is *Lepidurus apus*, which has existed for more than 200 million years, and is therefore presumably the oldest living species on Earth (e.g. Eder et al., 1997). In alluvial springbrooks, several species of the 'primitive' subfamilies of Diamesinae and Prodiamesinae (Diptera, Chironomidae) still occur. Similarly, the Salicaceae (willows) originally developed along dynamic rivers and subsequently began to occupy more stable habitats, although the family is still primarily associated with rivers (Karrenberg et al., 2002). Some morphological and phenological characteristics related to their life on the active zone (e.g. small seed size) are highly conservative and have not changed in response to altered selection pressures during the course of evolution (Karrenberg et al., 2002).

Aquatic and terrestrial communities

The fauna of dynamic floodplain rivers comprises a mix of obligate terrestrial species to obligate aquatic species, ranging from meiofauna in the hyporheos to mammals on vegetated islands. In braided rivers, the interplay between vertical and lateral connectivity results in diverse plant and faunal communities (see e.g. Robinson et al., 2002; Brunke et al., 2003). Lateral hydrological connectivity, for example, can be expressed as the duration of a water body connected at the surface to the main river channel. Along the Danube River (Alluvial Zone National Park, Austria), hydrological connectivity was the main determinant that explained the distribution of aquatic organisms across a river–floodplain gradient (Fig. 5; Tockner et al., 1998; Reckendorfer et al., 2005). Most species exhibited a distinct preference for a specific degree of connectivity. Based on species-specific 'connectivity preferences', the potential effect of river regulation and river–floodplain restoration on community composition can be predicted. Arscott et al. (2003) illustrated how differences in habitat stability, by comparing two contrasting headwater floodplains, influenced abundance and diversity patterns of caddis flies (Trichoptera). Beta diversity (spatial species turnover) was important for Trichoptera diversity in the highly dynamic braided river, whereas alpha diversity was high in a stable forested floodplain. This means that diversity is differently organized in braided rivers compared with more stable rivers. Similarly, Castella et al. (1991) illustrated how floodplain structure and dynamics influenced the scale at which invertebrates are organized. In the regulated Rhône River between-channel heterogeneity dominated, whereas in the more dynamic Ain River within-channel diversity dominated

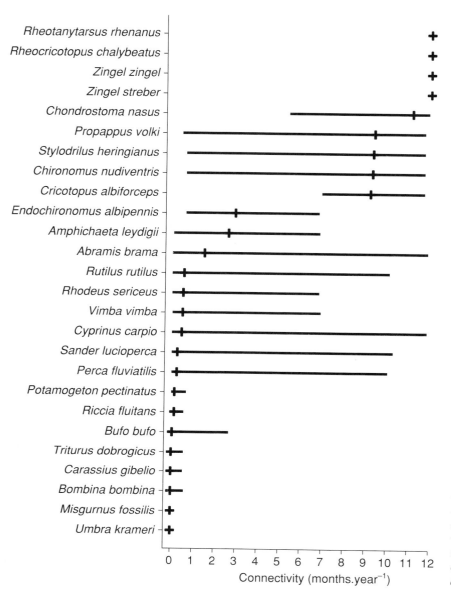

Fig. 5 Danube floodplain (Alluvial Zone National Park, Austria): species optima (different faunal groups) and tolerance ranges across a hydrological connectivity gradient. Connectivity is expressed as the number of months with an upstream hydrological surface connection to the main channel. (Modified after Reckendorfer et al., 2005.)

(i.e. higher alpha diversity within individual water bodies).

Distinct spatial differentiations of aquatic communities occur at even smaller scales. Around a large bar complex along the Fraser River (BC, Canada), Church et al. (2002) identified 13 aquatic habitat types on the basis of nine physical characteristics. Each of the habitats had a quite distinct fish assemblage, with eddy pools and channel nooks having the highest species diversity and fish density. Surprisingly, this demonstrated the relatively small spatial scale at which the fundamental connection is made between the biota and the physical system that supports them. The scale was set by the very local scale at which the animals differentiate their environment and adapt their behaviour. Another critical feature of habitat units is their limited occurrence at very high flows. At flows above 7000 $m^3 s^{-1}$ most open bar tops in Fraser River become submerged and large numbers of fish occupied the relatively slack water there. Above 9000 $m^3 s^{-1}$ (i.e. approximately mean annual flood), there were significant currents over most bar tops, and bar-edge habitats disappeared. Therefore, at these discharge levels fishes that normally occupy these units must seek refuge elsewhere. It is known that fish and

Table 2 Different reproduction strategies of the common toad (*Bufo b. bufo*) in relation to the dynamics of river–floodplain water bodies (After Kuhn, 1993)

Reproduction biology	Permanent water bodies	Temporary and dynamic floodplain water bodies
Spawning time	Early, fixed	Opportunistic
Spawning aggregation	Large (mostly)	Small
Male tactics	Search and fight	Mating calls
Selection of spawning sites	ND	Distinct
Spawning site fidelity	Distinct	ND

ND = no data available.

ditions in this zone compared with immigrants from the floodplain? All of these aspects are crucial for the development of restoration and conservation strategies for dynamic ecosystems.

Braided rivers are very dynamic systems that shift between dry and wet states seasonally or with a more stochastic rhythm. Depending on the organisms considered, drying or rewetting can trigger 'boom' and 'bust' periods (cf. Kingsford *et al.*, 1999). Terrestrial invertebrates, for example, rapidly colonize dry habitats, benefiting from stranding and dying aquatic organisms. Unfortunately, simultaneous investigations of aquatic and terrestrial assemblages have not yet been performed.

Trophic linkages across boundaries

Bar edges of braided rivers are inhabited by a diverse and abundant carnivorous arthropod community, primarily spiders, rove beetles, ground beetles and ants (Plachter & Reich, 1998; Sadler *et al.*, 2004). Large differences in productivity, together with diverse assemblages of generalist consumers, make gravel banks model systems to study spatial subsidies. Gut analyses of riparian ground beetles in a braided river (Isar, Germany) revealed a high proportion of aquatic insects in their diet (Hering & Plachter, 1996). Most recently, Paetzold *et al.* (2005) have proved that predation by ground dwelling arthropods is a quantitatively important pathway in the transformation of aquatic secondary production to the riparian food web. Ground beetles fed entirely on aquatic organisms; 80% of the diet of rove beetles, and ~50% of the diet of most abundant riparian spiders were of aquatic origin. Spiders and ground beetles can in turn significantly reduce the emergence of aquatic insects along the river shore, thereby controlling population dynamics of aquatic insects such as stoneflies and caddis flies (Paetzold & Tockner, 2005). Average consumption along the river edge by riparian arthropods can be as high as 40% of the total average aquatic insect emergence in the adjacent channel. Riparian arthropods along braided-river banks are generally highly mobile organisms, adapted to living in a highly dynamic environment (Plachter & Reich, 1998). Therefore, they can rapidly respond to pulsed subsidies by spatial redistributions. Mass emergence during a short, well-defined period of the year is common among aquatic insects and can be synchronized with long-term flood dynamics (Lytle & Poff, 2004). Peaks in surface drifting organisms occur during storms through accidental input of riparian invertebrates to the stream.

The efficiency of the transfer rate of energy across the aquatic–terrestrial boundary depends on a high ratio of shoreline length to stream area and the permeability of the boundary along braided banks. The alteration of riparian habitats may reduce the energy transfer between the channel and an adjacent gravel bar. Further, a decline in diversity of shoreline invertebrates is supposed to reduce the functional linkage between aquatic and terrestrial systems, although no data are available yet. Studies along 12 (formerly) braided rivers, which differed in their degree of hydrological and morphological modification, demonstrated that river regulation altered fundamentally the entire riparian arthropod community. Channelized sections that are also impacted by hydropeaking were almost completely devoid of terrestrial arthropods (A. Paetzold & K. Tockner, unpublished data).

Convex and concave islands

Lentic (standing) water bodies and vegetated islands are neglected habitats of braided rivers, mainly because of their absence in most extant rivers. They are very sensitive landscape elements that disappear as a consequence of river regulation, wood removal, and flow control. Today, ponds and islands are among the most endangered landscape features along river corridors (Homes et al., 1999; Gurnell & Petts, 2002; Hohensinner et al., 2004; Karaus et al., 2005). They can be used as ecosystem-level indicators of the environmental condition of river corridors.

River research has concentrated on the lotic channel. Only recently have lentic water bodies also been considered as integral habitats along river corridors (Ward et al., 2002). In particular **parafluvial ponds**, lentic water bodies within the active zone, are distinct habitats within the river mosaic. Investigation of pond heterogeneity and invertebrate diversity along the entire length of three braided Alpine river corridors exhibited peaks of parafluvial ponds in bar- and island-braided river reaches, with a maximum of 29 ponds per river-kilometre (Karaus et al., 2005). Numerous ponds were associated with vegetated islands. Indeed, the presence of vegetated islands enhanced the diversity of aquatic habitats (Arscott et al., 2000; Table 3). Ponds were absent in regulated and naturally constrained sections along each corridor. Although covering a small proportion of the total aquatic area (<3%), ponds contributed >50% to invertebrate diversity (U. Karaus & K. Tockner, unpublished data). Each pond harboured a distinct fauna, which supports the idea that ponds have an insular nature ('concave island') with a characteristic set of environmental properties. A high proportion of pond species were classified as rare, with a limited spatial distribution. Thermal characteristics (see Fig. 3) and water level fluctuations were the most important variables determining pond heterogeneity. Detailed investigations of aquatic invertebrates along an entire river corridor have confirmed the role of standing water bodies as foci of biodiversity (Arscott et al., 2005). Results suggested that heterogeneity in invertebrate assemblages within a habitat (e.g. a small pond) was equally as important as heterogeneity within or among reaches (i.e. lateral and longitudinal dimensions). The inclusion of lateral habitats such as backwaters and ponds also changed the interpretation of zoobenthic organization along the river continuum and illustrated that lateral habitats served as important sinks and sources for invertebrates. Overall, corridor water bodies provided redundancy and novelty (Arscott et al., 2005).

Vegetated islands are 'high energy landforms' (Osterkamp, 1998). Their formation requires: (i) a

Table 3 Biocomplexity in an island-braided and a bar-braided floodplain (Tagliamento River, NE Italy)

	Bar-braided	Island-braided	Reference
Width of active channel (m)	830	1000	Tockner et al. (2003)
Large wood (t ha^{-1})	15–73	102–158	Van der Nat et al. (2003a)
Channels (half-life expectancy; months)	4.1	7.7	Van der Nat et al. (2003b)
Aquatic habitat diversity (H')	1.6	2.0	Arscott et al. (2000)
Average number of ponds	7	22	Van der Nat et al. (2002)
Average shoreline length (km km^{-1})	13.7	20.9	Van der Nat et al. (2002)
Amphibian species	5	7	Klaus et al. (2002)
Carabid beetle species	34	47	Tockner et al. (2003)
Benthic invertebrates: α-diversity	30	27	Arscott et al. (2005)
Benthic invertebrates: β_2-diversity	10.5	21	Arscott et al. (2005)
Benthic invertebrates: γ-diversity	50	53	Arscott et al. (2005)

natural flood regime; (ii) an unconstrained river corridor; (iii) a sediment source; and (iv) a source of large woody debris. Such a combination of conditions is not present in highly managed river systems (Ward et al., 2002, Gurnell & Petts, 2002). For example, over 650 vegetated islands (>0.007 ha) occur along the corridor of the Fiume Tagliamento (Tockner et al., 2003; see Fig. 1B). Islands are, however, among the first landscape elements that disappear as a consequence of river regulation. For example, only six islands remain of approximately 2000 islands historically present in the Austrian Danube. There, the construction of hydropower plants led to a decrease in gravel bars and vegetated islands by 94% and 97%, respectively (Hohensinner et al., 2004). Similarly, the response of former braided rivers in the Great Plains (USA) to upstream dam construction was channel-narrowing and a reduction in geomorphological dynamics (Friedman et al., 1996). Channel narrowing was associated with an increase in native and exotic woody riparian species colonizing the former channel bed. Such increases in vegetation density can substantially alter channel geometry. Lateral mobility decreases, braiding intensity decreases and channel relief increases (Gran & Paola, 2001).

Ecologically, islands are pivotal landscape elements. They represent early successional stages, are colonized by a diverse and often endangered fauna and flora, are almost devoid of invasive species, have a high perimeter-to-area ratio, serve as stepping stones for migrating organisms such as small mammals, and serve as important natural retention structures along river corridors. Today, the dynamics and biodiversity of islands can be investigated only in large river beds with a whole set of islands of different size and different age. In Europe there are only a few possibilities left for such studies, as most large rivers have been modified: confined by dykes to protect arable land and housing, or used to produce hydraulic energy.

Floating organic matter links aquatic and terrestrial environments

Water movement links the riverine environments. As water moves across the river–floodplain system, it develops a hydrogeomorphological linkage that allows organisms to disperse and transport nutrients, sediments, and organic matter. Floating organic matter links aquatic with terrestrial compartments and upstream with downstream segments of river ecosystems (both energetically and as a vector for terrestrial and aquatic organisms). Given that some animals can survive to reach a downstream shore, upstream sections may serve as a source of colonists. In addition, canyons that separate alluvial valleys contain pockets of habitats that may operate as either refuges or population sinks periodically colonized by flood-borne animals.

The main driving factor in the cycling of floating organic debris and its associated fauna is the pulsing of flow. During flood events large amounts of organic material and organisms float downstream. During the decreasing limb of the hydrograph, organic material aggregates and accumulates in 'dead zones' and at retention structures along shoreline habitats, where it forms distinct 'drift lines'. With an increase in the water level, deposited material becomes resuspended and transported downstream (Fig. 9). Floating organic matter serves as a major dispersal vector for aquatic and terrestrial organisms along river corridors (Tenzer, 2003). Recent results from the Tagliamento River demonstrate that coarse particulate organic matter fractions such as wood, fruits or grass are exclusively transported at the water surface. Abundance and composition of invertebrates change rapidly between transport, accumulation and deposition phases. The number of organisms associated with floating organic debris is on average 20 times higher than that in the water column, and is primarily composed of terrestrial organisms. Many of the organisms are transported over long distances (tens of kilometres; Langhans, 2000). The removal of organic matter upstream of hydropower plants is considered to have a major impact on the ecology of river systems that significantly reduces local riparian species richness in the downstream reaches (Andersson et al., 2000). Floating material provides an indicator of the integrity of entire river corridors. However, basic research is required to establish and calibrate floating organic matter as an integrative indicator of connectivity along river corridors.

PROSPECT

Most braided rivers have been converted into incised single-thread channels (Surian, 1999; Piégay

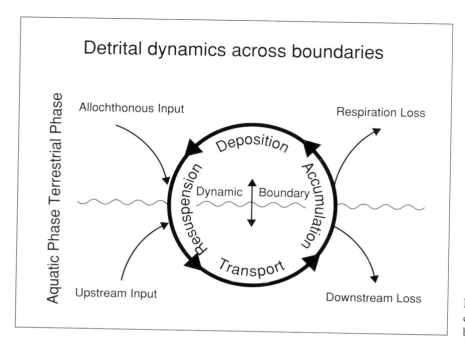

Fig. 9 Floating organic matter dynamics across aquatic–terrestrial boundaries.

et al., this volume, pp. 257–275). Remaining braided rivers are under enormous pressure from channelization, gravel exploitation and flood control measures. During recent decades river restoration activities have increased rapidly. Although a huge amount of detailed environmental knowledge has been compiled, up to 60% of all restoration projects are still ineffective (Tockner & Stanford, 2002). The main reasons for project failure include: (i) the 'missing-link' between natural and social science; and (ii) the lack of reference data from near-pristine ecosystems, which constrains our understanding and replication of ecosystem processes. More than for most other ecosystems, restoring braided rivers means restoring their underlying hydrogeomorphological dynamics. However, for the successful conservation and restoration of braided rivers it is necessary to document how the variable physical conditions of braided rivers govern species diversity and ecosystem processes. Future research on braided rivers has to focus on:

1 the ecological importance of the unsaturated zone;
2 groundwater–surface-water interactions and their role in controlling biodiversity and bioproduction;
3 thermal patch dynamics at varying spatial scales;
4 the relationship between habitat age diversity and species identity and richness;
5 adaptations and dispersal pathways in rivers;
6 the influence of the adjacent woodland on the active zone;
7 the importance of ponds and islands as foci for biodiversity;
8 how ponds and islands are physically and ecologically linked;
9 upscaling from experiments to entire river corridors in order to identify 'hot spots' and 'hot windows' of process activity.

In conclusion, it is clear that braided rivers are predetermined as model ecosystems for testing general ecological concepts.

ACKNOWLEDGEMENTS

The research presented in this paper is supported by a grant from the ETH-Forschungskommission (0-20572-98), by the EU-funded project 'tempQsim' (BBW Nr. 02.0072), and by a grant from the Rhône–Thur Project (EAWAG). Many thanks to David Arscott, Mark Lorang, Urs Uehlinger, Simone Langhans, Michael Döring, Dimitry van der Nat, Geoffrey Petts, Angela Gurnell, Peter Edwards, and James Ward for stimulating discussions during the Tagliamento project. Comments by Greg Sambrook Smith, Geoffrey Petts and an anonymous reviewer helped to improve this manuscript.

GLOSSARY

Adaptation Characteristics of an organism evolved as a consequence of natural selection in its evolutionary past and/or changes in the form or behaviour of an organism during its life as a response to environmental stimuli.

Biodiversity (or Biological Diversity) The variety of life in all its forms, levels and combinations. Includes ecosystem diversity, species diversity, and genetic diversity.

Community Populations of different species that interact with each other in an area.

Dispersal The spreading of individuals away from each other, e.g. of offspring from their parents and from regions of high density to regions of lower density.

Disturbance Any relatively discrete event in space and time that disrupts ecosystem, community, or population structure and changes resources, substrate, or the physical environment.

Ecological stability Ability of a property (e.g. population, community) of an ecosystem to return toward a steady-state equilibrium following a disturbance.

Ecosystem The interacting system of a biological community and its environmental surroundings which together form a recognizable self-contained entity.

Food web Representation of feeding relationships in a community that includes all the links revealed by dietary analysis.

Habitat Place or type of place where an organism, population, or community lives.

Hydrological connectivity Water-mediated transfer of matter, energy and/or organisms within or between elements of the hydrologic cycle.

Hyporheic zone A spatially fluctuating ecotone between the surface stream and the deep groundwater where important ecological processes and their requirements and products are influenced at a number of scales by water movement, permeability, substrate particle size, resident biota, and the physicochemical features of the overlying stream and adjacent aquifers.

Meiofauna Small benthic metazoans that pass through a 0.500 mm sieve and are retained on a 0.063 (or 0.045 mm) sieve. The majority of recognized phyla have meiofaunal representatives.

Metapopulation Set of local populations within some larger area, where typically dispersal from one local population to at least some other patches is possible.

Productivity The rate at which biomass is produced per unit area by any class of organisms.

Refugia Areas from which the recolonization after a disturbance event occurs.

Resilience Resilience, or relative stability, is a measure of the rate at which the property or system (see stability) approaches steady state following a disturbance.

r-Strategy Ecological strategy where organisms rely on high reproductive rates for continued survival within the community. Populations of r-strategists are subject to extreme fluctuations.

Sustainability Maintenance and/or improvement of the integrity of the life-support system on Earth.

REFERENCES

Adis, J. and Junk, W.J. (2002) Terrestrial invertebrates inhabiting lowland river floodplains of Central Amazonia and Central Europe: a review. *Freshwat. Biol.*, **47**, 711–731.

Andersson, E., Nilsson, C. and Johansson, M.E. (2000) Effects of river fragmentation on plant dispersal and riparian flora. *Regul. River. Res. Manag.*, **16**, 83–89.

Arscott, D.B., Tockner, K. and Ward, J.V. (2000) Aquatic habitat diversity along the corridor of an Alpine floodplain river (Fiume Tagliamento, Italy). *Arch. Hydrobiol.*, **149**, 679–704.

Arscott, D.B., Tockner, K. and Ward, J.V. (2001) Thermal heterogeneity along a braided floodplain river (Tagliamento River, northeastern Italy). *Can. J. Fish. Aquat. Sci.*, **58**, 2359–2373.

Arscott, D.B., Tockner, K., Van der Nat, D. and Ward, J.V. (2002) Aquatic habitat dynamics along a braided Alpine river ecosystem (Tagliamento River, Northeast Italy). *Ecosystems*, **5**, 802–814.

Arscott, D.B., Keller, B., Tockner, K. and Ward, J.V. (2003) Habitat structure and Trichoptera diversity in 2 headwater flood plains, N.E. Italy. *Int. Rev. Hydrobiol.*, **88**, 255–273.

Arscott, D.B., Tockner, K. and Ward, J.V. (2005) Lateral organization of aquatic invertebrates along the corridor of a braided floodplain river. *J. N. Am. Benthol. Soc.*, **24**, 934–954.

Baxter, C.V. and Hauer, F.R. (2000) Geomorphology, hyporheic exchanges, and selection of spawning habitat by bull trout (*Salvelinus confluentus*). *Can. J. Fish. Aquat. Sci.*, **57**, 1470–1481.

Best, J.L. and Bristow, C.S. (Eds) (1993) *Braided Rivers*. Special Publication 75, Geological Society Publishing House, Bath.

Bill, H.C., Spahn, P., Reich, M. and Plachter, H. (1997) Bestandsveränderungen und Besiedlungsdynamik der Deutschen Tamariske, Myricaria germanica (L.) Dessv., an der Oberen Isar (Bayern). *Z. Ökol. Naturs.*, **6**, 137–150.

Billi, P., Hey, R.D., Thorne, C.R. and Tacconi, P. (Eds) (1992) *Dynamics of Gravel-bed Rivers*. Wiley, Chichester.

Bonn, A. and Kleinwächter, M. (1999) Microhabitat distribution of spider and ground beetle assemblages (Araneae, Carabidae) on frequently inundated river banks of the River Elbe. *Z. Ökol. Naturs.*, **8**, 109–123.

Bormann, F.H. and Likens, G.E. (1979) *Pattern and Process in a Forested Ecosystem*. Springer, New York.

Boscaini, A., Franceschini, A. and Maiolini, B. (2000) River ecotones: carabid beetles as a tool for quality assessment. *Hydrobiologia*, **422/423**, 173–181.

Brunke, M. and Gonser, T. (1997) The ecological significance of exchange processes between rivers and groundwater. *Freshwat. Biol.*, **37**, 1–33.

Brunke, M., Hoehn, E. and Gonser, T. (2003) Patchiness of river-goundwater interactions within two floodplain landscapes and diversity of aquatic invertebrate communities. *Ecosystems*, **6**, 707–722.

Burgherr, P., Robinson, C.T. and Ward, J.V. (2002) Seasonal variation in zoobenthos across habitat gradients in an alpine glacial floodplain (Val Roseg, Swiss Alps). *J. N. Am. Benthol. Soc.*, **21**, 561–575.

Castella, E., Richardot-Coulet, M., Roux, C. and Richoux, P. (1991) Aquatic macroinvertebrate assemblages of two contrasting floodplains: the Rhône and Ain rivers, France. *Regul. River. Res. Manag.*, **6**, 289–300.

Church, M. and Jones, D. (1992) Channel bars in gravel-bed streams. In: *Gravel Bed Streams* (Eds R.D. Hey, J.C. Bathurst and C.R. Thorne), pp. 291–338. Wiley, Chichester.

Church, M., Ham, D. and Weatherly, H. (2002) *Gravel Management in the Lower Fraser River*. Report for the City of Chilliwack, BC, University of British Columbia, Vancouver, 104 pp.

Claret, C., Marmonier, P., Boissier, J.-M., Fontvieille, D. and Blanc, P. (1997) Nutrient inputs from parafluvial interstitial water to the river: importance of gravel bar heterogeneity. *Freshwat. Biol.*, **37**, 656–670.

Claret, C., Marmonier, P. and Bravard, J.-P. (1998) Seasonal dynamics of nutrient and biofilm in interstitial habitats of two contrasting riffles in a regulated large river. *Aquat. Sci.*, **60**, 33–55.

Dieterich, M. (1996) Methoden und erste Ergebnisse aus Untersuchungen zur Lebensraumfunktion von Schotterkörpern in Flussauen. *Verh. Ges. Ökol.*, **26**, 363–367.

Ebersole, J.L., Liss, W.J. and Frissell, C.A. (2003) Cold water patches in warm streams: physicochemical characteristics and the influence of shading. *J. Am. Wat. Resour. Assoc.*, **39**, 355–368.

Eder, E., Hödl, W. and Gottwald, R. (1997) Distribution and phenology of large branchiopods in Austria. *Hydrobiologia*, **359**, 13–22.

Fittkau, E.J. and Reiss, F. (1983) Versuch einer Rekonstruktion der Fauna europäischer Ströme und ihrer Alpen. *Arch. Hydrobiol.*, **97**, 1–6.

Fowler, R.T. and Death, R.G. (2001) The effect of environmental stability on hyporheic community structure. *Hydrobiologia*, **445**, 85–95.

Francis, R.A., Gurnell, A.M., Petts, G.E. and Edwards, P.J. (this issue) Riparian tree establishment on gravel bars: interactions between plant growth strategy and the physical environment. In: *Braided Rivers: Process, Deposits, Ecology and Management* (Eds G.H. Sambrook Smith, J.L. Best, C.S. Bristow and G.E. Petts), pp. 361–380. Special Publication 36, International Association of Sedimentologists. Blackwell, Oxford.

Friedman, J.M., Osterkamp, W.R. and Lewis, W.M. (1996) Channel narrowing and vegetation development following a Great Plains flood. *Ecology*, **77**, 2167–2181.

Gran, K. and Paola, C. (2001) Riparian vegetation controls on braided stream dynamics. *Water Resour. Res.*, **37**, 3275–328.

Gurnell, A.M. and Petts, G.E. (2002) Island-dominated landscapes of large floodplain rivers, a European perspective. *Freshwat. Biol.*, **47**, 581–600.

Gurnell, A.M., Petts, G.E., Hannah, D.M., Smith, B.P.G., Edwards, P.J., Kollmann, J., Ward, J.V. and Tockner, K. (2001) Island formation along the gravel-bed Fiume Tagliamento, Italy. *Earth Surf. Proc. Landf.*, **26**, 31–62.

Gurnell, A.M., Tockner, K., Edwards, P.J. and Petts, G. (2005) Effects of deposited wood on biocomplexity of river corridors. *Front. Ecol. Environ.*, **3**, 377–382.

Harner, M.J. and Stanford, J.A. (2003) Differences in cottonwood growth between a losing and a gaining reach of an alluvial floodplain. *Ecology*, **84**, 1453–1458.

Hering, D. and Plachter, H. (1996) Riparian ground beetles (Coleoptera, Carabidae) preying on aquatic invertebrates: a feeding strategy in alpine floodplains. *Oecologia*, **111**, 261–270.

Hohensinner, S., Habersack, H., Jungwirth, M. and Zauner, G. (2004) Reconstruction of the characteristics of a natural alluvial river–floodplain system and hydromorphological changes following human modifications: the Danube River (1812–1991). *Riv. Res. Appl.*, **20**, 25–41.

Holderegger, B. (1999) *Autökologie von Sphingonotus caerulans (Latreille, 1804) und Oedipoda caerulescens (Linné, 1758) (Orthoptera, Acrididae) in zwei unterschiedlichen Zonationstypen im Pfynwald (VS, Schweiz)*. Diploma thesis, Zoological Institute, University of Bern.

Homes, D., Hering, D. and Reich, M. (1999) The distribution and macrofauna of ponds in stretches of an alpine floodplain differently impacted by hydrological engineering. *Regul. River. Res. Manag.*, **15**, 405–417.

Huggenberger, P., Hoehn, E., Beschta, R. and Woessner, W. (1998) Abiotic aspects of channels and floodplains in riparian ecology. *Freshwat. Biol.*, **40**, 407–425.

Hughes, F.M.R. (1997) Floodplain biogeomorphology. *Progr. Phys. Geogr.*, **21**, 501–529.

Junk, W.J., Bayley, P.B. and Sparks, R.E. (1989) The flood pulse concept in river–floodplain systems. *Can. Fish. Aquat. Sci. Spec. Publ.*, **106**, 110–127.

Karaus, U., Alder, L. and Tockner, K. (2005) 'Concave islands': habitat heterogeneity and dynamics of parafluvial ponds in a gravel-bed river. *Wetlands*, **25**, 26–37.

Karrenberg, S., Edwards, P.J. and Kollmann, J. (2002) The life history of Salicaceae living in the active zone of flood plains. *Freshwat. Biol.*, **47**, 733–748.

Klaus, I., Baumgartner, C. and Tockner, K. (2002) Die Wildflusslandschaft des Tagliamento (Italien, Friaul) als Lebensraum einer artenreichen Amphibiengesellschaft. *Z. Feldherpetol.*, **8**, 21–30.

Kingsford, R.T., Porter, A.L. and Porter, J. (1999) Water flow on Cooper Creek in arid Australia determine 'boom' and 'bust' periods for waterbirds. *Biol. Conserv.*, **88**, 231–248.

Kollmann, J., Vieli, M., Edwards, P.J., Tockner, K. and Ward, J.V. (1999) Interactions between vegetation development and island formation in the Alpine river Tagliamento. *Appl. Veg. Sci.*, **2**, 25–36.

Kuhn, J. (1993) Fortpflanzungsbiologie der Erdkröte *Bufo b. Bufo* (L.) in einer Wildflussaue. *Z. Ökol. Naturs.*, **2**, 1–10.

Langhans, S. (2000) *Schwemmgut: Indikator der ökologischen Integrität einer Flussaue. Tagliamento, Italien*. Unpublished Diplomarbeit, ETH Zürich.

Lude, A., Reich, M. and Plachter, H. (1999) Life strategies of ants in unpredictable floodplain habitats of alpine rivers (Hymenoptera: Formicidae). *Entomol. Gener.*, **24**, 74–91.

Lytle, D.A. and Poff, N.L. (2004) Adaptation to natural flow regimes. *Trends Ecol. Evol.*, **19**, 94–100.

Malard, F., Tockner, K. and Ward, J.V. (1999) Shifting dominance of subcatchment water sources and flow paths in a glacial floodplain, Val Roseg, Switzerland. *Arc. Antarc. Alp. Res.*, **31**, 114–129.

Malard, F., Lafont, M., Burgherr, P. and Ward, J.V. (2001) A comparison of longitudinal patterns in hyporheic and benthic assemblages in a glacial river. *Arc. Antarc. Alp. Res.*, **33**, 457–466.

Malard, F., Tockner, K. Dole-Olivier, M.-J. and Ward, J.V. (2002) A landscape perspective of surface-subsurface hydrological exchanges in river corridors. *Freshwat. Biol.*, **47**, 621–640.

Malard, F., Uehlinger, U., Zah, R. and Tockner, K. (In press) Flood pulses and riverscape dynamics in a braided glacial river. *Ecology*.

Manderbach, R. and Plachter, H. (1997) Lebensstrategie des Laufkäfers Nebria picicornis (F. 1801) (Coleoptera, Carabidae) an Fliessgewässerufern. *Beitr. Ökol.*, **3**, 17–27.

Montgomery, D. R. (1999) Process domains and the river continuum. *J. Am. Wat. Resour. Assoc.*, **35**, 397–410.

Mosley, M.P. (1982) Analysis of the effect of changing discharge on channel morphology and instream uses in braided river, Ohau River, New Zealand. *Wat. Resour. Res.*, **18**, 800–812.

Muhar, S., Kainz, M., Kaufmann, M. and Schwarz, M. (1998) Erhebung und Bilanzierung flusstypspezifisch erhaltener Fliessgewässerabschnitte in Österreich. *Öst. Wass. Abfallw.*, **5/6**, 119–127.

Naugle, D.E., Jenks, J.A. and Kernohan, B.J. (1996) Use of thermal infrared sensing: to estimate density of white-tailed deer. *Wildl. Soc. Bull.*, **24**, 37–43.

Osterkamp, W.R. (1998) Processes of fluvial island formation, with examples from Plum Creek, Colorado and Snake River, Idaho. *Wetlands*, **17**, 530–545.

Paetzold, A. and Tockner, K. (2005) Quantifying the effects of riparian arthropod predation on aquatic insect emergence. *J. N. Am. Benthol. Soc.*, **24**, 395–402.

Paetzold, A., Schubert, C. and Tockner, K. (2005) Aquatic-terrestrial linkages along a braided river: Riparian arthropods feeding on aquatic insects. *Ecosystems*, **8**, 748–759.

Piégay, H., Grant, G., Nakamura, F. and Trustrum, N. (2006) Braided river management: from assessment of river behaviour to improved sustainable development. In: *Braided Rivers: Process, Deposits, Ecology and Management* (Eds G.H. Sambrook Smith, J.L. Best, C.S. Bristow and G.E. Petts), pp. 257–275. Special Publication 36, International Association of Sedimentologists. Blackwell, Oxford.

Plachter, H. and Reich, M. (1998) The significance of disturbances for populations and ecosystems in natural floodplains. *Proceedings of the International Symposium on River Restoration*, Tokyo, Japan, pp. 29–38.

Poiani, K.A., Richter, B.D. anderson, M.G. and Richter, H.E. (2000) Biodiversity conservation at multiple scales: functional sites, landscapes, and networks. *BioScience*, **50**, 133–146.

Poole, G.C. (2002) Fluvial landscape ecology: addressing uniqueness within the river discontinuum. *Freshwat. Biol.*, **47**, 641–660.

Poole, G.C., Stanford J.A., Frissell, C.A. and Running, S.W. (2002) Three-dimensional mapping of geomorphic controls on floodplain hydrology and connectivity from aerial photos. *Geomorphology*, **48**, 329–347.

Pugsley, C.W. and Hynes, H.B.N. (1986) Three-dimensional distribution of winter stonefly nymphs, *Allocapnia pygmea*, within the substrate of a Southern Ontario River. *Can. J. Fish. Aquat. Sci.*, **43**, 1812–1817.

Reckendorfer, W., Schmalfuss, R., Baumgartner, C., Habersack, H., Hohensinner, S., Jungwirth, M. and Schiemer, F. (2005) The Integrated River Engineering Project for the free-flowing Danube in the Austrian Alluvial Zone National Park: framework conditions, decision process and solutions. In: *Large Rivers. Arch. Hydrobiol.* (Suppl.), **155**, 169–186.

Reich, M. (1991) Grasshoppers (Orthoptera, Saltatoria) on alpine and dealpine riverbanks and their use as indicators for natural floodplain dynamics. *Regul. River. Res. Manag.*, **6**, 333–340.

Reich, M. (1994) Kies- und schotterreiche Wildflußlandschaften—primäre Lebensräume des Flußregenpfeifers (Charadrius dubius). *Vogel Umwelt*, **8**, 43–52.

Rempel, L.L., Richardson, J.S. and Healey, M.C. (1999) Flow refugia for benthic invertebrates during flooding of a large river. *J. N. Am. Benthol. Soc.*, **18**, 34–48.

Richards, K., Brasington, J. and Hughes, F. (2002) Geomorphic dynamics of floodplains: ecological implications and a potential modelling strategy. *Freshwat. Biol.*, **47**, 559–579.

Robinson, C.T., Tockner, K. and Ward, J.V. (2002) The fauna of dynamic riverine landscapes. *Freshwat. Biol.*, **47**, 661–677.

Sadler, J.P., Bell, D. and Fowles, A. (2004) The hydroecological controls and conservation value of beetles on exposed riverine sediments in England and Wales. *Biol. Conserv.*, **118**, 41–56.

Schatz, I., Steinberger, K.-H. and Kopf, T. (2002) *Auswirkungen des Schwellbetriebes auf uferbewohnende Arthropoden (Araneae; Inseceta: Coleoptera: Carabidae, Staphylinidae) am Inn im Vergleich zum Lech (Tirol, Österreich)*. Institut für Zoologie und Limnologie, Universty of Innsbruck.

Siepe, A. (1994) Das 'Flutverhalten' von Laufkäfern (Coloptera: Carabidae), ein Komplex von ökoethologischen Anpassungen in das Leben der periodisch überfluteten Aue—I: Das Schwimmverhalten. *Zool. Jb. Syst.*, **121**, 515–566.

Sempeski, P., Gaudin, P. and Herouin, E. (1998) Experimental study of young grayling (Thymallus thymallus) physical habitat selection factors in an artificial stream. *Arch. Hydrobiol.*, **141**, 321–332.

Stanford, J.A. (1998) Rivers in the landscape: introduction to the special issue on riparian and groundwater ecology. *Freshwat. Biol.*, **40**, 402–406.

Stanford, J.A., Ward, J.V. and Ellis, B.K. (1994) Ecology of the alluvial aquifers of the Flathead River, Montana. In: *Groundwater Ecology* (Eds J. Gibert, D.L. Danielopol, and J.A. Stanford), pp. 313–347. Aquatic Ecology Series, Academic Press, San Diego.

Stanford, J.A., Lorang, M.S. and Hauer, F.R. (2005) The shifting habitat mosaic of river ecosystems. *Verh. Internat. Verein. Limnol*, **29**, 123–136.

Stanley, E.H., Fisher, S.G. and Grimm, N.B. (1997) Ecosystem expansion and contraction in streams. *Bioscience*, **47**, 427–435.

Stelter, C., Reich, M., Grimm, V. and Wissel, C. (1997) Modelling persistence in dynamic landscapes: lessons from a metapopulation of the grasshopper *Bryodema tuberculata*. *J. Animal Ecol.*, **66**, 508–518.

Surian, N. (1999) Channel changes due to river regulation: the case of the Piave River, Italy. *Earth Surf. Process. Landf.*, **24**, 1135–1151.

Tanner, C. (1999) *Temperaturmessungen im Schnee, ein Projekt im Rahmen des koordinierten Lawinendienstes*. Eidgenössische Materialprüfungsanstalt, Dübendorf, Switzerland, 25 pp.

Tenzer, C. (2003) *Ausbreitung terrestrischer Wirbelloser durch Fliessgewässer*. Unpublished Doctoral thesis, University of Marburg (Germany), 184 pp.

Tockner, K. and Stanford, J. A. (2002) Riverine flood plains: present state and future trends. *Environ. Conserv.*, **29**, 308–330.

Tockner, K., Schiemer, F. and Ward, J.V. (1998) Conservation by restoration: the management concept for a river–floodplain system on the Danube River in Austria. *Aquat. Conserv.*, **8**, 71–86.

Tockner, K., Malard, F. and Ward, J.V. (2000) An extension of the Flood Pulse Concept. *Hydrol. Process.*, **14**, 2861–2883.

Tockner, K., Ward, J.V., Arscott, B.A., Edwards, P.J., Kollmann, J., Gurnell, A.M., Petts, G.E. and Maiolini, B. (2003) The Tagliamento River: A model ecosystem of European importance. *Aquat. Sci.*, **65**, 239–253.

Tockner, K., Klaus, I., Baumgartner, C. and Ward, J.V. (In press) Amphibian diversity and nestedness in a dynamic floodplain ecosystem (Tagliamento, NE Italy). *Hydrobiologia*.

Torgersen, C.E., Price, D.M., Li, H.W. and McIntosh, B.A. (1999) Multiscale thermal refugia and stream habitat associations of Chinook Salmon in Northeastern Oregon. *Ecol. Appl.*, **9**, 301–319.

Uehlinger, U., Malard, F. and Ward, J.V. (2003) Thermal patterns in the surface waters of a glacial river corridor (Val Roseg, Switzerland). *Freshwat. Biol.*, **48**, 284–300.

Van der Nat, D., Schmid, A.P., Tockner, K., Edwards, P.J. and Ward, J.V. (2002) Inundation dynamics in braided floodplains: Tagliamento River, northeast Italy. *Ecosystems*, **5**, 636–647.

Van der Nat, D., Tockner, K., Edwards, P.J. and Ward, J.V. (2003a) Large wood dynamics of complex Alpine river flood plains. *J. N. Am. Benthol. Soc.*, **22**, 35–50.

Van der Nat, D., Tockner, K., Edwards, P.J., Ward, J.V. and Gurnell, A.M. (2003b) Habitat change in braided flood plains (Tagliamento, NE-Italy). *Freshwat. Biol.*, **48**, 1799–1812.

Vischer, D. and Oplatka, M. (1998) Der Strömungswiderstand eines flexiblen Ufer- und Vorlandbewuchses. *Wasserwirtschaft*, **88**, 1–5.

Ward, J.V. (1992) *Aquatic Insect Ecology. Biology and Habitat*. Wiley, New York.

Ward, J.V. and Uehlinger, U. (Eds) (2003) *Ecology of a Glacial Floodplain*. Kluwer, Dordrecht.

Ward, J.V., Tockner, K. and Schiemer, F. (1999a) Biodiversity of floodplain river ecosystems: ecotones and connectivity. *Regul. River. Res. Manag.*, **15**, 125–139.

Ward, J.V., Tockner, K., Edwards, P.J., Kollmann, J., Bretschko, G., Gurnell, A.M., Petts, G.E. and Rossaro, B. (1999b) A reference system for the Alps: the 'Fiume Tagliamento'. *Regul. River. Res. Manag.*, **15**, 63–75.

Ward, J.V., Tockner, K., Uehlinger, U. and Malard, F. (2001) Understanding natural patterns and processes in river corridors as the basis for effective river restoration. *Regul. River. Res. Manag.*, **17**, 311–323.

Ward, J.V., Tockner, K., Arscott, D.B. and Claret, C. (2002) Riverine landscape diversity. *Freshwat. Biol.*, **47**, 517–539.

Winterbourn, M.J. (1978) Macroinvertebrate fauna of a New-Zealand forest stream. *N. Z. J. Zool.*, **5**, 157–169.

Wintersberger, H. (1996) Spatial resource utilisation and species assemblages of larval and juvenile fish. *Arch. Hydrobiol.(Supplement)*, **115**, 29–44.

Wohlgemuth-von Reiche, D., Griegel, A. and Weigmann, G. (1997) Reaktion terrestrischer Arthropodengruppen auf Überflutungen der Aue im Nationalpark Unteres Odertal. *Arbeitsber. Landschaftsökol. Münster*, **18**, 193–207.

Yoshimura, C., Omura, T., Furumai, H. and Tockner, K. (2005) Present state of rivers and streams in Japan. *River Res. Appl.*, **21**, 93–112.

Riparian tree establishment on gravel bars: interactions between plant growth strategy and the physical environment

ROBERT A. FRANCIS*, ANGELA M. GURNELL*, GEOFFREY E. PETTS[†] and PETER J. EDWARDS[‡]

*Department of Geography, King's College London, Strand, London, WC2R 2LS, UK (Email: robert.francis@kcl.ac.uk)
[†]School of Geography, Earth and Environmental Sciences, University of Birmingham, Edgbaston, Birmingham B15 2TT, UK
[‡]Geobotanical Institute, ETH Zentrum, GEO D 7, CH-8044 Zürich, Switzerland

ABSTRACT

This paper investigates the influence of propagule form (seedlings, vegetative fragments and living drift-wood) and the physical environment (relative elevation and sediment calibre of the location of deposition) on the establishment of two pioneer riparian tree species, *Populus nigra* L. and *Salix elaeagnos* Scop. Field observations suggest that vegetative reproduction by these species is a key process in the establishment of vegetation on bar surfaces, and also the formation of wooded islands. In a field experiment conducted along the braided Tagliamento River, Italy, during the 2002 growing season, cuttings of the two species were established in locations representing differing combinations of relative elevation (depth to water table) and sediment calibre. Survival and growth parameters were measured both during and at the end of the growing season. Measurements of seedling growth and photomonitoring of stem growth from living wood debris were also undertaken.

Initial results indicate that cuttings and seedlings of both species respond differently to combinations of elevation and sediment calibre, with cuttings performing best at lower elevations but showing no consistent preference for sediment calibre, while seedlings perform best at higher elevations and show a preference for either coarse (*S. elaeagnos*) or fine sediments (*P. nigra*). Of the two environmental parameters, elevation may have a greater importance for survival and growth of *P. nigra*, while sediment calibre may be the key variable for *S. elaeagnos*. Analysis of stem lengths from the various propagules suggests that growth rates can be substantially higher among fluvially deposited trees, followed by cuttings and then seedlings. Therefore, the topographical and sedimentary characteristics of bar formations in relation to propagule type and species deposited can exert a substantial influence upon vegetation establishment and, consequently, subsequent patterns and rates of sedimentation. This in turn may have a considerable impact on island development and river morphology.

Keywords *Populus nigra*, *Salix elaeagnos*, riparian vegetation, sedimentation, braided river, Tagliamento River.

INTRODUCTION

In many interpretations of river form and process, vegetation is seen as a passive factor, with its distribution, zonation and growth responding to the allogenic influences of river flow and sedimentation (Hupp & Osterkamp, 1985; Pautou & Decamps, 1985; Van Splunder et al., 1995; Robertson & Augspurger, 1999). Recently, it has been suggested that riparian vegetation may play a more active, autogenic role in modifying its environment, distribution, zonation and succession (Tooth & Nanson, 2000; Gurnell & Petts, 2002). A key example of this is the stabilization of bars leading to ridge and island development (Gurnell *et al.*, 2001), which may influence river pattern (Tooth & Nanson, 2000; Gran & Paola, 2001). The degree of the influence of riparian vegetation on its environment depends upon the rate and amount of biomass it produces.

Many pioneer species along rivers are opportunistic tree species such as the Salicaceae in the Northern Hemisphere, which reproduce in large numbers and have rapid above- and below-ground growth (Malanson, 1993; Braatne *et al.*, 1996). The Salicaceae form an important part of most temperate riparian vegetation communities (Malanson, 1993), with the distribution, zonation and vigour of members of the Salicaceae being strongly related to flow dynamics and water availability; although there are wide variations between species and sexes, species within this family being dioecious (Saachi & Price, 1992; Johnson, 1994; Douglas, 1995; Van Splunder *et al.*, 1995, Hughes *et al.*, 2000; Barsoum, 2002). Flood events are responsible for the hydrochorous distribution of both seeds and vegetative propagules (Murray, 1986; Johansson *et al.*, 1996; Goodson *et al.*, 2001). The timing and severity of a flood event, in conjunction with the quantity and physiological characteristics of the Salicaceae propagules, largely governs the location of propagule deposition along the river corridor.

Once deposition has occurred, an adequate water supply is vital for successful growth. Available water is largely dependent upon:

1 the input of water to the substrate;
2 the capacity of sediment within the aeration zone to retain moisture, whether of groundwater or precipitation origin;
3 the local topography.

These factors determine the depth to water table and the rate of water table decline during flow recession (Mahoney & Rood, 1992; Malanson, 1993; Segelquist *et al.*, 1993). The water retention capacity and hydraulic conductivity of the substrate is determined by its grain-size distribution and porosity (Ellis & Mellor, 1995).

Surfeit or deficit of water may both prove harmful to riparian vegetation. Drought conditions leading to desiccation have been recognized in several studies as the primary cause of mortality amongst Salicaceae seedlings in riparian environments (Sacchi & Price, 1992; Johnson, 1994, 2000; Douglas, 1995; Rood *et al.*, 1998). Desiccation usually occurs when seedlings are growing in free-draining substrates at relatively high elevations within the active zone, where they may be subject to rapid water-table decline after flood flows and to relatively deep water-table levels during low flow periods. However, some of these studies also found that low elevation sites, which were subject to frequent water inundation, were also associated with seedling mortality due to anoxia or hypoxia (Rood *et al.*, 1998; Johnson 2000). Therefore, relative height of substrate above mean water level, grain-size distribution and rate of groundwater decline all appear to be involved in determining water availability and thus the survival of young plants during the early growth phases. These factors all display a high heterogeneity within braided gravel-bed rivers (Church, 1996; Lewin, 1996; Petts *et al.*, 2000). Few detailed field-based studies have been performed into how these factors relate to the establishment of Salicaceae along rivers, and almost none have involved active field experiments (cf. Barsoum, 1998; Winfield & Hughes, 2002; Karrenberg *et al.*, 2003a).

Most studies of the relationships between moisture availability and the Salicaceae have focused on seedlings, as this has been considered the primary method of establishment within river ecosystems (Segelquist *et al.*, 1993; Johnson, 1994; Van Splunder *et al.*, 1996). Gurnell *et al.* (2001), however, suggested that vegetative reproduction, particularly from large pieces of drift wood, is also very important for the successful establishment of several Salicaeae species along the Tagliamento River in northeast Italy. Indeed, Gurnell *et al.* (2001) and Gurnell & Petts (2002) proposed that the mode of reproduction (sexual or vegetative) and whether vegetation is (A) dispersed across open bar surfaces, (B) aggregated in areas of fine sediment that have accumulated in the shelter of large pieces of driftwood, or (C) in the form of sprouting large wood pieces, control the rate of biomass production and the associated rate of aggradation of fine sediment around the growing biomass (see Fig. 1).

This paper presents initial results from ongoing field studies that are seeking to explore the growth trajectories hypothesized by Gurnell & Petts (2002), focusing on tree recruitment and growth (Fig. 1). A series of field experiments was initiated to investigate the impact of bar sediment calibre and surface

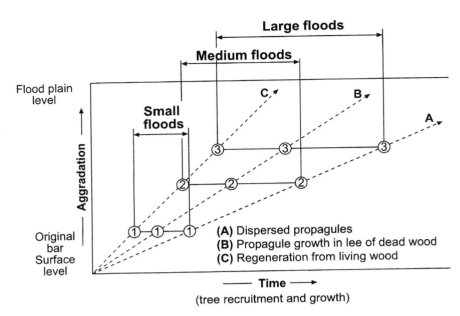

Fig. 1 Theoretical trajectories for tree recruitment and growth, and sediment aggradation, on bar surfaces within the active zone of the Tagliamento River. Trajectories are initiated by: (A) dispersed propagule establishment; (B) propagule establishment in the shelter of large wood pieces; (C) vegetation regeneration from living wood pieces. The rate of development differs between the trajectories over time, and so floods of varying magnitudes (indicated by circled numbers on the trajectories: small = 1, medium = 2, large = 3) will cause greater or lesser disturbance to the vegetation, depending upon the method of establishment. (Taken from Gurnell & Petts, 2002.)

elevation upon the survival and growth of two important species along the Tagliamento River: *Populus nigra* L. and *Salix elaeagnos* Scop. Results from pilot experiments are presented to evaluate the survival and growth of cuttings (i.e. representing Trajectory A, Fig. 1) within plots representing different sediment calibre and relative elevation, and to compare the growth rates of surviving cuttings with measurements taken of seedlings in geomorphologically comparable locations. A second series of observations is presented which explore comparative survival and growth rates associated with the initiation of Trajectories B and C (Fig. 1). These include the first summer survival rates of uprooted trees deposited by spring floods on bar surfaces, and also the stem growth rates in and around a small sample of flood-deposited trees over 18 months following their deposition. Comparison of all of these studies aims to:

1 evaluate the effects of substrate characteristics (grain-size distribution and permeability) and relative elevation on cutting and seedling establishment;
2 compare any differences in growth strategies that may be apparent between cuttings and seedlings;
3 relate cutting and seedling growth rates to those of living driftwood and fragments sheltered by dead driftwood, in order to provide estimates of relative rates of vegetation establishment during the initial stages of Trajectories A, B and C (Fig. 1) at the study site.

STUDY SITE

The Tagliamento River is situated in northeast Italy where it drains from its source in the dolomitic limestone Alps, through the Prealps and alpine foothills, to the Adriatic Sea (Fig. 2). The river has a catchment of 2850 km^2, and is approximately 170 km long. Despite engineering modifications along the most downstream 25 km, the natural morphology of the remainder of the main river is generally intact. Indirect impacts such as water abstraction from the headwaters and from groundwater for hydropower, irrigation and drinking water have decreased low flows, but major flood flows (up to 4000 m^3 s^{-1}) remain essentially unaffected (Ward *et al.*, 1999; Gurnell *et al.*, 2001). The main river is therefore a suitable subject for investigations into the links between hydrology, geomorphology and ecology within a large, braided, gravel-bed river (Ward *et al.*, 1999).

The flow regime of the Tagliamento River is 'extremely flashy' (Gurnell *et al.*, 2000), with high flows occurring predominantly in spring (snowmelt) and autumn (intense rain storms). Periods of low flow are most likely to occur during the winter (December to February) and during August. The river's active zone (the area subject to frequent modification by flood events and bordered laterally by permanent vegetation) is extensive, having a width of over 1.5 km in some reaches, and is

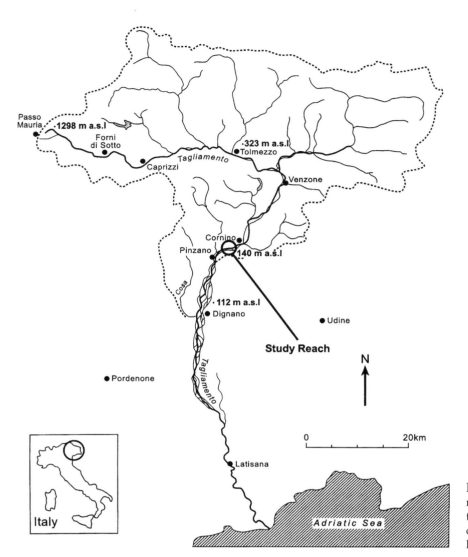

Fig. 2 The Tagliamento River, northeast Italy, showing location of the study reach. The catchment area of the river is indicated by the dashed line.

characterized by a heterogeneous mix of exposed riverine sediments and tree-covered bars (islands) (Petts *et al.*, 2000; Gurnell *et al.*, 2001). Riparian vegetation communities are dominated by the genera *Salix*, *Populus* and *Alnus*, with the three most frequently occurring species being *Salix elaeagnos*, *Populus nigra* and *Alnus incana* (Karrenberg *et al.*, 2003b). Of these three species, *A. incana* is present primarily in the upper reaches, while *P. nigra* is much more common in the central and downstream reaches. *Salix elaeagnos* is the most ubiquitous species, being present along the entire length of the river.

The data presented in this paper were collected in a braided reach, approximately 80 km from the river's source, at an altitude of approximately 140 m a.s.l. Here the river corridor is approximately 1150 m wide with an active zone of 750 m, and vegetated bars (islands) are common.

MATERIALS AND METHODS

Trajectory A—survival and growth rates of cuttings and seedlings growing on open gravel surfaces

A field experiment was conducted to investigate the survival and growth rates of *Salix elaeagnos* and *Populus nigra* cuttings and seedlings within plots selected to represent the range in sediment calibre (coarse, mixed, fine) and relative surface elevation (high, low) of unvegetated bars within the study reach. 'Coarse' sediments were classified as ranging from gravel to boulder size (2–256 mm), with 'fine' being sand or silt (<2 mm). 'Mixed' plots contained evidence of an abundance of both grain sizes. Six plots, (1) 'High Fine,' (2) 'Low Fine,' (3) 'High Mixed,' (4) 'Low Mixed,' (5) 'High Coarse,' and (6) 'Low Coarse', were planted with cuttings during April 2002, and cutting growth was measured during July and September 2002. Six plots of similar elevation and sediment calibre (7)–(12) were identified within which the growth of 1 (current) year and 2 (previous) year *S. elaeagnos* and *P. nigra* seedlings were measured during July 2002.

Relative elevation of the plots was measured using a Leica TCRA1105Plus Total Station and was then adjusted for floodplain slope. In each plot, surface sediment calibre was estimated using a photo-sieving technique (described by Petts *et al.*, 2000). Subsurface sediment calibre was determined from samples drawn from distinct layers within pits excavated to the water table or a depth of 0.5 m, whichever was closer to the surface, using a combination of dry-sieving and laser particle sizing (Malvern Instruments Mastersizer 2000). For each horizon an index of hydraulic conductivity (K) was calculated using sediment D_{10}, as this has been proposed as the effective particle size diameter controlling this factor:

$$K = D_{10}^2 \qquad (1)$$

where D_{10} is the grain size, with D in millimetres (Moreton *et al.*, 2002). This is utilized in this case as a surrogate for relative permeability of the substrate, to give an indication of the potential differences in available water between plots.

Table 1 provides information on the cutting plots (seedling plots were very similar to the corresponding cutting plots) and illustrates that although the 'High' plots were higher than the 'Low' plots there was some variability within these categories, reflecting difficulties in locating sufficiently large plots of relatively homogeneous sediment calibre. To place the relative elevations of the plots into the context of the relief of the entire active zone, Plot 3 was at the level of the vegetated surface of the nearest island, and the relief of the entire active zone (island surface to main channel bed) at the study site was approximately 3 m. The 'Fine' plots consisted of horizons that were predominantly sand, although thin layers of coarser material were present. Both the 'Mixed' and 'Coarse' plots were predominantly gravel, although the mixed plots did contain higher percentages of sand in most horizons and a smaller D_5, indicating a greater amount of very fine sediment. This is also reflected in the lower hydraulic conductivity of the 'Mixed' plots as opposed to the 'Coarse' plots, with the 'Fine' plots having the lowest of all. This indicates that water will drain more slowly through horizons with greater amounts of sand (such as the 'Fine' plots), leading to greater water availability, but also potentially longer periods of saturation. Clay was absent from almost all horizons.

Cutting methodology

Cuttings were taken from several individuals of *P. nigra* (10) and *S. elaeagnos* (12) situated on an island located close to the six plots. Each cutting was selected from old-growth branches and cut to a length of 40 cm, with care taken to ensure a diameter of ≥5 mm for at least three quarters of its length. This size was chosen because larger cuttings are high in carbohydrates and generally perform better in the early stages of plant development (Hartman & Kester, 1975), probably due to an increased capacity for root production (Richardson *et al.*, 1979; Burgess *et al.*, 1990). Indeed, Burgess *et al.* (1990) found that *Salix alba* cuttings with a length greater than 45 cm had 100% survival rates if diameter was greater than 6 mm, and an 83.3% survival rate if diameter was smaller than 6 mm. The cuttings were mixed before being planted to avoid any pronounced genetic or sexual bias within the characteristics of individual trees.

The cuttings were planted between 28 April and 2 May 2002, i.e. near the beginning of the growing season. Each plot was laid out as a 5 m × 5 m square and weeded to remove all growing vegetation. One hundred cuttings (50 *P. nigra* and 50 *S. elaeagnos*) were planted in alternate rows of 10 cuttings with a 0.5 m spacing between plants and rows. The cuttings were planted at an angle of approximately 45° to the ground surface, pointing downstream to roughly simulate hydrochorous deposition. This was achieved by using a metal dibber hammered into the gravel to make a planting hole that minimized sediment disturbance and cutting damage. The length of the cutting protruding from the substrate was always <10 cm. Each plot was watered daily for up to three days after planting to reduce the stress experienced by the cuttings and

Table 1 Sedimentary and elevational characteristics of cutting plots (1–6) set out along the Tagliamento River, northeast Italy

Plot	Horizon	Depth (cm)	D_5 (ϕ)	D_{50} (ϕ)	D_{95} (ϕ)	Gravel* (%)	Sand* (%)	Clay (%)	K†	Elevation (cm)‡
1 (High Fine)	Surface	0	(>−3)	(>−3)	(>−3)	0	100	ND	ND	−84
	A	0–29	4	1.9	0.2	2	98	0	0.016	
	B	30–40	2.6	0	<−1	44.9	55.1	0	0.053	
	C	41–50	2.6	1.4	−1	5	95	0	0.032	
2 (Low Fine)	Surface	0	(>−3)	(>−3)	(>−3)	0	100	ND	ND	−140
	A	0–21	3	1.6	0.1	0.8	99.2	0	0.02	
	B	22–50	2.3	<−1	<−1	72.4	27.1	0.5	0.168	
3 (High Mixed)	Surface	0	(>−3)	−3.3	−5.3	73	27	ND	ND	0
	A	0–15	3.8	<−1	<−1	70.9	29.1	0	0.022	
	B	16–50	1.9	<−1	<−1	61.1	38.9	0	0.125	
4 (Low Mixed)	Surface	0	(>−3)	(>−3)	−5.6	49	51	ND	ND	−137
	A	0–23	2.8	<−1	<−1	74.8	25.2	0	0.072	
	B	24–50	2	<−1	<−1	81.4	18.6	0	0.436	
5 (High Coarse)	Surface	0	(>−3)	−4.8	−6.5	97	3	ND	ND	−66
	A	0–22	1.3	<−1	<−1	79.7	20.3	0	0.865	
	B	23–50	1.7	<−1	<−1	79.2	20.8	0	0.578	
6 (Low Coarse)	Surface	0	(>−3)	−5	−6.9	98	2	ND	ND	−108
	A	0–35	1.9	<−1	<−1	80.5	19.5	0	0.25	
	B	35–50	1.9	<−1	<−1	74.1	25.9	0	0.336	

D_5, D_{50} and D_{95} represent percentages of the grain-size distribution that are finer than the diameter (ϕ) given; ND = not determined.
*Surface percentage of gravel and sand is based on an 8 mm diameter threshold to distinguish between the two sediment types (gravel > 8 mm, sand < 8 mm), as smaller grain sizes cannot be determined accurately from photographs.
†Index of permeability based on the D_{10} of the grain-size distribution (see Moreton et al., 2002).
‡Elevation is relative to the High Mixed plot, which was at the level of the vegetated surface of the nearest river island.

simulate the wet conditions that would occur after natural deposition during a flood.

Measurements of cutting growth parameters were taken between 12 and 16 July 2002 (approximately 75 days after planting) and between 25 and 27 September 2002 (approximately 150 days after planting). At the time of measurement cuttings were considered alive if they displayed any living stems. Cuttings that showed evidence of past regrowth but appeared dead at time of measurement, and those which were absent due to being dislodged or buried, were recorded as dead. Several measures of growth were recorded for the cuttings that were alive, including the number of stems produced by the cutting, the length of the longest stem, the total combined length of all stems, the number of leaves on the cutting, and the length of ten randomly selected leaves (from which an estimate of average leaf length was derived).

Seedling methodology

To compare the growth of naturally occurring seedlings of *S. elaeagnos* and *P. nigra*, six plots that were similar in their sediment calibre and elevation, and as close as possible to each of the cutting plots, were selected and marked out in an identical fashion to the cutting plots.

Measurements including seedling age, the number of stems, the total and longest stem length, and the number of leaves were taken for all *P. nigra* and *S. elaeagnos* seedlings found within the plots. The number of stems, stem length and number of leaves were assessed using the same methods as for cuttings. 'Species age' simply discriminated between new seedlings that had germinated in 2002 and second year seedlings that had germinated during 2001. This was assessed using two similar but not identical techniques for each of the two subject species. *Populus nigra* seedlings were aged by counting the winter bud scars present on the stem. These were often readily apparent, and new growth could usually be determined by a difference in stem colour (newer stem material being a dark red, older material a pale brown). *Salix elaeagnos* species were aged by counting leaf bud scars. These two different techniques were utilized because *P. nigra* has a monopodial growth form and so has an apical meristem, and leaves a more or less distinct winter bud scar. Growth of *S. elaeagnos*, like all willows, is sympodial, and it does not have an apical meristem; when the new growing season arrives, growth continues from one of the lateral shoots.

Seedlings were also identified by their erect stems (stems originating from buried wood often having an angled base), and by excavation of seedlings falling outside the plots to ensure that deposited wood fragments were absent. No seedlings of either species were judged to be older than 2 yr. When estimating the age of the seedlings in numbers of days, growth was arbitrarily assumed to have commenced on 1 May and, for year 2 seedlings, to have ceased between 1 October and 30 April inclusive.

Data analysis

Several of the measured properties of the cuttings and seedlings were found not to be normally distributed (after Anderson–Darling and Ryan–Joiner normality tests) even following log-transformation. Non-parametric tests (χ^2 contingency tables and Mann–Whitney tests with Bonferroni correction) were thus utilized to analyse the data.

Trajectories B and C—survival and growth rates of uprooted trees following their deposition on open gravel surfaces

During July 2001, all wood accumulations associated with an uprooted tree were located along nine transects across the active zone of the study site. Each accumulation was surveyed to assess its size (length), and whether it was alive (at least 10% of the branches displaying living shoots) or dead. These trees provide an indication of the proportion of tree-cored wood accumulations that can support vegetation development under Trajectories B (propagules sheltered by dead driftwood) and C (living driftwood).

On 26 April 2001, six trees that had been deposited on the open gravel during the previous winter were surveyed from both sides (i.e. to reveal the entire tree length) using scaled photographs. The same trees were resurveyed in exactly the same manner on 7 July 2001 (after 72 days), 2 September 2001 (129 days), 2 May 2002 (159 days assuming no growth from 1 October to 30 April inclusive) and 18 July 2002 (236 days). Measurements taken from the scaled photographs of the canopy height at five equally spaced points along each tree (from the lowest growth vertically to the top of the crown) provided estimates of average growth rates from the trees if they had sprouted, and/or from other propagules deposited around the trees. In addition, the maximum height of the canopy that sprouted from six wood pieces that were deposited by the river in the shelter of the root bole of two of the trees provided further estimates of growth rates from large vegetative fragments (branches), although as a result of their burial in fine sediment the exact length of these six wood pieces could not be determined. The height of sediment accumulated around the trees by the end of the photomonitoring period could also be measured from the photographs. All of these estimates of growth rates from photographed canopies could be compared with the measured growth of the longest shoots from each cutting and seedling in the plot experiments as an index of comparative biomass production under Trajectories A, B and C.

RESULTS

Trajectory A—survival and growth rates of cuttings and seedlings growing on open gravel surfaces

Cuttings

Figure 3 illustrates the number of survivors within the six cutting plots at the time of the July survey.

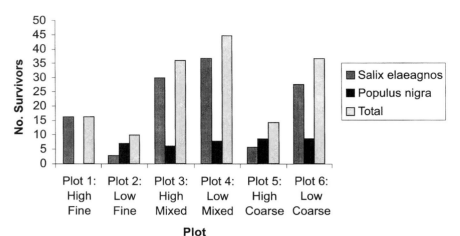

Fig. 3 Number of surviving cuttings within each plot, measured in July 2002.

Table 2 χ^2 analyses comparing differences in survival rates of *Salix elaeagnos* and *Populus nigra* within the cutting plots (Plots 1–6) during the July and September 2002 surveys. Bold entries indicate a significant difference ($P < 0.01$), with *Salix* > *Populus*

Plot	July 2002		September 2002	
	χ^2	P-value	χ^2	P-value
1 (High Fine)	**19.048**	**0.000**	**16.279**	**0.000**
2 (Low Fine)	1.778	0.182	ND	ND
3 (High Mixed)	**25**	**0.000**	**23.253**	**0.000**
4 (Low Mixed)	**33.980**	**0.000**	**31.818**	**0.000**
5 (High Coarse)	0.706	0.401	0.706	0.401
6 (Low Coarse)	**15.487**	**0.000**	**13.752**	**0.000**

ND = not determined.

Plot 2 (Low Fine) requires careful interpretation because the high mortality rate was in part attributable to cutting burial as a result of the partial inundation of the plot. Of the 600 cuttings planted during April 2002, 159 (26.5%) survived to July. These included 120 (75.5%) of *S. elaeagnos*, and 39 (24.5%) of *P. nigra*. Chi-square analyses indicated a significantly higher survival of *S. elaeagnos* over all six plots ($\chi^2 = 56.142$, $P < 0.001$) and also within plots 1, 3, 4 and 6 (Table 2).

At the end of September 2002, 146 (24.3%) of the original 600 cuttings were still alive (113 *S. elaeagnos* and 33 *P. nigra*) representing a further loss of seven *S. elaeagnos* and six *P. nigra* since the July survey. Of these further losses, three *S. elaeagnos* and seven *P. nigra* were located in Plot 2, which had been completely buried by sediment deposited during a further flood. One *P. nigra* cutting previously recorded as dead had re-sprouted in Plot 6 (Low Coarse), making the total loss six individuals despite seven dying in Plot 2; a 'dead' *S. elaeagnos* cutting had also re-sprouted in Plot 3 (High Mixed). In the remaining plots, losses of cuttings were negligible. Again, χ^2 analyses indicated a significant difference in survival between the two species over all six plots ($\chi^2 = 57.933$, $P < 0.001$) and also within Plots 1, 3, 4 and 6 (Table 2), again in favour of *S. elaeagnos*.

Table 3a gives statistics for each of the growth parameters measured in July 2002 for both species within each of the six cutting plots. On average, the highest total stem growth rates for *S. elaeagnos* were found in Plot 1 (High Fine), with Plot 3 (High Mixed) having the highest mean number of stems. Plot 6 (Low Coarse) had the highest mean stem length for *P. nigra*, with other parameters being similar (discounting Plot 1, High Fine, where all *P. nigra* died).

Table 3b provides statistics for each of the growth parameters measured in September 2002 for both species within Plots 1–6. The highest stem total growth rates for *S. elaeagnos* were now seen in Plot 6 (Low Coarse). Plot 3 (High Mixed) maintained its higher number of *S. elaeagnos* stems. For *P. nigra*, highest stem growth rates were now in Plot 4 (Low Mixed) instead of Plot 6 (Low Coarse).

For both of these sets of measurements (Table 3), there is some evidence of a positive relationship between growth of *P. nigra* and increasing sediment calibre. It is also apparent that the 'Low' plots almost always out-performed their corresponding 'High' plot. For *S. elaeagnos*, the Mixed plots often represented more extreme mean values, being higher or lower than their 'Fine' and 'Coarse' counterparts.

To evaluate the importance of relative elevation for each species, the growth data from the 'High' and 'Low' plots were pooled, and then Mann–Whitney tests performed with Bonferroni correction (Table 4a). This was then repeated for sediment calibre, pooling data for 'Fine,' 'Mixed' and 'Coarse' sites as appropriate (Table 4b). Variations in relation to relative elevation were most marked for *P. nigra*, where the majority of growth parameters are significantly larger in the lower sites. However, the longest stems and leaves also occurred on lower sites in the September survey for *S. elaeagnos*. In contrast, variations in relation to sediment calibre are most marked for *S. elaeagnos*, particularly in the July survey where the longest stems and largest number of stems occur on fine sediments, but longest leaves occur in relation to coarser sediments. The number of statistically significant differences obtained for growth parameters at varying elevations was higher over both periods of measurement for *P. nigra* cuttings than the number obtained for *S. elaeagnos* cuttings. The opposite is true for significant differences obtained due to variations in sediment calibre, with many more being found for *S. elaeagnos* cuttings than *P. nigra*. This indicates that differences in elevation (depth to water table) may be particularly important for *P. nigra*, whereas *S. elaeagnos* may be influenced more by the grain-size distribution and sorting of sediments (and therefore the retention of near-surface moisture).

Seedlings

Table 5 shows statistics for growth parameters measured in July 2002 for both species within the six seedling plots. No *S. elaeagnos* seedlings were found within the High Fine and High Mixed plots. For *S. elaeagnos* classified as 1 yr, the highest growth rates were in Plot 11 (High Coarse), while for 2 yr seedlings, Plot 12 (Low Coarse) performed best. For *P. nigra* seedlings, growth rates were highest in Plot 12 (Low Coarse) for both 1 and 2 yr seedlings. Table 5 suggests a positive relationship between growth rates and increasing sediment coarseness for many variables of both *S. elaeagnos* and *P. nigra*, with the exception of the 'Low' plots for *P. nigra*, where the relationships were generally negative. The 'Mixed' plots were often representative of higher or lower means than their 'Fine' and 'Coarse' counterparts.

Data were grouped according to elevation and sediment calibre and analysed using Mann–Whitney tests with Bonferroni correction to determine if there were any significant differences between growth rates of seedlings of different ages (1 or 2 yr) (Table 6). Although *S. elaeagnos* seedlings were not present in Plots 7 (High Fine) and 9 (High Mixed), those found on Plot 11 (High Coarse) that were aged 2 yr performed significantly better than those in the grouped 'Low' plots for all three measured growth parameters. The grouped *P. nigra* data also displayed a trend for significantly higher growth rates in the 'High' plots for 2 yr seedlings (Table 6). In relation to sediment calibre, *S. elaeagnos* seedlings displayed significantly higher growth rates in the 'Coarse' plots than in the other plots, while second-year *P. nigra* seedlings performed better on the 'Fine' plots.

Trajectories B and C—survival and growth rates of uprooted trees following their deposition on open gravel surfaces

Dead and living wood accumulations

Table 7 provides summary statistics describing the length in metres of a sample of 170 isolated trees and shrubs that had been deposited on open gravel at the study site. The length of trees that were alive

Table 3 Summary statistics for growth parameters of *S. elaeagnos* and *P.nigra* cuttings

(a) Cuttings measured in July 2002

Species	Variable	Plot 1 (High Fine)			Plot 2 (Low Fine)			Plot 3 (High Mixed)			Plot 4 (Low Mixed)			Plot 5 (High Coarse)			Plot 6 (Low Coarse)		
		Mean	N	SD	Mean	N	SD	Mean	N	SD	Mean	N	SD	Mean	N	SD	Mean	N	SD
Salix elaeagnos	Total stem length	458	16	235	193	3	85.8	280	30	141	294	37	166	213	28	82.5	273	6	157
	Longest stem length	181	16	48.3	105	3	42.7	72.7	30	28.9	114	37	61.8	117	28	45.3	118	6	64.7
	Number of stems	4.69	16	3.54	3.33	3	2.31	7.1	30	3.33	5.59	37	3.16	3.00	28	1.79	3.86	6	1.9
	Number of leaves	75.6	16	44.0	31.7	3	11.6	74.5	30	32.9	61.7	37	30.3	43.0	28	22.5	55.00	6	27.2
	Average leaf length	37.2	16	4.84	34.7	3	10.7	26.4	30	5.57	28.3	37	9.56	35.8	28	10.5	32.54	6	11.6
Populus nigra	Total stem length	0	0	0	150	7	160	78.7	6	35.5	150	8	75.3	100	9	35.5	197.4	9	84.0
	Longest stem length	0	0	0	82.1	7	52.3	71.2	6	22.3	108	8	51.1	93.3	9	36.7	138.6	9	55.1
	Number of stems	0	0	0	1.86	7	1.46	1.17	6	0.41	2.00	8	1.07	1.22	9	0.44	2.00	9	1.12
	Number of leaves	0	0	0	19.9	7	14.9	13.5	6	6.29	19.4	8	10.7	12.1	9	3.82	23.33	9	8.6
	Average leaf length	0	0	0	24.5	7	13.2	32	6	13.1	46.7	8	12.0	38.1	9	7.5	56.08	9	6.58

(b) Cuttings measured in September 2002

Species	Variable	Plot 1 (High Fine)			Plot 2 (Low Fine)			Plot 3 (High Mixed)			Plot 4 (Low Mixed)			Plot 5 (High Coarse)			Plot 6 (Low Coarse)		
		Mean	N	SD	Mean	N	SD	Mean	N	SD	Mean	N	SD	Mean	N	SD	Mean	N	SD
Salix elaeagnos	Total stem length	495	14	284	0	0	0	268	29	121	503	36	262	458	6	257	631	28	430
	Longest stem length	269	14	108	0	0	0	78.5	29	25.2	246	36	136	226	6	170	230	28	133
	Number of stems	3.79	14	2.08	0	0	0	6.62	29	2.64	4.61	36	2.39	4.17	6	1.47	5.04	28	2.52
	Number of leaves	73.9	14	47.5	0	0	0	58.4	29	23.0	68.2	36	36.7	63.0	6	29.9	83.0	28	48.9
	Average leaf length	35.5	14	6.99	0	0	0	25.5	29	4.79	36.6	36	10.1	37.7	6	5.3	35.3	28	9.97
Populus nigra	Total stem length	0	0	0	0	0	0	65.8	6	24.0	335	8	233	164	9	92.0	279	10	93.5
	Longest stem length	0	0	0	0	0	0	65.0	6	24.1	261	8	131	155	9	98.8	214	10	82.0
	Number of stems	0	0	0	0	0	0	1.17	6	0.41	1.75	8	1.04	1.22	9	0.44	1.78	10	0.83
	Number of leaves	0	0	0	0	0	0	9.67	6	3.62	10.7	8	7.59	16.6	9	4.93	26.2	10	8.63
	Average leaf length	0	0	0	0	0	0	31.6	6	12.6	51.5	8	16.2	47.9	9	7.72	62.6	10	11.3

Table 4 Mann–Whitney analyses comparing variations in growth parameters of surviving *Salix elaeagnos* and *Populus nigra* cuttings

(a) According to relative elevation. Italicized entries indicate a significant difference at the 0.05 level. Bold entries indicate a significant difference at the 0.01 level

Growth parameter	Salix elaeagnos		Populus nigra	
	P (adj for ties) High versus Low July 2002	P (adj for ties) High versus Low September 2002	P (adj for ties) High versus Low July 2002	P (adj for ties) High versus Low September 2002
Total stem length	0.2483	0.4590	**0.0073** (Low > High)	**0.0001** (Low > High)
Longest stem	0.4977	**0.0000** (Low > High)	0.1326	**0.0011** (Low > High)
Number of stems	0.1054	0.1544	*0.0220* (Low > High)	0.0584
Number of leaves	*0.0394* (High > Low)	0.1807	**0.0097** (Low > High)	0.1474
Average length of leaves	0.5478	**0.0001** (Low > High)	*0.0433* (Low > High)	**0.0021** (Low > High)

(b) According to sediment calibre. Bold entries indicate a significant difference between sediment calibres at the 0.05 level, with Bonferroni correction

Growth parameter	Salix elaeagnos		Populus nigra	
	Significant differences between sediment calibres July 2002	Significant differences between sediment calibres September 2002	Significant differences between sediment calibres July 2002	Significant differences between sediment calibres September 2002
Total stem length	None	**Coarse > Mixed**	None	None
Longest stem	**Fine > Mixed, Coarse**	**Fine > Mixed** **Mixed > Coarse**	None	None
Number of stems	**Fine > Mixed** **Mixed > Coarse**	None	None	None
Number of leaves	None	None	None	**Coarse > Mixed**
Average length of leaves	**Coarse > Fine**	None	**Fine, Coarse > Mixed**	None

Table 5 Summary statistics for growth parameters of *S. elaeagnos* and *P. nigra* seedlings measured in July 2002

(a) One-year Salix elaeagnos and Populus nigra seedlings July 2002

Species	Variable	Plot 7 (High Fine)			Plot 8 (Low Fine)			Plot 9 (High Mixed)			Plot 10 (Low Mixed)			Plot 11 (High Coarse)			Plot 12 (Low Coarse)		
		Mean	N	SD	Mean	N	SD	Mean	N	SD	Mean	N	SD	Mean	N	SD	Mean	N	SD
Salix elaeagnos	Total stem length	0	0	0	31.7	23	19.5	0	0	0	70.2	18	48.8	144	7	80.8	96.7	25	63.1
	Longest stem length	0	0	0	30.8	23	18.5	0	0	0	35.39	18	15.9	65.1	7	28.0	41.9	25	17.1
	Number of stems	0	0	0	1.04	23	0.21	0	0	0	2.61	18	1.65	4.08	7	1.66	2.86	25	1.57
	Number of leaves	0	0	0	10.6	23	3.73	0	0	0	26.2	18	16.3	37.4	7	14.5	25.6	25	12.3
Populus nigra	Total stem length	91.2	6	100	52.7	39	34.3	15.3	3	4.51	48.2	12	26.9	19.2	3	10.9	187	5	144
	Longest stem length	91.2	6	100	53.2	39	34.2	15.3	3	4.51	48.2	12	26.9	19.2	3	10.9	187	5	144
	Number of stems	1.00	6	0	1.00	39	0	1.00	3	0	1.00	12	0	1.00	3	0	1.00	5	0
	Number of leaves	10.0	6	5.33	8.49	39	2.37	6.67	3	2.52	9.00	12	1.76	5.4	3	0.89	17.3	5	8.08

(b) Two-year Salix elaeagnos and Populus nigra seedlings July 2002

Species	Variable	Plot 7 (High Fine)			Plot 8 (Low Fine)			Plot 9 (High Mixed)			Plot 10 (Low Mixed)			Plot 11 (High Coarse)			Plot 12 (Low Coarse)		
		Mean	N	SD	Mean	N	SD	Mean	N	SD	Mean	N	SD	Mean	N	SD	Mean	N	SD
Salix elaeagnos	Total stem length	0	0	0	192	21	205	0	0	0	166	10	93.0	371	11	233	541	24	280
	Longest stem length	0	0	0	84.2	21	52.3	0	0	0	71.7	10	35.2	127	11	44.5	135	24	39.7
	Number of stems	0	0	0	3.19	21	1.69	0	0	0	3.6	10	2.07	7.54	11	4.32	8.00	24	3.41
	Number of leaves	0	0	0	38.4	21	19.4	0	0	0	39.9	10	24.8	76.5	11	35.7	10	24	36.3
Populus nigra	Total stem length	273	32	116	124	12	78.7	80.5	47	34.5	96.8	38	51.1	59.9	8	25.5	390	8	159
	Longest stem length	257	32	107	124	12	78.7	79.5	47	33.5	96.8	38	51.1	59.9	8	25.5	355	8	112
	Number of stems	1.19	32	0.47	1.00	12	0	1.04	47	0.20	1.00	38	0	1.00	8	0	1.88	8	1.13
	Number of leaves	18.8	32	6.06	12.9	12	4.06	9.79	47	2.73	10.9	38	2.51	5.88	8	1.73	27.0	8	12.3

Table 6 Mann-Whitney analyses comparing variations in growth parameters of *Salix elaeagnos* and *Populus nigra* seelings aged 1 and 2 yr, surveyed in July 2002

(a) According to relative elevation. Italic entries indicate a significant difference at the 0.05 level. Bold entries indicate a significant difference at the 0.01 level

Growth parameter	*Salix elaeagnos*		*Populus nigra*	
	P (adjusted for ties) High versus Low age 1	P (adjusted for ties) High versus Low age 2	P (adjusted for ties) High versus Low age 1	P (adjusted for ties) High versus Low age 2
Total stem length	0.1785	**0.0001 (High > Low)**	0.9432	**0.0010 (High > Low)**
Longest stem	0.6362	*0.0264 (High > Low)*	0.9679	**0.0020 (High > Low)**
Number of leaves	0.3317	**0.0001 (High > Low)**	0.4553	**0.0010 (High > Low)**

(b) According to sediment calibre. Bold entries indicates a significant difference between sediment calibres at the 0.05 level, with Bonferroni correction

Growth parameter	*Salix elaeagnos*		*Populus nigra*	
	Significant differences between sediment calibres age 1	Significant differences between sediment calibres age 2	Significant differences between sediment calibres age 1	Significant differences between sediment calibres age 2
Total stem length	Mixed, Coarse > Fine Coarse > Mixed	Coarse > Fine, Mixed	None	Fine > Mixed
Longest stem	Coarse > Fine, Mixed	Coarse > Fine, Mixed	None	Fine > Mixed
Number of leaves	Mixed, Coarse > Fine	Coarse > Fine, Mixed	None	Fine > Mixed

Table 7 Summary statistics describing the lengths (m) of a sample of 170 dead and living trees deposited across the active zone of the study area in August 2001

Status	N	Mean	Median	Minimum	Maximum	Lower quartile	Upper quartile
Living trees	66	14.7	13.5	2.3	36.0	8.0	20.0
Dead trees	104	5.4	3.4	1.2	21.4	2.0	7.0

Table 8 Tree canopy and shrub growth rates and fine sediment accumulation associated with the deposition of uprooted trees on open gravel within the study area

Sample	Canopy height (trees)/ longest stem length (shrubs) (cm)			Growth rate of canopy (trees)/stems (shrubs) (mm day^{-1})	Canopy remaining after day 236 (%)	Depth of fine sediment deposited (m)
	72 days	129 days	236 days			
Tree 1 (*P. nigra*)	34	156	256	10.8	100	>0.6
Tree 2 (*P. nigra*)	36	26	66	2.8	10	0.4
Tree 3 (*S. elaeagnos*)	14	31	19	0.81	0	0.2
Tree 4 (*F. ornus*)	9	11	25	1.06	20	<0.1
Tree 5 (*A. incana*)	2	23	ND	1.78*	ND	ND
Tree 6 (*A. incana*—no shoots on day 72)	15	1	11	0.45	0	<0.1
Shrub 1 (*P. nigra*)	110	213	271	11.48	100	NA
Shrub 2 (*S. elaeagnos*)	110	154	175	7.41	100	NA
Shrub 3 (*P. nigra*)	77	110	ND	8.52*	ND	ND
Shrub 4 (*S. alba*)	67	83	138	5.85	100	NA
Shrub 5 (*S. elaeagnos*)	77	88	130	5.51	100	NA
Shrub 6 (*S. alba*)	76	88	121	5.13	100	NA

Growth rates, canopy remaining and depth of fine sediment after day 236 (*129). ND = no data. NA = not applicable.

($n = 66$) and dead ($n = 104$) ranged from 1.2 m to 36 m. After a $\log_{10}$ transformation to ensure that the variances of the two samples were not significantly different from one another, a Student *t*-test revealed that the living accumulations were significantly longer than the dead ones ($P < 0.0001$).

Growth rates from river-deposited trees and branches

Table 8 provides measurements of average canopy height of woody vegetation around six trees deposited during the winter of 2000–2001 on open gravel within the study area. The canopy heights include regrowth from the main tree and from any other propagules that may have accumulated around it. Three of the trees forming the core of wood accumulations (3, 5 and 6) were dead by the end of the observation period, although propagules within the accumulations were still alive, and so are representative of Trajectory B. One tree (1) was sprouting vigorously across its entire canopy and so is representative of Trajectory C. The remaining two trees (2 and 4) were still partly alive at the end of the observation period, and so may represent

Table 9 Cutting and seedling average growth rates estimated from the length of the longest stem

	Salix elaeagnos		Populus nigra	
	N	Longest stem length (daily increment) (mm)	N	Longest stem length (daily increment) (mm)
Cuttings (75 days growth)	120	111.7 (1.5)	39	102.2 (1.4)
Seedlings (75 days growth)	185	40.9 (0.6)	68	57.5 (0.8)
Cuttings (150 days growth)	113	200.7 (1.3)	33	182.3 (1.2)
Seedlings (230 days growth)	67	108.2 (0.5)	138	140.6 (0.6)

either Trajectory C or B in the medium term. Whichever trajectory is represented, by the end of the observation period all of the trees had become a focus for the accumulation of fine sediment and for the early growth of a wide variety of plants. Moreover, there is a clear positive relationship between the growth of the tree/shrub canopy and the depth of fine sediment around the central wood accumulation (Table 8).

Table 8 also gives the lengths of the longest shoots of six shrubs that could be consistently identified in the scaled photographs. Five were associated with Tree 1, and one (Shrub 3) was associated with Tree 4. All of these shrubs had sprouted from around the root bole of the central tree, and it is assumed that they had sprouted from large pieces of wood, such as branches that were trapped within the root bole and associated deposited sediment. In all cases the wood pieces from which the shrubs sprouted were partly buried in the sediment around the root bole and so their dimensions could not be measured.

The average growth rates (mm) in Table 8 can be compared with average growth rates (mm) observed from the seedling and cutting experiments (Table 9), illustrating substantial differences that are relevant to the three hypothesized trajectories (Fig. 1). It is apparent that, on average, seedlings grow more slowly than cuttings (at approximately half the rate of cutting growth); whereas trees and shrubs display a wide range of growth rates (Table 8), which can be many times that of cuttings, and over an order of magnitude higher than that of seedlings.

DISCUSSION

Interactions between the physical environment and propagule growth

The growth parameters examined here are important factors in determining the capability of riparian vegetation to survive within river environments, as well as the degree to which it can influence processes such as sedimentation and therefore island development. Surface biomass can provide substantial resistance to hydrological disturbance and increase surface roughness, while root biomass helps to reinforce substrates (Johnson, 1994; Gladwin & Roelle, 1998; Abernethy & Rutherford, 2001; Simon & Collison, 2002). Biomass may be a key factor in sedimentation, as shown by rates of accretion recorded around deposited trees and shrubs (Table 8; see also Fetherston et al., 1995; Gomi et al., 2001). Deposited driftwood is also less likely to be disturbed or entrained by flows than vegetative fragments due to its much greater mass, and so may also lead to aggradation in locations where smaller propagules would not be able to establish.

Vegetation density also influences sedimentation: this includes not only how individual plants are spaced within the community, but also the number of stems, leaves and the size of leaves produced by the plants (Thornton et al., 1997; Järvelä, 2002). This study suggests that greater sediment aggradation may be anticipated in patches of asexually produced vegetation than in patches produced by seed (in the short term, at least). This is because of the higher initial growth rates of cuttings compared with seedlings (approximately double; see Table 9) and the capacity of cuttings to produce multiple stems and therefore more leaves, potentially creating denser communities that are more effective at trapping sediment. These growth responses also maximize the chances of survival following sediment deposition, as longer stem lengths minimize the risk of complete burial, while greater numbers of stems may reduce the possibility of all stems being lost. Greater rates of above-ground growth are usually accompanied by a comparable increase in underground root growth within the first years of development (e.g. Pregitzer & Friend, 1996), and so vegetative propagules may also stabilize the bar substrate and prevent erosion more effectively than seedlings. Therefore, dense patches of vegetation resulting from both vegetative fragments and large trees, within a matrix of sparse vegetation and gravel, may lead to variations in both spatial patterns and rates of subsequent sediment deposition and erosion.

Trajectory A (Fig. 1) can be expected to contain large amounts of variability depending upon the species, type of propagule, number of individuals and location of deposition in terms of elevation and sediment calibre. The significantly higher ($P < 0.01$) survivorship of *S. elaeagnos* cuttings within many of the plots suggests that this species may produce more biomass and a denser community than vegetative propagules of *P. nigra*. Survival rates for seedlings are not known, and the lack or paucity of individuals of either species within specific plots is more likely to be the result of stochastic dispersal processes than the death of germinated seedlings.

Significant differences in growth rates both between type of propagule (cuttings versus seedlings) and species (*S. elaeagnos* versus *P. nigra*) were found in this study, although patterns were not always temporally consistent. For both species, seedlings appeared to grow faster at higher relative elevations, while cuttings performed better at lower elevations. One possible explanation for this is that seedlings may produce a single taproot that rapidly reaches the water table and allows them to remain at high elevations away from disturbance by high flows. Cuttings may produce larger numbers of adventitious roots that can utilize subsurface moisture but may not descend through the substrate as quickly. High root growth rates for both seedlings and cuttings of the Salicaceae have been documented (Pregitzer & Friend, 1996; Van Splunder *et al.*, 1996; Barsoum, 1998; Mahoney & Rood, 1998; Barsoum, 2002). As the second year seedlings had essentially experienced two growing seasons, the trend suggested for greater growth at higher elevations may in turn be demonstrated for cuttings once those cuttings at higher elevations have made a more substantial contact with the water table.

Contrasts between cuttings and seedlings are also observed from their growth responses to sediment calibre. Despite some observable trends relating to increasing sediment calibre (Table 3), statistically significant differences within the cuttings of both species showed little consistency, either temporally or in terms of plot dominance of the growth parameters. While the latter is also true for the first-year seedlings, by the second year an indication of the preference of the *S. elaeagnos* seedlings for coarse calibres, and of the *P. nigra* seedlings for fine calibres, is apparent. This is in contrast to several other studies, which have noted the preference of *Salix* spp. for sandy substrates, while *Populus* spp. prefer well-drained gravels (Dister *et al.*, 1989; Van Splunder *et al.*, 1995; Barsoum, 1998).

Despite the lack of replication, these results indicate that the high topographical and sedimentary heterogeneity that is present along the Tagliamento River helps to maintain a wide diversity of growth rates and morphologies for *S. elaeagnos* and *P. nigra*. Such variability in initial physical conditions at sites of propagule deposition may, through their inter-relations with propagule form and physiology, have an influence on subsequent patterns of sedimentation.

Relationships between propagule form and growth rates, and implications for the Tagliamento River

The data collected from the field experiments and photomonitoring allows some initial extrapolation of actual and estimated growth rates to compare with the conceptual growth trajectories described in Fig. 1. The growth trajectories for the average longest stem (Trajectory A) or canopy height (Trajectories B and C) of propagules are displayed in Fig. 4. Considerable variability can be observed among the individuals surveyed to represent Trajectories B and C, although general trends are apparent. Based upon these initial results, living driftwood (Trajectory C) has the most substantial role to play within vegetation establishment and

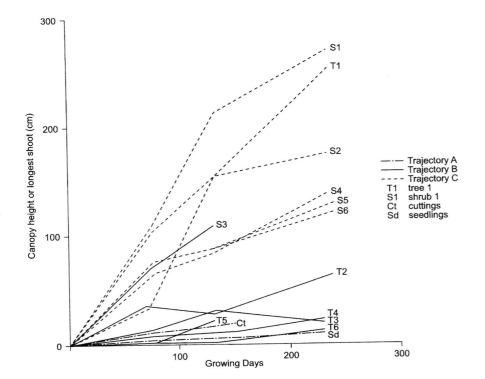

Fig. 4 Actual and estimated growth rates of the longest stem or canopy height of the various propagules investigated, over 230 growing days. Cuttings (Ct) and seedlings (Sd) are representative of Trajectory (A): dispersed propagule establishment. Trees 2–6 and Shrub 3 are representative of Trajectory (B): propagule establishment in the shelter of dead wood pieces. Tree 1 and Shrubs 1, 2, 4, 5 and 6 are representative of Trajectory (C): vegetation regeneration from living wood pieces.

island development, followed by propagules sheltered by dead wood accumulations (Trajectory B), diffuse smaller vegetative propagules and then seedlings (Trajectory A). This is because large wood accumulations can display a high rate of growth (up to ten times that of cuttings), and/or support the growth of sheltered propagules, which in combination with a large initial size can be expected to lead to correspondingly substantial sediment accumulation. This supports the initial concept defined in Fig. 1, and illustrates that riparian trees can have an active influence on their environment through their propagule form and physiology. Regrowth from driftwood, an important mode of vegetation establishment along the Tagliamento River, appears to be more important than in most other studies of vegetation establishment within the active zones of rivers, where growth from seedlings was observed to be the primary method of vegetation development (Niiyama, 1990; Johnson, 1994; Jones *et al.*, 1994).

The substantial numbers of deposited trees recorded within this section of one reach of the river, and observed in other reaches (Gurnell *et al.*, 2000, 2001; Van der Nat *et al.*, 2002, 2003), suggest that there are prolific opportunities for vegetation establishment via living and dead wood within the active zone of the Tagliamento River. The highly significant relationship between tree survival and length indicates the importance of the relatively undisturbed forested margin along the Tagliamento; these margins contain mosaics of mature trees that can be eroded and deposited during flood conditions (Karrenberg *et al.*, 2003b). This is likely to be a primary reason for the presence of river islands within many reaches of the river (Edwards *et al.*, 1999; Gurnell *et al.*, 2001). Along rivers that have been heavily modified and have lost their riparian forests, islands are much fewer in number or absent altogether (Gurnell & Petts, 2002).

The rates of biomass production and related sediment accumulation associated with cuttings and seedlings (Fig. 4, Trajectory A) are necessarily lower than those of large wood pieces. Furthermore, cuttings were also observed to have lower survival rates than deposited trees (26.5% to 38.8% respectively). This does not exclude Trajectory A from being a successful route to vegetation establishment and island development, however. Diffuse propagules may be particularly important under certain circumstances: in reaches where the riparian zone does not reach maturity, or where there are fewer trees, or where driftwood is removed; or during years when floods are limited, or do not deposit large amounts of woody debris. In these reaches, fine-scale variations in topography, sediment calibre, distribution and sorting, along with propagule form, distribution and species, will have a substantial influence on the establishment of riparian vegetation.

CONCLUSION

This paper has presented evidence that propagule form can play a substantial role in modifying the biomass production of riparian trees, and may therefore potentially influence rates and patterns of sedimentation. These initial results support the conceptual model of Gurnell & Petts (2002) (Fig. 1), although further work needs to be completed to validate the model. Important points to consider are:

1 Deposited trees and shrubs produce biomass more quickly, and in greater quantity, than smaller vegetative fragments (cuttings) and seedlings. Vegetative fragments (cuttings) grow at approximately twice the rate of seedlings during the early years of establishment. This suggests that, based on biomass production, trees and shrubs have the greatest capacity to actively influence their physical environment, while vegetative fragments have a much lower impact, and seedlings have the least influence.

2 If large pieces of wood are removed from rivers, or riparian habitat is lost or managed so that tree maturity is not achieved, the process of biomass production from living wood is removed from the system. This compromises the relationships between riparian trees and river form and processes, both ecological and hydrogeomorphological.

3 In the absence of large wood, riparian tree growth is limited to production from vegetative fragments and seeds. This study suggests that significant differences in growth may exist between propagule types of *Salix elaeagnos* and *Populus nigra*, with seedlings growing faster on higher elevation substrates and vegetative fragments at lower elevations, and between species. The influence of sediment calibre was also found to differ depending upon propagule type and species, although consistent patterns were lacking.

ACKNOWLEDGEMENTS

The research for this paper was supported by a University of Birmingham School Studentship to Robert Francis, and NERC Research Grant NER/B/S/2000/00298 to Angela Gurnell and Geoff Petts. Many thanks to Adam Bates, Lee Brown, Ian Morrissey, Lee Thornley and Jon Weller who assisted with various phases of the field experiment. The manuscript was improved by useful comments from Francine Hughes and David Gilvear.

REFERENCES

Abernethy, B. and Rutherford, I.D. (2001) The distribution and strength of riparian tree roots in relation to riverbank reinforcement. *Hydrol. Process.*, **15**, 63–79.

Barsoum, N. (1998) *A comparison of vegetative and non-vegetative regeneration strategies in* Populus nigra *and* Salix alba. Unpublished PhD thesis, Department of Geography, University of Cambridge, Cambridge, 369 pp.

Barsoum, N. (2002) Relative contributions of sexual and asexual regeneration strategies in *Populus nigra* and *Salix alba* during the first years of establishment on a braided gravel bed river. *Evol. Ecol.*, **15**, 255–279.

Braatne, J.H., Rood, S.B. and Heilman, P.E. (1996) Life history, ecology, and conservation of riparian cottonwoods in North America. In: *Biology of Populus and its Implications for Management and Conservation* (Eds R.F. Stettler, H.D. Bradshaw, Jr., P.E. Heilman and T.M. Hinckley), pp. 57–85. NRC Research Press, Ottawa, Ontario, Canada.

Burgess, D., Hendrickson, O.Q. and Roy, L. (1990) The importance of initial cutting size for improving the growth performance of *Salix alba* L. *Scand. J. Forest Res.*, **5**, 215–224.

Church, M. (1996) Channel morphology and typology. In: *River Flows and Channel Forms* (Eds G.E. Petts and P. Callow), pp. 185–202. Blackwell Science, Oxford.

Dister, E., Obrdlik, P., Schneider, E. and Wenger, E. (1989) Zur Ökologie und Gefährdung der Loire-Auen. *Nat. Landsch.*, **64**, 95–99.

Douglas, D.A. (1995) Seed germination, seedling demography, and growth of *Salix setchelliana* on glacial river gravel bars in Alaska. *Can. J. Bot.*, **73**, 673–679.

Edwards, P.J., Kollmann, J., Gurnell, A.M., Petts, G.E., Tockner, K. and Ward, J.V. (1999) A conceptual model of vegetation dynamics on gravel bars of a large Alpine river. *Wetlands Ecol. Manag.*, **7**, 141–153.

Ellis, S. and Mellor, A. (1995) *Soils and Environment*. Routledge, London, 346 pp.

Fetherston, K.L., Naiman, R.J. and Bilby, R.E. (1995) Large woody debris, physical processes, and riparian forest development in montane river networks of the Pacific Northwest. *Geomorphology*, **13**, 133–144.

Gladwin, D.H. and Roelle, J.E. (1998) Survival of plains cottonwood (*Populus deltoides* subsp. *Monilifera*) and saltcedar (*Tamarix ramosissima*) seedlings in response to flooding. *Wetlands*, **18**(4), 669–674.

Gomi, T., Sidle, R.C., Bryant, M.D. and Woodsmith, R.D. (2001) The characteristics of woody debris and sediment distribution in headwater streams, southeastern Alaska. *Can. J. For. Res.*, **31**, 1386–1399.

Goodson, J.M., Gurnell., A.M., Angold, P.G. and Morrissey, I.P. (2001) Riparian seed banks: structure, process and implications for future management. *Prog. Phys. Geogr.*, **25**(3), 301–325.

Gurnell, A.M. and Petts, G.E. (2002) Island-dominated landscapes of large floodplain rivers, a European perspective. *Freshwater Biol.*, **47**, 581–600.

Gurnell, A.M., Petts, G.E., Hannah, D.M., Smith, B.P.G., Edwards, P.J., Kollmann, J., Ward, J.V. and Tockner, K. (2000) Wood storage within the active zone of a large European gravel-bed river. *Geomorphology*, **34**, 55–72.

Gurnell, A.M., Petts, G.E., Hannah, D.M., Smith, B.P.G., Edwards, P.J., Kollman, J., Ward, J.V. and Tockner, K. (2001) Riparian vegetation and island formation along the gravel-bed Fiume Tagliamento, Italy. *Earth Surf. Proc. Landf.*, **26**, 31–62.

Gran, K. and Paola, C. (2001) Riparian vegetation controls on braided stream dynamics. *Water Resour. Res.*, **37**(12), 3275–3283.

Hartman, H.T. and Kester, D.E. (1975) *Plant Propagation: Principles and Practices*. Prentice-Hall, New Jersey, 662 pp.

Hughes, F.M.R., Barsoum, N., Richards, K.S., Winfield, M. and Hayes, A. (2000) The response of male and female black poplar (*Populus nigra* L. subspecies *betulifolia* (Pursh) W. Wettst.) cuttings to different water table depths and sediment types: implications for flow management and river corridor biodiversity. *Hydrol. Process.*, **14**, 3075–3098.

Hupp, C.R. and Osterkamp, W.R. (1985) Bottomland vegetation distribution along Passage Creek, Virginia, in relation to fluvial landforms. *Ecology*, **66**, 670–681.

Järvelä, J. (2002) Flow resistance of flexible and stiff vegetation: a flume study with natural plants. *J. Hydrol.*, **269**, 44–54.

Johansson, M.E., Nilsson, C. and Nilsson, E. (1996) Do rivers function as corridors for plant dispersal? *J. Veg. Sci.*, **7**, 593–598.

Johnson, W.C. (1994) Woodland expansion in the Platte River, Nebraska: patterns and causes. *Ecol. Monogr.*, **64**, 45–84.

Johnson, W.C. (2000) Tree recruitment and survival in rivers: influence of Hydrol. Process. *Hydrol. Process.*, **14**, 3051–3074.

Jones, R.H., Sharitz, R.R., Dixon, P.M., Segal, D.S. and Schneider, R.L. (1994) Woody plant regeneration in four floodplain forests. *Ecol. Monogr.*, **63**(3), 345–367.

Karrenberg, S., Blaser, S., Kollmann, J., Speck, T. and Edwards, P.J. (2003a) Root anchorage of saplings and cuttings of woody pioneer species in a riparian environment. *Funct. Ecol.*, **17**, 170–177.

Karrenberg, S., Kollmann, J., Edwards, P.J., Gurnell, A.M. and Petts, G.E. (2003b) Patterns in woody vegetation along the active zone of a near-natural Alpine river. *Basic Appl. Ecol.*, **4**(2), 157–166.

Lewin, J. (1996) Floodplain construction and erosion. In: *River Flows and Channel Forms* (Eds G.E. Petts, and P. Callow), pp. 203–220. Blackwell Science, Oxford.

Mahoney, J.M. and Rood, S.B. (1992) Response of a hybrid poplar to water table decline in different substrates. *For. Ecol. Manag.*, **54**, 141–156.

Mahoney, J.M. and Rood, S.B. (1998) Streamflow requirements for cottonwood seedling recruitment—an integrative model. *Wetlands*, **18**(4), 634–645.

Malanson, G.P. (1993) *Riparian Landscapes*. Cambridge University Press, Cambridge.

Moreton, D.J., Ashworth, P.J. and Best, J.L. (2002) The physical scale modelling of braided alluvial architecture and estimation of subsurface permeability. *Basin Res.*, **14**(3), 265–285.

Murray, D.R. (1986) Seed dispersal by water. In: *Seed Dispersal* (Ed. D.R. Murray), pp. 49–85. Academic Press, Sydney.

Niiyama, K. (1990) The role of seed dispersal and seedling traits in colonization and coexistence of *Salix* species in a seasonally flooded habitat. *Ecol. Res.*, **5**, 317–331.

Pautou, G. and Décamps, H. (1985) Ecological interactions between the alluvial forests and hydrology of the upper Rhône. *Arch. Hydrobiol.*, **104**(1), 13–37.

Petts, G.E., Gurnell., A.M., Gerrard, A.J., Hannah, D.M., Hansford, B., Morrissey, I., Edwards, P.J., Kollmann, J., Ward, J.V., Tockner, K. and Smith, B.P.G. (2000) Longitudinal variations in exposed riverine sediments: a context for the ecology of the Fiume Tagliamento, Italy. *Aquat. Conserv.*, **19**, 249–266.

Pregitzer, K.S. and Friend, A.L. (1996) The structure and function of *Populus* root systems. In: *Biology of Populus and its Implications for Management and Conservation* (Eds R.F. Stettler, H.D. Bradshaw, Jr., P.E. Heilman and T.M. Hinckley), pp. 331–350. NRC Research Press, Ottawa, Ontario, Canada.

Richardson, S.G., Barker, J.R., Crofts, K.A. and Van Epps, G.A. (1979) Factors affecting root of stem cuttings of salt desert shrubs. *J. Range Manag.*, **32**(4), 280–283.

Robertson, K.M. and Augspurger, C.K. (1999) Geomorphic processes and spatial patterns of primary forest succession on the Bogue Chitto River, USA. *J. Ecol.*, **87**, 1052–1063.

Rood, S.B., Kalischuk, A.R. and Mahoney, J.M. (1998) Initial cottonwood seedling recruitment following the flood of the century of the Oldman River, Alberta, Canada. *Wetlands*, **18**(4), 557–570.

Sacchi, C.F. and Price, P.W. (1992) The relative roles of abiotic and biotic factors in seedling demography of Arroyo Willow (*Salix Lasiolepis*: Salicaceae). *Am. J. Bot.*, **79**(4), 395–405.

Segelquist, C.A., Scott, M.L. and Auble, G.T. (1993) Establishment of *Populus deltoides* under simulated alluvial groundwater declines. *Am. Midl. Nat.*, **130**, 274–285.

Simon, A. and Collison, A.J.C. (2002) Quantifying the mechanical and hydrologic effects of riparian vegetation on streambank stability. *Earth Surf. Proc. Landf.*, **27**, 527–546.

Thornton, C.I., Abt, S.R. and Clary, W.P. (1997) Vegetation influence on small stream siltation. *J. Am. Water Resour. Assoc.*, **33**(6), 1279–1288.

Tooth, S. and Nanson, G.C. (2000) Anabranching rivers on the Northern Plains of arid central Australia. *Geomorphology*, **29**, 211–233.

Van der Nat, D., Tockner, K., Edwards P.J. and Ward, J.V. (2002) Quantification of large wood in large floodplain rivers: an area-based approach using differential GPS and GIS. *Verh. Int. Ver. Limnol.*, **28**, 332–335.

Van der Nat, D., Tockner, K., Edwards, P.J. and Ward, J.V. (2003) Large wood dynamics of complex Alpine river flood plains. *J. N. Am. Benthol. Soc.*, **22**(1), 35–50.

Van Splunder, I., Coops, H., Voesenek, L.A.C.J. and Blom, C.W.P.M. (1995) Establishment of alluvial forest species in floodplains: the role of dispersal timing, germination characteristics and water level fluctuations. *Acta Bot. Neerl.*, **44**, 269–278.

Van Splunder, I., Voesenek, L.A.C.J., Coops, H., De Vries, X.J.A. and Blom, C.W.P.M. (1996) Morphological responses of seedlings of four species of Salicaceae to drought. *Can. J. Bot.*, **74**, 1988–1995.

Ward, J.V., Tockner, K., Edwards, P.J., Kollmann, J., Bretschko, G., Gurnell, A.M., Petts, G.E. and Rossaro, B. (1999) A reference system for the Alps: the 'Fiume Tagliamento'. *Regul. River. Res. Manag.*, **15**, 63–75.

Winfield, M. and Hughes, F.M.R. (2002) Variation in *Populus nigra* clones: Implications for river restoration projects in the United Kingdom. *Wetlands*, **22**(1), 33–48.

Index

Note: page numbers in *italics* refer to figures, those in **bold** refer to tables

accretion, braid bars 20, *25*, 31
acoustic Doppler current profiling (ADCP) 1, 3
aeolian deposits 96
aerial photographs 14, 367
    flight height 122
    human impact assessment 330
    hydraulic model 312, 314
    modern rivers 78–9
afforestation 265, 266
aggradation
    bed elevation 305
    after Assam earthquake 295, *296*, 304
    cellular models 144
    engineering works 265–6
    gravel extraction 264–5
    human impact 335
    sediment 53
        supply 258, 260–1
    state of stream 219
    vegetation density 375
    Waimakariri River 127–8
Ain River (France) 261, 262, 345–6
airborne laser scanning (ALS) 312, 314
    data collection 315
    first-return intensity map 317–18
    ground-cover map 319
    image-relief maps 317–18
    use 324
    see also laser altimetry
alluvial architecture 81–98, *99*
alluvial dynamics recreation 270
Alnus (alder) 364
Alps, French, contracting braided systems 261–2
alternating-direction implicit factorization 156, 157–8
amphibians 350–1
ancient rivers
    analogues 75
    deposition 58
ants *347*, *348*, 351
    adaptations 350
aprons, bank protection 279, *280*, 281
aquatic communities 345–8
aquatic–terrestrial boundary
    energy transfer rate 351
    floating organic matter dynamics *354*

aquifers 52
    depositional models 41–2
    hydraulic conductivity 68
    porosity values 68
    sequential indicator simulation 66, 68
    stochastic simulation 65–6, **67**, 68
arthropods
    carnivorous 351
    riverine 347
        adaptations 350
        mass emergence 351
    terrestrial 344
Arve River (France) 261, 262
Assam earthquake (1950) 289, 291–308
    bed elevation 294–5, *296*
    braiding intensity changes 297–8
    river width effect 295–7
    sediment transport 291–2, 293–4
Avoca River (New Zealand) 141–2, **143**, 144–50

backwaters 322, *323*
    age 341
    habitat 352
bandals 281
bank erosion 116–17, 149
    channel widening 270
    conditions 282
    outer bank of bifurcation 245
bank protection 277–87
    measurement 283–4
    stabilized 282–3
    structures 282
    techniques 282–3
    see also engineering structures
bars
    alternate in shallow channels 153–72
    amplitude variance 154
    area 80, *83*
    asymptotic height 164
    axes 80–1, *82*, *84*, *85*
    celerity 169–71
    characteristic properties 164–71
    dimension measurement 6
    edges 351
    emergence 117
    equilibrium height 164
    evolution with time *167*
    fractal dimension estimates *85*
    front evolution 161–2

    gravel 361–78
    height 164–6, *167*, 179
    hierarchy 77–8
    inclination of front 162
    length 166–9, 171
    mode reduction 116–17
    modern rivers 78–82
    organization 153
    perimeter 80, *83*
    scaling 81–2
    seasonally flooded 281
    size 78
    stabilization 362
    submerged 346–7
    theoretical studies 29–30
    topography 179
    train stability 154
    tree establishment 361–78
    upstream migration 117
    width : length ratio 78
    see also migration, bars; *named types*
bathymetric surveys 315
Beam–Warming implicit scheme 156, 157
bed
    accretion in Meghna River 303, **304**
    alternate pattern 161
    bifurcations 242–4, *246–7*
    celerity of disturbance 292–3
    materials **212**, **214**
    movement inception 218–19
    numerical modelling 161, 162–3, 189
    planform morphology change 145
    profiles 242–4, *246–7*
    shear stresses 207, 322–3
    slope 138
        distribution of lateral 184, 193
        down-valley direction 119
        sediment transport 156
    stability 322–4
    topography 27–8, *163*
        modelling 30
    transverse inclination 245
    velocity behaviour 163
bed elevation *168*, 182, 189
    bed load transport *204*, 205
    bifurcations 244–5, *247–8*
    Brahmaputra–Padma–Lower Meghna river system 294–5, *296*
    deviations from median 193
    human impact 332–3
    Jamuna River 300–1

bed level
  bar form development 154
  changes 116–17, 124, 126, 140, 145
  flow field effects 154
  submerged 316
  Waitaki River 316
bed load transport 199–210, **211–14**, 267–8
  bed elevation *204*, 205
  braiding index 202–3
  channel width 205–9
  constant discharge/supply 202–3, *204*, 205, **212**
  data acquisition 201–2, *203*
  equilibrium state 202
  erosion *204*, 205
  experimental study 201–10, **211–14**
  flow depth 202–3, 205
  formula for braided rivers with limited width 205–9
  grain-size distribution 207
  hydrographs *204*, 205–6, 209, *210*, **213**
  input 201–2, 209–10
    reduction *204*, 205–6
  investigations 200–1, *203*
  microscale braided stream 217–30
  models 137, 149
  output 201–2, 209–10
  sheets 77
  threshold discharge (Q_0) 207–9
  topography 202–3, 205
bedform 77–8
  bar height 166
  migration 14
  Sagavanirktok River 16, 19
  scaling 81–2
bedsets **90**, **91**
  cross-plot ara *95*
  geometric means *96*, *98*, *99*
  sandstone 86–7
  thickness–width 89, 94, 96, *97*
  width 101
beetles 347–8, 351
  flood events 350
  phenological plasticity *349*
benthic organisms 344
bifurcations 233–5
  asymmetry 242
  bed elevation 244–5, *247–8*
  bed profiles 242–4, *246–7*
  BRT model 248, 250–4
  channel changes 239–40
  discharge
    partition 242
    ratios 253–4
  dynamic process 233–5
  equilibrium configurations 252, 253
  erosion of outer bank 245
  field data collection methods 236–41
  flow
    partition 237–8
    rating curve 238
  grain size 240–1
  hydraulic characteristics 241–5, *246–7*, 248
  inlet step 243–4, 251, 253
  input data 252
  morphology 241–5, *246–7*, 248
  sediment transport 239
  Shields stress 250, 251, 254
  site characterization 238–41
  topographic survey 238
  transverse bed inclination 245
  velocity measurement 237, 245, 248, *249*
  Y-shaped configuration 234
biodiversity 345
  foci 352
  habitat complexity 340
  maintenance 340
  surface–subsurface exchanges 344
biological productivity 258
biomass
  production rate 378
  root 373, 376
  surface 373
birds
  migratory 262–3
    roosting habitat creation 269–70
  shoreline 347
Bolla Pittaluga *et al* model *see* BRT model
borehole profiles, sedimentological models 54–5
bottom stresses 112
bounding surfaces, horizontally dominated 69
Brahmaputra–Jamuna River (Bangladesh) 1, 2, 20, 22–4, 25
  bank protection 277–87
  data collection 14
  morphological change 283–4, 291–2, *300*
  river training 277–87
  widening 291–2
Brahmaputra–Padma–Lower Meghna river system (Bangladesh) 289–308
  braiding intensity changes 297–8
  conceptual process–response model 299–303, **304**
  data sources/processing 294
  morphological changes 294–9
  width 295–7
braid bars 77
  accretion 20, *25*, 31
  Brahmaputra–Jamuna River 20, 22, *25*
  Calamus River 19
  cellular models 145
  meandering belt 128
  microhabitats 267, *268*
braid belt 108, 128
  active 127
  low-flow 127–8
braiding, causes 283–5, *286*
braiding index 6
  bed load transport 202–3
  Calamus River 19
  human impact 331–2, 333–4
  Jamuna River 301
  MP modelled river 188–9
braiding intensity
  Assam earthquake 304
  changes with Assam earthquake 297–8
  evolution 117
  human impact 331–2
  Jamuna River 301
  mechanism 305
  Meghna River 303
  Padma River 301
braidplain
  morphology 128, 314
  topography 127
Brenta River (Italy) 328–9
  channel adjustments 330–7
  human impacts 329–37
  hydrology **328**
  management 337
  physiography **328**
brick mattresses 279
BRT model 248, 250–4
Bufo bufo bufo (common toad) 350, **351**

caddis flies 349
Calamus River (Nebraska, US) 18–20, *21*
  channel migration 20, *21*
  data collection 14
canopy height 367–8, 373, **375**
canyons 353
celerity
  bars 169–71
  bed disturbance 292–3
  morphological change in Brahmaputra 298–9
cellular automata theory 184–5
cellular discretization 113, 114

cellular models 4–5, 137–50
  sensitivity 148
  water routing scheme validation 141–2
channel(s)
  abandonment periods 146, 148
  adjustments 328, 330–3
    causes 335–6
    magnitude 333–5
    temporal trends 335–6
  alternate pattern 161
  anastomosing 294, 306–7, 308
  anisotropy 117–18
  bifurcations
    asymmetry 243–4
    change 239–40
  cellular models of braided 143, 144–5
  change
    bifurcations 239–40
    estimates 126, **127**
    geometry 223–4
    spatial in form 145
    time scales 119
  curved 15, 19, 30
  degradation 264
  depth 80–1, 85
    storey thickness–width data 99, 100
    Waitaki River 315
  dynamics 146–8
  equilibrium topography 146
  geometry 12, 117–18, 223–4
  hierarchy 77–8
  migration in Calamus River 20, 21
  modelled morphology 142, 143, 144–6, 148–9
  narrowing 353
  nature of 14
  number with planform forcing 285, 286
  reflectance intensity 315
  Sagavanirktok River 15
  shallow and alternate bars 153–72
  simulated 160–1
  single-thread 306, 307, 333, 334–5, 353–4
  size 78
  spatial changes in form 145
  substrate mapping 324–5
  thalweg 141
  training works 267
  widening 270, 336
channel belts
  compound bar migration 41–2
  deposits 15

GPR reflection data 13–14
  modelling 30–1
channel fills 15–16
  deposits
    Calamus River 20
    Sagavanirktok River 15–16
  laboratory studies 28
  modelling 31, 34, 35, 36, 37, 39, 40
channel patterns 110
  human impact 332–3, 334
  morphological responses 306–8
  static analysis 119
channel width 84, 145
  Assam earthquake 304
  bar length 169
  bed load transport 206–9
  Brahmaputra River 295–7, 301
  human impact 330–1, 333–7
  invariance 190
  Meghna River 302–3
channel-belt width 75, 79–80
  prediction 99, 100
characteristics of braided rivers 195
  analysis methods 177–82
  spatial 195
  temporal 195
chars 281
Chézy coefficient 155
chute and lobe 234
chute cutoff 234
climate change 336
common midpoint method (CMP) for GPR 62–3
composite bars 77
compound braid bars 14–15
  bar head regions 31
  migration 14, 33–4, 41
  modelling 37, 39, 40
  preservation of truncated 41
  probability density function 41
computational fluid dynamics (CFD) 4–5
conceptual process–response model 299–303, **304**
  comparison with previous process-response models 304–5
  complex response 305–6
confluence scour 117
confluence scour zone 15
  flow 27–8
  modelling 30, 33, 36, 37
  sediment transport 27–8, 130
confluence–confluence spacing 181–2
confluence–diffluence unit bifurcation 234

connectivity preferences, species-specific 345, 346
conservation 340
contraction phase 258, 267–71
Courant number 115, 160
Cowlitz River (US) 305
critical stream power 220
cross strata 27
  Brahmaputra–Jamuna River 23–4
  Calamus River 20
  modelling 35, 39
  Sagavanirktok River 16
  sandy 35, 39
cross-beds 57
cross-dam, erodible 280
cross-stream direction 119
crustacea 345
curved channels
  Calamus River 19
  depositional models 30
  Sagavanirktok River 15

dams
  Brenta River 329
  channel adjustments 335–6
  erodible cross-dam 281
  Piave River 329, 330
  sediment regime alteration 330
  vegetated island impact 353
  Waitaki River 312–13
Danube River (Austria) 345, 346, 347
  vegetated islands 353
data loggers 284
degradation
  bed elevation after Assam earthquake 295, 296, 305
  state of stream 219
dendritic network development 127
deposition
  ancient rivers 58
  bar generation 161–2
  cellular models 144
  elements 57–8
  patterns 119
  point estimation weighting 123–4
  rate 41–2
  system-scale study 128–9
depositional models 11–13, 29–31, 53
  aquifer characterization 41–2
  channel bars 29–30
  gravel-bed rivers 31, 32, 32–5, 35, 38–9
  hydrocarbon reservoirs 41–2
  interpretation 100–2
  new 31–2, 32–5, 35, 36–8, 38–42
  sand-bed rivers 31, 36–8, 39–40

deposits 3–4
  Brahmaputra–Jamuna River 22–3, 24, 25
  Calamus River 20
  channel belts 15
  frozen rivers 12
  geometry 96, 98
  hydrogeological characterization 60–1, **62**, 63–5
  laboratory studies 28–9
  overbank 58
  Sagavanirktok River 14–16, 16, *17–18*
  scale invariance of subsurface 81–3
  statistical analysis 87
  studies 13–16, *17–18*, 18–20, *21*, 22–4, *25*, 26–8
  vertical accretion 23–4, *25*
depth sounders 14
dessication 362
diffluence–diffluence spacing 181–2
digital elevation models (DEMs) 1, 2–3, 14, 108, 110–11
  Avoca River 141–2, **143**, 144–50
  generated 129
  sediment transport 130
  system-scale studies 122–3, 129
  time-scaling 119–20
  Waimakariri River 122–30
  Waitaki River 314, 315–20
discharge
  measurements 236–41
  pulsing 341
  ratios at bifurcations 253–4
dissolved organic matter (DOM), upwelling 344
disturbance cascades 350
Drau River (Austria) 270
driftwood, tree regrowth 377
drill-cores
  data processing 63–5
  sedimentary structure types 60–1, **62**, 63–5, 70
  variogram computation 66, **67**, 68
Drôme River (France) 262, 266, 268–9
drought 362
  terrestrial arthropod recolonization 344
dry–wet zones 123
dunes 27, 77
  Brahmaputra–Jamuna River 20
  crest amplitude 60
  generation 155
  gravel 27, 60, 69–70
  migration 16, 60, 69–70
  Niobrara River 27

dykes 267, 270
dynamical systems approach 180
dynamics of braided rivers 1–3

earthquake, river system response 289, 291–308
echo-sounding 315
ecological value 258
  braiding intensity 258, 259
ecological windows 341–2
ecology of braided rivers 6–7, 262–4, 339–55
ecosystems
  quality improvement 269–71
  risk management balance 270
  stability 350
edge-detection techniques, substrate mapping 324
elevation change 119
embankment protection measures 262
energy
  slope 139
  transfer 351
engineering structures 267–8
  channel adjustments 335–6
  conceptual process–response model
  complex response 306
  costs 281–2
  effect mitigation 270
  flooding protection 265
  Waitaki River 312
  see also bank protection
environment
  adaptations 348–51
  extreme 345
erosion
  bar generation 161–2
  bed load transport *204*, 205
  cellular models 144
  human impact 336
  patterns 119
  point estimation weighting 123–4
  training river works 199
  see also bank erosion
Exner equation 116, 140, 155
  discretization 160
  mass balance 185
expansion phase 258
experimental studies 13
exploratory models 149, 177–96
  flume experiment 186, *187*, 188–90
  fractal methods 180–1
  laboratory river comparison 188–9
  scale-invariant properties 180–2
  see also Murray & Paola (MP) model

fauna
  endangered 340
  floating organic debris associated 353
  phenological plasticity 348–9
field data resolution 94
Fier River (France) 261
fish 263–4, 267, *268*
  refuges 347
  spawning 341
float tracking 283–4
flood pulse concept 6
flooding/flood events
  beetle activity 350
  changes 306
  dynamics 341
  protection 264–6
  riparian vegetation sensitivity 320, *321*, 322
  risk 267
  simulation 140
  Tagliamento River 363–4
  terrestrial arthropod recolonization 344
  wave migration speed 238
  willow seed/propagule distribution 362
floodplains, active
  biocomplexity **352**
  contraction/expansion dynamics 341–2
  permanent infrastructure building 258
  restoration 270
flora
  channel narrowing impact 353
  endangered 340
  phenological plasticity 348–9
flow
  acceleration 28
  angle 112–13
  contraction/expansion 110
  data acquisition 195
  divergence zones 28
  flashy regime 342, 364
  laboratory studies 28–9
  modelling 30
  partitioning 238
  processes
    modelling 116
    three-dimensional 110
  pulse 6, 341–2, 353
  rating curve 236, 238
  regime of Tagliamento River 342, 364
  routing law 109–10, 112–13, 150
  stage 6–7

flow (cont'd)
　studies 28
　system-scale study 117
　transverse displacement 245
　variability 128–9
flow depth 138
　bed load transport 202–3, 205
　hydraulic models 312
　sediment transport 222–9, *230*
　unit discharge 141, 142
flow velocity 163, 224–6
　bifurcations 238, 245, 248, *249*
　Brahmaputra–Jamuna River 20
　Calamus River 19
　hydraulic models 312
flow–sediment divergence magnitude 114
fluid dynamics models, computational 137
flume experiment
　bed load transport 200–1
　microscale 219, 228, *230*
　scales 186, *187*, 188–90
flushing-flow effectiveness 322–4
fluvial facies 81–98, *99*
fluvial reservoir models 76, 100–2
fluvial systems, universality 78
fractal methods 180–1
Fraser River (Canada) 263, 266–7
　aquatic communities 346–7
　braid bar microhabitats 267, *268*
friction sink term 131–2
Froude number 139, 322, *323*
frozen river deposits 12

genotypic flexibility 350–1
geographical information systems (GIS) 14
　channel bed topography 27–8
　digital elevation models 315–16
　human impact assessment 330
geometrical models 5
GEOSSAV software 52, 65
global positioning systems (GPS) 14, 120–1
　channel bed topography 27–8
　morphological change tracking 284
grain size 15
　bifurcations 240–1
　critical stream power index 222
　distribution 207
　measurements 238
　sediment transport 218–30
　velocity ratio 226
grasshoppers 347–8
　metapopulation formation 350

gravel(s)
　bimodal 55, *56*, 60, 70
　brown 56, 57
　coarse-grained sheets 58
　exposed sediments 342–3
　extraction 264–5, 267
　　regulation 267
　grey 55, *56*
　lobes 129
　long-term recruitment rates 267
　mining 329–30, 335–6
　mobility 323
　open-framework 16, 55, *56*, 60, 70
　removal of excess 264
　silty 56, 57
　texture types 55–6
gravel sheets
　geometry 58–9
　horizontally bedded 57–8
　simultaneous scour development 59–60
gravel-bed rivers 27
　bed load transport 199–210, **211–14**
　modelling 31, 32, *32*–5, 35, 38–9
gravelly braided river model 15
gravity, sediment discharge 171
ground cover classification 316–17, 319
ground penetrating radar (GPR) 3–4, 12
　ancient river depositional elements 58
　common midpoint method 62–3
　data
　　interpretation 70
　　processing 63–5
　　reflection 13
　high-resolution 53
　sedimentary structure types 60–1, 63–5
　sedimentological models 52, 54–5
　stratasets 41
　variogram computation 66, **67**, 68, 70
groundwater upwelling 343
groyne test structures 278, 279, *280*
gully formation 266

habitat(s)
　aquatic 267, 342–3
　availability 342
　complexity 340
　index of quality 342
　lateral 352
　management 267
　quality 342
　surface/subsurface 343–4

　terrestrial 342–3
　turnover 341
habitat conditions
　alteration 351–2
　braiding intensity 259
　diversity 312
　mapping/quantification 320, 322, *323*
　quantification *323*
hexagonal mesh 113
hierarchy 76, 77–8
humans
　impact on morphology 327–37
　needs balancing 267
hydraulic characteristics
　bifurcations 241–5, *246–7*, 248
　roughness 316–20
hydraulic conductivity of aquifers 68
hydraulic control, sinuosity differences 129–30
hydraulic models 141, 145, 146
　sediment transport 269
HYDRO2DE model 141, 145, 146
hydrocarbon reservoirs, depositional models 41–2
hydrodynamic modelling 154–6
　applications 320, *321*, 322–4
　two-dimensional 311–25
hydroelectric power 336
　floating organic debris removal 353
　see also dams
hydrogeomorphological dynamics restoration 354
hydrographs
　bed load transport *204*, 205–6, 209, *210*, **213**
　key parameters **212**
hydrological connectivity, lateral 345, *346*
hydro-operations, impact assessment 311–25
hydroworks, Waitaki River 312
hyporheic zone 344

incision periods 146, *148*
infrared cameras *see* IR-thermography
infrastructure construction 258, 261–2
inlet step 243–4, 251, 253
intermediate disturbance hypothesis 262
inundated areas
　low-flow 127
　system-scale measurement 122, 123, **125**
inundation
　polymodal patterns 342
　prediction 142

invertebrate communities 263–4
  adaptation **348**
  benthic 347
  biodiversity foci 352
  disturbance regime **348**
  fluvial style **348**
  heterogeneity 352
  organization 345–6
  rare 347
  refugia **348**
  terrestrial 351
IR-thermography 343
Ishikari River (Japan) 266
islands
  convex/concave 352–3
  development 362
  seasonally flooded 281
  tree impact 378
  vegetated 340, 352–3

Jacobian decomposition 158
Jamuna River (Bangladesh) 234
  anastomoses 306, 308
  bed elevation 300–1
  braiding intensity 301
  conceptual process–response model 300–1
  planform changes 306, *307*
  sediment load pulse 300–1
  widening 291–2, 296–7, 301
  see also Brahmaputra–Jamuna River (Bangladesh)

Kawerong River (Papua New Guinea) 199–200
Kayenta Formation (Colorado, US) 94, 96, *97*, *98*
Kolmogorov–Smirnov two-sample test 182, 184, 191, 193
Kurobe River (Japan) *340*

laboratory studies 28–9
  bed load transport 200–1
landscape visualization 324
landsliding, sediment inputs 266
land-use change 336
laser altimetry 107, 120–1
  inundated areas 122
  see also airborne laser scanning (ALS)
lentic areas 352
  biota 345
levees
  channel adjustments 329, 335
  raising 265–6
life cycle, asynchronic 349

Light Detection and Range (LIDAR)
  see airborne laser scanning (ALS); laser altimetry
linear stability analysis 284
lithofacies 65–6
  simulation 68
lithology
  Rhine River 65–6
  watershed 269
  Weise River 65–6
Lombach River (Switzerland) 206–7, **213**, **214**

McArdell & Faeh model 116–17
macroforms 77–8
management of braided rivers 257–72
  integrated 268–9
Manning friction law 112, 113, 131
mapping
  habitat conditions 322, *323*
  roughness 316–20
  substrate 324–5
  thermal heterogeneity 343
  topography 324
maps, historical 330
Markov chain geostatistics 52
mass conservation 114–15
  law of 185
mass emergence of arthropods 351
meandering channels 294
meandering stream migration 128
Meghna River (Bangladesh) 290–1
  anastomoses 306, 308
  bed accretion 303, **304**
  braiding intensity 303
  conceptual model 301–3, **304**
  planform changes 306–7
  width 302–3
    changes 291, 297
  see also Brahmaputra–Padma–Lower Meghna river system (Bangladesh)
mesoforms 77
metapopulation formation 350
microforms 77
migration
  bars 117
  channel 30
  bedform 14
  birds 262–3, 269–70
  channel 20, *21*
  compound bar
    braid 14, *33–4*, 41
    channel belts 41–2

dune 16
  gravel 60, 70
  flood wave speed 238
  meandering streams 128
  mobile terrestrial animals 349–50
  unit bars 15–16, *33–4*, 40
mobility 349, 351
modern rivers 76, 77
  aerial photographs 78–9
  analogues 75
  data 78–82
    combination with outcrop data 99, *100*
  satellite images 78–9
morphodynamics 179
  exploratory models 177
morphological change
  Brahmaputra–Jamuna River 283–4, 291–2, *300*
  Brahmaputra–Padma–Lower Meghna river system 294–9
  celerity for Brahmaputra 298–9
  patterns 195–6
  quantification 108, 120–4, **125**, 126–30
  timescale 128
morphological response model of Schumm 304–5
morphology
  bifurcations 241–5, *246–7*, 248
  changes 148
  development rate fluctuations 179
  human impact 327–37
  modelled 142, *143*, 144–6
  prediction methods 284
  quantification 108, 120–4, **125**, 126–30
  river system response to Assam earthquake 289–308
Mount St Helen (US) volcanic eruption 305
multibeam echo sounding (MBES) 1, 3
multispectral (MS) scanning, aerial 312, 314–15
  imagery classification 315–16
  substrate mapping 323–4
Murray & Paola (MP) model 109–17, 118
  concept 185
  evaluation 184–6, *187*, 188–96
  objectives 137–8, 149
  planform 194–5
  response to external forcing 194
  topography 186, 191, 194–5

Myricaria germanica (Tamaricaceae) 349

Navier–Stokes conservation 131
neotectonics, conceptual process–response model complex response 306
Niobrara River (Nebraska, US) 14, 24, 26
non-linear interactions 284
numerical modelling 4–5, 108–17
   alternate bars in shallow channels 153–72
   alternating-direction implicit factorization 156, 157–8
   Beam–Warming implicit scheme 156, 157
   cellular discretization 113, 114
   comparative perspective 117
   conceptual assessment 112–15
   discretization 113, 114, 115, 158
   equations 156–60
   exploratory models 149
   flow processes 116
   framework 160–1
   hexagonal mesh 113
   increased complexity 115–17
   Jacobian decomposition 158
   mesh resolution 115
   nomenclature 172–3
   numerical simulations 160–71
   post-positivist analysis 111–12
   process complexity increase 115
   reduced complexity approaches 109–10, 111
   reductionist approach 110, 137
   RHS discretization 159–60
   sediment transport 111–12
   synthesist approach 110, 137
   time-step treatment 115
   two-dimensional depth-averaged 284
   validity 110–12
   see also hydrodynamic modelling
nutrients, upwelling 344
Nyack River (Montana, US) floodplain 344

Ohau River (New Zealand) 341
optical imagery 130
optically stimulated luminescence (OSL) dating 14
organic matter, floating 353, *354*
outcrops **88, 90–3**
   data analysis 87
   geometric means 96, *97*
   hierarchy 76
   modern river data combination 99, *100*
   orientation 89, 94
   profiles 54–5
   sandstone bodies 85–6, 87
   scaling 76
   sedimentological models 54–5
   storeys 89
   two-dimensional 87, 89
overbank deposits 58

Padma River (Bangladesh) 290–1
   anastomoses 306, 308
   braiding intensity 301
   planform 301
      changes 306–7
   sediment transport 301
   width changes 297, 301
palaeochannel systems 76–7
palaeoflow, outcrop orientation 89
partition ratio 236
permafrost, GPR reflection data 13
phenotypic flexibility 350–1
photogrammetry 3, 107
   automated digital 120–1, 122–3
   digital 27–8
   inundated areas 122
   oblique terrestrial analytical 120
   substrate mapping 324
photography, scaled *see* aerial photographs
photo-sieving technique 365
Piave River (Italy) 328–9
   channel adjustments 330–7
   human impacts 329–37
   hydrology **328**
   management 337
   physiography **328**
planar strata, sandy 35, 39
planform
   characteristics 191
   definitions 306
   exploratory models 177, 190, 191
   forcing 285, *286*
   multi-temporal low-water 284, *284*
   Murray & Paola (MP) model 194–5
   Padma River 301
   predisturbance pattern 306
   sequential organization of patterns 180
   single-thread 306, 307
Platte River (USA) 262–3, 269–70
point bars 14–15, 77

ponds
   biodiversity foci 352
   parafluvial 340, 352
   shallow 342–3
pools 322, *323*
Populus (poplar) 364
   cuttings 364, 365–7
      survival/growth rate 367–9, **370**, **371**, 376–7
   seedlings 366, 367
      survival/growth rate 369, **372**, 374, 376–7
porosity values of aquifers 68
post-positivist analysis 111–12
predation by ground-dwelling arthropods 351
prediction methods 284
principle of twenty percent inefficiency 285
probability density function (PDF) of compound braid bars 41
processes of braided rivers, studies 13–28
process-imitating methods 52–3
propagules, vegetative
   growth 369, 374, 376–7
      form relationship 376–8
   hydrochorous distribution 362

reforestation 266, 336
refugia 344, 347, 350
   invertebrate communities **348**
Reisakubetsu River (Japan) 163, *165*
remote sensing 1, 2–3, 6–7, 283–4
   data
      collection 314–15
      reliability 324
   evaluation 323–4
   flight height 121–2
   hydro-operations impact assessment 311–25
   projected use 323–4
   strategy 314
   system-scale measurement 121–2
   see also satellite images
reproduction
   rate 349
   vegetative 362–3
restoration of rivers 270, 340, 354
revetment test structures 278–9, *280*, *281*
Rhine, River (France) 270
Rhine, River (Switzerland) *54*
   lithology 65–6
   sediments 55–71

Rhine, River (Switzerland) (cont'd)
  stochastic aquifer simulation 65–6, **67**, 68
Rhône River (Switzerland) 270, 342, 345–6
RHS discretization 159–60
Ridanna Creek (Italy) 235–54
  morphology 241–5, *246–7*, 248
  site characterization 238–41
ridges, development 362
riffles 322, *323*
ripples 77
risk management, ecological improvement balance 270
river(s)
  continuum concept 6
  expanding systems 260–1
  maintenance/management 336–7
  planform patterns 76–7
  reach widening 199–200
  see also ancient rivers; modern rivers; restoration of rivers
river training 277–87
  measurement 283–4
  techniques 282–3
riverscape aesthetics 271, *272*
root-mean-square error 318
Roubion Valley (France) 271, *272*
roughness mapping 316–20
r-strategists 349
runs 322, *323*

sabo dams 266
Sagavanirktok River (Alaska)
  depositional model *32–5*, 38–9
  deposits 14–16, 16, *17–18*
  GPR profile *17–18*
St Venant equation 131
Salicaceae (willow) 345, 362
  phenological plasticity 348–9
  vegetative reproduction 362–3
Salix elaeagnos (willow) 364
  cuttings 364, 365–7
    survival/growth rate 367–9, **370**, **371**, 376–7
  seedlings 366, 367
    survival/growth rate 369, **372**, 374, 376–7
salmon fisheries 258, 263, 267
salmonids, spawning 341
salt dilution measurement 237
sand 57
  lenses 56
  mobility 323
  overbank deposits 58
  plug 281

sand-bed rivers, depositional models 31, *36–8*, 39–40
sandbodies, fluvial 85–6
  dimension reconstruction 100–2
  genetic types 87
sandstone bodies 85–6, 87
satellite images
  Brahmaputra–Padma–Lower Meghna river system 294, *295*
  Meghna River 302–3
  modern rivers 78–9
  multi-temporal low-water planform 283–4
  see also remote sensing
scale(s) 76
  characteristic 184–6, *187*, 188–96
    modelled rivers *187*, 188–94
  computational grid raster structure 190
  flume experiment 186, *187*, 188–90
  modelled river 189
  spatial 190
scale invariance 5–6, 78, 82, 178, 179–82
  modelled rivers 192–4
  Murray & Paola model evaluation 184–6, *187*, 188–96
  subsurface fluvial deposits 81–3
  topographic patterns using transect data 182, *183*, 184
  topography 182
scaling
  anisotropic spatial 78, 119, 180
  bars 81–2
  bedform 81–2
  characteristics 117–20
  dynamic 118, 119
  dynamic anisotropy 180–1
  geometrical 111
  network link 180–1
  self-affinity 126–7
  spatial 117–20, 179, 195
  time-scale dependence 119–20
  width-based 127
scour depth topography 179
scour fill deposits 57, 58, *59*, 69
  simultaneous gravel sheet development 59–60
  trough-shaped 58
screens, floating 281, *281*
sediment(s)
  accumulation rate 378
  aggradation 52, 258, 260–1
  bimodal 56
  braiding intensity 258

delivery
  increase 267–9
  reduction 266–7
discharge 30
  gravity force 171
drill-cores 60–1, **62**, 63–5
entrainment thresholds 140
exposed riverine 347
extraction 329–30
flux 110, 227–8, 267
  input 224
GPR data 60–1, 63–5
landsliding 266
management 267
open-weave 56
reduced supply 146–7
routing model of fluvial deposition 30–1
saturated 344
sorting 154
  during flooding 69–70
structures 55–7
  types 56–7, 60–1
textures 55–7
trapping 266, 268
unsaturated 344
variogram computation 66, **67**, 68
vegetation density 375
wave propagation 292, 304
see also aggradation
sediment load
  aggradation 260–1
  Brahmaputra–Padma–Lower Meghna river system 294, *296*
  pulse in Jamuna River 300–1
  reduction 308
sediment transport
  Assam earthquake 291–2, 293–4
  bed slope 156
  bifurcation 239
  capacity 305
  cellular models 144
  confluence scour zone 130
  discretization 113–14
  efficiency 219, 227
    stream power index 223
  flow depth 222–9, *230*
  fluctuations 179
  grain size studies 220–30
  hydraulic models 269
  incipient motion conditions 222–5
  initiation 218–19
  laboratory studies 28–9
  lateral 111–12
  microscale braided stream 217–30

sediment transport (cont'd)
  modelling 30, 51–71, 116
    numerical models 111–12
    sinuosity differences 129–30
  motion inception 222
  non-linear law 110, 112–13
  Padma River 301
  rates 130, 139–40
  relations 219–20
  source factor evolution 228–9, 230
  stream length 222–9, 230
  studies 28
  system-scale study 117
  velocity ratio 226
sedimentary sequences in compound bar deposits 35
sedimentological model 51–71
  descriptive methods 53
  hydrogeological applications 53
  methods 54–5
  process-imitating methods 53
  Rhine River 55–71
  sinuosity differences 129–30
  structure-imitating methods 52–3
  Weise River 55–71
SEDSIM model 30–1
seeds, hydrochorous distribution 362
self-affinity 117–18
  computational models 180–1
  scaling 126–7
self-organizing systems 118
self-similarity 76, 81, 117
  deposit geometry 96, 98
  scaling parameters 118
semblance velocity analysis 61, 63
serial discontinuity concept 6
shallow water equations 112
  depth-averaged 131–2
shear stresses
  bed 207, 322–3
  bifurcations 250, 251
  critical 218, 222–3
  pattern 226
Shields stress 218, 222–3
  bifurcations 250, 251, 252, 254
shifting habitat mosaic (SHM) 341
shoot length 373, **375**
shoreline
  communities 340
  dead zones 347
  length 342
    stream area ratio 351
silt 57
  lenses 56
  overbank deposits 58

sinuosity 178–9
  differences 129–30
slope direction, down-valley 119
source factor evolution 228–9, 230
spatial redistribution 351
spatial scales 117–20, 178–9, 195
spawning areas 341
spiders 347, 351
state space approach, topography 182, 185
stereo-matching 122, 123
storeys 89, **92**, **93**
  cross-plot ara 95
  geometric means 98, 99, 102
  geometry prediction 98
  sandstone 86–7, 87
  thickness–width 89, 94, 95, 96, 97
    channel depth data 99, 100
  width 101, 102
storm-related events 108, 128
strata
  inclined 39–40
  low-angle 24
  see also cross strata
strataset
  geometry 19, 40–2
  Sagavanirktok River 16
  sandstone 86, 87
STRATSIM model 31
stream
  effectiveness coefficient 222, 224–5, 226
  length and sediment transport 222–9, 230
stream power 139
  critical 220
  index 222, 223
streamway concept 267
structure-imitating methods 52–3
submergence, relative 226–7
substrate mapping 323–4
subsurface–surface exchange processes 343–4
Sunwapta River (Canada) 182, 183, 184, 235–54
  morphology 241–5, 246–7, 248
  site characterization 238–41
  topography 192–4
superposition 77–8
surface–subsurface exchange processes 343–4
sustainable management 264–71
synoptic remote sensing 108
system-scale study 107–9
  deposition 129
  digital elevation models 129

error identification 122–3
flow 117
inundated areas 122, 123, **125**
methodological challenges 121
models 109–17
sediment transport 117
semi-empirical approaches 122
timescale for change 128
see also morphological change, quantification; morphology, quantification

Tagliamento River (Italy) 54–5, 328–9
  channel adjustments 330–7
  ecology 340–1
  flashy flow regime 342, 363
  floating organic debris 353
  flood events 363–4
  floodplain biocomplexity **352**
  flow regime 342, 363
  geometry 58, 59
  habitat mosaic 340
  human impacts 329–37
  hydrology **328**
  islands 353
  management 337
  morphological elements 58
  physiography **328**
  Salicaceae 362–3
  sediment sorting 69–70
  species diversity 347
  trees
    establishment on gravel bars 363–78
    growth rate/morphology 376–7
  vegetation 353, 364
temperature, ecosystem regulation 342–3
temporal evolution of braided rivers 262–4
temporal scales 178–9, 195, 258–60
terraces 146, 148
terrestrial communities 345–8
thalweg
  Calamus River 19
  flow concentration 141
  line 163
thermal heterogeneity 342–3
thermal patch dynamics 343
toad, common 350, **351**
topographic patterns 182
  scale invariance 182, 183, 184
topographic survey of bifurcations 238

topography
    bed load transport 202–3, 205
    braidplain 127
    characteristics 192–4
    data acquisition 195
    exploratory models 177
    flume experiment 188
    large spatial extent 110–11
    mapping 324
    Murray & Paola model 186, 191, 194–5
    remote sensing 7
    scale invariance 182
    scour depth 179
    state space approach 182, 185
    structural variability 179
    structure in modelled 193
    surface age change 140
    visualization 120, *121*
Toutle River (US) 305
transect graphs 191–2
transverse bar conversion 234
trees
    canopy height 367–8, 374, **375**
    cuttings 365–7
        survival/growth rate 364–5, **366**, 367–9, **370**, **371**, 374, **375**, 376–7
    deposited 376–8
    establishment on gravel bars 361–78
    growth 362–3
    recruitment 362–3
    regrowth from driftwood 377
    seedlings 366–7
        survival/growth rate 364–5, **366**, 369, **372**, 374, **375**, 376–7
    uprooted 367–8, 369, 374, **375**
trophic linkages across boundaries 351–2
trough fill deposits 57
turbulence phenomena 155

ultrasonic distance gauging (UDG) device 237

unit bars 14, 15, 27, 77
    Brahmaputra–Jamuna River 20
    migration 15–16, *33–4*, 40
    modelling 39
    Niobrara River 27
unit discharge
    flow depth 141, 142
    gamma distributions 145–6
    patterns *143*, 145
Ürümqi River (China) 217

Val Roseg (Switzerland) floodplain 341–2
validation 110–12
    high level 111–12
    water routing pathways 141–2
variogram computation 66, **67**, 68, 70
vegetation
    colonization 140, 262–3
    density 375
    flood-sensitivity 320, *321*, 322
    ground cover classification 316–17, 318
    pioneer 341
    riparian 320, *321*, 322, 362, 364
        encroaching 314
    submerged 320
    surface patterns 142, 144
    Tagliamento River 364
velocity analysis, semblance 61, 63
velocity measurement at bifurcations 238, 245, 248, *249*
velocity ratio of sediment transport 226–7
video imagery, dynamic scaling 118

Waiapu River (New Zealand) 260–1
Waiho River (New Zealand) 265–6
Waimakariri River (New Zealand) *121*
    aggradation 127–8

    bed load transport 200
    digital elevation models 122–30
    flooding protection 264
Waipaoa River (New Zealand) 260–1, 266
Waitaki River (New Zealand) 263–4, 312, *313*, 314
    digital elevation models 315–20
    flow diversion 314
    hydro-operations impact assessment 311–25
    restoration programme 270
water
    allocation to river restoration 270
    deficit 362
    discharge 30
    surfeit 362
water flow
    across-stream in Calamus River 19
    Brahmaputra–Jamuna River 20
water routing pathways 138, 139–41
    validation 141–2
water surface
    ground-based elevation 315
    width 145, **146**
water table, GPR reflection data 13
watershed, lithology 269
wave groups 154
weirs
    construction 267–8
    sediment trapping 268
Weise River (Switzerland) 54
    lithology 65–6
    sediments 55–71
    stochastic aquifer simulation 65–6, **67**, 68
wet–dry zones 123
wetlands, perifluvial 270
willow 345
    phenological plasticity 348–9
Wolman Count procedure 238